Partial Differential Equations

Equations

and **Boundary**

Value Problems

with

Mathematica

Second Edition

Partial Differential Equations and Boundary Value Problems with Mathematica

Second Edition

Prem K. Kythe
Pratap Puri
Michael R. Schäferkotter

CHAPMAN & HALL/CRC

A CRC Press Company

Boca Raton   London   New York   Washington, D.C.

Library of Congress Cataloging-in-Publication Data

Kythe, Prem K.
　　Partial differential equations and boundary value problems with Mathematica / by Prem
K. Kythe, Pratap Puri, Michael R. Schäferkotter.— 2nd ed.
　　　p. cm.
　　Rev. ed. of: Partial differential equations and Mathematica, c1997.
　　Includes bibliographical references and index.
　　ISBN 1-58488-314-6
　　　1. Differential equations, Partial. 2. Boundary value problems. 3. Mathematica
(Computer file) I. Puri, Pratap, 1938- II. Schuäferkotter, Michael R., 1955- III. Kythe,
Prem K. Partial differential equations and mathematica. IV. Title.

QA374 .K97 2002
515′.353—dc21　　　　　　　　　　　　　　　　　　　　　　　　　　2002074125

Visit the CRC Press Web site at www.crcpress.com

© 2003 by Chapman & Hall/CRC

No claim to original U.S. Government works
International Standard Book Number 1-58488-314-6
Library of Congress Card Number 2002074125
Printed in the United States of America 1 2 3 4 5 6 7 8 9 0
Printed on acid-free paper

To the Memory of Our Respective Brothers:

Kaushal

Vijay

David

Contents

Preface

This book on partial differential equations and boundary value problems with Mathematica provides a more accessible treatment of an important subject. There is a need to introduce technology into mathematics courses. The use of Mathematica is integrated in such a manner that it takes the interested student through this powerful tool to understand the subject even better.

Partial differential equations and initial and boundary value problems are encountered in a variety of applications in many fields including continuum mechanics, potential theory, geophysics, physics and biology, and mathematical economics.

Overview

The first edition of this book, 'Partial Differential Equations and Mathematica,' was published by CRC Press in 1997. That edition presented the theory and applications for solving initial and boundary value problems involving, in general, the first-order partial differential equations, and in particular, the second-order partial differential equations of mathematical physics. This basic theme is also reflected in the current revised second edition of the book. But the subject has been extended to include initial and boundary value problems not only from mathematical physics but also from continuum mechanics.

The highlights of the second edition are as follows:

1. Most of the chapters are revised, and additional problems added. Extreme care is taken to weed out typographical and other errors.

2. Some new material is added, namely, a new chapter (Chapter 8) involving more initial and boundary value problems of different types; wave propagation and dispersion; boundary layer flows; and the ill-posed problems in partial differential equations. A new section on fluctuating flows is added to Chapter 10.

3. The chapter on Green's functions is completely rewritten, to make it simpler for students to understand. A new section on numerical computation of Green's functions in the plane with related Mathematica codes is provided.

4. Mathematica sections are not only upgraded but expanded (more problems added), and moved toward the end of the chapter as a section called 'Mathematica Projects.' In the first edition these sessions were scattered throughout. This change provides the readers a choice to skip Mathematica.

5. The Mathematica support is provided by posting the Mathematica Packages and Notebooks on the CRC website *http://www.crcpress.com*, as well as on the following websites:
www.math.uno.edu/fac/pkythe.html, or *www.math.uno.edu/fac/ppuri.html*.
Look for the electronic products/download section on the CRC website.

These packages and notebooks require at least Mathematica 3.0. A glossary of Mathematica functions, which was included in the first edition as an appendix, is now available at the above websites as the file `Glossary.pdf`.

6. Mathematica is upgraded to its current version 4.0.

7. Five appendices are added; they deal with the Green's identities; some orthogonal polynomials; tables of transform pairs; Bessel functions; a list of Mathematica Notebooks and Packages available on the above wesites and a list of Mathematica Projects discussed in the book.

8. As in the first edition, the tradition of providing answers to every exercise is maintained. Useful hints are provided in some exercises, and complete solutions to certain difficult problems are included.

9. The number of solved examples is increased to a total of 171, exercises to a total of 239, and Mathematica projects to 44. Besides, available on the CRC website are several general purpose Mathematica Notebooks with explicit codes to solve related problems.

10. The bibliography is expanded to include references of cited literature as well as additional books on related topics for further reading.

11. The subject index is better organized and expanded to provide easy access to page numbers.

12. This book is oriented basically toward explaining different methods and problem solving rather than proving theorems. This approach provides the students with a thorough understanding of the subject of partial differential equations. Such a training is useful in both theoretical and technical applications.

To the Instructor

As in the first edition, most of the material in this textbook, especially the first six chapters, is developed for a beginner's course on partial differential equations. These chapters are designed primarily for junior/senior level undergraduate students in mathematics, physics, and engineering, who have completed at least the courses on multivariate calculus and ordinary differential equations, and possess some working knowledge of Mathematica in case they opt to use its versatility in symbolic manipulation and graphics capabilities. Adequate material on other topics from mathematical analysis is provided in the text as and when needed.

The book represents a two semester course. The first six chapters can be taught at the earliest after the completion of multivariate calculus and ordinary differential equations, while the remaining part definitely requires some degree of maturity. An important consideration at this point is the need for engineering, physics, and applied mathematics students to learn the subject at an earlier stage. In most cases they start using partial differential equations and their solutions prior to any formal training in the subject. As a result, their understanding of applied technical areas is hampered by the lack of familiarity with the theory and methods of partial differential equations. It is our hope that mathematics, physics, and engineering majors will take this course at the beginning of their junior/senior years and thus learn and enjoy other technical subjects with better understanding.

As mentioned above, the chapters present a balanced two-semester course material, which can be tailored to the needs of different levels of instructions. Although the following table outlines some suggested curricula at the three levels, the instructor is free to adapt whatever chapters or sections according to the limitations of the course.

Elementary/Juniors	Beginning Seniors	Beginning Graduates
Chapter 1	Chapter 1	Chapter 2
Chapter 3	Chapter 2	Chapter 7
Chapter 4	Chapter 3	Chapter 8
Chapter 5	Chapter 4	Chapter 9
Chapter 6	Chapter 5	Chapter 10
	Chapter 6	Chapter 11

It will be noticed that all chapters from Chapter 7 onward are more or less independent of each other. They can, therefore, be studied in any desired manner depending on the needs of the students.

The book provides a comprehensive and systematic coverage of the basic theory and applications that can readily be followed by undergraduate students at the junior or senior level.

One important aspect of the role of this textbook is that unlike most books on partial differential equations which are written primarily for mathematics graduate students, our book aims at providing a service course to students from varied disciplines. Our book fills in the need for a two-sequence course, involving juniors, seniors, and beginning graduate students who are drawn from engineering, physics, chemistry, geophysics, naval architecture, and mathematics.

To the Readers

Some information about the contents of the book is given below. More details can be found in the Table of Contents.

Chapter 1 provides some useful definitions, classification of second-order partial differential equations, some well-known equations, and the superposition principle. The method of characteristics for first- and second-order partial differential equations is studied in Chapter 2. This topic is usually ignored in most of the textbooks, or delayed toward the end of the book. However, it is our opinion and experience that it helps the students understand the nature of the solutions and form a guide for higher order partial differential equations, provided the topic is handled with clarity and ample geometrical presentations. Mathematica is found to be very useful in achieving this perspective. An old technique of inverse operators, borrowed from the theory of ordinary differential equations, has been used in Chapter 3 to solve homogeneous and nonhomogeneous partial differential equations with constant coefficients. Chapter 4 puts together the concepts of orthogonality, orthonormality, orthogonal polynomials, series of orthogonal functions, trigonometric Fourier series, eigenfunction expansions, and the Bessel functions. This material is needed in Chapter 5 which deals with the method of separation of variables for boundary and initial value problems. These problems involve the wave, heat, and Laplace equations, with homogeneous and nonhomogeneous boundary conditions in the Cartesian, cylindrical polar, and spherical geometries. The integral transforms, especially the Laplace and Fourier transforms, are presented in Chapter 6. These techniques are powerful tools to solve different types of boundary value problems with initial conditions.

The theory and applications of Green's functions, together with numerical computation of Green's functions in the plane are discussed in Chapter 7. A new chapter, Chapter 8, deals with wave propagation and dispersion, additional initial and boundary value problems, boundary layer flows, and ill-posed problems. The weighted residual methods based on the theory of the variational calculus are given in Chapter 9; this chapter prepares students who are interested in courses on the finite element and boundary element methods. Chapter 10 deals with the perturbation methods applied to problems involving partial differential equations; a new section on the fluctuating flows and applications is added for students interested in fluid dynamics. Finally, the numerical methods based on finite differences are studied in Chapter 11, where Mathematica unfolds the intricate details and the beauty of these methods.

The book begins with a general introduction to Mathematica (Chapter 0), which describes the Mathematica style, and some important concepts. There are five appendices which include useful material on Green's identities, certain orthogonal polynomials, tables of the Laplace and Fourier transform pairs, Bessel functions, a list of Mathematica Notebooks and Packages, and a list of Mathematica Projects discussed in the book.

From the First Edition

Why another introductory textbook on partial differential equations when so many are already available? The question is rightly asked, and the justification is in order.

We have found that every year an increasing number of students enter advanced courses involving boundary value problems, which deal mostly with numerical techniques, such as finite difference, finite element, or boundary element methods. At the same time they are introduced to partial differential equations as graduate students, although a few do manage to acquire some knowledge of the subject through other courses in engineering and physics. It is a pity that an opportunity to learn this subject at undergraduate level is lost, because the students encounter textbooks that are graded strictly for graduate level courses. Even when the textbooks are written for undergraduate students, quite often that may not be the case for a majority of students. Most of these textbooks, though written with quality material, are generally based on hard analysis. Our textbook, written for a two-semester course, is aimed at attracting junior and senior undergraduate students, so they get an early training in the subject and do not miss out on elementary techniques and simple beauty of the subject. In the pedagogical spirit of moderation, we have avoided the extreme situation where a beginner's course is so advanced and severe that it is likely to break the spirit of even mature students in an attempt to cover practically everything in the subject. On the other hand, one should encourage textbooks on this subject which in pace and thought are graded to undergraduate levels.

Accordingly, the authors have striven to produce a beginner's textbook that is mature, challenging, and instructive, and at the same time is reasonable in its demands. Certainly, it is not claimed that partial differential equations can become easy and effortless. However, the authors' combined classroom experience over a number of years justifies the effort that the subject can be made reasonably easy to understand despite its complexity, provided that the student has a thorough background in multivariate calculus and ordinary differential equations. It can impart understanding and profit even to the undergraduate juniors and seniors who take it only for one semester before their graduation. The goal, then, has been to produce a textbook that provides both the basic concepts and the methods for those who will take it only for a semester, and a textbook which also provides adequate training and encouragement for those who plan to continue their studies in the subject itself or in applied areas. The distinctive features and the scope of the book can be determined from the Table of Contents.

This textbook has evolved out of lecture notes developed while teaching the course to undergraduate seniors, and graduate students in mathematics, engineering, and physics at the University of New Orleans during the past three years. Much effort has gone into the organization of the subject matter in order to make the course attractive to students and the textbook easy to read. Although there is a large number of classical textbooks available on the subject, there has been a need for an introductory textbook with Mathematica. This publication, based on classroom experience, fulfills such a need. The mathematical contents of the book are simple enough for the average student to understand the methodology and the fine points of theory and techniques of partial differential equations.

The book provides a comprehensive and systematic coverage of the basic theory and applications that can readily be followed by undergraduate students at the junior or senior level.

Acknowledgments

The help provided by some of my colleagues and students is gratefully acknowledged. Thanks are also due to the editors at the CRC Press who offered valuable suggestions for editorial improvements. The authors thank the people at TechType Works, Inc., Gretna, Louisiana, for typesetting the manuscript.

Kythe, Puri, and Schäferkotter
New Orleans, Louisiana
August 2002

Notation

A list of the notations and abbreviations used in this book is given below.

a	thermal conductivity
$a_0,\ a_n,\ b_n$	Fourier coefficients, $n = 1, 2, \ldots$
Bi	Biot number
c	wave speed, or wave velocity
C	capacity of a conductor; cross-section of a pipe or channel
$d\lambda$	uniform length of a subinterval $(\lambda_{i-1}, \lambda_i)$, $i = 1, 2, \ldots$
D	domain; also, differential operator d/dx
$1/D$	inverse operator of D
$\mathrm{erf}(x)$	error function
$\mathrm{erfc}(x)$	complementary error function
Eq(s)	equation(s) (when followed by an equation number)
$\mathbf{E}$	electric field
$f(x)$	free term
$\|f\|$	norm of f, defined by $\langle f, f \rangle^{1/2}$
$\tilde{f}_s(n)$	finite Fourier sine transform
$\tilde{f}_c(n)$	finite Fourier cosine transform
$\mathcal{F}$	Fourier transform
$\mathcal{F}_c$	Fourier cosine transform
$\mathcal{F}_s$	Fourier sine transform
$\mathcal{F}^{-1}\tilde{f}(\alpha)$	inverse Fourier transform
$\mathcal{F}_c^{-1}\tilde{f}(\alpha)$	inverse Fourier cosine transform
$\mathcal{F}_s^{-1}\tilde{f}(\alpha)$	inverse Fourier sine transform
G	loss coefficient of a conductor
$G(\mathbf{x}, \mathbf{x}')$	Green's function, also written as $G(\mathbf{x} - \mathbf{x}')$
$G(\mathbf{x}, \mathbf{x}'; t, t')$	Green's function for a space-time operator

h	step size
$\hbar$	Planck's constant
$H(x)$	Heaviside unit step function
$H_n(x)$	Hermite polynomials of degree n
$H_n^{(1)}(x)$	Hankel function of the first kind and order n
$H_n^{(2)}(x)$	Hankel function of the second kind and order n
iff	if and only if
$\Im$	imaginary part of a complex quantity
$I_n(x)$	modified Bessel function of the first kind and order n
J	jacobian
$J_n(x)$	Bessel function of first kind and order n
k	thermal diffusivity
$K(s,x)$	kernel of an integral transform
$K_n(x)$	modified Bessel function of the second kind and order n
L	linear differential operator; induction coefficient of a conductor
L^*	adjoint operator of L
$L_n(x)$	Laguerre polynomials of degree n
$L_n^{(\alpha)}(x)$	generalized Laguerre polynomials
$\mathcal{L}$	Laplace transform, defined by $\mathcal{L}\{f(t)\} = F(s)$
$\mathcal{L}^{-1}$	inverse Laplace transform
$\mathbf{M}$	magnetic field
n, or $\mathbf{n}$	outward normal
p.v.	Cauchy's principal value of an integral
p	partial derivative u_x, or $\dfrac{\partial u}{\partial x}$; also, pressure
$p_n(x)$	polynomial of degree n
$P_n(x)$	Legendre polynomials of degree n
$P_n^{\alpha,\beta}$	Jacobi polynomials of degree n
q	partial derivative u_y, or $\dfrac{\partial u}{\partial y}$
Q	heat flux
r	partial derivative u_{xx}, or $\dfrac{\partial^2 u}{\partial x^2}$; also, residual
(r,θ)	polar coordinates
(r,θ,z)	cylindrical polar coordinates
R	resistance of a conductor
R^1	real line
R^n	Euclidean n-space
R^+	set of positive real numbers
$\Re$	real part of a complex quantity
s	variable of the Laplace transform; partial derivative u_{xy}, or $\dfrac{\partial^2 u}{\partial x \partial y}$
s_0	constant rate of suction
$\mathrm{Si}(x)$	sine-integral function

S	surface
t	partial derivative u_{yy}, or $\dfrac{\partial^2 u}{\partial y^2}$; also, time
t'	source point; singularity
T	tension; also, temperature
$T_n(x)$	Chebyshev polynomials of the first kind of degree n
$\langle f,g\rangle$	inner product of f and g
$\langle f,g\rangle_w$	inner product of u and v with a weight function w
u	dependent variable; displacement; temperature
$\mathbf{u}$	velocity field with components u,v,w
U_0	constant mean free-stream velocity
$U_n(x)$	Chebyshev polynomials of the second kind of degree n
V	volume
w	test function
$w(x)$	weight function
$\mathbf{x}$	a point $(x_1, x_2, \dots, x_n)$ in R^n; also, a field point
$\mathbf{x}'$	a source point; singularity
$Y_0(x)$	Weber's Bessel function of second kind of order zero
$Y_n(x)$	Bessel function of the second kind of order n
z	a complex number $z = x + iy$
α	variable of the Fourier transform
β	film coefficient
Γ	simple closed contour; also, circulation
$\Gamma(x)$	gamma function
δu	variation of u
$\delta(x,x')$	Dirac delta function; also denoted by $\delta(x-x')$
ε	perturbation parameter
ξ,η	characteristic coordinates
κ	damping coefficient
λ	eigenvalue
λ_n	eigenvalues
μ	coefficient of viscosity
ν	kinematic viscosity
ϕ_n	eigenfunctions
(ϕ_n,λ_n)	eigenpairs
(ρ,θ,ϕ)	spherical coordinates
$\vartheta(x,t)$	theta-function
ω	frequency
$\omega_x,\omega_y,\omega_z$	vorticity components along the axes
Ω	domain
$\boldsymbol{\omega}$	vorticity vector
∂D	boundary of the domain D
∇	grad, defined by $\nabla = \mathbf{i}\,\dfrac{\partial}{\partial x} + \mathbf{j}\dfrac{\partial}{\partial y} + \mathbf{k}\dfrac{\partial}{\partial z}$

∇_n	gradient along the outward normal n
∇^2	Laplacian $\dfrac{\partial^2}{\partial x^2} + \dfrac{\partial^2}{\partial y^2} + \dfrac{\partial^2}{\partial z^2}$
$\nabla^2 + k^2$	Helmholtz operator
∇^4	biharmonic operator, defined by $\nabla^4 u = \nabla^2(\nabla^2 u)$
$\dfrac{\partial}{\partial t} - k\nabla^2$	diffusion operator
$\dfrac{\partial}{\partial n}$	partial derivative with respect to n
$\square_c$	wave operator $\dfrac{\partial^2}{\partial t^2} - c^2\nabla^2$
$\blacksquare$	marker for the end of an example or a proof

0

Introduction to Mathematica

0.1. Introduction

Mathematica is a powerful mathematical programming environment. Mathematica
provides numerical, symbolic, and graphical tools in order to assist one in the mathematical aspects of problem solving. Significant uses have been found for Mathematica
in investigating and analyzing problems in engineering, mathematics, and physics, as
well as economics and other sciences. Mathematica can also be used as a high level
programming language. Mathematica will run on most of the major platforms, from
Cray supercomputers to desktop systems and laptops.

Mathematica is comprised of two parts, called the Front End and the Kernel. The
Kernel is the computation engine, which does all the calculations. The Front End
takes the form of either a notebook interface (advanced Front End) or a command
line interface, allowing the user to communicate with the Kernel.

Mathematica Notebooks allow one to add hyperlinks, notes, explanations, and
conclusions to the work in a similar fashion to a word processor. Mathematical
symbols and two-dimensional notations can be entered from the keyboard as well
as through the use of palettes. Version 4.0 supports a wide variety of text, graphics,
and sound formats through the use of the Import and Export commands. To find out
more about the Import and Export capabilities, use the Help Browser in Version 4.
Notebooks can be saved as HTML and TEX. Presentations can be prepared and also
documents for electronic publishing. Items can be cut and pasted within notebooks or
between notebooks in order to reuse or modify text, graphics, and calculations. Style

sheets facilitate the formatting of notebook cells and can be tailored to suit individual needs.

The Notebook Front End is a file that organizes text, graphics, and calculations in *cells*. Many different kinds of cells comprise a Mathematica Notebook. In Version 4, the Mathematica books are based on Mathematica symbolic expressions with over 500 user options. The Option Inspector is a tool that provides a listing of front end options and current settings. Commonly used options can also be set using menus or dialog boxes.

Input cells contain Mathematica commands and may be evaluated by pressing SHIFT-RETURN or simply RETURN when the cell is selected. Text cells contain text information and are not evaluated by the Kernel. Graphic cells contain pictures of plots and graphs.

Cells can be formatted with various attributes, such as the font to use in displaying text, font size, and color. The cells may be grouped in an outline fashion in order to organize a document into sections containing titles, headings, and subheadings.

A Mathematica Notebook can be transferred from one platform to another without losing information or formatting, though some consideration should be given in order to successfully port the notebooks. For example, one should use the Uniform Style command to eliminate font variation within cells when translating a Macintosh Notebook for Windows. For more information on this subject, see the note "Notebook Conversion Tips," which is available on MathSource (see section below). The Notebooks may also be sent via electronic mail since the notebooks are ASCII text files.

Most packages written in Version 2 or 3 will run unchanged under Version 4. The format of the Version 4 notebooks is significantly different from versions earlier than 2.2. When a notebook that has been created with Version 2.2 or earlier is opened in Version 4, a dialog box appears that allows one to automatically convert the notebook to Version 4.

0.2. Conventions

Reserved words in the Mathematica programming environment always begin with a capital letter. The arguments of functions are delimited by brackets ([ ]), while parentheses are used to effect grouping. Lists, which are the primary data structure of Mathematica are delimited with braces ({ }) with the elements of the list separated by commas.

Certain symbols should be pointed out. The multiplication symbol is represented by $*$ or by a space as in $a*b$ or $a\,b$. The symbol '$=$' stands for substitution, as in $t = 1$, while equal is denoted by '$==$', as in Equal[x,t], or $x == t$ yields True only if

x and t have the same value. The Mathematica command 'Not' can be rendered by !, as in $x! = t$, which is True if x and t do not have the same value. The last input is denoted %, while %n stands for In[n], which is the input cell number n. Finally, never type the prompt In[n]:= that begins each line. Mathematica automatically puts the In and Out prompts. Type only the text that follows the In prompt.

0.3. Getting Started

After the program is running, either a command line interface will appear or a Notebook will appear. To begin, one just types and Mathematica will put the characters into an *input* cell. With cursor in the input cell, press the ENTER key or SHIFT–RETURN keys together to evaluate the input cell and generate an output cell as shown below. The Mathematica Notebook Intro2Mma.nb, found on the CRC web server mentioned in the Preface, is a notebook for new users. The notebook provides explanation and examples in order to get started.

In[1]:   1+1

Out[1]:   2

In[2]:   x := Table[Sin[k],{k,1,5}]//N
          x

Out[2]:   {0.841471, 0.909297, 0.14112, −0.756802, −0.958924}

In[3]:   Plot[Sin[x],{x,0,2Pi}]

Out[3]:

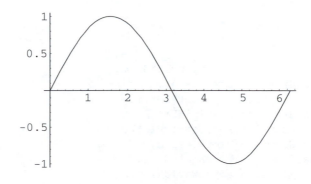

-Graphics-

Occasionally, it will be necessary to interrupt or abort a Mathematica calculation. On most systems there is a command key sequence to interrupt and abort. For GUI (Graphical User Interface) systems, there is a menu choice, under the Action menu, that will enable an interrupt or abort of a calculation.

0.4. File Manipulation

Mathematica can be used for file manipulation on many different computers as well as non-Unix systems. Mathematica has a simple set of file manipulation commands that is intended to work the same way with all operating systems. The advantage is that one never has to learn the file manipulation commands of different computer systems. A few of the commands for file and directory manipulation follow.

In[4]:
```
(* Give the current working directory *)
Directory[ ]
```
Out[4]:
```
Macintosh HD:Applications:Mathematica
```
In[5]:
```
(* List all files in the current working directory *)
FileNames[ ]
```
Out[5]:
```
{Mathematica, Mathematica Kernel, MathLive, Packages}
```
In[6]:
```
(* List all packages containing ''Plot'' in the two levels of sub-
directories below the current directory *)
FileNames["*Plot*.m", "*", 2]
```
In[7]:
```
{Packages:Graphics:ContourPlot3D.m,
Packages:Graphics:FilledPlot.m,
Packages:Graphics:ImplicitPlot.m,
Packages:Graphics:MultipleListPlot.m,
Packages:Graphics:ParametricPlot3D.m,
Packages:Graphics:PlotField.m,
Packages:Graphics:PlotField3D.m,
Packages:Miscellaneous:WorldPlot.m,
Packages:ProgrammingExamples:ParametricPlot3D.m}
```

0.5. Ordinary Differential Equations

Differential equations are used in many areas of natural science in order to study processes that are continuous in space or time. The Mathematica command DSolve computes solutions to ordinary differential equations, as well as systems of ordinary differential equations, and also first-order partial differential equations.

DSolve is a collection of algorithms that allows Mathematica to solve a wide range of equations. Mathematica can solve various types of equations including linear homogeneous and nonhomogeneous equations, second-order variable coefficient equations, second-order nonlinear equations, and first-order partial differential equations. A numerical approximation may be obtained using the NDSolve function.

In[8]: soln = DSolve[y'[x] == y[x], y[x], x]

Out[8]: {{y[x] -> E^x C[1]}}

The symbol C[1] is built into Mathematica. A rule can be used to set it.

In[9]: y[x]/.First[soln]/.C[1]->y0

Out[9]: E^x y0

If initial conditions are given, Mathematica uses them to determine the constants.

In[10]: DSolve[y'[x] == y[x], y[0]==y0, y'[0]==y0, y[x], x]

Out[10]: {{y[x] -> E^x y0}}

A question that very often appears at this stage is, "How can the results from DSolve be used in other calculations?" The answer is that the output from Solve,

NSolve, DSolve, or NDSolve is a list in which each element is a list of rules. One
can assign a name to the solution and easily check that the solution is correct.

In[11]: f[x_] := y[x]/.First[%]
          f[x]

Out[11]: $E^X C[1]$

In[12]: f'[x] == f[x]

Out[12]: True

0.6. To the Instructor

The Mathematics Department at the University of New Orleans realized the value
in using Mathematica for the teaching of Calculus. The authors recognized that the
symbolic, numerical, and graphical capabilities of Mathematica were well suited to
augment the teaching of partial differential equations. Subsequently, the first author
taught a course on Ordinary Differential Equations with Mathematica, and the first two
authors recently gave a course in Partial Differential Equations in which Mathematica
was used. The third author taught Calculus and Mathematica, and Vector Calculus
using Mathematica.

The notebook concept creates a manageable interface for the student, without
all the headaches of programming I/O (input and output), and creating sophisticated
graphics, as well as simplifying complicated algebraic expressions. It should be
understood by the students that, although Mathematica will provide the power to
perform mathematical tasks, Mathematica should not be used as a crutch to solve the
problems.

We suggest that the instructor work through the examples to discover what is
to be emphasized. We also ask that the instructor realize that Mathematica plays a
secondary role and that only representative examples are presented in the text material. Other examples and exercises can be worked out in an analogous fashion. The
instructor should also understand that developing more advanced programming and
solutions techniques is not the purpose nor the intent of this textbook. Sometimes
the nature of a problem may require some variation and/or modification of the given
Mathematica code. In all cases the student should be encouraged to explore possi-
bilities in Mathematica as long as there is no detriment to their learning the partial
differential equations material.

It is also suggested that Notebooks be downloaded from the World Wide Web at the CRC website *www.crcpress.com* (look for electronic products/download), and be made such that they cannot be erased or modified. To preserve the original Mathematica files that are downloaded from the above website, the students are advised to create a copy of the originals and modify only a copy. The instructor should also become aware of the hardware capabilities in terms of memory and speed. Mathematica is quite capable of using a substantial size of memory in evaluating expressions.

0.7. To the Student

The use of computer technology in the classroom continues to facilitate the learning of mathematics as well as other sciences. The technology, in this case, takes the form of Mathematica and the associated Notebooks. With this technology, the student can easily explore many of the graphical, numerical, and symbolic aspects of any number of problems. You will find that using Mathematica and the computer to learn partial differential equations can be both exciting and frustrating at the same time! A few suggestions will follow that will help you to maximize your experience.

Although Mathematica is a programming environment, you do not need to learn how to program Mathematica. You can learn by example and easily adapt the examples to solve most of the problems. We encourage you to copy, paste, and edit whenever possible. Besides, if you have a working example that can be slightly modified to solve your problem, then there will be less chance for a typing mistake if you let the computer do the typing by cutting and pasting. We recommend that you first cut and paste, and then modify the example.

Mathematica is capable of making mistakes. So the student should be able to verify and check some results by hand. At times, you will have to do the entire calculation by hand in order to verify that Mathematica is providing you with the correct answer to your problem.

The basic reference for Mathematica is Stephen Wolfram's *The Mathematica Book*, Fourth Edition (Mathematica Version 4, 1999). This book is the definitive reference for built-in functions and commands found in Mathematica.

A very useful facility in Mathematica is the Help Browser. The complete text of The Mathematica Book is available using the Help Browser, which can be utilized by choosing Help from the Help menu. After clicking a category in the Help Browser, a list of topics is displayed. Selecting a topic opens a list of subtopics. Selecting a particular command name yields the usage message associated with the command. An alphabetical index of information contained in all of the Help Browser categories can be found in the Master Index category.

Some of you will, no doubt, be interested in learning how to program Mathematica, or have a question about Mathematica. The student should understand that programming Mathematica is a means to an end and that Mathematica is used to motivate the student's learning of partial differential equations. The moderated newsgroup *comp.soft-sys.math.mathematica* is a forum which offers the opportunity to ask questions and receive answers regarding Mathematica related issues. Note that the standard rules of list netiquette apply.

0.8. MathSource

MathSource is an extensive, well-organized, and easily accessible online electronic library of Mathematica materials. A MathSource CD-ROM is available. For information on MathSource commands, send an email with Help Intro in the body to *mathsource@wri.com* or visit *http://www.mathsource.com* for valuable information.

1

Introduction

In many mathematical modeling formulations, partial derivatives are required to represent physical quantities. These derivatives always involve more than one independent variable, generally the space variables $x, y, \ldots$ and the time variable t. Such formulations have one or more dependent variables, which are the unknown functions of the independent variables. The resulting equations are called *partial differential equations*, which, together with the initial and/or boundary conditions, represent physical phenomena and are known as the initial or boundary value problems.

1.1. Notation and Definitions

Definitions about order, linearity, homogeneity, and solutions for partial differential equations are similar to those for the ordinary differential equations and are as follows: The *order* of a partial differential equation is the same as the order of the highest partial derivative appearing in the equation. The partial derivatives $\dfrac{\partial u}{\partial x}, \dfrac{\partial u}{\partial y}, \dfrac{\partial^2 u}{\partial x^2}, \dfrac{\partial^2 u}{\partial y \partial x}$, and $\dfrac{\partial^2 u}{\partial y^2}$ are sometimes denoted by u_x, u_y, u_{xx}, u_{xy}, and u_{yy} (or $p, q, r, s,$ and t), respectively. The most general first-order partial differential equation with two independent variables x and y has the form

$$F(x, y, u, p, q) = 0, \qquad p = u_x, \quad q = u_y. \tag{1.1}$$

The most general second-order partial differential equation is of the form

$$F(x, y, u, p, q, r, s, t) = 0, \qquad r = u_{xx}, \quad s = u_{xy}, \quad t = u_{yy}. \tag{1.2}$$

A partial differential equation is said to be *linear* if the unknown function u and all its partial derivatives appear in an algebraically linear form, i.e., of the first degree. For example, the equation

$$a_{11}\, u_{xx} + 2a_{12}\, u_{xy} + a_{22}\, u_{yy} + b_1\, u_x + b_2\, u_y + c_0\, u = f, \qquad (1.3)$$

where the coefficients $a_{11}, a_{12}, a_{22}, b_1, b_2$, and c_0 and the function f are functions of x and y, is a second-order linear partial differential equation in the unknown $u(x, y)$.

An operator L is a linear differential operator iff $L(\alpha u + \beta v) = \alpha L u + \beta L v$, where α and β are scalars, and u and v are any functions with continuous partial derivatives of appropriate order. A partial differential equation $Lu = 0$ is said to be *homogeneous*, whereas the equation $Lu = g$, where $g \neq 0$ is a given function of the independent variables, is said to be *nonhomogeneous*. For example,

$$(x + 2y)u_x + x^2 u_y = \cos(x^2 + y^2)$$

is a nonhomogeneous first-order linear equation, whereas the equation

$$(x + 2y)u_x + x^2 u_y = 0$$

is homogeneous. Thus, a linear homogeneous equation is such that whenever u is a solution of the equation, then cu is also a solution, where c is a constant. A function $u = \phi$ is said to be a *solution* of a partial differential equation if ϕ and its partial derivatives, when substituted for u and its partial derivatives occurring in the partial differential equation, reduce it to an identity in the independent variables. The *general* solution of a linear partial differential equation is a linear combination of all linearly independent solutions of the equation with as many arbitrary functions as the order of the equation; a partial differential equation of order k has k arbitrary functions. A *particular* solution of a partial differential equation is one that does not contain arbitrary functions or constants.

A partial differential equation is said to be *quasilinear* if it is linear in all the highest-order derivatives of the dependent variable. For example, the most general form of a quasilinear second-order equation is

$$A(x, y, u, p, q)\, u_{xx} + B(x, y, u, p, q)\, u_{xy} + C(x, y, u, p, q)\, u_{yy}$$
$$+ f(x, y, u, p, q) = 0. \qquad (1.4)$$

It is assumed that the reader is familiar with the theory and methods of ordinary differential equations. Since the subject of partial differential equations is broad, we will limit our discussion mainly to equations of the first order and certain well-known equations of the second order in the following chapters.

1.2. Initial and Boundary Conditions

A partial differential equation subject to certain conditions in the form of initial or boundary conditions is known as an initial value or a boundary value problem. The initial conditions, also known as *Cauchy conditions*, are the values of the unknown function u and an appropriate number of its derivatives at the initial point.

The boundary conditions fall into the following three categories:

(i) *Dirichlet boundary conditions* (also known as boundary conditions of the first kind), when the values of the unknown function u are prescribed at each point of the boundary ∂D of a given domain D.

(ii) *Neumann boundary conditions* (also known as boundary conditions of the second kind), when the values of the normal derivatives of the unknown function u are prescribed at each point of the boundary ∂D.

(iii) *Robin boundary conditions* (also known as boundary conditions of the third kind, or mixed boundary conditions), when the values of a linear combination of the unknown function u and its normal derivative are prescribed at each point of the boundary ∂D.

In heat transfer problems these three categories are also known as the isothermal, adiabatic, and outer heat conduction boundary conditions, respectively. The last category includes the effects of convection, radiation, and heat conduction into the surrounding medium, which is usually regarded negligible. The following problems are examples of each category:

$$u_t = k\,u_{xx}, \quad 0 < x < l,\ t > 0,$$
$$u(x,0) = f(x), \quad u_t(x,0) = g(x),\ 0 < x < l, \tag{1.5}$$
$$u(0,t) = T_1(t), \quad u(l,t) = T_2(t),\ t > 0;$$

$$u_t = k\,u_{xx}, \quad 0 < x < l,\ t > 0,$$
$$u(x,0) = f(x), \quad u_t(x,0) = g(x),\ 0 < x < l, \tag{1.6}$$
$$u_x(0,t) = T_3(t), \quad u_x(l,t) = T_4(t),\ t > 0;$$

$$u_t = k\,u_{xx}, \quad 0 < x < l,\ t > 0,$$
$$u(x,0) = f(x), \quad u_t(x,0) = g(x),\ 0 < x < l,$$
$$\left.\begin{array}{l} u(0,t) + \alpha\,u_x(0,t) = 0, \\ u(l,t) + \beta\,u_x(l,t) = 0, \end{array}\right\} \quad t > 0. \tag{1.7}$$

1.3. Classification of Second-Order Equations

If $f = 0$ in Eq (1.3), the most general form of a second-order homogeneous equation is

$$a_{11}\, u_{xx} + 2a_{12}\, u_{xy} + a_{22}\, u_{yy} + b_1\, u_x + b_2\, u_y + c_0\, u = 0. \qquad (1.8)$$

To show a correspondence of this equation with an algebraic quadratic equation, we replace u_x by α, u_y by β, u_{xx} by α^2, u_{xy} by $\alpha\beta$, and u_{yy} by β^2. Then the left side of Eq (1.8) reduces to a second-degree polynomial in α and β:

$$P(\alpha, \beta) = a_{11}\alpha^2 + 2a_{12}\alpha\beta + a_{22}\beta^2 + b_1\alpha + b_2\beta + c_0. \qquad (1.9)$$

It is known from analytical geometry and algebra that the polynomial equation $P(\alpha, \beta) = 0$ represents a *hyperbola, parabola,* or *ellipse* according as its discriminant $a_{12}^2 - a_{11}a_{22}$ is positive, zero, or negative. Thus, Eq (1.8) is classified as hyperbolic, parabolic, or elliptic according as the quantity $a_{12}^2 - a_{11}a_{22}$ is positive, zero, or negative.

An alternative approach to classify the types of Eq (1.8) is based on the following theorem:

Theorem 1.1. *The relation $\phi(x, y) = C$ is a general integral of the ordinary differential equation*

$$a_{11}\, dy^2 - 2a_{12}\, dx\, dy + a_{22}\, dx^2 = 0 \qquad (1.10)$$

iff $u = \phi(x, y)$ is a particular solution of the equation

$$a_{11}\, u_x^2 + 2a_{12}\, u_x\, u_y + a_{22}\, u_y^2 = 0. \qquad (1.11)$$

PROOF. Assume that the function $u = \phi(x, y)$ satisfies Eq (1.11). Then the equation

$$a_{11} \left(\frac{\phi_x}{\phi_y} \right)^2 - 2a_{12} \left(-\frac{\phi_x}{\phi_y} \right) + a_{22} = 0 \qquad (1.12)$$

holds for all x, y in the domain of definition of $u = \phi(x, y)$ with $\phi_y \neq 0$. In order that the relation $\phi(x, y) = C$ be the general solution of Eq (1.10), we must show that the function y defined implicitly by $\phi(x, y) = C$ satisfies Eq (1.10). Suppose that $y = f(x, C)$ is such a function. Then

$$\frac{dy}{dx} = - \left[\frac{\phi_x(x, y)}{\phi_y(x, y)} \right]_{y = f(x, C)}.$$

Hence, in view of Eq (1.12), we have

$$a_{11}\left(\frac{dy}{dx}\right)^2 - 2a_{12}\left(\frac{dy}{dx}\right) + a_{22}$$

$$= \left[a_{11}\left(-\frac{\phi_x}{\phi_y}\right)^2 - 2a_{12}\left(-\frac{\phi_x}{\phi_y}\right) + a_{22}\right]_{y=f(x,C)} = 0. \tag{1.13}$$

Thus, $y = f(x, C)$ satisfies Eq (1.11).

Conversely, let $\phi(x, y) = C$ be a general solution of Eq (1.10). We must show that for each point (x, y)

$$a_{11}\,\phi_x^2 + 2a_{12}\,\phi_x\,\phi_y + a_{22}\,\phi_y^2 = 0. \tag{1.14}$$

If we can show that Eq (1.14) is satisfied for an arbitrary point (x_0, y_0), then Eq (1.14) will be satisfied for all points. Since $\phi(x, y)$ represents a solution of Eq (1.10), we construct through (x_0, y_0) an integral of Eq (1.10), where we set $\phi(x_0, y_0) = C_0$, and consider the curve $y = f(x, C_0)$. For all points of this curve, we have

$$a_{11}\left(\frac{dy}{dx}\right)^2 - 2a_{12}\left(\frac{dy}{dx}\right) + a_{22}$$

$$= \left[a_{11}\left(-\frac{\phi_x}{\phi_y}\right)^2 - 2a_{12}\left(-\frac{\phi_x}{\phi_y}\right) + a_{22}\right]_{y=f(x,C_0)} = 0.$$

If we set $x = x_0$ in this equation, we get

$$a_{11}\,\phi_x^2(x_0, y_0) + 2a_{12}\,\phi_x(x_0, y_0)\,\phi_y(x_0, y_0) + a_{22}\,\phi_y^2(x_0, y_0) = 0,$$

where $y_0 = f(x_0, C_0)$. ∎

Eq (1.10) or (1.11) is called the *characteristic equation* of the partial differential equation (1.3) or (1.8); the related integrals are called the *characteristics*.

Eq (1.13), regarded as a quadratic equation in dy/dx, yields two solutions:

$$\frac{dy}{dx} = \frac{a_{12} \pm \sqrt{a_{12}^2 - a_{11}\,a_{22}}}{a_{11}}.$$

The expression under the radical sign determines the type of the differential equation (1.3) or (1.8). Thus, as before, Eq (1.3) or (1.8) is of the hyperbolic, parabolic, or elliptic type according as the quantity $a_{12}^2 - a_{11}a_{22} \gtreqless 0$.

EXAMPLE 1.1. The Tricomi equation $u_{xx} + xu_{yy} + u = 0$, for which $a_{12}^2 - a_{11}a_{22} = -x$, is hyperbolic if $x < 0$, parabolic if $x = 0$, and elliptic if $x > 0$. ∎

EXAMPLE 1.2. For the equation $(\sin x)\, u_{xx} + 2(\cos x)\, u_{xy} + (\sin x)\, u_{yy} = 0$, the quantity $a_{12}^2 - a_{11}\, a_{22} = \cos^2 x - \sin^2 x = \cos 2x$. Hence, the equation is hyperbolic for $-\pi/4 + n\pi < x < \pi/4 + n\pi$, parabolic for $x = \pi/4 + n\pi$, and elliptic for $\pi/4 + n\pi < x < 3\pi/4 + n\pi$, where n is an integer. ∎

EXAMPLE 1.3. Consider $u_{xx} - x^2 y u_{yy} = 0$, $y > 0$. Here $a_{12}^2 - a_{11} a_{22} = x^2 y > 0$, so the partial differential equation is hyperbolic. ∎

EXAMPLE 1.4. Consider $e^{xy} u_{xx} + u_{yy} \sinh x + u = 0$. Here $a_{12}^2 - a_{11} a_{22} = -e^{xy} \sinh x$, and the partial differential equation is hyperbolic if $x < 0$, parabolic if $x = 0$, and elliptic if $x > 0$. ∎

EXAMPLE 1.5. For the equation $u_{xx} + y\, u_{yy} = 0$, we have $a_{12}^2 - a_{11}\, a_{22} = -y$. Hence, the equation is hyperbolic for $y < 0$, parabolic for $y = 0$, and elliptic for $y > 0$. ∎

EXAMPLE 1.6. To find the region where the equation

$$u_{xx} + x^2\, u_{yy} = y\, u_y$$

is elliptic, note that $a_{12}^2 - a_{11}\, a_{22} = -x^2$. Hence, the given equation is elliptic in any region that does not contain the y-axis ($x = 0$). ∎

1.4. Some Known Equations

The following equations appear frequently in the study of physical problems.

1. *Heat equation* in R^1: $u_t = k\, u_{xx}$, where u denotes the temperature distribution and k the thermal diffusivity.

2. *Wave equation* in R^1: $u_{tt} = c^2 u_{xx}$, where u represents the displacement, e.g., of a vibrating string from its equilibrium position, and c the wave speed.

3. *Laplace equation* in R^2: $\nabla^2 u \equiv u_{xx} + u_{yy} = 0$, where $\nabla^2 = \nabla \cdot \nabla$ denotes the Laplacian.

4a. *Transport (Traffic) equation:* $u_t + a(u)\, u_x = 0$.

4b. *Transport equation* in R^1: $u_t + a\, u_x = 0$, where a is a constant.

5. *Burgers' equation* in R^1: $u_t + u\, u_x = 0$, which arises in the study of a stream of particles or fluid flow with zero viscosity.

6. *Eikonal equation* in R^2: $u_x^2 + u_y^2 = 0$, which arises in geometric optics.

7. *Poisson's equation* in R^n: $\nabla^2 u = f$, also known as the nonhomogeneous Laplace equation in R^n; it arises in various field theories and electrostatics.

8. *Helmholtz equation* in R^3: $\left(\nabla^2 + k^2\right) u = 0$, which arises, e.g., in underwater scattering.

9. *Klein-Gordon equation* in R^3: $u_{tt} - c^2\nabla^2 u + m^2 u = 0$, which arises in quantum field theory, where m denotes the mass.

10. *Telegraph equation* in R^3: $u_{tt} - c^2\nabla^2 u + \kappa u_t + u = 0$, where κ is the damping coefficient; it arises in the study of electrical transmission in telegraph cables when the current may leak to the ground.

11a. *Schrödinger equation* in R^3: $u_t = i[\nabla^2 u + V(x)u]$, where $V(x)$, $x \in R^3$, denotes the potential; it arises in quantum mechanics.

11b. *Cubic Schrödinger equation* in R^3: $u_t = i[\nabla^2 u + \sigma u |u|^2]$, where $\sigma = \pm 1$; this is a semilinear version of 11a.

12. *Sine-Gordon equation* in R^3: $u_{tt} - c^2\nabla^2 u + m^2 u = 0$, which arises in quantum field theory.

13. *Semilinear heat equation* in R^3: $u_t - k\nabla^2 u = f(x, t, u)$.

14a. *Semilinear wave equation* in R^3: $u_{tt} - c^2\nabla^2 u = f(x, t, u)$.

14b. *Semilinear Klein-Gordon equation* in R^3: $u_{tt} - c^2\nabla^2 u + m^2 u + \gamma u^p = 0$, where γ denotes a coupling constant, and $p \geq 2$ is an integer.

14c. *Dissipative Klein-Gordon equation* in R^3: $u_{tt} - c^2\nabla^2 u + \alpha u_t + m^2 u + u^p = 0$.

14d. *Dissipative sine-Gordon equation* in R^3: $u_{tt} - c^2\nabla^2 u + \alpha u_t + \sin u = 0$.

15. *Semilinear Poisson's equation* in R^3: $\nabla^2 u = f(x, u)$.

16. *Porous medium equation* in R^3: $u_t = k\nabla^2(u^a \nabla u)$, where $k > 0$ and $a > 1$ are constants; it is a quasilinear equation, and arises in seepage flows through porous media.

17. *Biharmonic equation* in R^3: $\nabla^4 u \equiv \nabla^2(\nabla^2 u) = 0$; it arises in elastodynamics.

18. *Korteweg de Vries (KdV) equation* in R^1: $u_t + c u u_x + u_{xxx} = 0$, which arises in shallow water waves.

19a. *Euler's equations* in R^3: $\mathbf{u}_t + (\mathbf{u} \cdot \nabla)\mathbf{u} + \dfrac{1}{\rho}\nabla p = 0$, where $\mathbf{u}$ denotes the velocity field, and p the pressure.

19b. *Navier-Stokes equations* in R^3: $\mathbf{u}_t + (\mathbf{u} \cdot \nabla)\mathbf{u} + \dfrac{1}{\rho}\nabla p = \nu\nabla^2\mathbf{u}$, where ν denotes the kinematic viscosity and ρ the density of the fluid.

20. *Maxwell's equations* in R^3: $\mathbf{E}_t - \nabla \times \mathbf{H} = 0$, $\mathbf{H}_t + \nabla \times \mathbf{E} = 0$, where $\mathbf{E}$ and $\mathbf{H}$ denote the electric and the magnetic field, respectively; they are a system of six equations in six unknowns.

Origins of these and other equations of mathematical physics are related to some interesting physical problems. We will derive some of them as examples, which will also bring out certain aspects of mathematical modeling of boundary value problems.

EXAMPLE 1.7. (*One-dimensional wave equation for vibrations of a string*)
Consider a stretched string of length l that is fixed at both ends. It is assumed that
(i) the string is thin and flexible, i.e., it offers no resistance to change of form except
a change in length, and (ii) the tension T in the string is much larger than the force
due to gravity acting on it so that the latter can be neglected. Let the string in its
equilibrium state be situated along the x-axis. Let $u(x,t)$ denote the displacement of
the string at time t from its equilibrium position. The shape of the string at a fixed t
is represented in Fig. 1.1.

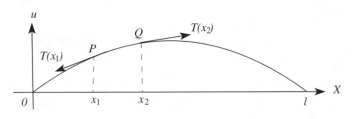

Fig. 1.1. Vibrations of a string.

We assume that the vibrations are small, which implies that the displacement
$u(x,t)$ and its derivative u_x are small enough so that their squares and products can
be neglected. As a result of vibrations, let a segment (x_1, x_2) of the string be deformed
into the segment PQ. Then at time t the length of the arc $\overset{\frown}{PQ}$ is given by

$$\int_{x_1}^{x_2} \sqrt{1 + u_x^2}\, dx \approx x_2 - x_1, \tag{1.15}$$

which simply means that under small vibrations the length of the segment of the
string does not change. By Hooke's law, the tension T at each point in the string is
independent of t, i.e., during the motion of the string any change in T can be neglected
in comparison with the tension in equilibrium. We will now show that the tension T
is also independent of x. In fact, it is evident from Fig. 1.1 that the x-component of
the resulting tension at the points P and Q must be in equilibrium, i.e.,

$$T(x_1) \cos \alpha(x_1) - T(x_2) \cos \alpha(x_2) = 0,$$

where $\alpha(x)$ denotes the angle between the tangent at a point x and the positive x-axis
at time t. Since the vibrations are small, we have

$$\cos \alpha(x) = \frac{1}{\sqrt{1 + \tan^2 \alpha(x)}} = \frac{1}{\sqrt{1 + u_x^2}} \approx 1, \tag{1.16}$$

which implies that $T(x_1) \approx T(x_2)$. Since x_1 and x_2 are arbitrary, the magnitude
of T is independent of x. Hence, if T_0 denotes the tension at equilibrium and T the
tension in the vibrating string, then $T \approx T_0$ for all x and t.

Now, the sum of the components of tension $T(x_1)$ at P and $T(x_2)$ at Q along the u-axis is given by

$$T_0 \left[\sin \alpha(x_2) - \sin \alpha(x_1) \right]$$

$$= T_0 \left[\frac{\tan \alpha(x_2)}{\sqrt{1 + \tan^2 \alpha(x_2)}} - \frac{\tan \alpha(x_1)}{\sqrt{1 + \tan^2 \alpha(x_1)}} \right]$$

$$= T_0 \left[\frac{u_{x_2}}{\sqrt{1 + u_{x_2}^2}} - \frac{u_{x_1}}{\sqrt{1 + u_{x_1}^2}} \right] \quad \text{by using (1.16)} \tag{1.17}$$

$$\approx T_0 \left[\frac{\partial u}{\partial x_2} - \frac{\partial u}{\partial x_1} \right] = T_0 \left[\frac{\partial u}{\partial x}\Big|_{x=x_2} - \frac{\partial u}{\partial x}\Big|_{x=x_1} \right]$$

$$= T_0 \int_{x_1}^{x_2} \frac{\partial^2 u}{\partial x^2}\, dx.$$

Let $g(x,t)$ denote the external force per unit length acting on the string along the u-axis. Then the component of $g(x,t)$ acting on the segment $\overset{\frown}{PQ}$ along the u-axis is given by

$$\int_{x_1}^{x_2} g(x,t)\, dx. \tag{1.18}$$

Let $\rho(x)$ be the density of the string. Then the inertial force on the segment $\overset{\frown}{PQ}$ is

$$-\int_{x_1}^{x_2} \rho(x)\, \frac{\partial^2 u}{\partial t^2}\, dx. \tag{1.19}$$

Hence, the sum of the components (1.17), (1.18), and (1.19) must be zero, i.e.,

$$\int_{x_1}^{x_2} \left[T_0 \frac{\partial^2 u}{\partial x^2} + g(x,t) - \rho(x)\frac{\partial^2 u}{\partial t^2} \right] dx = 0. \tag{1.20}$$

Since x_1 and x_2 are arbitrary, it follows from (1.20) that the integrand must be zero, which gives

$$\rho(x)\frac{\partial^2 u}{\partial t^2} = T_0 \frac{\partial^2 u}{\partial x^2} + g(x,t). \tag{1.21}$$

This represents the partial differential equation for the vibrations of the string. If $\rho = \text{const}$, then (1.21) reduces to

$$\frac{\partial^2 u}{\partial t^2} = c^2 \frac{\partial^2 u}{\partial x^2} + f(x,t), \tag{1.22}$$

where $c = \sqrt{T_0/\rho}$, and $f(x,t) = g(x,t)/\rho$. In the absence of external forces, Eq (1.22) becomes

$$\frac{\partial^2 u}{\partial t^2} = c^2 \frac{\partial^2 u}{\partial x^2}, \tag{1.23}$$

which is the wave equation for free vibrations (oscillations) of the string. ∎

EXAMPLE 1.8. (*Two-dimensional wave equation for oscillations of a membrane*) Suppose that a membrane which is a perfectly flexible, thin, stretched sheet occupies a region D in the xy-plane in its equilibrium state. Further, let the membrane be subjected to a uniform tension T applied on its boundary ∂D perpendicular to the lateral surface. This means that the force acting on an element ds of the boundary ∂D is equal to $T\,ds$. We will examine the transverse oscillations of the membrane, which move perpendicular to the xy-plane at each point in the direction of the u-axis. Thus, the displacement u at a point $(x, y) \in D$ is a function of x, y, and t. Assuming that the oscillations are small, i.e., the functions u, u_x, and u_y are so small that their squares and products can be neglected, let $A \in D$ denote an arbitrary area of the membrane, which is bounded by the curve L and lies in a state of equilibrium in the xy-plane. After the membrane is displaced from its equilibrium position, let the area A be deformed into an area A' bounded by a curve L' (see Fig. 1.2.), which at time t is defined by

$$A' = \iint_A \sqrt{1 + u_x^2 + u_y^2}\, dx\, dy \approx \iint_A dx\, dy = A.$$

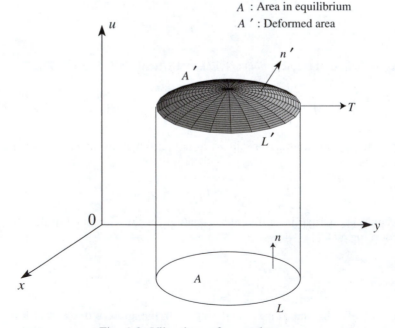

A : Area in equilibrium
A' : Deformed area

Fig. 1.2. Vibrations of a membrane.

Thus, we can neglect the change in A during the oscillations, and the tension in the membrane remains constant and equal to its initial value T.

Note that the tension T which is perpendicular to the boundary L' acts at all points in the tangent plane to the surface area A'. Let ds' denote an element of the boundary L'. Then the tension acting on this element is $T\,ds'$, and $\dfrac{\partial u}{\partial n} = \cos\alpha$, where α is the angle between the tension T and the u-axis, and n is the outward normal to the boundary L. The component of the tension acting on the element ds' in the direction of the u-axis is $T\dfrac{\partial u}{\partial n}\,ds'$. Hence, the component of the resultant force acting on the boundary L' along the u-axis is

$$T\int_{L'} \frac{\partial u}{\partial n}\,ds' \approx T\int_{L} \frac{\partial u}{\partial n}\,ds$$
$$= \iint_{A} \left(\frac{\partial^2 u}{\partial x^2} + \frac{\partial^2 u}{\partial y^2}\right) dx\,dy, \tag{1.24}$$

by Green's identity, where, in view of small oscillations, we have taken $ds' \approx ds$, and replaced L' by L. Let $g(x, y, t)$ denote an external force per unit area acting on the membrane along the u-axis. Then the total force acting on the area A' is given by

$$\iint_{A} g(x, y, t)\,dx\,dy. \tag{1.25}$$

Let $\rho(x, y, t)$ be the surface density of the membrane. Then the inertial force at all times t is

$$\iint_{A} \rho(x, y, t)\frac{\partial^2 u}{\partial t^2}\,dx\,dy. \tag{1.26}$$

Since the sum of the inertial force and the total force is equal and opposite to the resultant of the tension on the boundary L', we find from (1.24)–(1.26) that

$$\iint_{A} \left[\rho(x, y, t)\frac{\partial^2 u}{\partial t^2} - T\left(\frac{\partial^2 u}{\partial x^2} + \frac{\partial^2 u}{\partial y^2}\right) - g(x, y, t)\right] dx\,dy = 0,$$

or, since A is arbitrary,

$$\rho(x, y, t)\frac{\partial^2 u}{\partial t^2} = T\left(\frac{\partial^2 u}{\partial x^2} + \frac{\partial^2 u}{\partial y^2}\right) + g(x, y, t). \tag{1.27}$$

This is the partial differential equation for small oscillations of a membrane. If the density $\rho = \text{const}$, then Eq (1.27) in the absence of external forces reduces to

$$\frac{\partial^2 u}{\partial t^2} = c^2 \left(\frac{\partial^2 u}{\partial x^2} + \frac{\partial^2 u}{\partial y^2}\right), \quad c = \sqrt{T/\rho}. \; \blacksquare \tag{1.28}$$

EXAMPLE 1.9. (*Heat conduction equation for a uniform isotropic body*) Let $u(x, y, z, t)$ denote the temperature of a uniform isotropic body at a point (x, y, z) and time t. If different parts of the body are at different temperatures, then heat transfer takes place within the body. Consider a small surface element δS of a surface S drawn inside the body. Under the assumption that the amount of heat δQ passing through the element δS in time δt is proportional to $\delta S \, \delta t$, and the normal derivative is $\dfrac{\partial u}{\partial n}$, we get

$$\delta Q = -a \frac{\partial u}{\partial n} \, \delta S \, \delta t = -a \, \delta S \, \delta t \, \nabla_n u, \tag{1.29}$$

where a is the thermal conductivity of the body, which depends only on the coordinates (x, y, z) of points in the body but is independent of the direction of the normal to the surface S, and ∇_n denotes the gradient in the direction of the outward normal to the surface element δS. Let Q denote the *heat flux* which is the amount of heat passing through the unit surface area per unit time. Then Eq (1.29) implies that

$$Q = -a \frac{\partial u}{\partial n}. \tag{1.30}$$

Now, consider an arbitrary volume V bounded by a smooth surface S. Then, in view of (1.30), the amount of heat entering through the surface S in the time interval $[t_1, t_2]$ is given by

$$\begin{aligned} Q_1 &= -\int_{t_1}^{t_2} dt \iint_S a(x, y, z) \frac{\partial u}{\partial n} \, dS \\ &= \int_{t_1}^{t_2} dt \iiint_V \nabla \cdot (a \nabla u) \, dV, \end{aligned} \tag{1.31}$$

by divergence theorem, where n is the inward normal to the surface S. Let δV denote a volume element. The amount of heat required to change the temperature of this volume element by $\delta u = u(x, y, z, t + \delta t) - u(x, y, z, t)$ in time δt is

$$\delta Q_2 = [u(x, y, z, t + \delta t) - u(x, y, z, t)] \, c(x, y, z) \rho(x, y, z) \, \delta V, \tag{1.32}$$

where $c(x, y, z)$ and $\rho(x, y, z)$ are the specific heat and density of the body, respectively. Integrating (1.31) we find that the amount of heat required to change the temperature of the volume V by δu is given by

$$\begin{aligned} Q_2 &= \iiint_V [u(x, y, z, t + \delta t) - u(x, y, z, t)] \, c\rho \, dV \\ &= \int_{t_1}^{t_2} dt \iiint_V c\rho \frac{\partial u}{\partial t} \, dV. \end{aligned} \tag{1.33}$$

Now, we assume that the body contains heat sources, and let $g(x, y, z, t)$ denote the density of such heat sources. Then the amount of heat released by or absorbed in V in the time interval $[t_1, t_2]$ is

$$Q_3 = \int_{t_1}^{t_2} dt \iiint_V g(x, y, z, t) \, dV. \tag{1.34}$$

Since $Q_2 = Q_1 + Q_3$, we find from (1.31)–(1.34) that

$$\int_{t_1}^{t_2} dt \iiint_V \left[c\rho \frac{\partial u}{\partial t} - \nabla \cdot (a\nabla u) - g(x,y,z,t) \right] dV = 0,$$

or, since the volume V and the time interval $[t_1, t_2]$ are arbitrary, we get

$$\begin{aligned}
c\rho \frac{\partial u}{\partial t} &= \nabla \cdot (a\nabla u) + g(x,y,z,t) \\
&= \frac{\partial}{\partial x}\left(a\frac{\partial u}{\partial x}\right) + \frac{\partial}{\partial y}\left(a\frac{\partial u}{\partial y}\right) + \frac{\partial}{\partial z}\left(a\frac{\partial u}{\partial z}\right) + g(x,y,z,t),
\end{aligned} \tag{1.35}$$

which is the *heat conduction equation* for a uniform isotropic body. If c, ρ, and a are constant, Eq (1.35) becomes

$$\frac{\partial u}{\partial t} = k\left(\frac{\partial^2 u}{\partial x^2} + \frac{\partial^2 u}{\partial y^2} + \frac{\partial^2 u}{\partial z^2}\right) + f(x,y,z,t), \tag{1.36}$$

where $k = a/(c\rho)$ is known as the thermal diffusivity, and $f = g/(c\rho)$ denotes the heat source (sink) function. In the absence of heat sources or sinks (i.e., when $g(x,y,z,t) = 0$), Eq (1.36) reduces to the *homogeneous heat conduction equation*

$$\frac{\partial u}{\partial t} = k\nabla^2 u = k\left(\frac{\partial^2 u}{\partial x^2} + \frac{\partial^2 u}{\partial y^2} + \frac{\partial^2 u}{\partial z^2}\right). \tag{1.37}$$

In the case when the temperature distribution throughout the body reaches the steady state, i.e., when the temperature becomes independent of time, Eq (1.37) reduces to the *Laplace equation*

$$\nabla^2 u = \frac{\partial^2 u}{\partial x^2} + \frac{\partial^2 u}{\partial y^2} + \frac{\partial^2 u}{\partial z^2} = 0. \ \blacksquare \tag{1.38}$$

EXAMPLE 1.10. (*Heat conduction equation in $R^1 \times R^+$*) Consider a laterally insulated rod of uniform cross section with area A and constant density ρ, constant specific heat c, and constant thermal conductivity a. We assume that the temperature $u(x,t)$ is a function of x and t only, $t > 0$, and use the law of conservation of energy to derive the heat conduction equation. Consider a segment PQ of the rod, with coordinates x and $x + \Delta x$ (Fig. 1.3). Let R denote the rate at which the heat is accumulating on the segment PQ. Then, assuming that there are no heat sources or sinks in the rod, the rate R is given by

$$R = \int_x^{x+\Delta x} \frac{\partial c\rho A u(\xi, t)}{\partial t}\, d\xi.$$

Note that R can also be evaluated as the total flux across the boundaries of the segment PQ, which gives $R = aA[u_x(x + \Delta x, t) - u_x(x, t)]$. Now, using the mean-value theorem for integrals, we have

$$c\rho A\, u_t(x + h\Delta x, t)\, \Delta x = aA[u_x(x + \Delta x, t) - u_x(x, t)], \quad 0 < h < 1.$$

After dividing both sides by $c\rho A\Delta x$ and taking the limit as $\Delta x \to 0$, we get

$$u_t(x, t) = k\, u_{xx}(x, t). \ \blacksquare$$

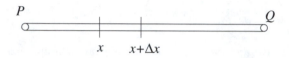

P 　　　　　　　　　　　　　　　　　　　　　　　　Q

x　　　$x+\Delta x$

Fig. 1.3. Segment PQ on a thin uniform rod.

EXAMPLE 1.11. (*One-dimensional traffic flow problem*) Let $\rho(x, t)$ denote the traffic density which represents the number of vehicles per mile at time t at an arbitrary yet fixed position x on a roadway. Let $q(x, t)$ denote the traffic flow which is a measure of number of vehicles per hour passing a fixed position x. Consider a section of the roadway bounded by the positions $x = x_1$ and $x = x_2$, and assume that there are no exits or entrances between these two positions. Then the number N of vehicles in the segment $[x_1, x_2]$ is given by $N = \int_{x_1}^{x_2} \rho(x, t)\, dt$. The rate of change of N with respect to time t is equal to the difference between the number of vehicles per unit time entering the position at $x = x_1$ and the number leaving at the position $x = x_2$, i.e.,

$$\frac{\partial N}{\partial t} = \frac{\partial}{\partial t} \int_{x_1}^{x_2} \rho(x, t)\, dx = q(x_1, t) - q(x_2, t). \tag{1.39}$$

As in the case of heat conduction (Example 1.9), Eq (1.39), which is also known as the integral representation of conservation of vehicles, can be written as

$$q(x_1, t) - q(x_2, t) = -\int_{x_1}^{x_2} \frac{\partial}{\partial x} q(x, t)\, dx = \frac{\partial}{\partial t} \int_{x_1}^{x_2} \rho(x, t)\, dx.$$

After interchanging the partial derivative with the integral on the right side and noting that x_1 and x_2 are arbitrary points, we obtain the partial differential equation

$$\frac{\partial \rho}{\partial t} + \frac{\partial q}{\partial x} = 0. \tag{1.40}$$

Let $u(x, t)$ denote the velocity of a vehicle. Then, since the number of vehicles per hour passing a given position is equal to the density of vehicles times the velocity of vehicles, we have

$$q(x, t) = \rho(x, t)u(x, t).$$

If we assume that the velocity u depends only on the density ρ, $u = u(\rho)$, i.e., the vehicles slow down as the traffic density increases, then $\dfrac{\partial u}{\partial \rho} \leq 0$. This inequality implies that the traffic flow depends only on the traffic density, i.e., $q = q(\rho)$. Then Eq (1.40) reduces to

$$\frac{\partial \rho}{\partial t} + \frac{\partial q}{\partial \rho}\frac{\partial \rho}{\partial x} = 0,$$

or

$$\frac{\partial \rho}{\partial t} + c(\rho)\frac{\partial \rho}{\partial x} = 0, \tag{1.41}$$

where $c(\rho) = \partial q/\partial \rho$. Eq (1.41) is a first-order homogeneous quasilinear partial differential equation. ∎

1.5. Superposition Principle

Let L denote a linear differential operator of any order and any kind. The superposition principle for homogeneous and nonhomogeneous linear differential equations is represented by the following two theorems:

Theorem 1.2. *Let $Lu = 0$ be a differential equation. Suppose u_1 and u_2 are two linearly independent solutions. Then $c_1 u_1 + c_2 u_2$ is also a solution.*

PROOF. By hypotheses $Lu_{1,2} = 0$. By definition $L(c_1 u_1 + c_2 u_2) = c_1 L u_1 + c_2 L u_2 = 0$. ∎

Theorem 1.3. *If $Lu = \sum_1^n c_i f_i$ is a nonhomogeneous linear differential equation and if $L g_k = f_k$, then $\sum_1^n c_i g_i$ is a solution of the above differential equation.*

PROOF. $L \sum_1^n c_i g_i = \sum_1^n c_i L g_i = \sum_1^n c_i f_i$. Thus $\sum_1^n c_i g_i$ satisfies the differential equation. ∎

It is obvious from these two theorems that if v is a solution of an equation $Lu = 0$ and if F is a solution of $Lu = f$, then $v + F$ is also a solution of $Lu = f$. A generalized superposition principle is as follows:

Theorem 1.4. *If each of the functions u_i, $i = 1, 2, \ldots$, is a solution of a linear homogeneous differential equation $L(u) = 0$, then the series $u = \sum_{i=1}^{\infty} C_i u_i$ is also a solution of this differential equation, provided that u and its derivatives appearing in $L(u)$ can be differentiated term-by-term.*

PROOF. If the derivatives of u appearing in $L(u) = 0$ can be differentiated term-by-term, we have

$$\frac{\partial^n u}{\partial x^m \partial t^{n-m}} = \sum_{i=1}^{\infty} C_i \frac{\partial^n u_i}{\partial x^m \partial t^{n-m}},$$

and since the equation $L(u) = 0$ is linear and a convergent series can be differentiated term-by-term, we can write

$$L(u) = L\left(\sum_{i=1}^{\infty} C_i u_i\right) = \sum_{i=1}^{\infty} C_i L(u_i) = 0.$$

The sufficient condition for term-by-term differentiability is the uniform convergence of the series $\displaystyle\sum_{i=1}^{\infty} C_i \frac{\partial^n u_i}{\partial x^m \partial t^{n-m}}$. ∎

1.6. Mathematica Projects

PROJECT 1.1. The classification of a given second-order partial differential equation into its type can be obtained by loading the Mathematica package Equation-Type.m, and the Notebook EquationType.nb found on the CRC website mentioned in the Preface.

PROJECT 1.2. Solutions of all exercises in the next section can be found in the Mathematica Notebook EquationType.nb on the CRC website. Although some of these exercises can be solved orally, the Mathematica files provide insight into the manipulative ease of Mathematica commands. Readers with advanced knowledge of Mathematica can skip this session.

1.7. Exercises

Classify the partial differential equation as hyperbolic, parabolic, or elliptic:

1.1.	$u_{xx} - 3u_{xy} + 2u_{yy} = 0$	ANS. Hyperbolic
1.2.	$4u_{xx} - 7u_{xy} + 3u_{yy} = 0$	ANS. Hyperbolic
1.3.	$u_{xx} + a^2 u_{yy} = 0,\ a \neq 0$	ANS. Elliptic
1.4.	$a^2 u_{xx} + 2a u_{xy} + u_{yy} = 0,\ a \neq 0$	ANS. Parabolic
1.5.	$4u_{tt} - 12u_{xt} + 9u_{xx} = 0$	ANS. Parabolic

1.6. $2u_{xt} + 3u_{tt} = 0$ Ans. Hyperbolic
1.7. $u_{xx} + 2u_{xy} + 5u_{yy} = 0$ Ans. Elliptic
1.8. $8u_{xx} - 2u_{xy} - 3u_{yy} = 0$ Ans. Hyperbolic

For what values of x and y are the following partial differential equations hyperbolic, parabolic, or elliptic?

1.9. $u_{xx} - xu_{yy} = 0$.
   Ans. Hyperbolic for $x > 0$, parabolic for $x = 0$, and elliptic for $x < 0$.

1.10. $u_{xx} - 2xu_{xy} + yu_{yy} = 0$.
   Ans. Hyperbolic for $x^2 > y$, parabolic for $x^2 = y$, and elliptic for $x^2 < y$.

1.11. $u_{xx} + 2xu_{xy} + (1 - y^2)u_{yy} = 0$.
   Ans. Hyperbolic for $x^2 + y^2 > 1$, parabolic for $x^2 + y^2 = 1$, and elliptic for $x^2 + y^2 < 1$.

1.12. $xu_{xx} + xu_{xy} + yu_{yy} = 0$.
   Ans. Hyperbolic for $x^2 > 4xy$, parabolic for $x^2 = 4xy$, and elliptic for $x^2 < 4xy$.

1.13. $(1 + y^2)u_{xx} + (1 + x^2)u_{yy} = 0$.
   Ans. Elliptic for all x and y.

1.14. $u_{xx} + xu_{xy} + yu_{yy} - xyu_y = 0$.
   Ans. Hyperbolic for $x^2 > 4y$, parabolic for $x^2 = 4y$, and elliptic for $x^2 < 4y$.

1.15. $u_{xx} + x^2u_{yy} = 0$.
   Ans. Elliptic for $x \neq 0$, and parabolic for $x = 0$.

1.16. $u_{xx} - 2\sin x\, u_{xy} - \cos^2 x\, u_y = 0$.
   Ans. Hyperbolic.

1.17. $yu_{xx} + u_{yy} = 0$.
   Ans. (Tricomi equation) Hyperbolic for $y < 0$, parabolic for $y = 0$, and elliptic for $y > 0$.

1.18. $u_{xx} - 2\cos x\, u_{xy} - (3 + \sin^2 x)\, u_{yy} - y\, u_y = 0$.
   Ans. Hyperbolic for all x.

1.19. $(1 + x^2)u_{xx} + (1 + y^2)u_{yy} + xu_x + yu_y = 0$.
   Ans. Elliptic for all x and y.

1.20. $x^2u_{xx} - y^2u_{yy} = 0$ $x > 0, y > 0$.
   Ans. Hyperbolic.

1.21. $y^2u_{xx} + x^2u_{yy} = 0$, $x > 0, y > 0$.
   Ans. Elliptic.

1.22. $x^2u_{xx} + 2xyu_{xy} + y^2u_{yy} = 0$.
   Ans. Parabolic for all nonzero x and y.

1.23. $(1 - x)u_{xx} - u_{yy} - u_x = 0$.
   Ans. Hyperbolic for $x < 1$, parabolic for $x = 1$, and elliptic for $x > 1$.

1.24. $x^2 u_{xx} - y^2 u_{yy} - 2y u_y = 0$.

    ANS. Hyperbolic for all nonzero x and y.

1.25. $x^2 u_{xx} - 2xy u_{xy} + y^2 u_{yy} + x u_x + y u_y = 0$.

    ANS. Parabolic for all nonzero x and y.

1.26. $u_{xx} - y u_{yy} - \dfrac{1}{2} u_y = 0$.

    ANS. Hyperbolic for $y > 0$, parabolic for $y = 0$, and elliptic for $y < 0$.

2

Method of Characteristics

It is customary in modern textbooks on partial differential equations to either completely ignore first-order partial differential equations or postpone their discussion to a later chapter. Since first-order partial differential equations are important from both physical and geometrical standpoints, their study is essential to understand the nature of solutions and form a guide to the solutions of higher-order partial differential equations.

First-Order Equations

First-order partial differential equations occur in a variety of physical problems. Some of the common ones are traffic flow, conservation laws, Mainardi-Codazzi relations in differential geometry, and shock waves. To solve first-order linear, quasilinear, or nonlinear partial differential equations, the method of characteristics is very useful. This method is explained in the next six sections. Second-order equations, confined to linear and quasilinear equations, are discussed in §2.7. We will limit our discussion to problems in R^2. An extension to higher dimensions, though routine, is more complicated.

2.1. Linear Equations with Constant Coefficients

The most general form of first-order linear partial differential equations with constant coefficients is

$$a\,u_x + b\,u_y + k\,u = f(x,y). \tag{2.1}$$

If $u(x,y)$ is a solution of Eq (2.1), then

$$du = u_x dx + u_y dy. \tag{2.2}$$

Comparing Eqs (2.1) and (2.2), we get the auxiliary system of equations

$$\frac{dx}{a} = \frac{dy}{b} = \frac{du}{f(x,y) - k\,u}. \tag{2.3}$$

The solution of the left pair is $bx - ay = c$. The other pair

$$\frac{dx}{a} = \frac{du}{f(x,y) - ku}$$

is reduced to an ordinary linear differential equation with u as the dependent variable and x as the independent variable. This equation is given by

$$\frac{du}{dx} + \frac{ku}{a} = \frac{f(x, (bx - c)/a)}{a},$$

for which the integrating factor is $e^{kx/a}$. This observation suggests that we introduce a new dependent variable $v = ue^{kx/a}$, which reduces Eq (2.1) to

$$a\,v_x + b\,v_y = f(x,y)e^{kx/a} \equiv g(x,y).$$

Note that a reduction can also be obtained by substituting $v = u\,e^{ky/b}$, which will lead to $av_x + bv_y = f(x,y)\,e^{ky/b}$. Thus, we need to consider only the formal reduced form

$$a\,u_x + b\,u_y = f(x,y). \tag{2.4}$$

The auxiliary system of equations for Eq (2.4) is

$$\frac{dx}{a} = \frac{dy}{b} = \frac{du}{f(x,y)}. \tag{2.5}$$

The solution of $\dfrac{dx}{a} = \dfrac{dy}{b}$ is $bx - ay = c$, or $ay - bx = c$, which, when solved for x, gives $x = \dfrac{ay - c}{b}$; a substitution of this value into $\dfrac{dy}{b} = \dfrac{du}{f(x,y)}$ yields

$$\frac{dy}{b} = \frac{du}{f\left(\dfrac{ay - c}{b}, y\right)},$$

which reduces to $du = F(y,c)dy$. Its solution is $u = G(y,c) + c_1$, where $G_y(y,c) = F(y,c)$. Thus, the general solution is obtained by replacing c_1 by $\phi(c)$ and c by $ay - bx$, thereby yielding

$$u(x,y) = G(y, ay - bx) + \phi(ay - bx),$$

and the solution of Eq (2.1) is

$$u(x, y) = \left[G(y, ay - bx) + \phi(ay - bx) \right] e^{-kx/a}.$$

Eqs (2.3) are known as the *equations of the characteristics*. These equations contain two independent equations, with two solutions of the form $F(x, y, u) = c_1$ and $G(x, y, u) = c_2$. Each of these solutions represents a family of surfaces. The curves of intersection of these two families of surfaces are known as the *characteristics* of the partial differential equation. The projections of these curves in the (x, y)-plane are called the *base characteristics*. These base characteristics are sometimes called characteristics for brevity when there is no ambiguity. But we will maintain this distinction between the characteristics and the base characteristics. The general solution represents a family of surfaces, and these surfaces are called *integral surfaces*.

Thus, the equation $ay - bx = c$ represents a family of planes. The intersection of any one of these planes with an integral surface is a curve whose projection in the (x, y)-plane is again given by $ay - bx = c$, but this time this equation represents a straight line and is the base characteristic. Thus, the solution u on a base characteristic $ay - bx = c$ is given by $u = G(y, c) + c_1$, and the general solution is the same as above.

An *alternative procedure* is to introduce a new set of coordinates

$$\xi = ay - bx, \quad \text{and} \quad \eta = ay + bx. \tag{2.6}$$

Their substitution into Eq (2.4) reduces it to

$$u_\eta = F(\xi, \eta), \quad \text{where} \quad F(\xi, \eta) = \frac{f\left((\eta - \xi)/(2b), (\eta + \xi)/(2a)\right)}{2ab}, \tag{2.7}$$

and the solution of (2.7) is

$$u(\xi, \eta) = \phi(\xi) + G(\xi, \eta), \tag{2.8}$$

where $G_\eta(\xi, \eta) = F(\xi, \eta)$. If $f(x, y) = 0$ in Eq (2.4), then the auxiliary system of equations is $dx/a = dy/b, du = 0$. The solutions of these equations are $ay - bx = c$, and $u = c_1 = \phi(c) = \phi(ay - bx)$. This procedure can also be regarded as a problem in rotation of axes (see Exercise 2.4).

Note that Eq (2.8) is

$$u(x, y) = \phi(ay - bx) + G\left(ay - bx, ay + bx\right).$$

If an initial condition $u(x, \psi(x)) = \mu(x)$ is prescribed, then

$$u\left(x, \psi(x)\right) = \mu(x) = \phi(a\psi(x) - bx) + G(a\psi(x) - bx, a\psi(x) + bx)$$

can be used to determine $\phi(x)$ uniquely. Thus, the existence of a unique solution u for the partial differential equation (2.1) subject to the above initial condition is established. For the existence and uniqueness of the solution, see also the end of §2.7.

EXAMPLE 2.1. Consider $2u_x - 3u_y = \cos x$. The auxiliary system of equations is

$$\frac{dx}{2} = \frac{dy}{-3} = \frac{du}{\cos x}.$$

The first solution is given by $3x + 2y = c$. The other equation is $\dfrac{dx}{2} = \dfrac{du}{\cos x}$. Its solution is $u = c_1 + \dfrac{1}{2}\sin x$. Noting that $c_1 = f(c)$ and $3x + 2y = c$, the general solution becomes

$$u = f(3x + 2y) + \frac{1}{2}\sin x.$$

Alternatively, the substitution $\xi = 3x + 2y$, $\eta = 3x - 2y$ reduces the given equation to

$$u_\eta = \frac{1}{12}\cos\frac{(\xi + \eta)}{6},$$

which yields

$$u = f(\xi) + \frac{1}{2}\sin\frac{(\xi + \eta)}{6}.$$

After replacing ξ and η by their values in x and y, we obtain the above general solution.

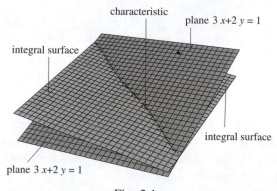

characteristic

plane 3 x+2 y = 1

integral surface

integral surface

plane 3 x+2 y = 1

Fig. 2.1.

In this problem the characteristics are given by the curves of intersection of the planes $3x + 2y = c$ and the integral surfaces $u = \dfrac{1}{2}\sin x + c_1$. The projections of these curves on the (x, y)-plane $u = 0$ are the base characteristics. Graphs of these characteristics are shown in Fig. 2.1 for $c = 1$ and $c_1 = 0$. If a linear partial differential equation is of the form $P(x, y)\, u_x + Q(x, y)\, u_y = 0$, then the base characteristics and the characteristics are the same curves. Now, we develop solutions for some specific conditions prescribed on initial curves. For example, if $u = 1$ on the initial curve $y = 0$, then

$$u = 1 + \frac{1}{2}\left(\sin x - \sin\frac{3x + 2y}{3}\right),$$

which is an integral surface denoted in Fig. 2.2 by S_1. Also, if $u = x^2$ on the initial curve $y = x$, then

$$u = \frac{1}{2}\sin x + \frac{(3x + 2y)^2}{25} - \frac{1}{2}\sin\frac{3x + 2y}{5},$$

which is another integral surface denoted by S_2. The graphs of the integral surfaces S_1 and S_2 and the characteristics are shown in Fig. 2.2. Note that an initial curve (or initial line) is a curve where an initial condition on u is prescribed. ∎

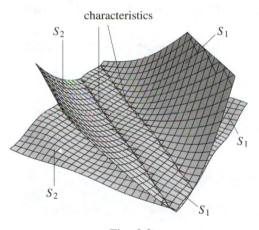

Fig. 2.2.

It is obvious that the solution of a first-order linear partial differential equation represents a surface and contains an arbitrary function and not an arbitrary constant. Clearly, the solution represents a family of surfaces. A unique surface is obtained if u is prescribed on an initial curve, which is not a characteristic. The reader can observe that the existence and uniqueness of the solution of a first-order equation are closely related to the existence and uniqueness of the solutions of the auxiliary system (2.5) of ordinary differential equations (see end of §2.7).

EXAMPLE 2.2. Consider $4u_x + u_y = x^2 y$. The auxiliary system of equations is

$$\frac{dx}{4} = \frac{dy}{1} = \frac{du}{x^2 y}.$$

The first solution is then given by $x - 4y = c$, and the solution, following the method of the previous example, is obtained as $x^2 y \, dx = 4 \, du$, which on using the first solution becomes $du = \dfrac{1}{16} \left(x^3 - cx^2 \right) \, dx$. On integration we have

$$u = c_1 + \frac{3x^4 - 4cx^3}{192} = f(c) + \frac{3x^4 - 4cx^3}{192}.$$

After replacing c by $x - 4y$, we get the general solution

$$u = f(x - 4y) + \frac{3x^4 - 4(x - 4y)x^3}{192} = f(x - 4y) - \frac{x^4}{192} + \frac{x^3 y}{12}. \quad \blacksquare$$

2.2. Linear Equations with Variable Coefficients

The general form of first-order linear partial differential equations with variable co-efficients is

$$P(x, y)u_x + Q(x, y)u_y + f(x, y)u = R(x, y). \tag{2.9}$$

Once again our attempt is to eliminate the term in u from Eq (2.9). This can be accomplished by substituting

$$u = ve^{-\zeta(x,y)},$$

where $\zeta(x, y)$ satisfies the equation

$$P(x, y)\, \zeta_x(x, y) + Q(x, y)\, \zeta_y(x, y) = f(x, y).$$

Hence, Eq (2.9) is formally reduced to

$$P(x, y)u_x + Q(x, y)u_y = R(x, y), \tag{2.10}$$

where P, Q, R in (2.10) are not the same as in (2.9). The method for solving these equations, known as *Lagrange's method*, is essentially the same as in the previous section except that now the auxiliary system of equations is

$$\frac{dx}{P} = \frac{dy}{Q} = \frac{du}{R}, \tag{2.11}$$

which becomes more complicated. This system has two solutions of the type

$$g(x, y, u) = c_1, \quad \text{and} \quad h(x, y, u) = c_2,$$

which represent general equations for two families of surfaces. The curves of intersection of these surfaces are the characteristics of Eq (2.10). The projection of a characteristic in the plane $u = 0$ is the base characteristic.

If $R(x, y) = 0$, then there is no difference in the base characteristics and the characteristics. Frequently the word 'base' is omitted from the term 'base characteristic.' These characteristics are clearly a one-parameter family of curves. In some cases it is convenient to introduce a parameter, say s, in the auxiliary system of equations, which are then expressed in the form

$$\frac{dx}{P} = \frac{dy}{Q} = \frac{du}{R} = ds. \tag{2.12}$$

EXAMPLE 2.3. Consider

$$3u_x + 4u_y + 14(x + y)u = 6x\, e^{-(x+y)^2}.$$

The first step is to find a function $\zeta(x, y)$ which satisfies the equation

$$3\zeta_x + 4\zeta_y = 14(x + y).$$

The characteristic equations are

$$\frac{dx}{3} = \frac{dy}{4} = \frac{d\zeta}{14(x + y)},$$

which are equivalent to $14(x + y)\, dx = 3\, d\zeta$ and $14(x + y)\, dy = 4\, d\zeta$. Adding these equations, we get $14(x + y)\, (dx + dy) = 7\, d\zeta$, which after integration gives a particular solution of the given equation as $\zeta(x, y) = (x + y)^2$. Then the substitution $u = ve^{-\zeta(x,y)}$ reduces the given equation to

$$3v_x + 4v_y = 6x,$$

with auxiliary equations

$$\frac{dx}{3} = \frac{dy}{4} = \frac{dv}{6x}.$$

The solution of $\dfrac{dx}{3} = \dfrac{dy}{4}$ is $4x - 3y = c_1$, and solution of $\dfrac{dx}{3} = \dfrac{dv}{6x}$ is $v = x^2 + c_2$.
Now, as before, v is easily found to be $v = x^2 + f(4x - 3y)$. On further examination we find that if we write $c_1 = 4x - 3y = g(x, y, v)$ and $c_2 = v - x^2 = h(x, y, v)$, and consider an expression of the form $F(g, h) = 0$, then

$$F_x = F_g(g_x + g_v v_x) + F_h(h_x + h_v v_x) = 0,$$

which reduces to

$$4F_g + (v_x - 2x)F_h = 0.$$

Similarly, the expression for F_y yields

$$-3F_g + v_y F_h = 0.$$

Eliminating F_g and F_h from these two equations, we get

$$\begin{vmatrix} 4 & v_x - 2x \\ -3 & v_y \end{vmatrix} = 0,$$

or

$$3v_x + 4v_y = 6x.$$

Thus, $F(g, h) = 0$ is also a solution of the differential equation. Of course, we must replace v by $u\,e^{(x+y)^2}$ to obtain the solution of the original problem. ∎

Solutions of the type $F(g, h) = 0$ or $g = F(h)$ or $h = F(g)$ are known as *general solutions*.

DEFINITION 2.1. Two C^1 functions g and h are said to be *functionally independent* if $\nabla g \times \nabla h \neq \mathbf{0}$.

We state two important results.

Theorem 2.1. *Let $g = c_1$ and $h = c_2$ be any two functionally independent solutions of the characteristic equations of Eq (2.9). Then $g = F(h)$, or $h = F(g)$, or $F(g, h) = 0$ represents the general solution of Eq (2.9).*

Corollary 2.1. *If $g = c_1$ is a solution of Eq (2.9), then $F(g) = 0$ is also a solution of Eq (2.9).*

EXAMPLE 2.4. Let us further consider the reduced equation $3v_x + 4v_y = 6x$ from Example 2.3. The auxiliary equations in the parametric form are

$$\frac{dx}{3} = \frac{dy}{4} = \frac{dv}{6x} = ds.$$

The solutions are

$$x = 3s + c_1, \quad \text{and} \quad y = 4s + c_2, \tag{2.13}$$

and, therefore,

$$v = 9s^2 + 6c_1 s + c_3. \tag{2.14}$$

The last solution is obtained by first substituting $x = 3s + c_1$ in $\dfrac{dv}{6x} = ds$. The values of x, y, v in terms of s represent the parametric form of the equation of the characteristics of the given partial differential equation. To find a specific characteristic, we must have an initial condition, e.g., $x = x_0$, $y = y_0$, $v = v_0$ at $s = s_0$. By eliminating s from (2.13) and (2.14), we get

$$4x - 3y = 4c_1 - 3c_2 = \alpha \quad \text{and} \quad v = x^2 - c_1^2 + c_3 = x^2 + \beta,$$

from which we get

$$v = x^2 + f(4x - 3y). \blacksquare$$

Note that, except for singular cases, a unique characteristic will, in general, pass through a point in space. Thus, if a continuous initial curve is prescribed, then a unique characteristic passes through every point of the initial curve. The locus of these characteristics will form the integral surface. Hence, if the initial curve is a characteristic itself, then the existence of an integral surface cannot be guaranteed (see end of §2.7).

EXAMPLE 2.5. Consider

$$u_x + e^x u_y = y, \quad u(0, y) = 1 + y.$$

The auxiliary system of equations is

$$\frac{dx}{1} = \frac{dy}{e^x} = \frac{du}{y}. \tag{2.15}$$

The system (2.15) is equivalent to two equations: $dy = e^x \, dx$, and $du = y \, dx$. The first equation has the solution $y = e^x + c$, which reduces the second equation to $du = (e^x + c) \, dx$, whose solution is $u = e^x + cx + c_1$. Hence,

$$u = e^x + cx + f(c), \quad \text{or} \quad u = e^x + (y - e^x)x + f(y - e^x).$$

Now, in view of the initial condition $u(0, y) = 1 + f(y - 1) = 1 + y$, we get $f(y) = y + 1$. Hence, the solution is

$$u(x, y) = e^x + (y - e^x)x + y - e^x + 1 = 1 + y + xy - xe^x. \blacksquare$$

EXAMPLE 2.6. Consider

$$2y \, u_x + u_y = x, \quad u(0, y) = f(y).$$

The auxiliary system of equations is

$$\frac{dx}{2y} = \frac{dy}{1} = \frac{du}{x}.$$

In this example a slightly different procedure will be demonstrated. We will first find the equation of the characteristics through an arbitrary point (x_0, y_0, u_0) on the integral surface. The left pair in the auxiliary equations is $dx = 2y \, dy$, whose solution is the family of surfaces

$$S_1 : x = y^2 + c.$$

When a surface of the family S_1 passes through the point (x_0, y_0, u_0), its equation becomes $x = y^2 + x_0 - y_0^2$. If the initial curve is $x = 0$, where $u(0, y) = f(y)$, then the value of y on the initial curve is given by $\hat{y}_0^2 = y_0^2 - x_0$, where $\hat{y}_0$ is the value of y on the integral surface through (x_0, y_0, u_0) at $x = 0$. The right pair of auxiliary equations is $du = x \, dy$, which also represents a family of surfaces. Let us denote the integral surface through $(0, \hat{y}_0, f(\hat{y}_0))$ by S_2. The curve of intersection of S_1 and S_2 is a characteristic (see Fig. 2.3). The differential equation of this characteristic is given by $du = (y^2 + x_0 - y_0^2) \, dy$. Its solution is

$$u = \frac{y^3}{3} + (x_0 - y_0^2)y + c_1.$$

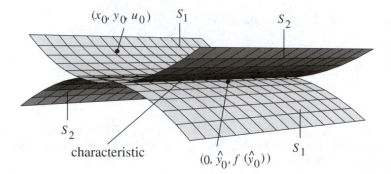

(x_0, y_0, u_0) S_1 S_2

S_2

characteristic $(0, \hat{y}_0, f(\hat{y}_0))$ S_1

Fig. 2.3.

At $x = 0$,

$$u(0, \hat{y}_0) = \frac{\hat{y}_0^3}{3} + (x_0 - y_0^2)\,\hat{y}_0 + c_1 = f(\hat{y}_0),$$

and substituting the value of $\hat{y}_0$, we get

$$c_1 = f((y_0^2 - x_0)^{1/2}) + \frac{2}{3}(y_0^2 - x_0)^{3/2}.$$

Using this value of c_1 in u, the value u_0 of u at (x_0, y_0) is given by

$$u_0 \equiv u(x_0, y_0) = \frac{2}{3}(y_0^2 - x_0)^{3/2} - \frac{2}{3}y_0^3 + x_0 y_0 + f((y_0^2 - x_0)^{1/2}).$$

Since (x_0, y_0, u_0) is an arbitrary point, the expression for $u(x_0, y_0)$ is generalized to

$$u(x, y) = \frac{2}{3}(y^2 - x)^{3/2} - \frac{2}{3}y^3 + xy + f((y^2 - x)^{1/2}). \; \blacksquare$$

EXAMPLE 2.7. Consider

$$(x + 2y)u_x + (y - x)u_y = y.$$

The auxiliary equations are

$$\frac{dx}{x + 2y} = \frac{dy}{y - x} = \frac{du}{y},$$

which are equivalent to two equations:

$$\frac{dy}{dx} = \frac{y - x}{x + 2y}, \quad \text{and} \quad dx + dy = 3\,du.$$

The first of these equations is a homogeneous ordinary differential equation of the first order. A standard substitution for such problems is $y = vx$, which leads to a first-order ordinary differential equation with separable variables

$$\frac{(1 + 2v)dv}{1 + 2v^2} = -\frac{dx}{x}.$$

This equation can be solved to give

$$\sqrt{2}\tan^{-1}\frac{\sqrt{2}y}{x} + \ln(x^2 + 2y^2) = c_1.$$

The other solution can be obtained by solving the second equation, which gives $u = \frac{1}{3}(x + y) + c_2$. Hence, the general solution becomes

$$u = \frac{1}{3}(x + y) + f\left[\sqrt{2}\tan^{-1}\frac{\sqrt{2}y}{x} + \ln(x^2 + 2y^2)\right]. \blacksquare$$

2.3. First-Order Quasilinear Equations

If the coefficients P, Q, and R in Eq (2.9) are functions of x, y, and u, but not of u_x and u_y, where P, Q, and R may be nonlinear in u, then the equation is known as *quasilinear*. However, in these equations the first-order derivatives occur only in the first degree, although the equation need not be linear in u. Such equations occur in shock waves of various kinds, e.g., traffic flow, water waves. The basic technique is the same as for first-order linear equations. The starting point is still the system of auxiliary equations (2.11).

EXAMPLE 2.8. Consider the quasilinear equation

$$u_x + u\,u_y = 0, \quad u(0, y) = f(y).$$

The system of auxiliary equations is

$$\frac{dx}{1} = \frac{dy}{u}, \quad du = 0.$$

Here $du = 0$ implies $u = c$. Using this value of u in $\frac{dx}{1} = \frac{dy}{u}$ we get $y = cx + c_1$. Thus, the characteristics are given by the curves of intersection of the surfaces $y - ux = c_1$ and $u = c$, and the general solution can be expressed as $u = g(y - ux)$. Applying the initial condition, we get $f(y) = g(y)$. Hence, the solution to the problem is $u = f(y - ux)$.

If $f(y) = y$, the solution becomes $u = \dfrac{y}{1 + x}$.

If $f(y) = y^2$, then we have $u = (y - ux)^2$, which after some algebraic simplification yields

$$u = \frac{1 + 2xy \pm \sqrt{1 + 4xy}}{2x^2}.$$

A careful examination by checking the limit as $x \to 0$ shows that the valid solution is

$$u = \frac{1 + 2xy - \sqrt{1 + 4xy}}{2x^2}. \quad \blacksquare$$

EXAMPLE 2.9. Consider

$$u_x + g(u)u_y = 0, \quad u(0, y) = f(y).$$

The system of auxiliary equations is

$$\frac{dx}{1} = \frac{dy}{g(u)}, \quad du = 0.$$

As in Example 2.8, $u = c$, and $y = g(c)x + c_1$. The general solution is $u = h(y - xg(u))$. After applying the initial condition, we get $f(y) = h(y)$. Hence, the solution to the problem is $u = f(y - xg(u))$. $\blacksquare$

EXAMPLE 2.10. Consider

$$u_x + u_y = u^2 + 1, \quad u(0, y) = f(y).$$

The auxiliary equations are

$$dx = dy = \frac{du}{u^2 + 1},$$

whose solutions are $y = x + c$, and $\tan^{-1} u = x + c_1$. Thus, the general solution is $u = \tan(x + g(y - x))$, and the particular solution for the problem is

$$u = \frac{\tan x + f(y - x)}{1 - f(y - x) \tan x}. \; \blacksquare$$

EXAMPLE 2.11. Consider the partial differential equation

$$2y\, u_x + u_y = 1,$$

where $u = 1$ is prescribed on the initial curve $y = 0$ (x-axis) for $0 \le x \le 1$. The auxiliary system (2.11) for this equation is

$$\frac{dx}{2y} = dy = du.$$

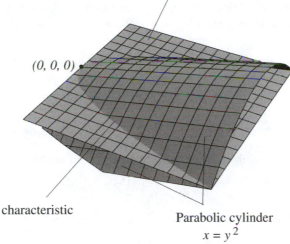

u = y

(0, 0, 0)

characteristic

Parabolic cylinder
$x = y^2$

Fig. 2.4.

The solution of $dx = 2y\, dy$ is the parabolic cylinder $x = y^2 + A$, while that of $du = dy$ is the plane $u = y + B$, where the parameters A and B are constant for each characteristic. For any point (x_0, y_0, u_0) we find that $A = x_0 - y_0^2$, and $B = u_0 - y_0$. If, for example, the point (x_0, y_0, u_0) is taken as the origin of the coordinate system, then

$$x = y^2, \quad \text{and} \quad u = y. \tag{2.16}$$

The equation of the characteristic is the intersecting curve (in this case the parabola) of the solution (2.16), as shown in Fig. 2.4.

The characteristic through the point $(1, 0, 1)$ is the intersection of the parabolic cylinder $x = y^2 + 1$ and the plane $u = y + 1$. Moreover, if $u = x^2$ on the initial curve $y = 0$, then $u = y + (x - y^2)^2$ represents the integral surface. ∎

EXAMPLE 2.12. Consider

$$u_x + 2uu_y = 1.$$

The auxiliary equations are

$$dx = \frac{dy}{2u} = du.$$

The solutions are $u = x + c$, and $u^2 = y + c_1$. Hence, the general solution is

$$u = x + f(u^2 - y), \quad \text{or} \quad u^2 = y + g(u - x).$$

It is important to choose the appropriate general solutions for the given initial conditions. Thus, for example, if the initial curve (line) is the line $x = y$, and the value of u on this initial line is $u(y, y) = y$, then from the first solution we get $u = x$. But the second solution gives $y^2 = y + g(0)$, which does not yield a value for the function g.

If the initial curve is $y = x$, and $u(y, y) = y^2$, then the second solution yields

$$u^2 = y + (u - x)^2 + 2(u - x) \pm (u - x)\sqrt{1 + 4(u - x)},$$

where the plus sign corresponds to $x > \dfrac{1}{2}$, $y > \dfrac{1}{2}$, and the minus sign corresponds to $x < \dfrac{1}{2}$, $y < \dfrac{1}{2}$. Substituting $y = x$ and $u = y^2$ in the second solution, we get $y^4 = y + g(y^2 - y)$. Now, let $z = y^2 - y$. Then

$$y = \frac{1}{2}\left(1 \pm \sqrt{1 + 4z}\right), \quad \text{and} \quad y^4 - y = z^2 + 2z \pm z\sqrt{1 + 4z} = g(z),$$

which gives the solution of the equation. But if we use the first solution we get $y^2 = y + f(y^4 - y)$, which is more difficult to resolve. ∎

EXAMPLE 2.13. Consider

$$(y + u)\, u_x + y\, u_y = x - y.$$

In this problem we will demonstrate the use of the auxiliary equations in parametric form (2.12). Thus, we have

$$\frac{dx}{y + u} = \frac{dy}{y} = \frac{du}{x - y} = ds,$$

which is rewritten as

$$\frac{dx}{ds} = y + u, \quad \frac{dy}{ds} = y, \quad \frac{du}{ds} = x - y.$$

The solution of the middle equation is $y = Ae^s$. Addition of the first and third equations yields

$$\frac{d(u + x)}{ds} = u + x.$$

Its solution is $u + x = Be^s$. Subtracting the first equation from the third results in

$$\frac{d(u - x)}{ds} = x - u - 2y,$$

which can be expressed as

$$\frac{d(u - x)}{ds} + (u - x) = -2Ae^s.$$

This equation is linear in $u - x$. Its solution is

$$u - x = Ce^{-s} - Ae^s.$$

Replacing e^s by $\dfrac{y}{A}$ in the two solutions, we get

$$u + x = \frac{B}{A}y, \quad \text{and} \quad u - x + y = \frac{CA}{y}.$$

Noting that B/A and CA can be replaced by c_1 and c_2, and that $c_2 = f(c_1)$, we have

$$(u - x + y)y = f\left(\frac{u + x}{y}\right) \tag{2.17}$$

as the general solution of the given equation. ∎

EXAMPLE 2.14. Consider

$$x\,y\,u\,u_x + (x^2 - u^2)\,u_y = x^2 y.$$

The auxiliary equations are

$$\frac{dx}{xyu} = \frac{dy}{x^2 - u^2} = \frac{du}{x^2 y}.$$

From

$$\frac{dx}{xyu} = \frac{du}{x^2 y},$$

we get $x^2 - u^2 = c_1$, and then using this solution in

$$\frac{dy}{x^2 - u^2} = \frac{du}{x^2 y},$$

we obtain

$$y\,dy = \frac{c_1\,du}{u^2 + c_1}.$$

This equation yields

$$y^2 = \begin{cases} 2a\tan^{-1}\dfrac{u}{a} + c_2, & \text{if } c_1 = a^2, \\[2ex] a\ln\dfrac{u-a}{u+a} + c_2, & \text{if } c_1 = -a^2. \end{cases}$$

The general solution can now be found. ∎

2.4. First-Order Nonlinear Equations

The general form of a first-order nonlinear equation is

$$F(x, y, u, u_x, u_y) = 0. \tag{2.18}$$

Consider the two-parameter family of surfaces

$$f(x, y, u, a, b) = 0. \tag{2.19}$$

Then

$$f_x + f_u u_x = 0, \quad \text{and} \quad f_y + f_u u_y = 0. \tag{2.20}$$

Equations (2.19) and (2.20) form a set of three equations in the two parameters a and b. If a and b are eliminated from these equations, we obtain an equation of the type (2.18). Therefore, it is reasonable to assume that the solution of (2.18) is of type (2.19). It is clear that any envelope* of this family will also be a solution of Eq (2.18). At this point we state the *Cauchy problem*: Determine $u(x, y)$ such that u and its partial derivatives satisfy

$$F(x, y, u, u_x, u_y) = 0,$$

subject to the condition $u(0, y) = \phi(y)$. In this case $u(x, y)$ is prescribed on the initial line $x = 0$ (y-axis). Initial data can, however, be prescribed on any simple

---

*An envelope of the family of surfaces $f(x, y, u, a, b) = 0$ is a surface which touches some member of this family at every point.

curve which is not a characteristic of Eq (2.18). Thus, for example, if the parametric form of the curve is $x = x(s), y = y(s)$, then the initial data can be written as

$$u(x(s), y(s)) = \psi(s).$$

We will now distinguish between different kinds of solutions of Eq (2.18).

A two-parameter family of solutions of the type $f(x, y, u, a, b) = 0$ is known as a *complete integral*.

If $b = g(a)$, where g is an arbitrary function, and an envelope of the family of solutions of the complete integral is found, then the envelope whose equation contains an arbitrary function is known as the *general integral* corresponding to the solution (2.19).

If the two-parameter family of solutions (2.19) has an envelope, then the equation of this envelope is known as the *singular integral* of Eq (2.18).

2.4.1. Cauchy's Method of Characteristics. This method is similar to the method of characteristics discussed earlier for linear and quasilinear partial differential equations.

Consider a first-order nonlinear equation (2.18). For convenience, we will use the notation

$$u_x = p \quad \text{and} \quad u_y = q. \tag{2.21}$$

If $u = u(x, y)$ is a solution of (2.18), then u and its partial derivatives satisfy Eq (2.18). But the total derivative of u is given by

$$du = p \, dx + q \, dy. \tag{2.22}$$

Differentiating (2.22) and (2.18) with respect to p, we get

$$dx + \frac{dq}{dp} \, dy = 0, \tag{2.23}$$

and

$$F_p + F_q \frac{dq}{dp} = 0. \tag{2.24}$$

Equations (2.23) and (2.24) yield

$$\frac{dx}{F_p} = \frac{dy}{F_q} = \frac{p \, dx + q \, dy}{p \, F_p + q \, F_q} = \frac{du}{p \, F_p + q \, F_q} = dt, \tag{2.25}$$

where t is a parameter. These are the *auxiliary equations* (also known as the equations of the characteristics) of Eq (2.18). It can be easily verified that they reduce to the characteristic equations of a linear or a quasilinear partial differential equation according as Eq (2.18) is linear or quasilinear. The parameter t introduced in (2.25) is such that

$$\frac{dx}{dt} = F_p, \quad \frac{dy}{dt} = F_q \quad \text{and} \quad \frac{du}{dt} = p\,F_p + q\,F_q. \tag{2.26}$$

Since p is a function of t, then

$$\frac{dp}{dt} = p_x \frac{dx}{dt} + p_y \frac{dy}{dt} = p_x F_p + p_y F_q = p_x F_p + q_x F_q, \quad (q_x = p_y). \tag{2.27}$$

Differentiating Eq (2.18) with respect to x, we have

$$F_x + F_u u_x + F_p p_x + F_q q_x = 0.$$

Using this equation and inserting the value of $F_p p_x + F_q q_x$ in (2.27), we get

$$\frac{dp}{dt} = -(F_x + F_u u_x) = -(F_x + pF_u). \tag{2.28}$$

Similarly,

$$\frac{dq}{dt} = -(F_y + qF_u). \tag{2.29}$$

Combining equations (2.25), (2.28), and (2.29) we find that

$$\frac{dx}{F_p} = \frac{dy}{F_q} = \frac{du}{pF_p + qF_q} = \frac{dp}{-(F_x + pF_u)} = \frac{dq}{-(F_y + qF_u)} = dt. \tag{2.30}$$

This system of auxiliary equations is then used to solve a nonlinear equation.

The difference between the auxiliary equations (2.25) and the corresponding equations (2.11) for a linear partial differential equation is that the equations (2.25) contain p and q explicitly and, therefore, in order to solve them we need additional equations which are included in (2.30). The solution is found by eliminating p and q from the solutions of (2.30) and the given equation. The eliminant will, in general, contain two arbitrary constants and represent a complete integral of the equation. We demonstrate the method by the following examples.

EXAMPLE 2.15. Consider

$$u = 4p\,q.$$

The auxiliary system (2.30) for this equation is

$$\frac{dx}{dt} = -4q, \quad \frac{dy}{dt} = -4p, \quad \frac{du}{dt} = -8pq, \quad \frac{dp}{dt} = -p, \quad \frac{dq}{dt} = -q. \tag{2.31}$$

Since

$$\frac{dx}{dt} = 4\frac{dq}{dt}, \quad \text{and} \quad \frac{dy}{dt} = 4\frac{dp}{dt},$$

we get

$$x + c_1 = 4q, \quad \text{and} \quad y + c_2 = 4p. \tag{2.32}$$

Substituting these values of p and q into the given equation, we get the complete solution as

$$u = \frac{1}{4}(x + c_1)(y + c_2). \tag{2.33}$$

However, if we demand that the solution pass through a given curve, then (2.33) may or may not yield the required solution. For example, if we require that $u(0, y) = y^2$ is the initial condition, then the solution given by (2.33) fails to yield any solution satisfying this initial condition. To avoid this situation, we follow a different approach. The initial values for x, y, and u at $t = 0$ can be taken to be $x_0 = 0$, $y_0 = \nu$, and $u_0 = \nu^2$, where ν is a parameter. Let the initial value of q be $q_0 = \left(\dfrac{\partial u}{\partial y}\right)_0$. Thus,

$$q_0 = \left(\frac{\partial u}{\partial y}\right)_0 = 2y_0 = 2\nu,$$

and since $u_0 = 4p_0 q_0$, we find that the initial value of p is $p_0 = \dfrac{\nu}{8}$. But from equations (2.31), we note that p and q are solved in terms of t as

$$p = Ae^{-t}, \quad \text{and} \quad q = Be^{-t}. \tag{2.34}$$

Substituting the initial values into (2.32) and (2.34), we find that

$$c_1 = 8\nu, \quad c_2 = -\nu/2, \quad A = \nu/8, \quad B = 2\nu.$$

Then, from (2.32) and the given equation, x, y, and u are expressed in terms of ν and t as

$$x = 8\nu(e^{-t} - 1), \quad y = \frac{\nu}{2}(e^{-t} + 1), \quad u = \nu^2 e^{-2t}. \tag{2.35}$$

The required solution is now found by eliminating ν and t from (2.35) as

$$u = \left(\frac{x}{16} + y\right)^2.$$

It is of interest to explore the relationship between the above solution and the complete integral (2.33). Let $c_1 = -16c_2$. Then Eq (2.33) becomes

$$4u = (x - 16c_2)(y + c_2). \tag{2.36}$$

The envelope of this family of surfaces is obtained by eliminating c_2 from (2.36) and the equation

$$4\frac{\partial u}{\partial c_2} = -16(y + c_2) + x - 16c_2 = 0,$$

resulting in $u = \left(\dfrac{x}{16} + y\right)^2$. Hence, $u = \left(\dfrac{x}{16} + y\right)^2$ is an envelope of (2.33). ∎

Now, we consider some special cases of Eq (2.18).

EXAMPLE 2.16. Consider

$$u = px + qy + f(p,q), \quad \text{or} \quad F = px + qy + f(p,q) - u = 0. \tag{2.37}$$

This is a special type of a nonlinear equation. It always has a complete solution which is obtained in a simple manner. The last two of the characteristic equations are

$$dp = 0, \quad \text{and} \quad dq = 0,$$

which yield $p = a$ and $q = b$, where a and b are arbitrary constants. Substituting these values into (2.37), we get $u = ax + by + f(a,b)$ as the complete solution. Equations of the type (2.37) are known as *Clairaut equations*. There are other special types of partial differential equations which yield the complete solution in a relatively easy manner. ∎

EXAMPLE 2.17. Consider

$$f(p,q) = 0. \tag{2.38}$$

Note that the characteristic equations yield $dp = 0$ and $dq = 0$. So $p = a$, and solution of $f(p,q) = 0$ for q gives $q = g(a)$. Then observing that

$$du = a\,dx + g(a)\,dy,$$

we get

$$u = ax + g(a)y + c.$$

As an example, let $f(p,q) = p^2 + q^2 - 1 = 0$. The auxiliary equations (2.30) are

$$\frac{dx}{dt} = p, \quad \frac{dy}{dt} = q, \quad du = dt, \quad dp = 0, \quad dq = 0.$$

Using $dp = 0$, we get $p = a$ and $q = \sqrt{1 - a^2}$, and these two combined with $du = p\,dx + q\,dy$ yield

$$u = ax + y\sqrt{1 - a^2} + c.$$

This is a complete solution. Another complete solution is obtained by using $p = a$ in $\dfrac{dx}{dt} = p$ and noting that $du = dt$, thus yielding $du = \dfrac{dx}{a}$, which gives

$$u = \frac{x}{a} + \alpha.$$

Similarly, $du = \dfrac{dy}{\sqrt{1 - a^2}}$ implies that

$$u = \frac{y}{\sqrt{1 - a^2}} + \beta.$$

Thus, $au = x + a\,\alpha$, and $u\sqrt{1 - a^2} = y + \beta\sqrt{1 - a^2}$. Replacing $a\alpha$ and $\beta\sqrt{1 - a^2}$ by $-c$ and $-d$, respectively, and eliminating a, we get

$$u^2 = (x - c)^2 + (y - d)^2,$$

which is another complete solution. ∎

EXAMPLE 2.18. Consider

$$F(u, p, q) = 0. \tag{2.39}$$

The last three terms of the system (2.30) yield

$$\frac{dp}{-pF_u} = \frac{dq}{-qF_u} = dt, \quad \text{or} \quad \frac{dp}{p} = \frac{dq}{q},$$

i.e., $p = A\,q$, where A is an arbitrary constant. This equation together with (2.39) is solved for p and q, and then we proceed as in the previous example. Thus, let $F(u, p, q) = u^2 + pq - 4 = 0$. Then $p = A\,q$, where we take A as a positive constant, say, $A = a^2$, and get

$$q = \pm\frac{1}{a}\sqrt{4 - u^2} \quad \text{and} \quad p = \pm a\sqrt{4 - u^2}.$$

Therefore, since $du = p\,dx + q\,dy$, we get

$$du = \pm\sqrt{4 - u^2}\left(a\,dx + \frac{1}{a}\,dy \right),$$

or

$$\frac{du}{\sqrt{4 - u^2}} = \pm\left(a\,dx + \frac{1}{a}\,dy \right),$$

which gives

$$\sin^{-1}\frac{u}{2} = \pm\left(a\,x + \frac{1}{a}\,y + c \right).$$

Hence,

$$u = \pm 2\sin\left(a\,x + \frac{1}{a}\,y + c \right).$$

If $p = -a^2 q$, then the solution is $u = \pm 2 \cosh(ax + y/a + c)$. ∎

If $F(x, y, u, p, q) = 0$ is independent of u and can be expressed as $\phi(x, p) = \psi(y, q)$, then each of these functions must be constant. Thus, if $\phi(x, p) = c$ and $\psi(y, q) = c$ are solved for p and q, then a complete integral is obtained. Some examples are as follows.

EXAMPLE 2.19. Consider

$$F(x, y, u, p, q) = p^2(1 - x^2) - q^2(4 - y^2) = 0.$$

Then

$$p^2(1 - x^2) = q^2(4 - y^2) = a^2,$$

which gives

$$p = \frac{a}{\sqrt{1 - x^2}}, \quad \text{and} \quad q = \frac{a}{\sqrt{4 - y^2}},$$

where we have ignored the negative solutions. Now, since $du = p\, dx + q\, dy$, we have

$$du = \frac{a}{\sqrt{1 - x^2}}\, dx + \frac{a}{\sqrt{4 - y^2}}\, dy,$$

which after integration gives

$$u = a\left(\sin^{-1} x + \sin^{-1}\frac{y}{2}\right) + b. \;∎$$

EXAMPLE 2.20. Consider

$$F(x, y, u, p, q) = 2pqy - pu - 2a = 0, \tag{2.40}$$

for which

$$F_x = 0, \quad F_y = 2pq, \quad F_u = -p, \quad F_p = 2qy - u, \quad \text{and} \quad F_q = 2py.$$

The auxiliary system of equations (2.30) are

$$\frac{dx}{2qy - u} = \frac{dy}{2py} = \frac{du}{2pqy - pu + 2pqy} = \frac{du}{pu + 4a} = \frac{dp}{p^2} = \frac{dq}{-pq}.$$

The last pair reduces to

$$\frac{dp}{p} + \frac{dq}{q} = 0,$$

which gives $pq = \alpha$. Using this value of pq in (2.40), we get

$$2\alpha y - pu - 2a = 0,$$

which yields

$$pu = 2\alpha y - 2a, \quad \text{and} \quad q = \frac{\alpha u}{2(\alpha y - a)}.$$

The latter equation is equivalent to $\dfrac{2}{u}\dfrac{\partial u}{\partial y} = \dfrac{\alpha}{\alpha y - a}$, which yields

$$2\ln u = \ln(\alpha y - a) + \ln \phi(x), \quad \text{or} \quad u^2 = (\alpha y - a)\,\phi(x),$$

where $\phi(x)$ is an arbitrary function. Hence,

$$2u\,\frac{\partial u}{\partial x} = (\alpha y - a)\,\phi'(x).$$

Comparing this with $pu = 2\alpha y - 2a$, we have $2pu = 4(\alpha y - a) = (\alpha y - a)\,\phi'(x)$, which gives $\phi'(x) = 4$, i.e., $\phi(x) = 4(x - \beta)$. Finally, after replacing $\phi(x)$ in the above expression for u^2, the solution is

$$u^2 = 4(\alpha y - a)(x - \beta). \ \blacksquare$$

EXAMPLE 2.21. We will find a complete solution of $u = \dfrac{1}{p} + \dfrac{1}{q}$ by two methods, and discuss the relationship between the two solutions.

METHOD 1. Since

$$F_x = F_y = 0, F_u = -1, F_p = -1/p^2, F_q = -1/q^2,$$

the auxiliary equations (2.30) are

$$\frac{dx}{dt} = -1/p^2, \quad \frac{dy}{dt} = -1/q^2, \quad \frac{du}{dt} = -\frac{1}{p} - \frac{1}{q} = -u,$$

$$\frac{dp}{dt} = p, \quad \frac{dq}{dt} = q.$$

Noting that $\dfrac{dp}{p\,dt} = 1$, and $\dfrac{dq}{q\,dt} = 1$, we can express the first two equations as $dx = -\dfrac{dp}{p^3}$, $dy = -\dfrac{dq}{q^3}$. Solving these we get $p^2 = \dfrac{1}{2x + a}$, and $q^2 = \dfrac{1}{2y + b}$. Taking the positive square roots, we get $u = \sqrt{2x + a} + \sqrt{2y + b}$ as a solution.

METHOD 2. The solutions of the characteristic equations can be expressed as $u = Ae^{-t}, p = Be^t, q = Ce^t$. Hence, $up = AB = A_1, uq = AC = A_2$. Using these values in the given equation, we get

$$u = \frac{1}{p} + \frac{1}{q}, \quad \text{which implies} \quad \frac{1}{A_1} + \frac{1}{A_2} = 1.$$

Integrating $up = AB = A_1, uq = AC = A_2$, we get $u^2 = 2A_1 x + g_1(y)$, and $u^2 = 2A_2 y + g_2(x)$. Comparing these two values of u^2, we find that

$$u^2 = 2(A_1 x + A_2 y + A_3) = 2\left(A_1 x + \frac{A_1}{A_1 - 1} y + A_3\right).$$

Now, we discuss the relationship between the two solutions

$$u = \sqrt{2x + a} + \sqrt{2y + b}, \quad \text{and} \quad u^2 = \left(Ax + \frac{A}{A - 1} y + B\right),$$

where we suppress the subscripts and replace A_3 by B. Define

$$b = -a(A - 1) + B(A - 1)/A.$$

Then

$$\phi(x, y, u) = -u + \sqrt{2x + a} + \sqrt{2y - a(A - 1) + B(A - 1)/A} = 0,$$

and

$$\phi_a = \frac{1}{2\sqrt{2x + a}} - \frac{A - 1}{2\sqrt{2y - a(A - 1) + B(A - 1)/A}} = 0.$$

Thus,

$$(A - 1)\sqrt{2x + a} = \sqrt{2y - a(A - 1) + B(A - 1)/A},$$

which yields $u = \sqrt{2x + a} + (A - 1)\sqrt{2x + a}$. Hence, $u = A\sqrt{2x + a}$, or $u^2 = 2A^2 x + A^2 a$. Now, solving $(A - 1)\sqrt{2x + a} = \sqrt{2y - a(A - 1) + B(A - 1)/A}$ for a and using the value so obtained in the expression for u^2, we get

$$u^2 = 2\left(Ax + \frac{A}{A - 1} y + B\right).$$

Obviously, the second solution is the envelope of the first. ∎

2.5. Geometrical Considerations

Before we discuss the geometrical interpretation of the partial differential equation (2.18), i.e., $F(x, y, u, p, q) = 0$, let us recall the geometrical interpretation of a first-order ordinary differential equation $y' = f(x, y)$. Here $f(x, y)$ represents the slope of any integral curve at the point (x, y). This slope is unique at every point. If we plot $f(x, y) = c$, then the curve so obtained is known as an *isocline* or a curve of constant slope. Of course, the curve itself does not have constant slope, but every integral curve which intersects $f(x, y) = c$ has the slope c at the point of intersection. Since the correspondence between integral curves and the points of an isocline is one-to-one, the number of integral curves is, in general, equal to the number of points on the isocline, i.e., there exists a single infinity of them. However, the exception to this rule occurs when isoclines intersect at a point, in which case the point is a singular point of the differential equation, or when the isocline is also a solution curve, in which case the isocline is a straight line with slope c and the isocline is an envelope of the integral curves, except for the equation $\dfrac{dy}{dx} = \dfrac{y}{x}$ whose isoclines and the integral curves are the same.

The situation for a partial differential equation is somewhat complicated. In this case the values of p and q are not unique at a fixed point (x, y, u). If an integral surface is $g(x, y, u) = 0$, then p and q represent the slopes of the curves of intersection of the surface with the planes $y = \text{const}$ and $x = \text{const}$, respectively. Moreover, p, q, -1 represent the direction ratios of the normal to the surface at the point (x, y, u). The derivatives p and q are constrained by Eq (2.18). Obviously, at a fixed point, p and q can be represented by a single parameter. Hence, there are infinitely many possible normals and consequently infinitely many integral surfaces passing through any fixed point. So, unlike the case of ordinary differential equations, we cannot determine a unique integral surface by making it pass through a point.

Cauchy established that a unique integral surface can be obtained by making it pass through a continuous twisted space curve, also known as an *initial curve*, except when the curve is a characteristic of the differential equation.

The infinity of normals passing through a fixed point generates a cone known as the *normal cone*. The corresponding tangent planes to the integral surfaces envelope a cone known as the *Monge cone*. In the case of a linear or a quasilinear equation, the normal cone degenerates into a plane since each normal is perpendicular to a fixed line. Consider the equation $a\,p + b\,q = c$, where a, b, and c are functions of x, y, and u. Then the direction $p, q, -1$ is perpendicular to the direction ratios a, b, c. This direction is fixed at a fixed point. The Monge cone then degenerates into a coaxial set of planes known as the *Monge pencil*. The common axis of the planes is the line through the fixed point with direction ratios a, b, c. This line is known as the *Monge axis*.

2.6. Some Theorems on Characteristics

Suppose $u(x, y) = f(x, y)$ is an integral surface S of the partial differential equation (2.18). Then the set of numbers

$$\left(x_0, y_0, \ u_0 = u(x_0, y_0), \ p_0 = \left(\frac{\partial f}{\partial x}\right)_{(x_0, y_0)}, \ q_0 = \left(\frac{\partial f}{\partial y}\right)_{(x_0, y_0)}\right),$$

which represents a plane with normal $(p_0, q_0, -1)$ and passing through the point (x_0, y_0, u_0), is called a *plane element*. If the point (x_0, y_0, u_0) lies on the surface S, then the element $(x_0, y_0, u_0, p_0, q_0)$ satisfies Eq (2.18) and is called an *integral element* of the surface. Let R denote a neighborhood of (x_0, y_0) in the plane $u = 0$. If the functions f_x and f_y are continuous in R, then the element $(x_0, y_0, u_0, p_0, q_0)$ is called a *tangent element* of S.

A curve Γ with parametric equations $x = x(t), y = y(t), u = u(t)$ lies on the surface S if $u(t) = f(x(t), y(t))$ for all admissible values of t. If a point P_0 on Γ corresponds to the value t_0 of the parameter t, then the direction of the tangent line is given by $\left(\dfrac{dx}{dt}, \dfrac{dy}{dt}, \dfrac{du}{dt} \right)_{t=t_0}$. This direction is normal to $\left(p_0 = \left(\dfrac{\partial u}{\partial x} \right)_{t_0}, q_0 = \left(\dfrac{\partial u}{\partial y} \right)_{t_0}, -1 \right)$ if

$$\left(\frac{du}{dt} \right)_{t_0} = p_0 \left(\frac{dx}{dt} \right)_{t_0} + q_0 \left(\frac{du}{dt} \right)_{t_0}.$$

Thus, a set of five functions $x(t), y(t), u(t), p(t), q(t)$, which satisfy the condition

$$\frac{du}{dt} = p(t) \frac{dx}{dt} + q(t) \frac{dy}{dt},$$

defines a *strip* on the curve Γ. If this strip is an integral element, then it is an integral strip of the partial differential equation. If this integral strip at each point touches a generator of the Monge cone, then the integral strip is a *characteristic strip*.

Now, we state some theorems on the characteristics. Their proofs can be found in the references cited below.

Theorem 2.2. *A necessary and sufficient condition for a surface to be an integral surface of a partial differential equation is that at each point its tangent element must touch its elementary cone (tangent cone or Monge cone).*

Theorem 2.3. *The function $F(x, y, u, p, q)$ is constant along every characteristic strip of the equation $F(x, y, u, p, q) = 0$.*

Theorem 2.4. *If a characteristic strip contains at least one integral element of $F(x, y, u, p, q) = 0$, it is an integral of the equation $F(x, y, u, u_x, u_y) = 0$.*

For the linear partial differential equation $a\,p + b\,q = c$, we have

Theorem 2.5. *Every surface generated by a one-parameter family of characteristic curves is an integral surface of the partial differential equation.*

Theorem 2.6. *Every characteristic curve which has one point in common with an integral surface lies entirely on the integral surface.*

Theorem 2.7. *Every integral surface is generated by a one-parameter family of characteristic curves.*

Theorem 2.8. *If*

$$D = \frac{dx}{ds}\frac{dy}{dt} - \frac{dy}{ds}\frac{dx}{dt} = a\frac{dy}{dt} - b\frac{dx}{dt} \neq 0$$

everywhere on an initial curve C, then the initial value problem has one and only one solution. If, however, $D = 0$ everywhere along C, the initial value problem cannot be solved unless C is a characteristic curve, and then the problem has an infinity of solutions.

Proofs of Theorems 2.2, 2.3, and 2.4 are given in Sneddon (1957, pp. 62–64) and of Theorems 2.5, 2.6, 2.7, and 2.8 in Courant and Hilbert (1965, pp. 64–66).

Second-Order Equations

2.7. Linear and Quasilinear Equations

For a linear or quasilinear partial differential equation of second or higher order, the characteristic equation is determined by the highest-order terms in the partial differential equation. These terms are known as the *principal part* of the partial differential equation. While the solution of the characteristic equation leads to the solution of the first-order partial differential equation, the solution of the characteristic equation of a second-order partial differential equation leads to a coordinate transformation which, when applied, reduces the second-order partial differential equation to a simpler form. This simpler form is called the *canonical form*. Consider a second-order partial differential equation

$$a_{11}\,u_{xx} + 2a_{12}\,u_{xy} + a_{22}\,u_{yy} + F(x,y,u,u_x,u_y) = 0, \tag{2.41}$$

where $a_{11}, a_{12},$ and a_{22} are functions of x and y only, and F is a function of $x, y, u, u_x,$ and u_y. The different canonical forms of Eq (2.41) are

$$\text{Elliptic}: u_{\xi\xi} + u_{\eta\eta} + G(\xi,\eta,u,u_\xi,u_\eta) = 0,$$

$$\text{Hyperbolic}: \begin{cases} u_{\xi\xi} - u_{\eta\eta} + G(\xi,\eta,u,u_\xi,u_\eta) = 0, \\ u_{\xi\eta} + G(\xi,\eta,u,u_\xi,u_\eta), \end{cases} \tag{2.42}$$

$$\text{Parabolic}: u_{\xi\xi} + G(\xi,\eta,u,u_\xi,u_\eta) = 0.$$

To reduce Eq (2.41) to a canonical form, we introduce a reversible transformation

$$\xi = \xi(x,y), \quad \text{and} \quad \eta = \eta(x,y), \tag{2.43}$$

with the condition that the Jacobian

$$J = \frac{\partial(\xi, \eta)}{\partial(x, y)} = \xi_x \, \eta_y - \eta_x \, \xi_y \neq 0. \tag{2.44}$$

The functions ξ and η are known as the *characteristic coordinates*. Using the transformation (2.43) and noting that

$$
\begin{aligned}
u_x &= u_\xi \, \xi_x + u_\eta \, \eta_x, \\
u_y &= u_\xi \, \xi_y + u_\eta \, \eta_y, \\
u_{xx} &= u_{\xi\xi} \, \xi_x^2 + 2u_{\xi\eta} \, \xi_x \, \eta_x + u_{\eta\eta} \, \eta_x^2 + u_\xi \, \xi_{xx} + u_\eta \, \eta_{xx}, \\
u_{xy} &= u_{\xi\xi} \, \xi_x \, \xi_y + u_{\xi\eta} \, (\xi_x \, \eta_y + \xi_y \, \eta_x) + u_{\eta\eta} \, \eta_x \, \eta_y + u_\xi \, \xi_{xy} + u_\eta \, \eta_{xy}, \\
u_{yy} &= u_{\xi\xi} \, \xi_y^2 + 2u_{\xi\eta} \, \xi_y \, \eta_y + u_{\eta\eta} \, \eta_y^2 + u_\xi \, \xi_{yy} + u_\eta \, \eta_{yy},
\end{aligned}
$$

Eq (2.41) reduces to

$$A_{11} \, u_{\xi\xi} + 2A_{12} \, u_{\xi\eta} + A_{22} \, u_{\eta\eta} + G(\xi, \eta, u, u_\xi, u_\eta) = 0, \tag{2.45}$$

where G is a function of ξ, η, u, u_ξ, and u_η, and A_{11}, A_{12}, and A_{22} are functions of ξ and η, given by

$$
\begin{aligned}
A_{11} &= a_{11} \, \xi_x^2 + 2a_{12} \, \xi_x \, \xi_y + a_{22} \, \xi_y^2, \\
A_{12} &= a_{11} \, \xi_x \, \eta_x + a_{12} \, (\xi_x \, \eta_y + \xi_y \, \eta_x) + a_{22} \, \xi_y \, \eta_y, \\
A_{22} &= a_{11} \, \eta_x^2 + 2a_{12} \, \eta_x \, \eta_y + a_{22} \, \eta_y^2.
\end{aligned}
\tag{2.46}
$$

The function G is linear or nonlinear according as F is linear or nonlinear.

If we now choose ξ and η such that both satisfy the condition

$$a_{11} \, \xi_x^2 + 2a_{12} \, \xi_x \, \xi_y + a_{22} \, \xi_y^2 = 0, \tag{2.47}$$

then $A_{11} = A_{22} = 0$, and Eq (2.41) reduces to

$$2A_{12} \, u_{\xi\eta} + G(\xi, \eta, u, u_\xi, u_\eta) = 0.$$

It can be verified (after some tedious algebra) that

$$A_{12}^2 - A_{11} A_{22} = \left(a_{12}^2 - a_{11} a_{22}\right) \left(\xi_x \, \eta_y - \xi_y \, \eta_x\right)^2. \tag{2.48}$$

This means that the sign of the quantity $(a_{12}^2 - a_{11} a_{22})$ is invariant under the reversible transformation (2.43), and the quantity itself is invariant if $|J| = 1$. An important consequence of this result is that the partial differential equation does not change its classification under nonsingular transformations (see §1.3, Theorem 1.1, where it is proved that if $\xi = \phi(x, y)$ is a solution of Eq (2.47), then $\phi(x, y) = c$ is a solution of Eq (1.10)). Both Eqs (1.10) and (2.47) are called characteristic equations of the partial differential equation (2.41). Equation (1.10) has two solutions given by

$$\frac{dy}{dx} = \frac{a_{12} \pm \sqrt{a_{12}^2 - a_{11}a_{22}}}{a_{11}}. \tag{2.49}$$

If the partial differential equation is hyperbolic, then there are two solutions, resulting in two characteristics for the partial differential equation. If the partial differential equation is parabolic, then there is only one real solution, and hence only one characteristic. In the elliptic case there are no real solutions, and so there are no characteristics.

In order to transform the partial differential equation to its canonical form, we introduce two independent variables $\xi = \xi(x, y) = c_1$ and $\eta = \eta(x, y) = c_2$, where ξ and η (called the characteristic coordinates) are solutions of Eq (2.45). In the case of a parabolic equation we have only one solution, so η is chosen arbitrarily except that it must satisfy the condition (2.44). In the case of an elliptic equation, the solutions are complex conjugates, and we can use the real and imaginary parts of the solutions as the new independent variables. It is shown in Chapter 5 that canonical forms are frequently necessary to solve partial differential equations by the method of separation of variables.

Note that the first-order terms in (2.45) can be eliminated by a transformation of the type $u = v \, e^{\alpha x + \beta y}$ with an appropriate choice of α and β.

Now, we demonstrate by some examples the effectiveness of this technique for reducing second-order partial differential equations to canonical forms.

EXAMPLE 2.22. Transform the partial differential equation

$$y^2 \, u_{xx} - 4x \, y \, u_{xy} + 4x^2 \, u_{yy} + (x^2 + y^2) \, u_x + u_y = 0$$

to its canonical form. This equation is parabolic, and in view of Theorem 1.1, its characteristic equation is given by

$$y^2 \, (dy)^2 + 4x \, y \, dx \, dy + 4x^2 \, (dx)^2 = 0,$$

which has only one solution

$$\frac{dy}{dx} = -\frac{2x}{y}, \quad \text{which yields} \quad 2x^2 + y^2 = c.$$

In this case there is only one characteristic curve, and so we make the substitution

$$\xi = 2x^2 + y^2, \quad \text{and} \quad \eta = x,$$

where ξ and η are the characteristic coordinates. The substitution for η is arbitrary in this situation, the only condition being that the Jacobian should be nonsingular. Thus, we have

$$
\begin{aligned}
u_x &= 4x\,u_\xi + u_\eta, \\
u_y &= 2y\,u_\xi, \\
u_{xx} &= 16x^2\,u_{\xi\xi} + 8x\,u_{\xi\eta} + u_{\eta\eta} + 4u_\xi, \\
u_{xy} &= 8x\,y\,u_{\xi\xi} + 2y\,u_{\xi\eta}, \\
u_{yy} &= 4y^2\,u_{\xi\xi} + 2u_\xi.
\end{aligned}
$$

Substitution of these values into the partial differential equation leads to the canonical form

$$
\left(\xi - 2\eta^2\right) u_{\eta\eta} + \left[4\xi + 4\eta(\xi - \eta^2) + 2\sqrt{\xi - 2\eta^2}\right] u_\xi + \left(\xi - \eta^2\right) u_\eta = 0. \ \blacksquare
$$

EXAMPLE 2.23. Reduce the partial differential equation

$$
y^2\,u_{xx} - 4x\,y\,u_{xy} + 3x^2\,u_{yy} - \frac{y^2}{x}\,u_x - \frac{3x^2}{y}\,u_y = 0
$$

to the canonical form. The principal part in this partial differential equation is similar to the previous example, but it is hyperbolic, and in view of Theorem 1.1, its characteristic equation is

$$
y^2\,(dy)^2 + 4x\,y\,dx\,dy + 3x^2\,(dx)^2 = 0,
$$

which has two independent solutions:

(i) $\quad \dfrac{dy}{dx} = -\dfrac{3x}{y}, \quad$ which yields $\quad 3x^2 + y^2 = c_1;$

(ii) $\quad \dfrac{dy}{dx} = -\dfrac{x}{y}, \quad$ which yields $\quad x^2 + y^2 = c_2.$

Hence, the new independent variables are

$$
\xi = 3x^2 + y^2, \quad \text{and} \quad \eta = x^2 + y^2.
$$

The partial derivatives of u with respect to the new variables are given by

$$
\begin{aligned}
u_x &= 6x\,u_\xi + 2x\,u_\eta, \\
u_y &= 2y\,u_\xi + 2y\,u_\eta, \\
u_{xx} &= 36x^2\,u_{\xi\xi} + 24x^2\,u_{\xi\eta} + 4x^2\,u_{\eta\eta} + 6u_\xi + 2u_\eta, \\
u_{xy} &= 12x\,y\,u_{\xi\xi} + 16x\,y\,u_{\xi\eta} + 4x\,y\,u_{\eta\eta}, \\
u_{yy} &= 4y^2\,(u_{\xi\xi} + 2u_{\xi\eta} + u_{\eta\eta}) + 2\,(u_\xi + u_\eta).
\end{aligned}
$$

Substituting these values into the given partial differential equation, we get the canonical form

$$(\xi - \eta)(\xi - 3\eta)\, u_{\xi\eta} = 0,$$

whose solution is

$$u = f(3x^2 + y^2) + g(x^2 + y^2). \ \blacksquare$$

We mention here an important property of hyperbolic partial differential equations. They are capable of transporting a discontinuity in the initial data along a characteristic. The solutions that are in C^2 are called *strict solutions*, whereas those with discontinuity in the function u or its first-order derivatives are called *weak* or *generalized solutions*. We demonstrate this idea by the following example. For more details, the reader is referred to John (1982) and Courant and Hilbert (1965).

EXAMPLE 2.24. Solve the partial differential equation

$$u_{tt} - u_{xx} = 0, \quad 0 < x < \infty,$$

subject to the conditions

$$u(0, t) = H(t)e^{-t}, \quad u(x, 0) = u_t(x, 0) = 0,$$

where $H(t)$ is the Heaviside unit step function. By introducing the characteristic coordinates $\xi = x + t, \eta = x - t$, the partial differential equation is reduced to

$$u_{\xi\eta} = 0.$$

Its solution is given by

$$u = f(\xi) + g(\eta) = f(x + t) + g(x - t).$$

The term $f(x + t)$ represents a wave traveling with a negative velocity coming from infinity. Since there are no sources or boundaries at infinity, it is not possible for a wave to either emanate or be reflected from infinity. This is also known as Sommerfeld's radiation condition. Therefore, the function $f(x + t)$ must be taken to be zero. Thus, we have from the boundary condition

$$g(-t) = H(t)e^{-t},$$

which yields

$$u = H(t - x)e^{-(t-x)}.$$

The initial conditions are then automatically satisfied. In this case the discontinuity in u propagates along the characteristic $x = t$. $\ \blacksquare$

EXAMPLE 2.25. Transform the partial differential equation

$$y^2 u_{xx} - 4x y u_{xy} + 8x^2 u_{yy} = 0$$

to the canonical form. In this case the partial differential equation is of the elliptic type. In view of Theorem 1.1, the characteristic equation is

$$y^2 (dy)^2 + 4x y \, dx \, dy + 8x^2 (dx)^2 = 0,$$

and its solution is

$$\frac{dy}{dx} = -(1 \pm i) \frac{2x}{y},$$

which yields

$$y^2 + 2(1 \pm i) x^2 = c_{1,2}.$$

In this situation we define the new independent coordinates as

$$\xi = 2x^2 + y^2, \quad \text{and} \quad \eta = 2x^2.$$

Then, as in Examples 2.23 and 2.24, we get the canonical form

$$2\eta(\xi - \eta)(u_{\xi\xi} + u_{\eta\eta}) + (\xi + \eta) u_\xi + (\xi - \eta) u_\eta = 0. \blacksquare$$

It can be seen from these examples that canonical forms, though more of theoretical interest, also provide, in some cases, the general solution of the partial differential equations.

The well-known Cauchy-Kowalewsky theorem guarantees the uniqueness and existence of quasilinear partial differential equations under certain specific conditions. A statement of this theorem for two independent variables is as follows:

Theorem 2.9. *Consider a quasilinear second-order partial differential equation which can be solved for u_{xx}, i.e.,*

$$u_{xx} = F(x, y, u_x, u_y), \tag{2.50}$$

where F is an analytic function of x, y, u_x, and u_y in a domain $\Omega \subset R^2$. Let the Cauchy data on a curve $x = x_0$ be

$$u(x_0, y) = f(y), \quad \text{and} \quad u_x(x_0, y) = g(y),$$

where f and g are analytic functions in a neighborhood of a point (x_0, y_0). Then the Cauchy problem has an analytic solution in some neighborhood of the point (x_0, y_0), and this solution is unique in the class of analytic functions.

Simply stated, this theorem guarantees a unique solution $u(x, y)$ in the form of a Taylor's series in a neighborhood of the point (x_0, y_0). The above statement is true if the second-order partial differential equation can be solved for u_{yy} or u_{xy}. A similar statement holds for first-order partial differential equations. There is, however, an exception. If the Cauchy data is prescribed on a characteristic, a unique solution may not exist. For example, consider $u_x = 0$. Its solution is $u = f(y)$. If the Cauchy data is $u(x,0) = \phi(x)$, where $\phi(x)$ is not constant, then no solution can be found.

For a general statement of this theorem for higher-order partial differential equations, see Courant and Hilbert (1965) and Petrovskii (1967). It should be pointed out that the form of complete solutions and general solution is not unique. Also, in the case of nonlinear problems there may be more than one valid solution satisfying the initial conditions.

2.8. Mathematica Projects

PROJECT 2.1. A complete Mathematica solution for Example 2.1 and the individual plots of the integral surfaces S_1 and S_2 are available in the Notebook Example2.1.nb.

PROJECT 2.2. The plots of the surfaces $y = e^x + c$ and $u = e^x + (y - e^x)x$ for $c = 1$ appearing in Example 2.5 are available in the Notebook Example2.5.nb. These plots show that the intersection of these surfaces is a characteristic.

PROJECT 2.3. The plots of the surfaces

$$\sqrt{2}\tan^{-1}\frac{\sqrt{2}y}{x} + \ln(x^2 + 2y^2) = 1, \quad \text{and} \quad u = \frac{1}{3}(x + y)$$

appearing in Example 2.7 are available in the Notebook Example2.7.nb. Note that the intersection of these two surfaces gives a particular characteristic.

PROJECT 2.4. For Example 2.8, the following plots are given in the Notebook Example2.8.nb:

(i) The graphs of $y - cx = c_1$ for $c = 1, 2, 3$ and $c_1 = 0, 1, 2$. These curves are characteristics and also base characteristics.

(ii) The graph of $u = \dfrac{y}{1+x}$ represents an integral surface.

(iii) The graph of $u = \dfrac{1 + 2xy - \sqrt{1 + 4xy}}{2x^2}$ represents another integral surface.

PROJECT 2.5. For Example 2.15, the Notebook Example2.15.nb contains the plot of the solution which represents an integral surface.

PROJECT 2.6. For Example 2.18, the Notebook Example2.18.nb contains the plots of

$$u = 2\sin\left(ax + \frac{1}{a}y + c\right)$$

for $c = 0, 1, 2$ and $a = 1, 2$; these plots show some particular integral surfaces.

PROJECT 2.7. For Example 2.19, the Notebook `Example2.19.nb` contains the plots of the solution for $a = 1$ and $b = 0, 1$; they represent some particular integral surfaces.

PROJECT 2.8. For Example 2.22, plots of the characteristic curves are available in the Notebook `Example2.22.nb`.

PROJECT 2.9. For Example 2.23, plots of the characteristic curves are available in the Notebook `Example2.23.nb`.

PROJECT 2.10. For Example 2.25, the computations of the canonical form are given in the Notebook `Example 2.25.nb`.

PROJECT 2.11. For Exercise 2.4, see the Notebook `Exercise2.4.nb`.

2.9. Exercises

Note that particular solutions are unique only for linear equations, but the form of general solutions and complete integrals is not.

2.1. Find the general solution of the equation $4p + 7q = 2u$.
    ANS. $u^2 = e^x f(7x - 4y)$.

2.2. Find the general solution of the equation $5p - 2q = 3$.
    ANS. $5u = 3x + f(2x + 5y)$.

2.3. Find the general solution of the equation $p + 2q = u + 3$.
    ANS. $u = -3 + e^x f(2x - y)$.

2.4. Find the general solution of the equation $3\,u_x + 4\,u_y - 5\,u = 10$, subject to the initial condition $u(x, 0) = x$.

SOLUTION. In this solution we demonstrate an alternative approach based on the rotation of axes. We first rotate the x, y axes through an angle θ; thus, the new axes ξ, η are related to the old axes x, y by the relations

$$\xi = x \cos\theta + y \sin\theta, \qquad x = \xi \cos\theta - \eta \sin\theta,$$
$$\eta = -x \sin\theta + y \cos\theta, \qquad y = \xi \sin\theta + \eta \cos\theta.$$

Hence, we have

$$u(x, y) = u(\xi \cos\theta - \eta \sin\theta, \xi \sin\theta + \eta \cos\theta) = w(\xi, \eta),$$

$$\frac{\partial}{\partial x} \equiv \frac{\partial \xi}{\partial x} \frac{\partial}{\partial \xi} + \frac{\partial \eta}{\partial x} \frac{\partial}{\partial \eta} = \cos\theta \frac{\partial}{\partial \xi} - \sin\theta \frac{\partial}{\partial \eta},$$

$$\frac{\partial}{\partial y} \equiv \frac{\partial \xi}{\partial y} \frac{\partial}{\partial \xi} + \frac{\partial \eta}{\partial y} \frac{\partial}{\partial \eta} = \sin\theta \frac{\partial}{\partial \xi} + \cos\theta \frac{\partial}{\partial \eta}.$$

The given partial differential equation then becomes

$$(3\cos\theta + 4\sin\theta)\frac{\partial w}{\partial \xi} + (4\cos\theta - 3\sin\theta)\frac{\partial w}{\partial \eta} - 5\,w = 10.$$

The coefficient of $\dfrac{\partial w}{\partial \eta}$ in the above equation vanishes if $\tan\theta = 4/3$, i.e., $\sin\theta = 4/5$, and $\cos\theta = 3/5$. With these values, the above equation reduces to

$$\frac{\partial w}{\partial \xi} - w = 2,$$

which has the general solution

$$w(\xi, \eta) = -2 + g(\eta)\,e^{\xi},$$

or

$$u(x, y) = -2 + g\left(\frac{3y - 4x}{5}\right)e^{(3x+4y)/5}.$$

Note that this is the general formal solution of the given partial differential equation. Now, to find the particular solution subject to the initial condition $u(x, 0) = x$, we have $\eta = -\dfrac{4x}{5}$ when $y = 0$, and then the above equation gives

$$x = -2 + g(\eta)\,e^{3x/5},$$

thus,

$$g(\eta) = (x + 2)\,e^{-3x/5} = (2 - 5\eta/4)\,e^{3\eta/4}.$$

Substituting this value of $g(\eta)$ into the above general solution, we obtain

$$u(x, y) = -2 + \left(2 - \frac{4}{5}\eta\right)e^{3\eta/4}\,e^{(3x+4y)/5}$$

$$= -2 + \left[2 - \frac{5}{4}\left(\frac{3y - 4x}{5}\right)\right]e^{3(3y-4x)/20+(3x+4y)/5}$$

$$= -2 + \left[2 + \frac{1}{4}(4x - 3y)\right]e^{5y/4},$$

which is the unique solution of the given partial differential equation subject to the given initial condition $u(x, 0) = x$.

2.5. Find the general solution of $3\,u_x + 4\,u_y - 2\,u = 1$. Using the method of characteristics as well as the method of rotation of axes, find the particular solution subject to the initial condition $u(x, 0) = x^2$.

ANS. The general solution is

$$u(x, y) = -\frac{1}{2} + g\left(\frac{3y - 4x}{5}\right)e^{2(3x+4y)/25},$$

and the particular solution is

$$u = -\frac{1}{2} + \left(\frac{1}{2} + \frac{(3y - 4x)^2}{16}\right) e^{y/2}.$$

2.6. Find the solution of $u_x - u_y + u = 1$, such that $u(x,0) = \sin x$.

SOLUTION. $\tan\theta = -1$, thus $\theta = -\pi/4, 3\pi/4$, and

$$\frac{\partial w}{\partial \xi} - \frac{1}{\sqrt{2}} w = -\frac{1}{\sqrt{2}},$$

whose general solution is $w = 1 + g(\eta)\, e^{\xi/\sqrt{2}}$, or

$$u(x,y) = 1 + g\left(-\frac{x+y}{\sqrt{2}}\right) e^{(y-x)/\sqrt{2}}.$$

Using the initial condition, we get $\sin x = 1 + g(-x/\sqrt{2})\, e^{-x/\sqrt{2}}$, so that

$$g(\eta) = -(\sin\sqrt{2}\eta + 1)\, e^{-\eta/\sqrt{2}}.$$

Then

$$u(x,y) = 1 - (\sin\sqrt{2}\eta + 1)\, e^{-\eta/\sqrt{2}}\, e^{\xi/\sqrt{2}} = 1 - [1 - \sin(x+y)]\, e^y. \ \blacksquare$$

2.7. Solve $u_x + u_y - u = 0$, subject to the initial condition $u(x,0) = h(x)$.

SOLUTION. Here $\tan\theta = 1$, thus $\theta = \pi/4$, and $\sqrt{2}\dfrac{\partial w}{\partial \xi} = w$, whose general solution is $w = g(\eta)\, e^{\xi/\sqrt{2}}$, or

$$u(x,y) = g(\eta)\, e^{\xi/\sqrt{2}}.$$

The initial condition yields

$$h(x) = g(-x/\sqrt{2})\, e^{x/2} = g(\eta)\, e^{\eta/\sqrt{2}},$$

or $g(\eta) = h(-\sqrt{2}\eta)\, e^{\eta/\sqrt{2}}$. Hence

$$u(x,y) = h(-\sqrt{2}\eta)\, e^{\xi/\sqrt{2}} = h(x-y)\, e^y.$$

2.8. Solve the equation $p + q + 3u = e^{-3x}\sin(x+2y)$, with the initial condition $u(x,0) = 0$.

ANS. $u(x,y) = (1/3)e^{-3x}[\cos(x-y) - \cos(x+2y)]$.

2.9. Solve the equation $2p + q + 2(2x - y)u = 6x^2 e^{y^2 - x^2}$.

ANS. $2y - x = c_1$ and $u e^{x^2 - y^2} - x^3 = c_2$;
general solution: $F(2y - x, u e^{x^2 - y^2} - x^3) = 0$.

2.10. Solve the equation $p + q\sqrt{1 - y^2} = 0$, with initial conditions (a) $u(0, y) = y$,
(b) $u(x, 0) = x^2$.
ANS. (a) $u(x, y) = y \cos x - \sqrt{1 - y^2} \sin x$; (b) $u = (\sin^{-1} y - x)^2$.

2.11. Find a general solution of the equation $xp - yq = x$.
ANS. $u = x + f(xy)$, or $xy = g(u - x)$.

2.12. Find a general solution of the equation $x^2 p + q = xu$.

ANS. $u = xf\left(\dfrac{1 + xy}{x}\right)$.

2.13. Find the solution of the equation $F = px + qy - 2u = 0$,
(a) subject to the initial condition $u = ay^2 - b$ at $x = 1$;
(b) $u = a(1 + y^2) + 2by$, at $x = 1$.
ANS. (a) $u = ay^2 - bx^2$; (b) $u = a(x^2 + y^2) + 2bxy$.

2.14. Find a general solution of the equation $F = pxy + qy^2 - 2uy - 4q = 0$. Also
find the particular solution subject to the initial condition $x = 1$, $au = 4 + b - y^2$.
ANS. $f\left(\dfrac{y^2 - 4}{x^2}, \dfrac{u}{x^2}\right) = 0$, $au - bx^2 + y^2 = 4$.
Note that both general and complete solutions are not unique.

2.15. Find a general solution of the equation $F = 2px + qy - u - x = 0$.
ANS. $y = (u - x)f\left(\dfrac{y^2}{x}\right)$, or $u = x + y f\left(\dfrac{y^2}{x}\right)$.

2.16. Solve the equation $p + q = \dfrac{1}{3}u^{-2}$, with the initial condition $u(0, y) = \sin y$.
ANS. $u^3(x, y) = x + \sin^3(y - x)$.

2.17. Find the general solution of the equation $2p + uq = \dfrac{u^2}{y}$.

ANS. $\dfrac{u}{y} = c_1$, and $\dfrac{2y \ln y - xu}{u} = c_2$; and the general solution is

$$F\left(\dfrac{u}{y}, \dfrac{2y \ln y - xu}{u}\right) = 0.$$

2.18. Find two functionally independent solutions of $(y - u)p + (u - x)q = x - y$.
ANS. $x + y + u = c_1$, and $x^2 + y^2 + u^2 = c_2$.

2.19. Find a general solution of the equation $F = pu - aq - x = 0$.
ANS. $f[(u + x)e^{y/a}, (u - x)e^{-y/a}] = 0$. Discuss the solution in the neighborhood
of $x = 0$.

2.20. Find the solution of $(y + u)p + (y + 1)q = x - y$, which passes through $x = 0, u = y$.

ANS. $(u - 1)^2 = (x - y)^2 + \dfrac{y + 1 - 3x - 3u}{y + 1 - x - u}$.

2.21. Find a solution of $(1 + q^2)u - px = 0$, which passes through the curve $2u = x^2, y = 0$.

ANS. $4u^2 = x^2(4y + x^2)$.

2.22. Find a solution of $F = p^2 + q^2 - 4u = 0$, which passes through $y = 0, u = x^2 + 1$.

ANS. $u = x^2 + (y + 1)^2$.

2.23. Find a solution of $F = u - p^2 + q^2 = 0$, which passes through $y = 0$, $4u + x^2 = 0$.

ANS. $4u = -(x \pm \sqrt{2}y)^2$.

2.24. Find a complete solution of the equation $F = pqy - pu/2 - cq = 0$.

ANS. $\pm(a^2y^2 - 2cu)^{3/2} - a^3y^3 = 3ac(2cx - uy) - b$.

Investigate if a solution exists at $x = 0$, and determine if this solution is a function of y such that both this solution and its partial derivative q vanish for $y = 0$.

ANS. Yes, and $b = 0$ gives this solution.

2.25. Find a complete solution of the equation
$F = 2px^2y + 2qxy^2 + pq - 4uxy = 0$.

ANS. $u = ax^2 + by^2 + ab$.

2.26. Find a complete solution of the equation
$F = 1 + upx + uqy - u^2 = 0$.

ANS. $u^2 = 1 + ax^2 + by^2$.

2.27. Find a complete solution of the equation $F = u^2(p^2 + q^2 + 1) - a^2 = 0$.

ANS. $u^2 + \dfrac{(\alpha x + y - \beta)^2}{1 + \alpha^2} = a^2$.

In problems (2.28)–(2.35), find the characteristics where possible and reduce the partial differential equation to its canonical form. Find the solution if possible.

2.28. $(a^2 + x^2)u_{xx} + (a^2 + y^2)u_{yy} + xu_x + yu_y = 0$. ANS. $u_{\xi\xi} + u_{\eta\eta} = 0$.

HINT: Solve the characteristic equation to determine that the substitution is

$$\xi = \log\left(x + \sqrt{a^2 + x^2}\right), \quad \text{and} \quad \eta = \log\left(y + \sqrt{a^2 + y^2}\right).$$

2.29. The Tricomi equation $u_{yy} - yu_{xx} = 0$ for $y > 0$.

ANS. $u_{\xi\eta} - \dfrac{u_\xi - u_\eta}{6(\xi - \eta)} = 0$.

2.30. $(1 + \sin x)u_{xx} - 2\cos x u_{xy} + (1 - \sin x)u_{yy} + \dfrac{(1 + \sin x)^2}{2\cos x}u_x +$
$\dfrac{1}{2}(1 - \sin x)u_y = 0$.

Ans. $4u_{\eta\eta} + u_\eta = 0, \xi, \eta = y \pm \log(1 + \sin x)$. The reduced form can now be solved to yield $u = f(\xi) + e^{-\eta/4}g(\xi)$.

2.31. $e^{2x}u_{xx} - 2e^{x+y}u_{xy} + e^{2y}u_{yy} + e^{2x}u_x + e^{2y}u_y = 0$.
Ans. $u_{\eta\eta} = 0, \xi = e^{-x} + e^{-y}, \eta = e^{-x} - e^{-y}, u = f(\xi) + \eta\,g(\xi)$.

2.32. $e^{2x}u_{xx} - 5e^{x+y}u_{xy} + 4e^{2y}u_{yy} + e^{2x}u_x + 4e^{2y}u_y = 0$.
Ans. $u_{\xi\eta} = 0, \xi = e^{-x} + e^{-y}, \eta = 4e^{-x} + e^{-y}, u = f(\xi) + g(\eta)$.

2.33. $9u_{xx} - 12u_{xy} + 4u_{yy} + 12u_x - 8u_y + 4u = 0$.
Ans. $36u_{\eta\eta} + 12u_\eta + u = 0, u = [f(\xi) + \eta g(\xi)]e^{-\eta/6}$, where $\xi, \eta = 2x \pm 3y$.

2.34. $3u_{xx} - 7u_{xy} + 2u_{yy} + 3u_x - u_y = 0$.
Ans. $5u_{\xi\eta} - u_\xi = 0, u = f(\eta) + g(\xi)\,e^{\eta/5}, \xi = 2x + y, \eta = x + 3y$.

2.35. $2u_{xx} + 6u_{xy} + 9u_{yy} + 2u_x + 3u_y - 2u = 0$.
Ans. $9(u_{\xi\xi} + u_{\eta\eta}) + 6u_\eta - 4u = 0, \xi = y - \dfrac{3}{2}x, \eta = \dfrac{3}{2}x$.

3

Linear Equations with Constant Coefficients

We will use the inverse operator method to solve homogeneous and nonhomogeneous partial differential equations with constant coefficients. This method, although basically developed and frequently used for solving ordinary differential equations, becomes useful for finding general solutions of partial differential equations with constant coefficients. The problem of finding the general solutions of second-order partial differential equations with constant coefficients and determining their particular solutions under auxiliary (initial) conditions is also discussed in §3.3. Before we discuss the partial differential equations with constant coefficients, we first review in §3.1 the technique of inverse operators from the theory of ordinary differential equations. This review will prove useful in discussing the homogeneous and nonhomogeneous partial differential equations with constant coefficients.

3.1. Inverse Operators

If D represents $\dfrac{d}{dx}$, then $\dfrac{1}{D}$ is defined as the inverse operator of D, i.e.,

$$\frac{1}{D}\,\phi(x) = \int \phi(x)\,dx.$$

If $f(D)$ represents a polynomial in D with constant coefficients, then $f(D)$ is a linear differential operator, and we define its inverse as $1/f(D)$. Thus,

$$f(D)\left[\frac{1}{f(D)}\phi(x)\right] = \phi(x),$$

where $\dfrac{1}{f(D)}$ is an inverse operator that, when operating on $\phi(x)$, gives the particular solution of the equation $f(D)\,y = \phi(x)$, which contains no solution of the equation $f(D)\,y = 0$. Note that $\dfrac{1}{f(D)}\,[f(D)\phi(x)]$ is not necessarily equal to $\phi(x)$. But if $\left[\dfrac{1}{f(D)}\phi(x)\right] = \psi(x)$, then $\psi(x)$ contains arbitrary constants, and $\psi(x) = \phi(x)$ for some values of these arbitrary constants. In the sequel we will ignore arbitrary constants. Some formulas for the operator pair $f(D)$ and $\dfrac{1}{f(D)}$ are listed below:

1. $f(D)\left[\dfrac{1}{f(D)}\phi(x)\right] = \phi(x).$

2. $\dfrac{1}{f_1(D)f_2(D)}\,\phi(x) = \dfrac{1}{f_1(D)}\left(\dfrac{1}{f_2(D)}\,\phi(x)\right) = \dfrac{1}{f_2(D)}\left(\dfrac{1}{f_1(D)}\,\phi(x)\right).$

3. $\dfrac{1}{f(D)}\,[c_1\phi_1(x) + c_2\phi_2(x)] = c_1\dfrac{1}{f(D)}\phi_1(x) + c_2\dfrac{1}{f(D)}\phi_2(x).$

4. $\dfrac{1}{f(D)}\,e^{ax} = \dfrac{1}{f(a)}e^{ax}$, provided that $f(a) \neq 0.$

5. $f(D)\phi(x)e^{ax} = e^{ax}\,f(D+a)\phi(x).$

6. $\dfrac{1}{f(D)}\,\phi(x)e^{ax} = e^{ax}\,\dfrac{1}{f(D+a)}\phi(x).$

7. $\dfrac{1}{(D-a)^m}\,e^{ax} = \dfrac{x^m e^{ax}}{m!}.$

8. $\dfrac{1}{(D-a)^m f(D)}\,e^{ax} = \dfrac{x^m}{m!f(a)}e^{ax}, \quad f(a) \neq 0.$

9. $\dfrac{1}{D^2 + a^2}\left\{\begin{array}{c}\cos\\\sin\end{array}\right. bx = \dfrac{\left\{\begin{array}{c}\cos\\\sin\end{array}\right. bx}{a^2 - b^2}, \quad |a| \neq |b|.$

10. $\dfrac{1}{f(D^2)}\left\{\begin{array}{c}\cos\\\sin\end{array}\right. ax = \dfrac{\left\{\begin{array}{c}\cos\\\sin\end{array}\right. ax}{f(-a^2)}$, provided that $f(-a^2) \neq 0.$

11. $\dfrac{1}{(D^2 + a^2)}\left\{\begin{array}{c}\cos\\\sin\end{array}\right. ax = \pm\dfrac{\left\{\begin{array}{c}x\sin\\x\cos\end{array}\right. ax}{2a}.$

12. $\dfrac{1}{aD^2 + bD + c}\left\{\begin{array}{c}\cos\\\sin\end{array}\right. \omega x = \dfrac{(c - a\omega^2)\left\{\begin{array}{c}\cos\\\sin\end{array}\right.\omega x \pm b\omega\left\{\begin{array}{c}\sin\\\cos\end{array}\right.\omega x}{(c - a\omega^2)^2 + b^2\omega^2}.$

13. $\dfrac{1}{f(D)} x^n = \dfrac{1}{a_0[1+g(D)]} x^n$

$\qquad = \dfrac{1}{a_0} \left[1 - g(D) + g^2(D) - g^3(D) + \cdots + g^n(D) + \cdots\right] x^n,$

where the terms g^k for $k \geq n+1$ are ignored.

14. $\dfrac{1}{f(D)} \{x\phi(x)\} = x\,\dfrac{1}{f(D)} \{\phi(x)\} - f'(D)\dfrac{1}{[f(D)]^2} \{\phi(x)\}.$

PROOF OF FORMULA 4. If $\phi(x) = e^{ax}$, we know that

$$De^{ax} = ae^{ax}, \quad D^2 e^{ax} = a^2 e^{ax}, \quad \ldots \quad D^n e^{ax} = a^n e^{ax}.$$

Thus,

$$f(D)e^{ax} = f(a)e^{ax}.$$

If we take $\dfrac{1}{f(D)}$ as the inverse operator of $f(D)$, then obviously

$$\frac{1}{f(D)}e^{ax} = \frac{1}{f(a)}e^{ax}, \quad \text{provided that } f(a) \neq 0.$$

Therefore, the particular integral y_p of the equation $f(D)y = Ae^{ax}$ is given by

$$y = \frac{A}{f(a)}e^{ax}, \quad f(a) \neq 0.$$

EXAMPLE 3.1. Consider $\left(D^2 + D + 1\right) y = e^{2x}$. Then

$$y_p = \frac{1}{D^2 + D + 1}e^{2x} = \frac{1}{2^2 + 2 + 1}e^{2x} = \frac{1}{7}e^{2x}. \ \blacksquare$$

EXAMPLE 3.2. Consider $\left(D^4 + 8\right) y = e^x$. Then

$$y_p = \frac{1}{D^4 + 8}e^x = \frac{1}{1^4 + 8}e^x = \frac{1}{9}e^x. \ \blacksquare$$

AN APPLICATION OF FORMULA 6. If $\phi(x) = V(x)e^{ax}$, where $V(x)$ is a function of x of the type x^n, $\cos bx$, $\sin bx$, or $x^n \cos bx$, $x^n \sin bx$, we first give a heuristic justification for formula 6. Note that

$$D\left(Ve^{ax}\right) = e^{ax}DV + aVe^{ax} = e^{ax}(D+a)V,$$

which yields

$$D^2\left(Ve^{ax}\right) = D\left[e^{ax} \cdot (D+a)V\right] = e^{ax}(D+a)^2 V.$$

Similarly,

$$D^n \left(V e^{ax} \right) = e^{ax} (D + a)^n V,$$

which, in general, gives

$$f(D) \left(V e^{ax} \right) = e^{ax} f(D + a) V.$$

We can, therefore, assume that

$$\frac{1}{f(D)} \left(V e^{ax} \right) = e^{ax} \frac{1}{f(D + a)} V.$$

We will use this formula to obtain the particular integral in this case. Since for the ordinary differential equation $f(D)y = \phi(x) = V e^{ax}$, we have

$$y_p = \frac{1}{f(D)} V e^{ax} = e^{ax} \frac{1}{f(D + a)} V,$$

our problem reduces to finding $\dfrac{1}{f(D + a)} V$. If V is of the form x^n, $\cos ax$ or $\sin bx$, then we use formula 13 or 10 to obtain the solution; if V is of the form $x^n \cos bx$ or $x^n \sin bx$, then we use Euler's formula and write

$$e^{ibx} = \cos bx + i \sin bx,$$

or

$$\cos bx = \Re \left\{ e^{ibx} \right\}, \quad \sin bx = \Im \left\{ e^{ibx} \right\},$$

so that if $V = x^n \cos bx$, we write $V = \Re \left\{ x^n e^{ibx} \right\}$, and if $V = x^n \sin bx$, we write $V = \Im \left\{ x^n e^{ibx} \right\}$. Thus,

$$\frac{1}{f(D + a)} V = \begin{cases} \Re \left\{ \dfrac{1}{f(D + a)} x^n e^{ibx} \right\} \\ \text{or} \\ \Im \left\{ \dfrac{1}{f(D + a)} x^n e^{ibx} \right\}. \end{cases}$$

Now, $f(D + a)$ is another polynomial in D, and we can write $f(D + a) = f_1(D)$. We consider

$$\frac{1}{f(D + a)} x^n e^{ibx} = \frac{1}{f_1(D)} x^n e^{ibx},$$

and using the formula 6 we get

$$\frac{1}{f_1(D)} x^n e^{ibx} = e^{ibx} \frac{1}{f_1(D + ib)} x^n,$$

where $\dfrac{1}{f_1(D+ib)}x^n$ is evaluated by the use of formula 13. This discussion also covers the case when $f(x) = x^n \cos ax$, or $x^n \sin ax$.

EXAMPLE 3.3. Consider $\left(D^4 + D^3 - 3D^2 - D + 2\right) y = 4e^x$. Then

$$
y_p = \frac{4e^x}{D^4 + D^3 - 3D^2 - D + 2} = 4\frac{1}{D^4 + D^3 - 3D^2 - D + 2}e^x
$$

$$
= 4\frac{1}{(D-1)^2}\frac{1}{D^2 + 3D + 2}e^x = 4\frac{1}{(D-1)^2}\frac{1}{1 + 3 + 2}e^x
$$

$$
= \frac{4}{6}\frac{1}{(D-1)^2}e^x = \frac{2}{3}e^x\frac{1}{(D+1-1)^2}\cdot 1
$$

$$
= \frac{2}{3}e^x\frac{1}{D^2}\cdot 1 = \frac{2}{3}e^x\frac{x^2}{2} = \frac{1}{3}x^2e^x. \ \blacksquare
$$

EXAMPLE 3.4. When $\phi(x) = e^{ax}$ and $f(a) = 0$, the ordinary differential equation is of the form

$$
f(D)y = Ae^{ax}.
$$

Since $f(a) = 0$, $(D - a)$ must be a factor of $f(D)$. Then $f(D) = (D - a)^n f_1(D)$, where $f_1(a) \neq 0$, and

$$
y_p = A\frac{1}{(D-a)^n f_1(D)}e^{ax} = A\frac{1}{(D-a)^n}\left(\frac{1}{f(D)}e^{ax}\right)
$$

$$
= A\left(\frac{1}{(D-a)^n}\right)\frac{1}{f_1(D)}e^{ax}
$$

$$
= \frac{A}{f_1(a)}\frac{1}{(D-a)^n}(e^{ax} \cdot 1),
$$

and, using formula 6, we write

$$
\frac{A}{f_1(a)}\frac{1}{(D-a)^n}(e^{ax} \cdot 1) = \frac{A}{f_1(a)}e^{ax}\frac{1}{D^n}(1) = \frac{A}{f_1(a)}e^{ax}\left(\frac{x^n}{n!}\right).
$$

If $f(D)y = \sin ax$, or $\cos ax$ and if $f(D)$ becomes zero by letting $D^2 = -a^2$, then it is convenient to consider $\cos ax$ and $\sin ax$ as real and imaginary parts of e^{iax} and deal with the problem as for e^{ax}. $\blacksquare$

AN APPLICATION OF FORMULA 10. If $\phi(x) = \sin ax$ or $\cos ax$, then we know that

$$
D^2\begin{Bmatrix} \cos ax \\ \sin ax \end{Bmatrix} = -a^2\begin{Bmatrix} \cos ax \\ \sin ax, \end{Bmatrix}
$$

and

$$D^{2n} \begin{Bmatrix} \cos ax \\ \sin ax \end{Bmatrix} = (-1)^n a^{2n} \begin{Bmatrix} \cos ax \\ \sin ax. \end{Bmatrix}$$

Thus,

$$f(D^2) \begin{Bmatrix} \cos ax \\ \sin ax \end{Bmatrix} = f(-a^2) \begin{Bmatrix} \cos ax \\ \sin ax. \end{Bmatrix}$$

Even if the operator $f(D)$ is not an even function, we can still use the above formula; for example,

$$\begin{aligned}
\left(D^3 + 2D^2 + 3D + 1\right) \cos ax &= \left[(-a^2) D + 2(-a^2) + 3D + 1\right] \cos ax \\
&= \left[(3 - a^2) D + 1 - 2a^2\right] \cos ax \\
&= -(3 - a^2) a \sin ax + (1 - 2a^2) \cos ax.
\end{aligned}$$

Therefore, we notice that when an operator of the type $f(D)$ is applied to $\cos ax$ or $\sin ax$, we can set $D^2 = -a^2$, and reduce $f(D)$ to a linear operator of first order. We will use this observation to find particular integrals in this case.

If $f(D)$ happens to be of the form $f(D^2)$, i.e., if the ordinary differential equation is

$$f(D^2) y = \begin{cases} \cos ax \\ \sin ax, \end{cases}$$

then

$$y = \frac{A}{f(D^2)} \begin{cases} \cos ax \\ \sin ax, \end{cases} = \frac{A}{f(-a^2)} \begin{cases} \cos ax \\ \sin ax, \end{cases}$$

provided that $f(-a^2) \neq 0$.

If $f(D)$ is, in general, a polynomial containing both even and odd degree terms, then we let $D^2 = -a^2$, and reduce $f(D)$ to a linear operator in D of the form $\alpha D + \beta$, so that if

$$f(D)y = A \begin{cases} \cos ax \\ \sin ax, \end{cases}$$

then

$$y = \frac{A}{f(D)} \begin{cases} \cos ax \\ \sin ax. \end{cases}$$

Then we let $D^2 = -a^2$, which reduces $f(D)$ to $\alpha D + \beta$. Thus,

$$y = \frac{A}{\alpha D + \beta} \begin{cases} \cos ax \\ \sin ax, \end{cases}$$

$$= \frac{A(\alpha D - \beta)}{(\alpha D + \beta)(\alpha D - \beta)} \begin{cases} \cos ax \\ \sin ax, \end{cases}$$

$$= \frac{A(\alpha D - \beta)}{\alpha^2 D^2 - \beta^2} \begin{cases} \cos ax \\ \sin ax, \end{cases}$$

$$= \frac{A}{-\alpha^2 a^2 - \beta^2} \begin{cases} -\alpha a \sin ax - \beta \cos ax \\ \alpha a \cos ax - \beta \sin ax. \end{cases}$$

EXAMPLE 3.5. Let $\left(D^4 + D^2 + 1\right) y = 2 \sin x$. Then

$$y_p = \frac{2}{D^4 + D^2 + 1} \sin x.$$

Now, if we let $D^2 = -1$, then $D^4 + D^2 + 1 = 1$, and we get $y_p = 2 \sin x$. ∎

EXAMPLE 3.6. Let $\left(D^2 + 1\right) y = 2 \sin x$. Then

$$y_p = \frac{2}{D^2 + 1} \sin x.$$

Now, if we let $D^2 = -1$, then $D^2 + 1 = 0$, and the above method is not applicable. In this case we use the fact that $\sin x = \Im \left\{ e^{ix} \right\}$ and use formulas 4 and 6, as follows: Since $\left(D^2 + 1\right) y = 2 \Im \left\{ e^{ix} \right\}$, we have

$$y_p = \frac{1}{D^2 + 1} 2 \Im \left\{ e^{ix} \right\}$$

$$= 2 \frac{1}{(D - i)(D + i)} \Im \left\{ e^{ix} \right\} = 2 \Im \left\{ \frac{1}{(D - i)(D + i)} e^{ix} \right\}$$

$$= 2 \Im \left\{ \frac{1}{D - i} \frac{1}{2i} e^{ix} \right\} \quad \text{(formula 4)}$$

$$= \frac{1}{i} \Im \{ e^{ix} \} \frac{1}{D + i - i} \cdot 1 \quad \text{(formula 6)}$$

$$= -i \Im \{ e^{ix} \} \frac{1}{D} \cdot 1 = -\Im \{ ix\, e^{ix} \}$$

$$= -\Im \{ ix \left(\cos x + i \sin x \right) \} = -x \cos x. \quad ∎$$

EXAMPLE 3.7. Let $\left(D^3 + 2D^2 + D + 1\right) y = 3 \cos 2x$. Then

$$y_p = \frac{3}{-2^2 D + 2\left(-2^2\right) + D + 1} \cos 2x$$

$$= \frac{3}{-3D - 7} \cos 2x = \frac{-3(3D - 7)}{(3D + 7)(3D - 7)} \cos 2x$$

$$= \frac{-3(3D - 7)}{9D^2 - 49} \cos 2x = \frac{-3}{9\left(-2^2\right) - 49}(3D - 7) \cos 2x$$

$$= -\frac{3}{85}[6 \sin 2x + 7 \cos 2x]. \quad \blacksquare$$

EXAMPLE 3.8. Consider $\left(D^3 - D^2 + 4D - 4\right) y = 2 \sin 2x$. Then

$$y_p = \frac{1}{D^3 - D^2 + 4D - 4} 2 \sin 2x = 2 \frac{1}{D^3 - D^2 + 4D - 4} \sin 2x.$$

If we put $D^2 = -4$, we get $f(D) = 0$, and therefore, we write

$$y_p = 2\frac{1}{D^3 - D^2 + 4D - 4} \sin 2x = 2\Im \left\{ \frac{1}{D^3 - D^2 + 4D - 4} e^{2ix} \right\}$$

$$= 2\Im \left\{ \frac{1}{(D - 1)\left(D^2 + 4\right)} e^{2ix} \right\} = 2\Im \left\{ \frac{1}{(D - 2i)(D - 1)(D + 2i)} e^{2ix} \right\}$$

$$= 2\Im \left\{ \frac{1}{D - 2i} \left(\frac{1}{(D - 1)(D + 2i)} e^{2ix} \right) \right\}$$

$$= 2\Im \left\{ \frac{1}{(D - 2i)} \left(\frac{1}{(2i - 1)(4i)} e^{2ix} \right) \right\}$$

$$= \frac{1}{2} \Im \left\{ \frac{1}{(D - 2i)} \frac{1}{(-2 - i)} e^{2ix} \right\}$$

$$= -\frac{1}{2} \Im \left\{ \frac{1}{(2 + i)} \frac{1}{(D - 2i)} \left(e^{2ix} \cdot 1 \right) \right\}$$

$$= -\frac{1}{2} \Im \left\{ \frac{2 - i}{\left(2^2 + 1\right)} e^{2ix} \right\} \frac{1}{D}(1) \quad \text{(using formula 6)}$$

$$= -\frac{1}{2} \Im \left\{ \frac{2 - i}{5} e^{2ix} \right\} \frac{1}{D}(1) = -\frac{1}{10} \Im \left\{ (2 - i)e^{2ix} \right\} \cdot x$$

$$= -\frac{1}{10} \Im \left\{ x(2 - i)\left(\cos 2x + i \sin 2x \right) \right\} = \frac{x}{10} \left(\cos 2x - 2 \sin 2x \right). \quad \blacksquare$$

AN APPLICATION OF FORMULA 13. If $\phi(x) = x^n$, then

$$f(D) = a_0 D^n + a_1 D^{n-1} + \cdots + a_{n-1}D + a_n$$

$$= a_n \left[1 + \frac{a_{n-1}}{a_n}D + \frac{a_{n-2}}{a_n}D^2 + \cdots + \frac{a_0}{a_n}D^n \right],$$

which gives

$$\frac{1}{f(D)} = \frac{1}{a_n(1 + g(D))},$$

where

$$g(D) = \frac{1}{a_n}\left(a_{n-1}D + a_{n-2}D^2 + \cdots + a_0 D^n\right),$$

and

$$\frac{1}{f(D)} = \frac{1}{a_n}(1 + g(D))^{-1}.$$

Now, to find the particular integral y_p of $f(D)y = Ax^n$, we apply the inverse $\dfrac{1}{f(D)}$ of $f(D)$ to the ordinary differential equation and get

$$\begin{aligned}
y &= \frac{1}{f(D)}(Ax^n) \\
&= A \cdot \frac{1}{a_n}(1 + g(D))^{-1} x^n \\
&= \frac{A}{a_n}\left[1 - g(D) + (g(D))^2 - (g(D))^3 + \cdots\right]x^n,
\end{aligned}$$

where terms of degree $n + 1$ and higher in D are ignored in the above expansion on the right side.

EXAMPLE 3.9. Consider $\left(D^2 + D + 2\right)y = x^4$. Then

$$\begin{aligned}
y_p &= \frac{1}{D^2 + D + 2}x^4 = \frac{1}{2\left(1 + \frac{1}{2}D + \frac{1}{2}D^2\right)}x^4 \\
&= \frac{1}{2}\left[1 + \left(\frac{1}{2}D + \frac{1}{2}D^2\right)\right]^{-1}x^4 \\
&= \frac{1}{2}\left[1 - \left(\frac{1}{2}D + \frac{1}{2}D^2\right) + \left(\frac{1}{2}D + \frac{1}{2}D^2\right)^2 - \left(\frac{1}{2}D + \frac{1}{2}D^2\right)^3 \right. \\
&\quad \left. + \left(\frac{1}{2}D + \frac{1}{2}D^2\right)^4 - \cdots\right]x^4 \\
&= \frac{1}{2}\left[1 - \frac{1}{2}D - \frac{1}{2}D^2 + \frac{1}{4}D^2 + \frac{1}{2}D^3 + \frac{1}{4}D^4 - \frac{1}{8}D^3 - \frac{3}{8}D^4 \right. \\
&\quad \left. + \frac{1}{16}D^4 + O(D^5)\right]x^4,
\end{aligned}$$

where $O(D^5)$ means terms containing D^5 and higher powers in D. Thus,

$$\begin{aligned}
y_p &= \frac{1}{2}\left[x^4 - 2x^3 - 6x^2 + 3x^2 + 12x + 6 - 3x - 9 + \frac{3}{2}\right] \\
&= \frac{1}{2}\left[x^4 - 2x^3 - 3x^2 + 9x - \frac{3}{2}\right]. \quad\blacksquare
\end{aligned}$$

EXAMPLE 3.10. Consider $\left(D^4 + 2D^3 + D^2\right) y = x^3$. Then

$$y_p = \frac{x^3}{D^2 \left(D^2 + 2D + 1\right)} = \frac{x^3}{D^2 \left(1 + D\right)^2}$$

$$= \frac{1}{D^2} \left(1 + D\right)^{-2} x^3$$

$$= \frac{1}{D^2} \left[1 - 2D + 3D^2 - 4D^3 + O\left(D^4\right)\right] x^3$$

$$= \frac{1}{D^2} \left[x^3 - 6x^2 + 18x - 24\right]$$

$$= \frac{1}{20} x^5 - \frac{1}{2} x^4 + 3x^3 - 12x^2. \ \blacksquare$$

3.2. Homogeneous Equations

Let L be a linear partial differential operator with constant coefficients in two variables x and y. Then $Lu = f(x, y)$ is a partial differential equation with constant coefficients. If we define $D_x = \dfrac{\partial}{\partial x}$ and $D_y = \dfrac{\partial}{\partial y}$, and use the notation $D_x^i = \dfrac{\partial^i}{\partial x^i}$, then

$$L = \sum_{i,j=0}^{m,n} A_{ij} D_x^i D_y^j. \tag{3.1}$$

First, we discuss the homogeneous equation $Lu = 0$, and limit ourselves to the case where L can be expressed as a product of linear factors in D.

Theorem 3.1. *If $u_1, u_2, \ldots, u_n$ are solutions of $Lu = 0$, where L is a linear partial differential operator, then $\sum_{i=1}^{n} c_i u_i$ is also a solution of $Lu = 0$.*

PROOF. Since $u_1, u_2, \ldots, u_n$ are solutions of $Lu = 0$, we have $Lu_1 = Lu_2 = \cdots = Lu_n = 0$. But then $L\left(\sum_1^n c_i u_i\right) = \sum_1^n c_i Lu_i = 0$. $\blacksquare$

Theorem 3.2. *If the operator L of order n can be factored into n linearly independent factors of the type $a_i D_x + b_i D_y + c_i$, where a_i, b_i, and c_i are constants, then the general solution of $Lu = 0$ is given by*

$$u = \sum_{i=1}^{n} f_i(a_i y - b_i x) e^{-c_i x / a_i}. \tag{3.2}$$

PROOF. It was established in §2.1 that the solution of $(a_i D_x + b_i D_y + c_i)u_i = 0$ is $u_i = \phi(a_i y - b_i x)\, e^{-c_i x/a_i}$. Then obviously u_i are n linearly independent solutions of $Lu = 0$ for $i = 1, 2, \ldots, n$. Hence, by Theorem 3.1, $\sum_1^n u_i$ is a solution of $Lu = 0$.

Theorem 3.3. *If $aD_x + bD_y + c$ is a factor of multiplicity k, then the corresponding solution is*

$$\sum_{i=0}^{k-1} x^i f_i(ay - bx)\, e^{-cx/a}. \tag{3.3}$$

A proof can be established by substitution.

EXAMPLE 3.11. Solve $(4D_x^2 - 16D_x D_y + 15D_y^2)u = 0$. Note that this equation can be written as $(2D_x - 3D_y)(2D_x - 5D_y)u = 0$, so from (3.2) the solution is $u = f_1(3x + 2y) + f_2(5x + 2y)$. ∎

EXAMPLE 3.12. Solve $(2D_x^2 - D_x D_y - 6D_y^2 + 4D_x - 8D_y)u = 0$. Note that this equation can be written as $(D_x - 2D_y)(2D_x + 3D_y + 4)u = 0$, so from (3.2) the solution is $u = f_1(2x + y) + e^{-2x} f_2(2y - 3x)$. ∎

EXAMPLE 3.13. Solve $(3D_x + 7D_y)(2D_x - 5D_y + 3)^3 u = 0$. The equation is already in the factored form, and its solution from (3.2) and (3.3) is

$$u = f(3y - 7x) + e^{-3x/2}[f_0(5x + 2y) + x f_1(5x + 2y) + x^2 f_2(5x + 2y)]. \quad ∎$$

Note that the functions f in (3.2) and (3.3) can also be taken as $f(b_i x - a_i y)$ and $f(bx - ay)$, respectively.

3.3. Nonhomogeneous Equations

If the partial differential equation is $Lu = f(x)$, then we find the general solution u_c (complementary function) to the homogeneous equation $Lu = 0$ and search for any function $g(x)$ that satisfies $Lu = f(x)$, where $g(x)$ is also known as the particular solution and sometimes denoted by u_p. Thus, the solution of the partial differential equation is $u = u_c + g(x)$. In this section we discuss some methods for finding the particular solutions.

The operator technique for finding the particular solutions for ordinary differential equations is applicable for the cases where the method of undetermined coefficients for ordinary differential equations is used. This technique is useful in finding particular solutions of partial differential equations. Thus, if $f(D_x, D_y)$ is a linear partial differential operator with constant coefficients, then the corresponding inverse operator is defined as $\dfrac{1}{f(D_x, D_y)}$. We state some obvious results:

$$f(D_x, D_y)\left[\frac{1}{f(D_x, D_y)}\phi(x, y)\right] = \phi(x, y), \tag{3.4}$$

$$\frac{1}{f_1(D_x, D_y)f_2(D_x, D_y)}\phi(x, y) = \frac{1}{f_1(D_x, D_y)}\left[\frac{1}{f_2(D_x, D_y)}\phi(x, y)\right]$$

$$= \frac{1}{f_2(D_x, D_y)}\left[\frac{1}{f_1(D_x, D_y)}\phi(x, y)\right], \tag{3.5}$$

$$\frac{1}{f(D_x, D_y)}\left[c_1\,\phi_1(x, y) + c_2\,\phi_2(x, y)\right] = c_1\frac{1}{f(D_x, D_y)}\phi_1(x, y)$$

$$+ c_2\frac{1}{f(D_x, D_y)}\phi_2(x, y), \tag{3.6}$$

$$\frac{1}{f(D_x, D_y)}e^{ax+by} = \frac{1}{f(a, b)}e^{ax+by}, \quad f(a, b) \neq 0, \tag{3.7}$$

$$f(D_x, D_y)\phi(x, y)e^{ax+by} = e^{ax+by}f(D_x + a, D_y + b)\,\phi(x, y), \tag{3.8}$$

$$\frac{1}{f(D_x, D_y)}\phi(x, y)\,e^{ax+by} = e^{ax+by}\frac{1}{f(D_x + a, D_y + b)}\phi(x, y)$$

$$= e^{ax}\frac{1}{f(D_x + a, D_y)}\,e^{by}\,\phi(x, y) = e^{by}\frac{1}{f(D_x, D_y + b)}\,e^{ax}\,\phi(x, y), \tag{3.9}$$

$$f(D_x^2, D_y^2)\cos(ax + by) = f(-a^2, -b^2)\cos(ax + by), \tag{3.10}$$

$$f(D_x^2, D_y^2)\sin(ax + by) = f(-a^2, -b^2)\sin(ax + by). \tag{3.11}$$

EXAMPLE 3.14. For a particular solution of the equation

$$\left(3D_x^2 + 4D_xD_y - D_y\right)u = e^{x-3y},$$

note that

$$u_p = \frac{1}{\left(3D_x^2 + 4D_xD_y - D_y\right)}e^{x-3y}$$

$$= \frac{1}{[3 + 4(-3) - (-3)]}e^{x-3y} = -\frac{1}{6}e^{x-3y}. \ \blacksquare \tag{3.12}$$

EXAMPLE 3.15. For a particular solution of partial differential equation

$$(3D_x^2 - D_y)u = \sin(ax + by),$$

we have

$$u_p = \frac{1}{(3D_x^2 - D_y)}\sin(ax + by) = \frac{1}{(-3a^2 - D_y)}\sin(ax + by)$$

$$= -\frac{D_y - 3a^2}{D_y^2 - 9a^4}\sin(ax + by) = \frac{b\cos(ax + by) - 3a^2\sin(ax + by)}{b^2 + 9a^4}. \ \blacksquare \tag{3.13}$$

EXAMPLE 3.16. To find a particular solution for the equation

$$(3D_x^2 - D_y)u = e^x \sin(x+y),$$

we have

$$u_p = \frac{1}{(3D_x^2 - D_y)} e^x \sin(x+y) = e^x \frac{1}{(3(D_x+1)^2 - D_y)} \sin(x+y)$$

$$= e^x \frac{1}{(3D_x^2 + 6D_x + 3 - D_y)} \sin(x+y)$$

$$= e^x \frac{1}{(3(-1) + 6D_x + 3 - D_y)} \sin(x+y)$$

$$= e^x \frac{1}{(6D_x - D_y)} \sin(x+y) = e^x \frac{(6D_x + D_y)}{(36D_x^2 - D_y^2)} \sin(x+y)$$

$$= e^x \frac{7 \cos(x+y)}{-35}$$

$$= -\frac{1}{5} e^x \cos(x+y). \ \blacksquare \tag{3.14}$$

EXAMPLE 3.17. To solve $u_{tt} - c^2 u_{xx} = 0$, such that $u(x,0) = e^{-x}$, $u_t(x,0) = 1 + x$, note that the partial differential equation can be written as $(D_t - cD_x)(D_t + cD_x)u = 0$, which gives the solution as $u = f(x+ct) + g(x-ct)$. This solution is known as the d'Alembert's solution (see Eq (5.24)). Applying the initial conditions, we get

$$f(x) + g(x) = e^{-x}, \tag{3.15}$$

$$cf'(x) - cg'(x) = 1 + x. \tag{3.16}$$

On integrating (3.16) with respect to x, we get

$$f(x) - g(x) = \frac{1}{c}(x + \frac{x^2}{2}) + c_1. \tag{3.17}$$

Eqs (3.15) and (3.17) yield

$$f(x) = \frac{1}{2} e^{-x} + \frac{1}{2c}(x + \frac{x^2}{2}) + \frac{c_1}{2},$$

$$g(x) = \frac{1}{2} e^{-x} - \frac{1}{2c}(x + \frac{x^2}{2}) - \frac{c_1}{2}.$$

Hence

$$u = \frac{1}{2} e^{-x}(e^{ct} + e^{-ct}) + (x+1)t. \ \blacksquare$$

A general scheme for initial value problems for the wave equation is as follows: Solve $u_{tt} = c^2 u_{xx}$, subject to the conditions $u(x,0) = \phi(x)$ and $u_t(x,0) = \psi'(x)$. Then, as in the above example,

$$f(x) + g(x) = \phi(x), \quad cf'(x) - cg'(x) = \psi'(x).$$

Consequently,

$$f(x) = \frac{1}{2}\left[\phi(x) + \frac{1}{c}\psi(x) + c_1\right],$$

$$g(x) = \frac{1}{2}\left[\phi(x) - \frac{1}{c}\psi(x) - c_1\right],$$

which yields

$$u(x,t) = \frac{1}{2}\left[\phi(x+ct) - \phi(x-ct)\right] + \frac{1}{2c}[\psi(x+ct) - \psi(x-ct)].$$

EXAMPLE 3.18. It is interesting to note that we can solve the Laplace equation by the above method. To solve

$$u_{xx} + u_{yy} = 0,$$

such that $u(x,0) = \phi(x)$ and $u_y(x,0) = \psi'(x)$, we express $u_{xx} + u_{yy} = 0$ as

$$(D_y + iD_x)(D_y - iD_x)u = 0,$$

and, therefore, its general solution is

$$u = f(x + iy) + g(x - iy).$$

Applying the initial conditions, we get $f(x) + g(x) = \phi(x)$, and $if'(x) - ig'(x) = \psi'(x)$. Hence,

$$f(x) = \frac{1}{2}\left[\phi(x) - i\psi(x) + c\right], \quad g(x) = \frac{1}{2}\left[\phi(x) + i\psi(x) - c\right].$$

Thus,

$$u(x,y) = \frac{1}{2}\left[\phi(x+iy) + \phi(x-iy)\right] + \frac{i}{2}\left[\psi(x-iy) - \psi(x+iy)\right].$$

The final value of u is real. If $\phi(x) = e^{-x}$, and $\psi' = \dfrac{1}{1+x^2}$, the solution is given by

$$u(x,y) = \frac{1}{2}\left[e^{-(x+iy)} + e^{-(x-iy)}\right] + \frac{i}{2}\left[\tan^{-1}(x-iy) - \tan^{-1}(x+iy)\right].$$

Set $\alpha = \tan^{-1}(x - iy) - \tan^{-1}(x + iy)$. Then

$$\tan\alpha = \frac{x - iy - (x + iy)}{1 + x^2 + y^2} = iz,$$

where $z = -\dfrac{2y}{1+x^2+y^2}$. If we simplify $\tan\alpha = iz$ by using

$$\sin\alpha = \frac{e^{i\alpha} - e^{-i\alpha}}{2i} \quad \text{and} \quad \cos\alpha = \frac{e^{i\alpha} + e^{-i\alpha}}{2},$$

we get $e^{2i\alpha} = (1-z)/(1+z)$, or

$$\alpha = \frac{i}{2}\,\ln\frac{1+z}{1-z} = \frac{i}{2}\,\ln\frac{x^2 + (1-y)^2}{x^2 + (1+y)^2},$$

after substituting the above value of z. Hence,

$$u(x,y) = e^{-x}\cos y + \frac{i\alpha}{2} = e^{-x}\cos y - \frac{1}{4}\,\ln\left[\frac{x^2 + (1-y)^2}{x^2 + (1+y)^2}\right]. \quad \blacksquare$$

EXAMPLE 3.19. Consider

$$u_{tt} - c^2\,u_{xx} = e^{-|x|}\,\sin t, \quad u(x,0) = 0, \quad u_t(x,0) = e^{-|x|}.$$

The particular solution is given by

$$u_p = -\frac{1}{1+c^2}\,e^{-|x|}\,\sin t,$$

and the complete solution by

$$u(x,t) = f(x - ct) + g(x + ct) - \frac{1}{1+c^2}\,e^{-|x|}\,\sin t.$$

After applying the initial conditions, we have $u(x,0) = f(x) + g(x) = 0$, which implies that $g(x) = -f(x)$, and

$$u_t(x,0) = -cf'(x) - cf'(x) - \frac{1}{1+c^2}\,e^{-|x|} = e^{-|x|}.$$

Thus, $f'(x) = -\dfrac{2+c^2}{2c\,(1+c^2)}\,e^{-|x|}$. Hence,

$$f(x) = \left[H(x) - H(-x)\right]\frac{2+c^2}{2c\,(1+c^2)}\,e^{-|x|},$$

where $H(x)$ is the Heaviside unit step function, and the solution is

$$u(x,t) = \left[H(x - ct) - H(ct - x)\right]\frac{2+c^2}{2c\,(1+c^2)}\,e^{-|x-ct|}$$

$$- \left[H(x + ct) - H(-(x + ct))\right]\frac{2+c^2}{2c\,(1+c^2)}\,e^{-|x+ct|} - \frac{1}{1+c^2}\,e^{-|x|}\,\sin t. \quad \blacksquare$$

Other techniques from ordinary differential equations such as the method of undetermined coefficients and the variation of parameters technique can also be extended to find the particular solution corresponding to the nonhomogeneous term in the partial differential equations.

3.4. Mathematica Projects

PROJECT 3.1. Mathematica Notebook `InverseOperator.nb` can be used to solve examples of the above type as well as the exercises given below.

3.5. Exercises

Evaluate (use the inverse operator method of §3.1):

3.1. $(D-3)^{-1}(x^3+3x-5)$.   ANS. $-(9x^3+9x^2+33x-34)/27$.

3.2. $(D-1)^{-1}(2x)$.   ANS. $-2x-2$.

3.3. $(D-1)^{-1}(x^2)$.   ANS. $-(x^2+2x+2)$.

3.4. $(4D^2-5D)^{-1}(x^2\,e^{-x})$.   ANS. $e^{-x}(81x^2+234x+266)/729$.

3.5. $(D^2-3D+2)^{-1}\sin 2x$.   ANS. $\dfrac{3\cos 2x-\sin 2x}{20}$.

3.6. $D^{-2}(2\sin 2x)$.   ANS. $-\dfrac{1}{2}\sin 2x.$.

3.7. $D^{-3}x$.   ANS. $x^4/24$.

3.8. $D^{-2}(3e^{3x})$.   ANS. $e^{3x}/3$.

3.9. $D^{-1}(2x+3)$.   ANS. x^2+3x.

3.10. $(D^3-D^2)^{-1}(2x^3)$.   ANS. $-2\left(\dfrac{x^5}{20}+\dfrac{x^4}{4}+x^3+3x^2\right)$.

3.11. $(D^2+3D+2)^{-1}(e^{ix})$.   ANS. $\dfrac{1-3i}{10}e^{ix}$.

3.12. $(D^2-3D+2)^{-1}(3\sin x)$.   ANS. $\dfrac{3}{10}(\sin x+3\cos x)$.

3.13. $(D^2+3D+2)^{-1}(8+6e^x+2\sin x)$.   ANS. $4+e^x+(\sin x-3\cos x)/5$.

3.14. $(D^5+2D^3+D)^{-1}(2x+\sin x+\cos x)$.   ANS. $x^2+x^2(\cos x-\sin x)/8$.

Find the general solution of the following partial differential equations:

3.15. $(3D_x^2-2D_xD_y-5D_y^2)u=3x+y+e^{x-y}$.
ANS. $u=f(5x+3y)+g(x-y)+\dfrac{11}{54}x^3+\dfrac{1}{6}x^2y+\dfrac{1}{8}xe^{x-y}$.

3.16. $(D_x^4-10D_x^2D_y^2+9D_y^4)u=135\sin(3x+2y)$.
ANS. $f_1(3x+y)+f_2(3x-y)+g_1(x+y)+g_2(x-y)-\sin(3x+2y)$.

3.17. $(D_x-2D_y)^3u=125e^x\sin y$.
ANS. $f_1(2x+y)+xf_2(2x+y)+x^2f_3(2x+y)-e^x(2\cos y+11\sin y)$.

3.18. Find the particular solution for the following partial differential equations:

(a) $(D_x^2 - D_y)u = 17e^{x+y}\sin(x - 2y)$.
ANS. $-e^{x+y}\{\sin(x - 2y) + 4\cos(x - 2y)\}$.

(b) $(D_x^2 + D_y^2)u = 6xy + 25e^{3x+4y}$.
ANS. $x^3y + e^{3x+4y}$.

(c) $(D_x^2 + D_y^2 - D_x)u = 37e^{5y}\cos(3x + 4y)$.
ANS. $e^{5y}\sin(3x + 4y)$.

3.19. Show that $u = f(ay - bx)\,e^{-cy/b}$ is also a solution of $(aD_x + bD_y + c)\,u = 0$.

3.20. Prove formula 14 of §3.1.

HINT. Use $D^n\{x\,\phi(x)\} = x\,D^n\phi(x) + n\,D^{n-1}\phi(x)$, which implies that
$$f(D)\{x\phi(x)\} = xf(D)\{\phi(x)\} + f'(D)\{\phi(x)\}.$$

3.21. Solve $u_{tt} - c^2u_{xx} = 0$, subject to the conditions $u(x,0) = \ln(1 + x^2)$ and $u_t(x,0) = e^{-x}$.

ANS. $u(x,t) = \dfrac{1}{2}[\ln\{1 + (x + ct)^2\} + \ln\{1 + (x - ct)^2\}] + \dfrac{1}{c}e^{-x}\cosh ct$.

4

Orthogonal Expansions

Unlike ordinary differential equations, the general solution of a partial differential equation consists of one or more arbitrary functions. It is not easy to determine the particular form of these functions from the prescribed boundary and initial conditions even if the general solution is known. However, it is often possible to solve a specific boundary value or initial value problem in the form of an infinite series of functions known as eigenfunctions or characteristic functions. This chapter is devoted to developing orthogonal series, trigonometric Fourier series, eigenfunction expansions, and the Bessel functions. Orthogonal expansions are important for the method of separation of variables, which is discussed in the next chapter.

4.1. Orthogonality

The *inner product* of two real-valued functions f_1 and f_2 on an interval $a \leq x \leq b$ is defined by

$$\langle f_1, f_2 \rangle = \int_a^b f_1(x) f_2(x)\, dx, \qquad (4.1)$$

provided the integral in (4.1) exists.

DEFINITION 4.1. The functions f_1 and f_2 are said to be *orthogonal* on the interval $a \leq x \leq b$ if $\langle f_1, f_2 \rangle = 0$, i.e., the integral in (4.1) vanishes.

DEFINITION 4.2. A set of real-valued functions $\{f_1(x), f_2(x), \dots\}$ defined on the interval $a \leq x \leq b$ is said to be an *orthogonal set of functions* on this interval if for each m and n, $m \neq n$,

$$\langle f_m, f_n \rangle = \int_a^b f_m(x) f_n(x) \, dx = 0, \quad m, n = 1, 2, \ldots, \tag{4.2}$$

provided each integral exists. Thus, for example, the orthogonality relations for the function $\cos(n\pi x / l)$ are

$$\int_0^l \cos \frac{n\pi x}{l} \cos \frac{m\pi x}{l} \, dx = \begin{cases} 0, & n \neq m, \\ \dfrac{l}{2}, & n = m \neq 0, \\ l, & n = m = 0. \end{cases} \tag{4.3}$$

DEFINITION 4.3. The *norm* of the functions $f_n(x)$ is denoted by $\|f_n\|$ and defined by

$$\|f_n\| = \sqrt{\langle f_n, f_n \rangle} = \left(\int_a^b f_n^2(x) \, dx \right)^{1/2} \geq 0. \tag{4.4}$$

DEFINITION 4.4.   An orthogonal set of functions $\{f_n(x)\}$ is called an *orthonormal set of functions* on the interval $a \leq x \leq b$ if $\|f_n\| = 1$ for all $n = 1, 2, \ldots$.

Thus, if $\{f_n(x)\}$ is an orthonormal set of functions, then

$$\langle f_m, f_n \rangle = \int_a^b f_m(x) f_n(x) \, dx = \delta_{mn}, \tag{4.5}$$

where

$$\delta_{mn} = \begin{cases} 1 & \text{if } m = n \\ 0 & \text{if } m \neq n \end{cases}$$

is the Kronecker delta. If an orthogonal set of functions $\{f_n(x)\}$ is defined on the interval $a \leq x \leq b$ with $\|f_n\| \neq 0$, we can always construct an orthonormal set of functions $\{g_n(x)\}$ by the formula

$$g_n(x) = \frac{f_n(x)}{\|f_n\|}, \quad a \leq x \leq b. \tag{4.6}$$

In fact, in view of (4.5),

$$\langle g_m, g_n \rangle = \int_a^b \frac{f_m(x)}{\|f_m\|} \frac{f_n(x)}{\|f_n\|} \, dx = \frac{1}{\|f_m\| \|f_n\|} \langle f_m, f_n \rangle = \delta_{mn}, \tag{4.7}$$

and hence $\|g_n\| = 1$ for all n.

EXAMPLE 4.1. The functions $f_n(x) = \sin nx$, $n = 1, 2, \ldots$, form an orthogonal set of functions on the interval $-\pi \leq x \leq \pi$, since

$$
\begin{aligned}
\langle f_m, f_n \rangle &= \int_{-\pi}^{\pi} \sin mx \sin nx \, dx \\
&= \frac{1}{2} \int_{-\pi}^{\pi} [\cos(m-n)x - \cos(m+n)x] \, dx \\
&= \frac{1}{2} \left[\frac{\sin(m-n)x}{m-n} - \frac{\sin(m+n)x}{m+n} \right]_{-\pi}^{\pi} = 0, \; m \neq n,
\end{aligned}
\tag{4.8}
$$

and

$$
\begin{aligned}
\langle f_n, f_n \rangle &= \int_{-\pi}^{\pi} \sin^2 nx \, dx = \frac{1}{2} \int_{-\pi}^{\pi} (1 - \cos 2nx) \, dx \\
&= \frac{1}{2} \left[x - \frac{\sin 2nx}{2n} \right]_{-\pi}^{\pi} = \pi.
\end{aligned}
$$

Thus, $\|f_n\| = \sqrt{\pi}$, and the orthonormal set of functions is given by

$$
\{g_n(x)\} = \left\{ \frac{\sin nx}{\sqrt{\pi}} \right\}, \quad n = 1, 2, \ldots . \; \blacksquare
$$

EXAMPLE 4.2. The set of functions

$$
\left\{ \frac{1}{\sqrt{2\pi}}, \frac{\cos x}{\sqrt{\pi}}, \frac{\sin x}{\sqrt{\pi}}, \frac{\cos 2x}{\sqrt{\pi}}, \frac{\sin 2x}{\sqrt{\pi}}, \ldots \right\}
$$

forms an orthogonal set on the interval $-\pi \leq x \leq \pi$. In Example 4.1, we have seen that $\dfrac{\sin nx}{\sqrt{\pi}}$ is orthonormal on the interval $-\pi \leq x \leq \pi$. Now, to verify the orthonormality of other functions, we have

$$
\int_{-\pi}^{\pi} \frac{\cos mx}{\sqrt{\pi}} \frac{\cos nx}{\sqrt{\pi}} \, dx
$$
$$
= \begin{cases} \frac{1}{2\pi} \int_{-\pi}^{\pi} [\cos(m-n)x + \cos(m+n)x] \, dx = 0, & \text{for } m \neq n, \\ \frac{1}{2\pi} \int_{-\pi}^{\pi} [1 + \cos 2mx] \, dx = 1, & \text{for } m = n. \end{cases}
$$

Also,

$$
\int_{-\pi}^{\pi} \frac{\sin mx}{\sqrt{\pi}} \frac{\cos nx}{\sqrt{\pi}} \, dx = \frac{1}{2\pi} \int_{-\pi}^{\pi} [\sin(m-n)x + \sin(m+n)x] \, dx = 0.
$$

Finally,

$$\int_{-\pi}^{\pi} \frac{1}{\sqrt{2\pi}} \frac{\sin nx}{\sqrt{\pi}}\, dx = -\frac{1}{\pi\sqrt{2}} \frac{\cos nx}{n} \Big|_{-\pi}^{\pi} = 0,$$

$$\int_{-\pi}^{\pi} \frac{1}{\sqrt{2\pi}} \frac{\cos nx}{\sqrt{\pi}}\, dx = \frac{1}{\pi\sqrt{2}} \frac{\sin nx}{n} \Big|_{-\pi}^{\pi} = 0,$$

$$\int_{-\pi}^{\pi} \left(\frac{1}{\sqrt{2\pi}}\right)^2 dx = \frac{x}{2\pi} \Big|_{-\pi}^{\pi} = 1. \ \blacksquare$$

4.2. Orthogonal Polynomials

DEFINITION 4.5. The *weighted inner product* of two functions f and g, with weight $w > 0$, is defined by

$$\langle f, g \rangle_w = \int_a^b w(x) f(x) g(x)\, dx. \tag{4.9}$$

Two nonzero functions $f(x)$ and $g(x)$ are said to be *orthogonal* with respect to the weight function w if $\langle f, g \rangle_w = 0$. A sequence of functions $\{f_n(x)\}$, (or polynomials $\{p_n(x)\}$), $n = 1, 2, \ldots$, is orthogonal with respect to $w(x)$ if every pair of them is orthogonal. Some well-known classes of orthogonal polynomials are defined for different values of a, b, and $w(x)$ for $n = 0, 1, 2, \ldots$, as follows:

Orthogonal Polynomials	a	b	$w(x)$
Chebyshev (1st kind) $T_n(x)$	-1	1	$(1 - x^2)^{-1/2}$
Chebyshev (2nd kind) $U_n(x)$	-1	1	$(1 - x^2)^{1/2}$
Hermite $H_n(x)$	$-\infty$	∞	e^{-x^2}
Jacobi $P_n^{(\alpha,\beta)}(x)$	-1	1	$(1 - x)^\alpha (1 - x)^\beta$
Laguerre $L_n(x)$	0	∞	e^{-x}
Legendre $P_n(x)$	-1	1	1

For more details, see Appendix B.

EXAMPLE 4.3. The Chebyshev polynomials of the first kind $T_n(x)$ and of the second kind $U_n(x)$ have the weight $w(x) = 1/\sqrt{1 - x^2}$, and are the solutions of the differential equation

$$(1 - x^2)\, y'' - x\, y' + n^2\, y = 0, \quad n = 0, 1, 2, \ldots .$$

The polynomials $T_n(x)$ are defined on the interval $-1 < x < 1$ by

$$T_n(x) = \cos\left(n \cos^{-1} x\right) = x^n - \binom{n}{2} x^{n-2} \left(1 - x^2\right) + \binom{n}{4} x^{n-4} \left(1 - x^2\right)^2 - \cdots,$$

and their orthogonality relations are

$$\int_{-1}^{1} \frac{T_m(x)T_n(x)}{\sqrt{1 - x^2}} \, dx = \begin{cases} 0 & \text{if } m \neq n, \\ \pi/2 & \text{if } m = n \neq 0, \\ \pi & \text{if } m = n = 0. \end{cases}$$

The polynomials $U_n(x)$ are defined by $U_n(x) = \dfrac{1}{n} \sqrt{1 - x^2} \, \dfrac{d}{dx} T_n(x)$, and their orthogonality relations are

$$\int_{-1}^{1} \frac{U_m(x)U_n(x)}{\sqrt{1 - x^2}} \, dx = \begin{cases} 0 & \text{if } m \neq n, \\ \pi/2 & \text{if } m = n \neq 0, \\ 0 & \text{if } m = n = 0. \end{cases} \blacksquare$$

EXAMPLE 4.4. The Legendre polynomials $P_n(x)$ are the solutions of the Legendre equation

$$\left(1 - x^2\right) y'' - 2x \, y' + n(n+1) \, y = 0.$$

These polynomials are also called the zonal harmonics of the first kind. The orthogonality relation for the Legendre polynomials $P_n(x)$ is

$$\int_{-1}^{1} P_n(x) \, P_m(x) \, dx = \begin{cases} 0 & \text{if } n \neq m, \\ \dfrac{2}{2n + 1} & \text{if } n = m. \end{cases} \blacksquare$$

4.3. Series of Orthogonal Functions

Some important types of series expansions are obtained from orthogonal sets of functions.

DEFINITION 4.6. Let $\{g_1(x), g_2(x), \dots\}$ be an orthogonal set of functions on an interval $a \leq x \leq b$, and let a function $f(x)$ be represented in terms of the functions $g_n(x)$, $n = 1, 2, \dots$, by a convergent series

$$f(x) = \sum_{n=1}^{\infty} c_n \, g_n(x). \tag{4.10}$$

This series is called a *generalized Fourier series* of $f(x)$, and the coefficients c_n, $n = 1, 2, \ldots$, are called the *Fourier coefficients* of $f(x)$ with respect to the orthogonal set of functions $g_n(x)$, $n = 1, 2, \ldots$.

In view of Definition 4.4, it is easy to determine the coefficients c_n. If we multiply both sides of (4.10) by $g_m(x)$ for a fixed m, integrate with respect to x over the interval $[a, b]$, and assume term-by-term integration, which is justified in the case of uniform convergence, we obtain

$$
\langle f, g_m \rangle = \int_a^b f g_m \, dx = \int_a^b \left(\sum_{n=1}^{\infty} c_n \, g_n(x) \right) g_m(x) \, dx
$$

$$
= \sum_{n=1}^{\infty} c_n \, \langle g_n(x), g_m(x) \rangle.
\tag{4.11}
$$

Since, in view of the relations (4.6) and (4.7), $\langle g_n, g_m \rangle = \delta_{nm}$, we find that $\langle g_n, g_m \rangle = \|g_n\|^2$ for $n = m$, and then (4.11) yields

$$
\langle f, g_n \rangle = c_n \|g_n\|^2,
$$

or

$$
c_n = \frac{\langle f, g_n \rangle}{\|g_n\|^2} = \frac{1}{\|g_n\|^2} \int_a^b f(x) g_n(x) \, dx.
\tag{4.12}
$$

EXAMPLE 4.5. Using the orthogonal set of functions from Example 4.2 and formula (4.12), the representation (4.10) becomes

$$
f(x) = a_0 + \sum_{n=1}^{\infty} \left(a_n \, \cos nx + b_n \, \sin nx \right),
\tag{4.13}
$$

where, since $\|g_0\| = \dfrac{1}{\sqrt{2\pi}}$, $\|g_n\| = \dfrac{1}{\sqrt{\pi}}$ for $n = 1, 2, \ldots$, the coefficients are given by

$$
a_0 = \frac{1}{2\pi} \int_{-\pi}^{\pi} f(x) \, dx,
$$

$$
a_n = \frac{1}{\pi} \int_{-\pi}^{\pi} f(x) \cos nx \, dx,
\tag{4.14}
$$

$$
b_n = \frac{1}{\pi} \int_{-\pi}^{\pi} f(x) \sin nx \, dx,
$$

for $n = 1, 2, \ldots$. ∎

In the above example we have introduced the *trigonometric Fourier series* of a periodic function $f(x)$ of period 2π, under the assumption that the series (4.13) converges and represents the function $f(x)$. The coefficients a_0, a_n, and b_n for $n = 1, 2, \ldots$ are called the *Fourier coefficients*. The trigonometric Fourier series as well as the Fourier sine and cosine series for a p-periodic and a $2L$-periodic, piecewise continuous function $f(x)$ together with the associated theorems are discussed in the following section.

4.4. Trigonometric Fourier Series

A function $f(x)$ is said to be *periodic* of period p if $f(x+p) = f(x)$, $c \leq x \leq c+p$, where c is a constant. For example, the functions $\sin x$ and $\cos x$ have period 2π. More generally, each of the functions $\sin \dfrac{2n\pi x}{p}$ and $\cos \dfrac{2n\pi x}{p}$ is periodic of period p, where n is a positive integer. Hence, if the infinite series

$$\frac{1}{2}a_0 + \sum_{n=1}^{\infty} \left(a_n \cos \frac{2n\pi x}{p} + b_n \sin \frac{2n\pi x}{p} \right) \tag{4.15}$$

is convergent, then it represents a function of period p.

Theorem 4.1. (Fourier Theorem I for periodic functions) *Let $f(x)$ be a single-valued, piecewise continuous, periodic function of period p on a finite interval $I = [c, c+p]$, where c is a constant. Then the series (4.15) converges to $f(x)$ at all points of continuity and to*

$$\frac{1}{2} \left[f(x+) + f(x-) \right] \tag{4.16}$$

at the points of discontinuity (and also at all points of continuity). The coefficients a_0, a_n, and b_n are given by

$$\begin{aligned} a_n &= \frac{2}{p} \int_c^{c+p} f(x) \cos \frac{2n\pi x}{p}\, dx, \quad n = 0, 1, 2, \ldots, \\ b_n &= \frac{2}{p} \int_c^{c+p} f(x) \sin \frac{2n\pi x}{p}\, dx, \quad n = 1, 2, \ldots. \end{aligned} \tag{4.17}$$

If we set $p = 2L$ and $c = -L$ in (4.17), then these formulas become

$$\begin{aligned} a_n &= \frac{1}{L} \int_{-L}^{L} f(x) \cos \frac{n\pi x}{L}\, dx, \quad n = 0, 1, 2, \ldots, \\ b_n &= \frac{1}{L} \int_{-L}^{L} f(x) \sin \frac{n\pi x}{L}\, dx, \quad n = 1, 2, \ldots. \end{aligned} \tag{4.18}$$

EXAMPLE 4.6. Let $f(x) = |x|$, $-\pi \le x \le \pi$ (Fig. 4.1). Note that $f(x)$ is an even function, since $f(-x) = |-x| = |x| = f(x)$. Then, from (4.18), with $L = \pi$, we have

$$a_0 = \frac{1}{\pi} \int_{-\pi}^{\pi} |x| \, dx = \frac{2}{\pi} \int_0^{\pi} x \, dx = \pi,$$

$$a_n = \frac{2}{\pi} \int_0^{\pi} f(x) \cos nx \, dx = \frac{2}{\pi} \int_0^{\pi} x \cos nx \, dx$$

$$= \frac{2}{\pi} \left[\frac{\cos nx}{n^2} + \frac{x \sin nx}{n} \right]_0^{\pi}$$

$$= \begin{cases} -\dfrac{4}{n^2 \pi}, & n \text{ odd}, \\ 0, & n \text{ even}, \end{cases}$$

and $b_n = 0$ for $n = 1, 2, \dots$. Hence, the Fourier series (4.15) for $f(x) = |x|$ is

$$|x| = \frac{\pi}{2} - \frac{4}{\pi} \sum_{n=1}^{\infty} \frac{\cos(2n-1)x}{(2n-1)^2}, \quad -\pi \le x \le \pi. \ \blacksquare$$

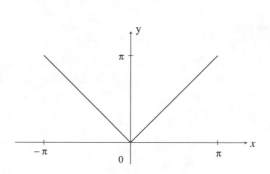

Fig. 4.1. Graph of $f(x) = |x|$, $-\pi \le x \le \pi$.

EXAMPLE 4.7. Consider

$$f(x) = \begin{cases} 2, & x = 0, \\ 3, & 0 < x < 2, \\ 2, & x = 2, \\ 1, & 2 < x < 4, \end{cases}$$

such that $f(x+4) = f(x)$. This function is piecewise continuous and is of period 4 (see Fig. 4.2). Note that $f(2-) = 3$ and $f(2+) = 1$. In this example, $c = 0, p = 4$. Then, using (4.17), we get

$$a_0 = \frac{1}{2} \left(\int_0^2 3 \, dx + \int_2^4 1 \, dx \right) = 4,$$

$$a_n = \frac{1}{2}\left(\int_0^2 3\cos\frac{n\pi x}{2}\,dx + \int_2^4 1\cos\frac{n\pi x}{2}\,dx\right) = 0,$$

$$b_n = \frac{1}{2}\left(\int_0^2 3\sin\frac{n\pi x}{2}\,dx + \int_2^4 1\sin\frac{n\pi x}{2}\,dx\right)$$

$$= \frac{3 - 2\cos n\pi - \cos 2n\pi}{n\pi} = \begin{cases} 0, & \text{if } n \text{ is even}, \\ \dfrac{4}{n\pi}, & \text{if } n \text{ is odd}. \end{cases}$$

Hence, the Fourier series for $f(x)$ is

$$f(x) = 2 + \frac{4}{\pi}\sum_{n=1}^{\infty}\frac{1}{2n-1}\sin\frac{(2n-1)\pi x}{2}.$$

Also, in view of (4.16), for example, $f(4) = \frac{1}{2}\left[f(4+) + f(4-)\right] = \frac{1}{2}(3+1) = 2.$ ∎

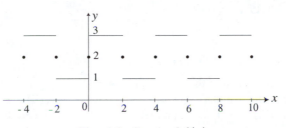

Fig. 4.2. Graph of $f(x)$.

EXAMPLE 4.8. Consider the wave equation $u_{tt} = c^2\,u_{xx}$, where c is the wave velocity, subject to the boundary conditions $u(0,t) = 0 = u(l,t)$ for $t \geq 0$, and the initial conditions $u(x,0) = x$, and $u_t(x,0) = 0$. We assume the d'Alembert's solution of the wave equation (see Example 3.17 and (5.24)), which is

$$u(x,t) = f(x+ct) + g(x-ct). \tag{4.19}$$

From the boundary condition $u(0,t) = 0$, we find that

$$f(ct) + g(-ct) = 0,$$

or, setting $ct = z$, we get $f(z) = -g(-z)$. Using the other boundary condition $u(l,t) = 0$, we have

$$f(l+ct) + g(l-ct) = 0,$$

which implies that $f(ct+l) - f(ct-l) = 0$, i.e., $f(z) = f(z+2l)$. This last equation means that f is a periodic function of period $2l$. Thus, the solution reduces to

$$u(x,t) = f(ct+x) - f(ct-x).$$

If we apply the initial conditions, we get

$$f(x) - f(-x) = x, \quad \text{and} \quad f'(x) - f'(-x) = 0,$$

which means that $f'(x)$ is an even function, and therefore $f(x)$ an odd function, i.e., $f(x) = -f(-x)$. Hence, $2f(x) = x$. Since $f(x)$ is an odd periodic function of period $2l$, it can be expressed as a Fourier sine series

$$f(x) = \frac{x}{2} = \frac{l}{\pi} \sum_{n=1}^{\infty} \frac{(-1)^{n+1}}{n} \sin \frac{n\pi x}{l},$$

which yields

$$u(x,t) = \frac{l}{\pi} \sum_{n=1}^{\infty} \frac{(-1)^{n+1}}{n} \left[\sin \frac{n\pi(ct+x)}{l} - \sin \frac{n\pi(ct-x)}{l} \right]$$

$$= \frac{2l}{\pi} \sum_{n=1}^{\infty} \frac{(-1)^{n+1}}{n} \sin \frac{n\pi x}{l} \cos \frac{n\pi ct}{l}.$$

This solution can be compared with (5.23). ∎

EXAMPLE 4.9. Consider the wave equation $u_{tt} = c^2 u_{xx}$, where c is the wave velocity, subject to the boundary conditions (i) $u_x(0,t) = 0$, (ii) $u(l,t) = 0$, and the initial conditions (iii) $u(x,0) = x$, and (iv) $u_t(x,0) = 0$. Using the d'Alembert's solution (4.19), we find that

$$\frac{\partial u(x,t)}{\partial x} = \frac{\partial f(x+ct)}{\partial(x+ct)} \frac{\partial(x+ct)}{\partial x} + \frac{\partial g(x-ct)}{\partial(x-ct)} \frac{\partial(x-ct)}{\partial x}.$$

If we set $ct = z$, then, in view of the boundary condition (i), this solution yields $u_x(0,t) = f'(z) + g'(-z) = 0$, i.e., $f'(z) = -g'(-z)$, which upon integration with respect to z yields $f(z) = g(-z) - c_1$. Hence,

$$g(z) = f(-z) + c_1. \tag{4.20}$$

Condition (ii) gives $f(l+ct) + g(l-ct) = 0$, which, by using the value of g from (4.20), becomes

$$f(l+ct) + f(ct-l) + c_1 = 0.$$

Condition (iii), in view of (4.20), gives

$$f(x) + f(-x) + c_1 = x,$$

whereas condition (iv) gives

$$cf'(x) + cf'(-x) = 0,$$

i.e., $f'(x) = -f'(-x)$. This implies that f' is an odd function, and, therefore, $f(x)$ is an even function, i.e., $f(-x) = f(x)$. Hence, $2f(x) = x - c_1$. If we define

$$\psi(x) = f(x) + \frac{c_1}{2},$$

then we have

$$\psi(ct + l) + \psi(ct - l) = 0,$$

or, by taking $ct - l = v$, we get

$$\psi(v) + \psi(v + 2l) = 0.$$

If we set $v = \zeta + 2l$, then

$$\psi(\zeta + 4l) = -\psi(\zeta + 2l) = \psi(\zeta).$$

Hence, ψ is a periodic function of period $4l$. Also, $\psi(x) + \psi(-x) = x$. Let

$$\psi(x) = A_n \sin \frac{n\pi x}{2l} + B_n \cos \frac{n\pi x}{2l}.$$

Since $0 \le x \le l$, we cannot integrate from 0 to $4l$. Thus, it is not possible to use formula (4.14) to determine the coefficients A_n and B_n. Therefore, we proceed as follows:

$$\begin{aligned}
u(x,t) &= \psi(ct + x) + \psi(ct - x) \\
&= A_n \left[\sin \frac{n\pi}{2l}(ct + x) + \sin \frac{n\pi}{2l}(ct - x) \right] \\
&\quad + B_n \left[\cos \frac{n\pi}{2l}(ct + x) + \cos \frac{n\pi}{2l}(ct - x) \right] \\
&= 2 \left[A_n \sin \frac{n\pi ct}{2l} \cos \frac{n\pi x}{l} + B_n \cos \frac{n\pi ct}{2l} \cos \frac{n\pi x}{2l} \right].
\end{aligned}$$

Using the initial condition (iv), we find that $u_t(x,0) = 0$, which gives $A_n = 0$. Again, condition (ii) yields $2B_n \cos \frac{n\pi ct}{2l} \cos \frac{n\pi}{2} = 0$. Thus, $B_n = 0$ if $n = 2m$, and $B_n \ne 0$ if $n = 2m - 1$, where m is a positive integer. This gives

$$u(x,t) = \sum_{m=1}^{\infty} 2B_{2m-1} \cos \frac{(2m-1)\pi ct}{2l} \cos \frac{(2m-1)\pi x}{2l}.$$

Since

$$u(x,0) = \sum_{m=1}^{\infty} 2B_{2m-1} \cos \frac{(2m-1)\pi x}{2l} = x,$$

we find that

$$
\begin{aligned}
B_{2m-1} &= \frac{1}{l} \int_0^l x \cos \frac{(2m-1)\pi x}{2l} \, dx \\
&= \frac{1}{l} \left\{ \left[x \sin \frac{(2m-1)\pi x}{2l} \right]_0^l \frac{2l}{(2m-1)\pi} \right. \\
&\quad \left. - \frac{2l}{(2m-1)\pi} \int_0^l \sin \frac{(2m-1)\pi x}{2l} \, dx \right\} \\
&= \frac{2(-1)^{m-1} l}{(2m-1)\pi} - \frac{4l}{(2m-1)^2 \pi^2} . \blacksquare
\end{aligned}
$$

DEFINITION 4.7. A function f defined on an interval $[a, b]$ is said to satisfy the *Lipschitz condition* if there exists a constant $M > 0$ such that

$$
|f(x) - f(y)| < M |x - y|. \tag{4.21}
$$

Obviously, such a function is uniformly continuous and bounded.

Theorem 4.2. (Fourier Theorem II) *Let $f(x)$ denote a periodic function of period 2π, which is integrable on $-\pi < x < \pi$. If this integral is improper, let it be absolutely convergent. Then at each point x which is interior to an interval on which f is piecewise continuous, the Fourier series for the function f converges to the average value (4.16).*

The asymptotic behavior of the Fourier coefficients of a periodic function $f(x)$ is given by the following theorem:

Theorem 4.3. *As $n \to \infty$, the Fourier coefficients a_n and b_n always approach zero at least as rapidly as α/n, where α is a constant independent of n. If the function $f(x)$ is piecewise continuous, then either a_n or b_n, and in general both, decrease no faster than α/n. In general, if $f(x)$ and its first $(k-1)$ derivatives satisfy the conditions of the Fourier theorems I and II, then the Fourier coefficients a_n and b_n approach zero as $n \to \infty$ at least as rapidly as α/n^{k+1}. Moreover, if $f^{(k)}(x)$ is not continuous everywhere, then either a_n or b_n, and in general both, approach zero no faster than α/n^{k+1}.*

This theorem implies that the smoother the function f is, the faster its Fourier series converges. Note that the analysis of the Fourier series for functions in R^2 and R^3 is similar. Proofs of these theorems are available in Davis (1963), Churchill and Brown (1978), and Walker (1988).

DEFINITION 4.8. Let f be a function defined on the interval $0 \le x \le L$ such that the integrals

$$
\int_0^L f(x) \sin \frac{n\pi x}{L} \, dx, \quad n = 1, 2, \dots,
$$

exist. Then the series

$$\sum_{n=1}^{\infty} b_n \sin \frac{n\pi x}{L}, \tag{4.22}$$

where

$$b_n = \frac{2}{L} \int_0^L f(x) \sin \frac{n\pi x}{L} \, dx, \quad n = 1, 2, \dots, \tag{4.23}$$

is called the *Fourier sine series* of f on the interval $0 \le x \le L$.

Note that the series expansion (4.22) is identical to the trigonometric Fourier series (4.15) of an *odd* function defined on the interval $-L \le x \le L$, which coincides with $f(x)$ on the interval $0 \le x \le L$.

DEFINITION 4.9. Let f be a function defined on the interval $0 \le x \le L$ such that the integrals

$$\int_0^L f(x) \cos \frac{n\pi x}{L} \, dx, \quad n = 0, 1, 2, \dots,$$

exist. Then the series

$$\frac{a_0}{2} + \sum_{n=1}^{\infty} a_n \cos \frac{n\pi x}{L}, \tag{4.24}$$

where

$$a_n = \frac{2}{L} \int_0^L f(x) \cos \frac{n\pi x}{L} \, dx, \quad n = 0, 1, 2, \dots, \tag{4.25}$$

is called the *Fourier cosine series* of f on the interval $0 \le x \le L$.

Note that the series expansion (4.24) is identical to the trigonometric Fourier series (4.15) of an *even* function defined on the interval $-L \le x \le L$, which coincides with $f(x)$ on the interval $0 \le x \le L$.

EXAMPLE 4.10. To find the Fourier cosine series of period 2π which represents $f(x) = x$ on the interval $0 < x < \pi$, let $L = \pi$ in (4.25). Then

$$a_0 = \frac{2}{\pi} \int_0^\pi x \, dx = \pi,$$

$$a_n = \frac{2}{\pi} \int_0^\pi x \cos nx \, dx = \frac{2}{\pi} \left[\frac{x \sin nx}{n} + \frac{\cos nx}{n^2} \right]$$

$$= \frac{2(\cos n\pi - 1)}{\pi n^2} = \begin{cases} 0 & \text{if } n \text{ is even,} \\ -\dfrac{4}{\pi n^2} & \text{if } n \text{ is odd.} \end{cases}$$

Hence, the Fourier cosine series is

$$x = \frac{\pi}{2} - \frac{4}{\pi} \sum_{n-1}^{\infty} \frac{1}{(2n-1)^2} \cos(2n-1)x.$$

The graph of the right side of this series is given in Fig. 4.3. Note that at $x = 0$, the right side of the series equals

$$\frac{\pi}{2} - \frac{4}{\pi} \sum_{1}^{\infty} \frac{1}{(2n-1)^2} = \frac{\pi}{2} - \frac{4}{\pi} \frac{\pi^2}{8} = 0,$$

since $\sum_{1}^{\infty} \frac{1}{n^2} = \frac{\pi^2}{6}$. At $x = \pi$, the right side of the above series is obviously equal

to $\frac{\pi}{2} + \frac{4}{\pi} \frac{\pi^2}{8} = \pi$. ∎

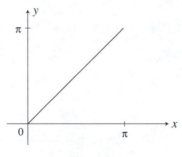

Fig. 4.3.

EXAMPLE 4.11. In the case of a jump discontinuity the Fourier series leads to what is known as the *Gibbs phenomenon*. For example, consider the Fourier sine series for

$$f(x) = \begin{cases} 1, & 0 < x < \pi, \\ -1, & -\pi < x < 0. \end{cases}$$

Then the coefficients b_n are determined by (4.18), and we get

$$f(x) = \frac{4}{\pi} \sum_{n=0}^{\infty} \frac{1}{2n+1} \sin(2n+1)x.$$

The partial sums

$$M_k = \frac{4}{\pi} \sum_{n=0}^{k} \frac{1}{2n+1} \sin(2n+1)x$$

define the k harmonics, which approximate the jump as shown in Fig. 4.4. Notice the sharp peaks in the harmonics M_1, M_2, M_{10}, M_{20}, and M_{40} near 0, which is the discontinuity of $f(x)$. Gibbs showed that the height or *overshoot* of these peaks is greater than $f(0+)$ by about 9%. The width of the overshoot goes to zero as $k \to \infty$, but the height remains at 9% both at the top and the bottom such that

$$\lim_{k \to \infty} \max |f(x) - M_k(x)| \neq 0.$$

This phenomenon does not go away even when the number of harmonics is increased. See Exercises 4.23 and 4.24 for more on this phenomenon. ∎

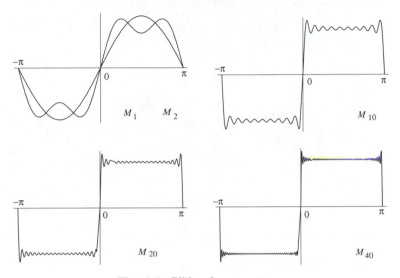

Fig. 4.4. Gibbs phenomenon.

EXAMPLE 4.12. Consider

$$u_{tt} = c^2 u_{xx}, \quad 0 < x < 1,$$

subject to the conditions $u(0, t) = u(l, t) = 0$, for $t \geq 0$, and $u(x, 0) = x$, $u_t(x, 0) = 0$. The general solution is

$$u = f(x + ct) + g(x - ct).$$

From the boundary conditions we find that

$$f(ct) + g(-ct) = 0, \quad \text{or} \quad f(z) + g(-z) = 0,$$

which yields $f(z) = -g(-z)$. Also $f(l + ct) + g(l - ct) = 0$ is equivalent to

$$f(ct + l) - f(ct - l) = 0,$$

which in turn gives $f(z) = f(z + 2l)$. This last equation implies that the function $f(x)$ is a periodic function of period $2l$. Thus, the solution reduces to

$$u = f(ct + x) - f(ct - x).$$

Applying the initial conditions, we get

$$f(x) - f(-x) = x, \quad \text{and} \quad f'(x) - f'(-x) = 0,$$

i.e., $f'(x)$ is an even function, which means that $f(x)$ is an odd function, i.e., $f(x) = -f(-x)$. Hence $2f(x) = x$. Since $f(x)$ is an odd periodic function of period $2l$, it can be expressed as a Fourier sine series. Thus,

$$f(x) = \frac{x}{2} = \frac{l}{\pi} \sum_0^\infty \frac{(-1)^{n+1}}{n} \sin \frac{n\pi x}{l},$$

which yields

$$u(x,t) = \frac{l}{\pi} \sum_0^\infty \frac{(-1)^{n+1}}{n} [\sin \frac{n\pi}{l}(ct + l) - \sin \frac{n\pi}{l}(ct - l)]$$

$$= \frac{2l}{\pi} \sum_0^\infty \frac{(-1)^{n+1}}{n} \sin \frac{n\pi x}{l} \cos \frac{n\pi ct}{l}. \quad \blacksquare$$

4.5. Eigenfunction Expansions

The Sturm-Liouville problems arise in the solution of boundary value problems when the method of separation of variables is used. This method, discussed in detail in the next chapter, is one of the most useful methods in solving boundary value problems involving partial differential equations. A Sturm-Liouville problem consists of the Sturm-Liouville equation

$$\frac{d}{dx}\left[p(x)\frac{dy}{dx}\right] + [q(x) + \lambda w(x)]\, y = 0, \tag{4.26}$$

which is a linear second-order ordinary differential equation defined on a given interval $a \le x \le b$ and satisfies the boundary conditions of the form

$$a_1\, y(a) + b_1\, y'(a) = 0,$$
$$a_2\, y(b) + b_2\, y'(b) = 0, \tag{4.27}$$

where λ is a real parameter, and a_1, a_2, b_1, and b_2 are given real constants such that a_1 and b_1, or a_2 and b_2 are both not zero. It is obvious that the system (4.26)–(4.27) always has a trivial solution $y = 0$. The nontrivial solutions of this problem are called the *eigenfunctions* $\phi_n(x)$ and the corresponding values of λ the *eigenvalues* λ_n of the problem. The pair (ϕ_n, λ_n) is known as the *eigenpair*.

DEFINITION 4.10. The boundary conditions of the type

$$y(a) = y(b), \quad y'(a) = y'(b) \tag{4.28}$$

are known as the *periodic boundary conditions*. In this case the solution is of period $(b - a)$.

Theorem 4.4. *Let the functions p, q, w, and p' in Eq (4.26) be real-valued and continuous on the interval $a \leq x \leq b$. Let $\phi_m(x)$ and $\phi_n(x)$ be the eigenfunctions of the problem (4.26)–(4.27) with corresponding eigenvalues λ_m and λ_n, respectively, such that $\lambda_m \neq \lambda_n$. Then*

$$\int_a^b \phi_m(x)\, \phi_n(x)\, w(x)\, dx = 0, \quad m \neq n, \tag{4.29}$$

i.e., the eigenfunctions ϕ_m and ϕ_n are orthogonal with respect to the weight function $w(x)$ on the interval $a \leq x \leq b$.

Proof of this theorem can be found in many textbooks on ordinary differential equations, e.g., Ross (1964) and Boyce and DiPrima (1992).

The eigenfunction expansion of an arbitrary function $f(x)$ in the interval $a \leq x \leq b$ is given by

$$f(x) = \sum_{n=1}^{\infty} c_n\, \phi_n(x), \tag{4.30}$$

where ϕ_n are the eigenfunctions, with corresponding eigenvalues λ_n for the Sturm-Liouville (or eigenvalue) problem (4.26), subject to the boundary conditions (4.27) or (4.28), where the coefficients c_n are determined by (4.12).

EXAMPLE 4.13. The set $\{1, \cos x, \sin x, \cos 2x, \sin 2x, \dots\}$ of orthogonal functions in Example 4.3 consists of the eigenfunctions of the eigenvalue problem

$$y'' + \lambda\, y = 0, \quad y(-\pi) = y(\pi), \quad y'(-\pi) = y'(\pi). \tag{4.31}$$

The corresponding eigenvalues are $\lambda_n = n^2$ for $n = 1, 2, \dots$. ∎

EXAMPLE 4.14. For the eigenvalue problem $y'' + \lambda\, y = 0$, where $0 < x < L$, and (a) subject to the Dirichlet boundary conditions $y(0) = 0 = y(L)$, the eigenpair is

$$\phi_n(x) = \sin \sqrt{\lambda_n}\, x, \quad \lambda_n = \left(\frac{n\pi}{L}\right)^2, \quad n = 1, 2, \dots;$$

and (b) subject to the Neumann boundary conditions $y'(0) = 0 = y'(L)$, the eigenpair is

$$\phi_n(x) = \cos \sqrt{\lambda_n}\, x, \quad \lambda_n = \left(\frac{n\pi}{L}\right)^2, \quad n = 0, 1, 2, \dots. ∎$$

Tables 4.1 and 4.2 at the end of this chapter provide the data for the solution of the eigenvalue problems with three types of boundary conditions defined in §1.2 in the Cartesian and the polar coordinates, respectively.

4.6. Bessel Functions

The Bessel functions of order ν are the solutions of the ordinary differential equation

$$x^2 \frac{d^2y}{dx^2} + x \frac{dy}{dx} + (x^2 - \nu^2)\, y = 0, \quad \nu \geq 0, \tag{4.32}$$

which is known as the Bessel equation of order ν. Its regular solutions for each $\nu \geq 0$ are the Bessel functions of the first kind

$$J_\nu(x) = \sum_{n=0}^{\infty} \frac{(-1)^n}{n!(n+\nu)!} \left(\frac{x}{2}\right)^{2n+\nu}, \quad -\infty < x < \infty. \tag{4.33}$$

These power series solutions are obtained by the Frobenius method, details of which are available in textbooks on ordinary differential equations, e.g., Ross (1964) and Boyce and DiPrima (1992). The infinite series (4.33) is uniformly convergent and can be differentiated or integrated term-by-term. The differentiation and integration formulas are as follows:

$$J_\nu(-x) = (-1)^\nu\, J_\nu(x),$$

$$\frac{d}{dx}[x^\nu\, J_\nu(x)] = x^\nu\, J_{\nu-1}(x),$$

$$\frac{d}{dx}[x^{-\nu}\, J_\nu(x)] = -x^{-\nu}\, J_{\nu+1}(x),$$

$$x\, J_\nu'(x) = \nu\, J_\nu(x) - x\, J_{\nu+1}(x), \tag{4.34}$$

$$x\, J_\nu'(x) = -\nu\, J_\nu(x) + x\, J_{\nu-1}(x),$$

$$\int_0^x t^\nu\, J_{\nu-1}(t)\, dt = x^\nu\, J_\nu(x).$$

In particular,

$$J_0'(x) = -J_1(x), \quad \text{and} \quad \int_0^x t\, J_0(t)\, dt = x J_1(x).$$

The integral representation for $J_\nu(x)$ is

$$J_\nu(x) = \frac{1}{\pi} \int_0^\pi \cos(x \sin\theta - \nu\theta)\, d\theta, \tag{4.35}$$

where for $\nu = 0$ we have

$$J_0(x) = \frac{1}{\pi} \int_0^\pi \cos(x \sin\theta)\, d\theta = \frac{2}{\pi} \int_0^{\pi/2} \cos(x \cos\theta)\, d\theta. \tag{4.36}$$

Note that J_{2k} is an even function and J_{2k+1} an odd function. Moreover, $J_0(0) = 1$, but $J_\nu(0) = 0$ for $\nu \geq 1$. In fact, J_ν has a zero of multiplicity ν at $x = 0$. The integral representation (4.35) shows that $\|J_\nu(x)\| \leq 1$ for all real x and $\nu \geq 0$.

From the graphs of J_0, J_1, J_2, J_3 (Fig. 4.5) and their derivatives, it is found that each J_ν decays for large x, and their zeros are almost evenly spread and interlaced. In fact, for real ν the functions $J_\nu(x)$ and $J_\nu'(x)$ each have countably many real zeros, all of which are simple except $x = 0$. For nonnegative ν, let the n-th positive zero of these functions be denoted by $\alpha_{\nu,n}$ and $\alpha_{\nu,n}'$; then the zeros interlace according to the inequalities

$$\alpha_{\nu,1} < \alpha_{\nu+1,1} < \alpha_{\nu,2} < \alpha_{\nu+1,2} < \alpha_{\nu,3} < \cdots ,$$
$$\nu < \alpha_{\nu,1}' < \alpha_{\nu+1,1}' < \alpha_{\nu,2}' < \alpha_{\nu+1,2}' < \alpha_{\nu,3}' < \cdots ,$$

i.e., each J_ν possesses an increasing unbounded sequence of positive zeros. In fact, all zeros of J_ν are real for $\nu \geq -1$. But for $\nu < -1$ and ν not an integer, the number of complex zeros of J_ν is twice the integer part of $(-\nu)$. If the integer part of $(-\nu)$ is odd, two of these zeros lie on the imaginary axis. If $\nu \geq 0$, all zeros of J_ν are real (for details, see Abramowitz and Stegun, 1965, p. 372).

By taking the limit of the integral representation (4.36) as $x \to \infty$, it can be shown that

$$\lim_{x \to \infty} J_\nu(x) = 0.$$

In fact, in view of the Riemann-Lebesgue lemma which states that

$$\lim_{\kappa \to \infty} \int_{-\pi}^{\pi} F(x) \cos \kappa x \, dx = \lim_{\kappa \to \infty} \int_{-\pi}^{\pi} F(x) \sin \kappa x \, dx = 0,$$

where $F(x)$ is Lebesgue-integrable on $[-\pi, \pi]$, we get

$$\lim_{x \to \infty} J_0(x) = \lim_{x \to \infty} \frac{2}{\pi} \int_0^{\pi/2} \cos(x \cos \theta) \, d\theta$$

$$= -\lim_{x \to \infty} \frac{2}{\pi} \int_0^1 \frac{\cos(xy)}{\sqrt{1 - y^2}} \, dy = 0.$$

Similarly,

$$J_\nu(x) = \frac{1}{\pi} \int_0^\pi \cos(x \sin \theta - \nu\theta) \, d\theta$$

$$= \frac{1}{\pi} \int_0^\pi [\cos \nu\theta \, \cos(x \sin \theta) + \sin \nu\theta \, \sin(x \sin \theta)] \, d\theta$$

tends to zero as $x \to \infty$.

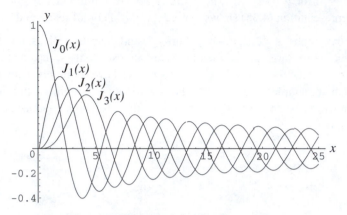

Fig. 4.5. Graphs of J_0, J_1, J_2, and J_3.

Note that the Bessel equation in the polar coordinates

$$\frac{d^2y}{dr^2} + \frac{1}{r}\frac{dy}{dr} + \left(\lambda^2 - \frac{\nu^2}{r^2}\right)y = 0, \quad 0 \le r \le a, \quad \nu \ge -\frac{1}{2}, \tag{4.37}$$

is a Sturm-Liouville equation.

EXAMPLE 4.15. To find the solution of the Bessel equation for $\nu = 0$, note that in this case Eq (4.32) reduces to

$$x^2\frac{d^2y}{dx^2} + x\frac{dy}{dx} + x^2y = 0, \tag{4.38}$$

and its solution from (4.33) is

$$J_0(x) = \sum_{n=0}^{\infty} \frac{(-1)^n}{(n!)^2}\left(\frac{x}{2}\right)^{2n}, \quad -\infty < x < \infty. \tag{4.39}$$

However, since Eq (4.38) is a second-order ordinary linear differential equation, it should have two linearly independent solutions. A second solution is found by assuming that y is of the form $y(x) = J_0(x)\ln x + \sum_{n=1}^{\infty} A_n x^n$. If we substitute this solution into Eq (4.38), and note that $x^2 J_0''(x) + x J_0'(x) + x^2 J_0(x) = 0$, then we get

$$2x J_0'(x) + \sum_{n=1}^{\infty} n(n-1) A_n x^n + \sum_{n=1}^{\infty} n A_n x^n + \sum_{n=1}^{\infty} A_n x^{n+2} = 0,$$

where $J_0'(x)$ is obtained from (4.39). Thus, combining all the summations in the above equation, we have

$$A_1 x + 4A_2 x^2 + \sum_{n=3}^{\infty}\left(n^2 A_n + A_{n-2}\right)x^n + \sum_{n=1}^{\infty}\frac{(-1)^n}{n!\,(n-1)!\,2^{2n-2}}x^{2n} = 0.$$

Obviously, $A_1 = 0$; the x^2-terms give $A_2 = 1/4$, and for odd n we have the recursive formula

$$n^2 A_n + A_{n-2} = 0, \quad n \text{ odd.}$$

This yields $A_{2n+1} = 0$ for $n > 0$. Also, the terms in x^{2n} yield the recursive relation

$$(2n)^2 A_{2n} + A_{2n-2} + \frac{(-1)^n}{n!\,(n-1)!\,2^{2n-2}} = 0 \quad \text{for } n \geq 2,$$

which, by iteration, gives

$$A_{2n} = \frac{(-1)^{n+1}}{2^{2n}\,(n!)^2} \left(1 + \frac{1}{2} + \frac{1}{3} + \cdots + \frac{1}{n} \right), \quad n \geq 1.$$

If we denote the series $1 + \dfrac{1}{2} + \dfrac{1}{3} + \cdots + \dfrac{1}{n}$ by $g(n)$, then the other solution is given by

$$y(x) = J_0(x) \ln x + \sum_{n=1}^{\infty} \frac{(-1)^{n+1}\,g(n)}{(n!)^2} \left(\frac{x}{2} \right)^{2n}. \tag{4.40}$$

This solution is called the *Neumann's Bessel function of the second kind and of order zero.* A linear combination of $J_0(x)$ and the solution (4.40), of the form $a\,J_0(x) + b\,y(x)$, becomes the general solution of Eq (4.38). If we take

$$a = A + \frac{2B}{\pi}\,(\gamma - \ln n), \quad b = \frac{2}{\pi}\,B,$$

where A and B are arbitrary constants, and γ is the Euler's constant defined by

$$\gamma = \lim_{n \to \infty} \left[g(n) - \ln n \right],$$

then the general solution of Eq (4.38) is

$$y(x) = A\,J_0(x) + B\,Y_0(x), \tag{4.41}$$

where

$$Y_0(x) = \frac{2}{\pi} \left\{ J_0(x) \left[\ln \left(\frac{x}{2} \right) + \gamma \right] + \sum_{n=1}^{\infty} \frac{(-1)^{n+1}\,g(n)}{(n!)^2} \left(\frac{x}{2} \right)^{2n} \right\}. \tag{4.42}$$

The solution $Y_0(x)$ is known as the *Weber's Bessel function of the second kind and of order zero.*

Alternatively, to obtain the other linearly independent solution $Y_0(x)$, we use the method of reduction of order and take $y = J_0(x)\phi(x)$, where $\phi(x)$ is an unknown function to be determined. Then, after solving the modified differential equation, we get

$$y(x) = J_0(x) \int \frac{1}{x J_0^2(x)} \, dx,$$

which, after some computations, yields

$$y(x) = J_0(x) \ln|x| + \frac{x^2}{4} - \frac{3x^4}{128} + \frac{11x^6}{13824} - \cdots .$$

However, in practice, we use a linear combination of this solution and the Bessel function $J_0(x)$, and denote it by

$$Y_0(x) = \frac{2}{\pi} \left[J_0(x) \ln|x| + \sum_{n=1}^{\infty} \frac{(-1)^{n+1} x^{2n}}{(n!)^2 \, 2^{2n}} \, g(n) + (\gamma - \ln 2) J_0(x) \right]. \quad \blacksquare$$

EXAMPLE 4.16. Consider the polar form of the Laplace operator in R^1:

$$L \equiv \frac{d^2}{dr^2} + \frac{1}{r} \frac{d}{dr}.$$

Then the corresponding eigenvalue problem is $L\phi + \lambda^2 \phi = 0$. It can be shown that the radially symmetric eigenfunctions ϕ_k of the Laplace equation subject to the Dirichlet boundary condition $\phi(1) = 0$ on the unit disk $U = \{r \leq 1\}$ are

$$\phi_k = J_0(\lambda_k r), \quad k = 1, 2, \ldots ,$$

such that

$$\lambda_1 < \lambda_2 < \lambda_3 < \cdots$$

are the positive zeros of J_0. These eigenfunctions form a complete orthogonal set, that is,

$$\langle J_0(\lambda_m r), J_0(\lambda_n r) \rangle = \int_0^1 J_0(\lambda_m r) \, J_0(\lambda_n r) \, r \, dr = 0 \quad \text{for } m \neq n.$$

Then, for any function f with the norm

$$\|f\|^2 = \int_0^1 [f(r)]^2 \, r \, dr < +\infty,$$

we have the *Fourier-Bessel expansion*

$$f(r) = \sum_{n=1}^{\infty} c_n \, J_0(\lambda_n r), \qquad (4.43)$$

where

$$c_n = \frac{\langle f, J_0(\lambda_n r)\rangle}{\|J_0(\lambda_n r)\|^2}, \quad \|J_0(\lambda_n r)\|^2 = \frac{J_1^2(\lambda_n)}{2}. \quad\blacksquare \tag{4.44}$$

In general, we have

Theorem 4.5. *The eigenfunctions* $\phi_n = J_\nu(\lambda_n r)$, $n = 1, 2, \ldots$, *form a complete orthogonal set of radially symmetric square-integrable functions such that for a square-integrable function f the Fourier-Bessel expansion*

$$f(r) = \sum_{n=1}^{\infty} c_n J_\nu(\lambda_n r) \tag{4.45}$$

holds, where

$$c_n = \frac{\langle f, J_\nu(\lambda_n r)\rangle}{\|J_\nu(\lambda_n r)\|^2}, \tag{4.46}$$

λ_n *are the positive zeros of J_ν for $n = 1, 2, \ldots$, and*

$$\|J_\nu(\lambda_n r)\|^2 = \int_0^1 J_\nu^2(\lambda_n r)\, r\, dr$$
$$= \frac{[J_\nu'(\lambda_n)]^2}{2} = \frac{J_{\nu+1}^2(\lambda_n)}{2} = \frac{J_{\nu-1}^2(\lambda_n)}{2}. \tag{4.47}$$

The last two relations in (4.47) are obtained by using the following formulas:

$$\int_0^1 r J_\nu^2(\lambda_n r)\, dr = \frac{1}{2}\left[J_\nu'(\lambda_n)\right]^2 + \frac{1}{2}\left(1 - \frac{n^2}{\lambda_n^2}\right)\left[J_\nu(\lambda_n)\right]^2,$$

$$\lambda_n J_\nu'(\lambda_n) = \lambda_n J_{\nu-1}(\lambda_n) - \nu J_\nu(\lambda_n),$$

$$\lambda_n J_\nu'(\lambda_n) = \nu J_\nu(\lambda_n) - \lambda_n J_{\nu+1}(\lambda_n),$$

and recalling that $J_\nu(\lambda_n) = 0$.

Theorem 4.6. *The eigenfunctions for the Laplace operator with zero Dirichlet boundary condition on the unit disk $U = \{r \le 1\}$ are*

$$J_0(\lambda_n r), \quad J_\nu(\lambda_n r)\cos\nu\theta, \quad J_\nu(\lambda_n r)\sin\nu\theta, \tag{4.48}$$

for $n = 1, 2, \ldots$, and $\nu \ge 1$, with the eigenvalues λ_n that are the positive zeros of $J_\nu(\lambda r)$. These eigenfunctions form a complete orthogonal basis in the space $L^2(U)$ of square-integrable functions on U.

Proofs of these theorems can be found in Watson (1944).

For the three types of boundary conditions (§1.2) the eigenpairs for Eq (4.37) are defined in Table 4.2 at the end of this chapter. Thus, a function $f(r, \theta)$ defined on the set $U \times (0, 2\pi)$ has an eigenfunction expansion of the form

$$f(r, \theta) = \sum_{n=0}^{\infty} \left[f_n(r)\cos n\theta + g_n(r)\sin n\theta\right], \tag{4.49}$$

where each of the functions f_n and g_n has an expansion of the form (4.45).

4.7. Mathematica Projects

PROJECT 4.1. The Mathematica Notebook `Orthonormality.nb` is available on the CRC web server. It can be used to verify the orthonormality of sets of functions.

PROJECT 4.2. We will present a Mathematica session to evaluate the integral

$$\int_{-1}^{1} P_n(x)\, P_m(x)\, dx$$

for $n, m = 0, \dots, 6$, which appears in Example 4.4.

In[1]: `legendre = Table[LegendreP[n,x],{n, 0 , 10}];`
          `legendre[[Range[6]]]`

Out[1]:
1

x

$\dfrac{-1 + 3\,x^2}{2}$

$\dfrac{-3\,x + 5\,x^3}{2}$

$\dfrac{3 - 30\,x^2 + 35\,x^4}{8}$

$\dfrac{15\,x - 70\,x^3 + 63\,x^5}{8}$

The plots of these Legendre polynomials are given below.

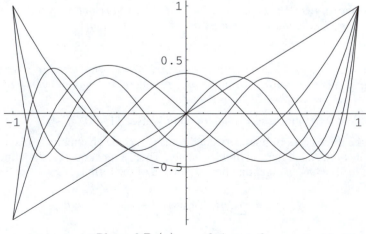

Plots of $P_n(x)$, $n = 0, 1, \dots, 6$.

-Graphics-

In[2]:
```
orthogonality =
Table[Integrate[legendre[[i]] legendre[[j]],
    {x, -1, 1}], {i, 1, 10}, {j, 1 , 10}];

MatrixForm[orthogonality]
```
Out[2]:

2	0	0	0	0	0	0	0	0	0
0	2/3	0	0	0	0	0	0	0	0
0	0	2/5	0	0	0	0	0	0	0
0	0	0	2/7	0	0	0	0	0	0
0	0	0	0	2/9	0	0	0	0	0
0	0	0	0	0	2/11	0	0	0	0
0	0	0	0	0	0	2/13	0	0	0
0	0	0	0	0	0	0	2/15	0	0
0	0	0	0	0	0	0	0	2/17	0
0	0	0	0	0	0	0	0	0	2/19

PROJECT 4.3. The following Mathematica session illustrates Example 4.6.

In[3]:
```
Clear[f,a0,a,b,fourier];
f[x_]:= Abs[x]

Integrate[Abs[x],{x,-Pi,Pi}]
```
Out[5]: π^2

Observe that the function $f(x) \cos nx$ is even and $f(x) \sin nx$ is odd. So we proceed as follows:

In[6]: `a0 = (2/Pi) Integrate[x,{x,0,Pi}]`

Out[6]: π

In[7]: `a = Table[2 Integrate[x Cos[n x],{x,0,Pi}]/Pi,{n,10}]`

Out[7]: $\left\{ \dfrac{-4}{Pi}, 0, \dfrac{-4}{9\,Pi}, 0, \dfrac{-4}{25\,Pi}, 0, \dfrac{-4}{49\,Pi}, 0, \dfrac{-4}{81\,Pi}, 0 \right\}$

In[8]: `b = Table[(Integrate[-x Sin[n x],{x,-Pi,0}]`
`            + Integrate[x Sin[n x],{x,0,Pi}])/Pi,{n,10}]`

Out[8]: $\{0,\ 0,\ 0,\ 0,\ 0,\ 0,\ 0,\ 0,\ 0,\ 0\}$

In[9]:   ```
fourier[x_] = a0/2 +
    Sum[a[[n]] Cos[n x]+ b[[n]] Sin[n x],{n,1,10}]
```

*Out[9]:*   $\left\{\dfrac{Pi}{2} - \dfrac{4\,Cos[x]}{Pi} - \dfrac{4\,Cos[3\,x]}{9\,Pi} - \dfrac{4\,Cos[5\,x]}{25\,Pi} - \dfrac{4\,Cos[7\,x]}{49\,Pi} - \dfrac{4\,Cos[9\,x]}{81\,Pi}\right\}$

*In[10]:*   `Plot[{f[x],fourier[x]},{x,-0.5,0.5}]`

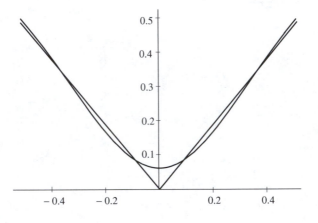

PROJECT 4.4.   The following Mathematica session illustrates Example 4.7.

*In[11]:*   ```
Clear[f,a0,a,b,fourier];
f[x_]:= If[x==0,2, If[0<x<2,3, If[x==2,2,
If[2<x<4,1]]]]
Integrate[f[x],{x,0,4}]
```

Out[13]:   ```
Integrate[If[x == 0, 2, If[0 < x < 2, 3, If[x== 2, 2,
      If[2 < x < 4, 1]]]], {x, 0, 4}]
```

*In[14]:*   `a0 = (1/2) (Integrate[3,{x,0,2}] + Integrate[1,{x,2,4}])`

*Out[14]:*   `4`

*In[15]:*   ```
a = Table[1/2 (Integrate[3 Cos[n Pi x/2],{x,0,2}] +
      Integrate[Cos[n Pi x/2],{x,2,4}]),{n,16}]
```

Out[15]: $\{0,0,0,0,0,0,0,0,0,0,0,0,0,0,0,0\}$

In[16]:   ```
b = Table[1/2 (Integrate[3 Sin[n Pi x/2],{x,0,2}]+
      Integrate[Sin[n Pi x/2],{x,2,4}]),{n,16}]
```
```

Out[16]:

$$\left\{\frac{4}{\text{Pi}}, 0, \frac{4}{3\,\text{Pi}}, 0, \frac{4}{5\,\text{Pi}}, 0, \frac{4}{7\,\text{Pi}}, 0, \frac{4}{9\,\text{Pi}}, 0, \frac{4}{11\,\text{Pi}}, 0, \frac{4}{13\,\text{Pi}}, 0, \frac{4}{15\,\text{Pi}}, 0\right\}$$

In[17]:   `fourier[x_]:= a0/2 + Sum[b[[n]] Sin[n Pi x/2],{n,1,16}]`

```
Plot[{f[x],fourier[x]},{x,-0.5,4},
     PlotRange -> {{-1,5},{-1,3.5}}]
```

Out[18]:

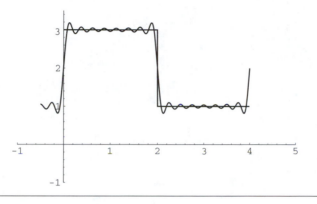

PROJECT 4.5. `plotfourier.nb`, available in the Mathematica Notebook, generates a table of the elements of a trigonometric Fourier series and plots their graphs for a given function (see Example 4.10).

PROJECT 4.6. `eigenpair.nb`, available in the Mathematica Notebook, can be used to obtain the eigenvalues and eigenfunctions for a given boundary value problem (see Example 4.13).

PROJECT 4.7. Plots of the Bessel functions $J_\nu(x)$ and their derivatives $J'_\nu(x)$ for $\nu = 0, 1, 2, 3$ are available in `bessel.nb` in the Mathematica Notebook. For definitions of these functions, see §4.6, and Appendix D.

4.8. Exercises

Show that each set of functions in Exercises 4.1–4.7 is orthogonal on the given interval, and determine the corresponding orthonormal set of functions:

4.1. $\left\{\sin\dfrac{n\pi x}{l}\right\}, n = 1, 2, 3, \ldots ; \quad -l \le x \le l.$

4.2. $\left\{\cos\dfrac{2n\pi x}{l}\right\}, n = 0, 1, 2, \ldots ; \quad -l \le x \le l.$

4.3. $\{\sin 2nx\}, n = 1, 2, 3, \dots;$　$0 \le x \le \pi.$

4.4. $\{\cos 2nx\}, n = 0, 1, 2, 3, \dots;$　$0 \le x \le \pi.$

4.5. $\{\sin 3nx\}, n = 1, 2, \dots;$　$-\pi \le x \le \pi.$

4.6. $\{\cos 3nx\}, n = 0, 1, 2, 3, \dots;$　$-\pi \le x \le \pi.$

4.7. $\{\sin 2nx, \cos 2nx\}, n = 0, 1, 2, 3, \dots;$　$|x| \le \pi.$

4.8. Find the Fourier series for each of the following function $f(x)$ of period 2π:

(a) $f(x) = \begin{cases} 1 & \text{if } -\pi/2 < x < \pi/2, \\ -1 & \text{if } \pi/2 < x < 3\pi/2. \end{cases}$

(b) $f(x) = x, \quad -\pi < x < \pi.$

(c) $f(x) = x^2, \quad -\pi < x < \pi.$

Ans. (a) $\dfrac{4}{\pi}\left(\cos x - \dfrac{1}{3}\cos 3x + \dfrac{1}{5}\cos 5x - \cdots\right).$

(b) $2\left(\sin x - \dfrac{1}{2}\sin 2x + \dfrac{1}{3}\sin 3x - \dfrac{1}{4}\sin 4x + \cdots\right).$

(c) $\dfrac{2\pi^2}{3} - 4\left(\cos x - \dfrac{1}{4}\cos 2x + \dfrac{1}{9}\cos 3x - \dfrac{1}{16}\cos 4x + \cdots\right).$

4.9. Find the Fourier series of each of the periodic functions $f(x)$ of period T:

(a) $f(x) = \begin{cases} -1 & \text{if } -1 < x < 0 \\ 1 & \text{if } 0 < x < 1, \quad T = 2. \end{cases}$

(b) $f(x) = 1 - x^2, \quad -1 < x < 1, \quad T = 2.$

Ans. (a) $\dfrac{4}{\pi}\left(\sin \pi x + \dfrac{1}{3}\sin 3\pi x + \dfrac{1}{5}\sin 5\pi x + \cdots\right).$

(b) $\dfrac{2}{3} + \dfrac{4}{\pi^2}\left(\cos \pi x - \dfrac{1}{4}\cos 2\pi x + \dfrac{1}{9}\cos 3\pi x - \cdots\right).$

4.10. Find the trigonometric Fourier series of the function $f(x) = x,\ -4 \le x \le 4.$

Ans. Since $f(x)$ is odd, we have $a_n = 0$. Then, with $L = 4$,

$$b_n = \frac{1}{4}\int_0^4 x \sin\frac{n\pi x}{4}\,dx = -\frac{8}{n\pi}\cos n\pi = \frac{8(-1)^{n+1}}{n\pi}, \quad n = 1, 2, \dots,$$

and the required series is

$$x = \frac{8}{\pi}\sum_{n=1}^{\infty}\frac{(-1)^{n+1}}{n}\sin\frac{n\pi x}{4}.$$

4.11. Find the trigonometric Fourier series of the function

$$f(x) = \begin{cases} \pi, & -\pi \le x < 0, \\ x, & 0 \le x \le \pi. \end{cases}$$

ANS. Note that $L = \pi$. The function f is neither odd nor even. Then

$$a_0 = \frac{1}{\pi} \int_{-\pi}^{\pi} f(x)\, dx = \frac{3\pi}{2},$$

$$a_n = \frac{1}{\pi} \int_{-\pi}^{\pi} f(x) \cos nx\, dx = \frac{1}{\pi}\left[\frac{\cos n\pi - 1}{n^2}\right]$$

$$= \begin{cases} -\dfrac{2}{\pi n^2} & \text{for odd } n, \\ 0 & \text{for even } n, \end{cases}$$

$$b_n = \frac{1}{\pi} \int_{-\pi}^{\pi} f(x) \sin nx\, dx = -\frac{1}{n}, \quad n = 1, 2, \ldots .$$

The required series is

$$f(x) = \frac{3\pi}{4} + \sum_{n=1}^{\infty}\left[\frac{(-1)^n - 1}{\pi n^2}\cos nx - \frac{1}{n}\sin nx\right].$$

4.12. Find the trigonometric Fourier series of the function

$$g(x) = \begin{cases} \pi, & -\pi \leq x < 0, \\ \dfrac{\pi}{2}, & x = 0, \\ x, & 0 \leq x \leq \pi. \end{cases}$$

HINT. The function $g(x)$ is the same as $f(x)$ in Exercise 4.11, except at $x = 0$. Since these two functions have the same values at all points except a finite number (only one in this case) in the same interval, the function $g(x)$ has the same Fourier series as that for $f(x)$ in Exercise 4.11.

4.13. Find the trigonometric Fourier cosine series of the function $f(x) = a^2 - x^2$, $0 \leq x \leq a$.

ANS. $a^2 - x^2 = \dfrac{2}{3}a^2 - \dfrac{4a^2}{\pi^2}\displaystyle\sum_{n=1}^{\infty}\frac{(-1)^n}{n^2}\cos\frac{n\pi x}{a}.$

4.14. Express $\sin x$ as a Fourier cosine series and show that

$$\frac{1}{2} = \frac{1}{1\cdot 3} + \frac{1}{3\cdot 5} + \frac{1}{5\cdot 7} + \cdots .$$

ANS. $\sin x = \dfrac{2}{\pi}\left(1 - \dfrac{2}{1\cdot 3}\cos 2x - \dfrac{2}{3\cdot 5}\cos 4x - \cdots\right)$. Take $x = 0$ and obtain the series for $1/2$.

4.15. Show that the extrema of the partial sums of the Fourier series

$$S_{2n-1} = \sin x + \frac{1}{3}\sin 3x + \cdots + \frac{1}{2n - 1}\sin(2n - 1)x$$

occur at equal distances, and they lie between the extrema of S_{2n+1}, except at $x = \pi/2$, which is common to all partial sums.

HINT. The extrema are determined by the solution of the equation

$$0 = \cos x + \cos 3x + \cdots + \cos(2n-1)x$$

$$= \Re\left\{e^{ix} + e^{3ix} + \cdots + e^{(2n+1)ix}\right\} = -\frac{\sin 2nx}{\sin x}, \quad x \neq 0, \pi,$$

which gives $x = \left\{\dfrac{\pi}{2n}, \dfrac{2\pi}{2n}, \ldots, \dfrac{(2n-1)\pi}{2n}\right\}$.

4.16. Show that

$$\int_0^{2\pi} \sin^{2n} \theta \, d\theta = \frac{2\pi(2n)!}{(n!2^n)^2}.$$

HINT. Use induction, or the following result:

$$\int_0^{\pi/2} \cos^m \theta \sin^n \theta \, d\theta = \frac{\Gamma\left(\dfrac{m+1}{2}\right)\Gamma\left(\dfrac{n+1}{2}\right)}{2\Gamma\left(\dfrac{m+n}{2}+1\right)}.$$

4.17. Show that $e^{x(t-1/t)/2} = J_0(x) + \displaystyle\sum_{n=1}^\infty J_n(x)\left[t^n + (-1)^n t^{-n}\right].$

HINT. Multiply the series expansions for $e^{xt/2}$ and $e^{-x/(2t)}$.

4.18. Show that $\cos x = J_0(x) + 2\displaystyle\sum_{n=1}^\infty (-1)^n J_{2n}(x),$

$$\sin x = 2\sum_{n=0}^\infty (-1)^n J_{2n+1}(x).$$

HINT. Set $t = i$ in Exercise 4.17.

4.19. Show that $\cos(x \sin \theta) = J_0(x) + 2\displaystyle\sum_{n=1}^\infty J_{2n}(x) \cos 2n\theta,$

$$\sin(x \sin \theta) = 2\sum_{n=0}^\infty J_{2n+1}(x) \sin(2n+1)\theta.$$

HINT. Set $t = e^{i\theta}$ in Exercise 4.17.

4.20. From the results of Exercise 4.19, deduce that

$$J_{2n}(x) = \frac{1}{\pi} \int_0^\pi \cos(x \sin \theta) \cos 2n\theta \, d\theta,$$

$$J_{2n+1}(x) = \frac{1}{\pi} \int_0^\pi \sin(x \sin \theta) \sin(2n+1)\theta \, d\theta.$$

4.21. Show that $\displaystyle\int_0^a J_\nu'^2(y)\, y\, dy = \frac{(a^2 - \nu^2)\, J_\nu^2(a) + a^2\, J_\nu'^2(a)}{2}.$

HINT. Use the identity

$$x\, J_\nu'(x)\, (x\, J_\nu'(x))' = J_\nu'(x)[x^2 J_\nu''(x) + x J_\nu'(x)] = (\nu^2 - x^2) J_\nu(x)\, J_\nu'(x).$$

4.22. (a) Show that the partial sums M_k in Example 4.11 can be approximated by $\dfrac{2}{\pi}\, \mathrm{Si}(z)$, where $\mathrm{Si}(z)$ is the sine-integral function of $z = 2(k+1)x$. Compute $\lim\limits_{k\to\infty}\, \lim\limits_{x\to 0}\, M_k$ and $\lim\limits_{x\to 0}\, \lim\limits_{k\to\infty}\, M_k$, and conclude that the Gibbs phenomenon occurs whenever a discontinuity is approximated.

(b) Use the Fourier sine series for $f(x) = 1$ in Example 4.11 and show that

(i) $\dfrac{\pi}{4} = 1 - \dfrac{1}{3} + \dfrac{1}{5} + \dfrac{1}{7} + \cdots;$

(ii) $\dfrac{\pi^2}{8} = 1 + \dfrac{1}{3^2} + \dfrac{1}{5^2} + \dfrac{1}{7^2} + \cdots;$

(iii) $\dfrac{\pi^2}{32} = 1 - \dfrac{1}{3^2} + \dfrac{1}{5^2} - \dfrac{1}{7^2} + \cdots;$

(iv) $\dfrac{\pi^4}{96} = 1 + \dfrac{1}{3^4} + \dfrac{1}{5^4} + \dfrac{1}{7^4} + \cdots.$

(v) $\dfrac{\pi}{4}\left(\dfrac{\pi}{2} - x\right) = \cos x + \dfrac{1}{3^2}\cos 3x + \dfrac{1}{5^2}\cos 5x + \cdots;$

(vi) $\dfrac{\pi}{8}\left(\dfrac{\pi x^2}{2} - \dfrac{x^3}{3}\right) = 1 - \cos x + \dfrac{1}{3^4}(1 - \cos 3x) + \dfrac{1}{5^4}(1 - \cos 5x) + \cdots;$

(vii) $\dfrac{\pi}{8}\left(\dfrac{\pi x^3}{2} - \dfrac{\pi x^2}{2} + \dfrac{x^3}{3}\right) = \cos x + \dfrac{1}{3^4}\cos 3x + \dfrac{1}{5^4}\cos 5x + \cdots.$

SOLUTION. (a) Starting with the partial sums M_k, we get

$$M_k = \frac{4}{\pi} \sum_{n=0}^{k} \frac{1}{2n+1}\, \sin(2n+1)x = \frac{4}{\pi} \sum_{n=0}^{k} \int_0^x \cos(2n+1)t\, dt$$

$$= \frac{2}{\pi} \int_0^x \sum_{n=0}^{k} \left\{ e^{(2n+1)it} + e^{-(2n+1)it} \right\} dt$$

$$= \frac{2}{\pi} \int_0^x \left\{ e^{it}\, \frac{1 - e^{2(k+1)it}}{1 - e^{2it}} + e^{-it}\, \frac{1 - e^{-2(k+1)it}}{1 - e^{-2it}} \right\} dt$$

$$= \frac{2}{\pi} \int_0^x \frac{\sin 2(k+1)t}{\sin t}\, dt \qquad\qquad (4.50)$$

$$\approx \frac{2}{\pi} \int_0^x \frac{\sin 2(k+1)t}{t}\, dt, \quad \text{since } \sin t \approx t \text{ for sufficiently small } t$$

$$= \frac{2}{\pi} \int_0^z \frac{\sin u}{u}\, du, \quad \text{where } 2(k+1)t = u \text{ and } z = 2(k+1)x$$

$$= \frac{2}{\pi}\, \mathrm{Si}(z).$$

Note that the sine-integral function $\mathrm{Si}(z)$ has the power series expansion $\mathrm{Si}(z) = z - \dfrac{1}{3}\dfrac{z^3}{3!} + \dfrac{1}{5}\dfrac{z^5}{5!} + \cdots$, and $\mathrm{Si}(\infty) = \pi/2$ (see Fig. 4.6); these graphs are obtained by using the Mathematica commands:

```
Plot[SinIntegral[z], {z, 0, 10π}, PlotRange -> All]
Plot[SinIntegral[z], {z, 0, 100π}, PlotRange -> All]
```

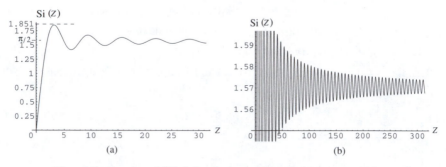

Fig. 4.6. Graphs of $\mathrm{Si}(z)$ for $0 < z < 10\pi$ and $0 < z < 100\pi$.

From (4.50) we conclude that if for finite k we set $x = 0$, then $z = 0$ and $M_k = 0$, and then if we let $k \to \infty$, then

$$\lim_{k\to\infty} \lim_{x\to 0} M_k = 0.$$

But if for $x > 0$ we first let $k \to \infty$, then $z \to \infty$, and

$$\lim_{x\to 0} \lim_{k\to\infty} M_k = \frac{2}{\pi} \int_0^\infty \frac{\sin u}{u}\, du = \frac{2}{\pi}\frac{\pi}{2} = 1.$$

Hence, these two limits are not interchangeable.

From the graphs in Fig. 4.6 we find that for small values of u, where $\sin u$ can be approximated by u, the values of $\mathrm{Si}(z)$ are proportional to z, but for large values of z the values of $\mathrm{Si}(z)$ approach $\pi/2$ asymptotically. In between these values, the function $\mathrm{Si}(z)$ oscillates successively with decreasing amplitudes and has maxima and minima at $z = \pi, 2\pi, 3\pi, \ldots$ (or, at $\ldots, x_n = \dfrac{\pi}{2(n+1)}, x_{n+1} = \dfrac{\pi}{2(2n+2)}, \ldots$), where the first absolute maximum occurs at the point $(\pi, 1.851)$ at which, in view of (4.50), the approximations $M_k, M_{k+1}, \ldots$ have the fixed value $M = \dfrac{2}{\pi}(1.851) = 1.18$. This value of M exceeds the value of $f(x) = 1$ by about 18%, which is known as an *overshoot*, i.e., this is the upper limit of the range of values given by (4.50); its lower limit is -1.18 which is obtained by approaching zero from the negative side in the sequence $\{\ldots, -x_k, -x_{k+1}, \ldots\}$. This behavior of approximating functions M_k, defined by (4.50), where the values of M exceed the range of discontinuity ± 1 is the Gibbs phenomenon that occurs whenever a discontinuity is approximated. This is an example of nonuniform convergence of the approximation process.

(b) (i) Take $x = \pi/2$ in the series for $f(x) = 1$.
(ii) Restrict the series for $f(x)$ to the interval $(0, \pi)$ and integrate from 0 to x:

$$\frac{\pi}{4} x = 1 - \cos x + \frac{1}{3^2}\left(1 - \cos 3x\right) + \frac{1}{5^2}\left(1 - \cos 5x\right) + \cdots ; \qquad (4.51)$$

then set $x = \pi/2$.
(iii) Subtract (4.51) from b(ii) and integrate from 0 to x:

$$\frac{\pi}{8}\left(\pi x - x^2\right) = \sin x + \frac{1}{3^3}\sin 3x + \frac{1}{5^3}\sin 5x + \cdots ; \qquad (4.52)$$

then set $x = \pi/2$.
(iv) Integrate (4.52) with respect to x and set $x = \pi/2$.
(v) Subtract (4.51) from b(ii).
(vi) Integrate b(v) from 0 to x.
(vii) Integrate (4.52) with respect to x and set $x = \pi/2$.

4.23. Find the Fourier sine series for $f(x) = 10 + x$ for the interval $0 < x < l$, and discuss the Gibbs phenomenon from the graphs of its partial sums.
    SOLUTION. The Fourier sine series is

$$10 + x = \frac{2}{\pi} \sum_{n=1}^{\infty} \left[\frac{10\left(1 - (-1)^n\right)}{n} + \frac{(-1)^{n+1} l}{n}\right] \sin\frac{n\pi x}{l}.$$

This series is valid everywhere except the end points where it becomes zero. The graphs of the partial sums S_n are given in Fig. 4.7 for $n = 1, 5, 10, 20, 40$. These graphs show that the series converges to $10 + x$ for $0 < x < l$. For these finitely many partial sums S_n the solution starts from zero at $x = 0$, shoots up beyond $10 + x$, and then reduces to zero at $x = l$ (we have taken $l = 10$ in this figure). This overshoot, as seen in Fig. 4.7, exhibits the Gibbs phenomenon.

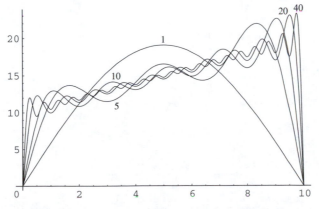

Fig. 4.7. Graphs of partial sums S_n for $n = 1, 5, 10, 20, 40$.

4.24. (a) Determine the eigenvalues and eigenfunctions for the Sturm-Liouville problem $y'' + \lambda^2 y = 0$, $y(0) = 0$, $y'(\pi) = \alpha\, y'(\pi)$.

(b) Show that the eigenfunctions form an orthogonal set, i.e.,

$$\int_0^\pi \sin \lambda_n x \, \sin \lambda_m x \, dx = \begin{cases} 0, & \text{if } n \neq m, \\ \dfrac{\pi}{2}\left[1 - \dfrac{\sin \lambda_n \pi \, \cos \lambda_n \pi}{\lambda_n \pi}\right], & \text{if } n = m, \end{cases}$$

where λ_n, $n = 1, 2, \dots$, are the consecutive positive roots of the transcendental equation $\tan \lambda \pi = \alpha \lambda$, α being a constant (Fig. 4.8).

(c) Expand an arbitrary function $f(x)$ in terms of the eigenfunctions as

$$f(x) = \sum_{n=1}^\infty B\,(\lambda_n)\, \sin \lambda_n x,$$

and show that

$$B\,(\lambda_n) = \frac{2}{\pi} \int_0^\pi \frac{f(x)\,\sin \lambda_n x \, dx}{1 - \dfrac{\sin 2\lambda_n \pi}{2\lambda_n \pi}}.$$

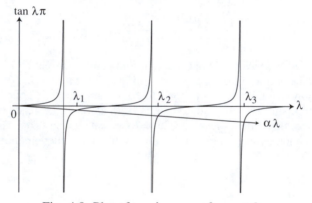

Fig. 4.8. Plot of $\tan \lambda x = \alpha x$ for $\alpha < 0$.

HINT. Note that

$$\int \sin \lambda_n x \, \sin \lambda_m x \, dx = \frac{1}{2} \int \left[\cos\left(\lambda_n - \lambda_m\right)x - \cos\left(\lambda_n + \lambda_m\right)x\right] dx$$

$$= \frac{1}{2}\left[\frac{\sin\left(\lambda_n - \lambda_m\right)x}{\lambda_n - \lambda_m} - \frac{\sin\left(\lambda_n + \lambda_m\right)x}{\lambda_n + \lambda_m}\right]$$

$$= \frac{\lambda_n \cos \lambda_n x \, \sin \lambda_m x - \lambda_m \cos \lambda_n x \, \sin \lambda_m x}{\lambda_n^2 - \lambda_m^2}$$

$$= \frac{\lambda_n \lambda_m}{\lambda_n^2 - \lambda_m^2} \cos \lambda_n x \, \cos \lambda_m x \left[\frac{\tan \lambda_n x}{\lambda_n} - \frac{\tan \lambda_m x}{\lambda_m}\right].$$

Hence,

$$\int_0^\pi \sin \lambda_n x \, \sin \lambda_m x \, dx$$

$$= \frac{\lambda_n \lambda_m}{\lambda_n^2 - \lambda_m^2} \cos \lambda_n x \, \cos \lambda_m x \left[\frac{\tan \lambda_n x}{\lambda_n} - \frac{\tan \lambda_m x}{\lambda_m} \right] \Bigg|_0^\pi = 0,$$

because $\tan \lambda \pi = \alpha \lambda$. If $n = m$, then

$$\int_0^\pi \sin^2 \lambda_n x \, dx = \frac{1}{2} \int_0^\pi \left[1 - \cos 2\lambda_n x \right] dx$$

$$= \frac{1}{2} \left[x - \frac{\sin 2\lambda_n x}{2\lambda_n} \right]_0^\pi = \frac{1}{2} \left[\pi - \frac{\sin \lambda_n \pi \, \cos \lambda_n \pi}{\lambda_n} \right]. \; \blacksquare$$

Compare this problem with Example 5.9 and Fig.5.3.

Table 4.1: Eigenvalue Problem in Cartesian Coordinates

$$\frac{d^2y}{dx^2} + \lambda^2 y = 0,\ 0 < x < L,\ \text{subject to the boundary conditions } a_1 y(0) + b_1 y'(0) = 0,\ a_2 y(L) + b_2 y'(L) = 0.$$

Eigenfunction expansion: $f(x) = \sum_{n=1}^{\infty} c_n \phi_n(x)$, $\quad c_n = \dfrac{1}{\|\phi_n\|^2} \displaystyle\int_0^L f(x)\phi_n(x)\,dx$, $\quad h_i = \dfrac{a_i}{b_i}$, $(i = 1, 2)$.

	Boundary conditions		ϕ_n	$\|\phi_n\|^2$	λ_n are the roots of
	at $x = 0$	at $x = L$			
1.	Dirichlet	Dirichlet	$\sin \lambda_n x$	$\dfrac{L}{2}$	$\sin \lambda L = 0$, i.e.,
	$a_1 \neq 0, b_1 = 0$	$a_2 \neq 0, b_2 = 0$			$\lambda_n = \dfrac{n\pi}{L}$, $n = 0, 1, 2, \ldots$
2.	Dirichlet	Neumann	$\sin \lambda_n x$	$\dfrac{L}{2}$	$\cos \lambda L = 0$, i.e.,
	$a_1 \neq 0, b_1 = 0$	$a_2 = 0, b_2 \neq 0$			$\lambda_n = \dfrac{(2n-1)\pi}{2L}$, $n = 1, 2, \ldots$
3.	Dirichlet	Robin	$\sin \lambda_n x$	$\dfrac{\lambda_n L - \sin \lambda_n L \cos \lambda_n L}{2\lambda_n}$	$\lambda + h_2 \tan \lambda L = 0^{*}$
	$a_1 \neq 0, b_1 = 0$	$a_2 \neq 0, b_2 \neq 0$			
4.	Neumann	Dirichlet	$\cos \lambda_n x$	$\dfrac{L}{2}$	$\cos \lambda L = 0$, i.e.,
	$a_1 = 0, b_1 \neq 0$	$a_2 \neq 0, b_2 = 0$			$\lambda_n = \dfrac{(2n-1)\pi}{L}$, $n = 1, 2, \ldots$
5.	Neumann	Neumann	$\cos \lambda_n x$	$\dfrac{L}{2}\,^{**}$	$\sin \lambda L = 0$, i.e.,
	$a_1 = 0, b_1 \neq 0$	$a_2 = 0, b_2 \neq 0$			$\lambda_n = \dfrac{n\pi}{L}$, $n = 0, 1, 2, \ldots$

TABLES 119

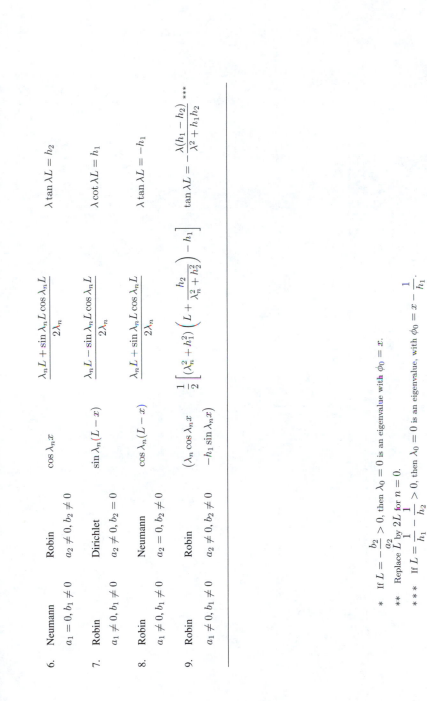

6. Neumann $a_1 = 0, b_1 \neq 0$ / Robin $a_2 \neq 0, b_2 \neq 0$	$\cos\lambda_n x$	$\dfrac{\lambda_n L + \sin\lambda_n L\cos\lambda_n L}{2\lambda_n}$	$\lambda\tan\lambda L = h_2$
7. Robin $a_1 \neq 0, b_1 \neq 0$ / Dirichlet $a_2 \neq 0, b_2 = 0$	$\sin\lambda_n(L-x)$	$\dfrac{\lambda_n L - \sin\lambda_n L\cos\lambda_n L}{2\lambda_n}$	$\lambda\cot\lambda L = h_1$
8. Robin $a_1 \neq 0, b_1 \neq 0$ / Neumann $a_2 = 0, b_2 \neq 0$	$\cos\lambda_n(L-x)$	$\dfrac{\lambda_n L + \sin\lambda_n L\cos\lambda_n L}{2\lambda_n}$	$\lambda\tan\lambda L = -h_1$
9. Robin $a_1 \neq 0, b_1 \neq 0$ / Robin $a_2 \neq 0, b_2 \neq 0$	$(\lambda_n\cos\lambda_n x \\ -h_1\sin\lambda_n x)$	$\dfrac{1}{2}\left[(\lambda_n^2+h_1^2)\left(L+\dfrac{h_2}{\lambda_n^2+h_2^2}\right)-h_1\right]$	$\tan\lambda L = -\dfrac{\lambda(h_1-h_2)}{\lambda^2+h_1 h_2}$ ***

*   If $L = -\dfrac{b_2}{a_2} > 0$, then $\lambda_0 = 0$ is an eigenvalue with $\phi_0 = x$.

**   Replace L by $2L$ for $n = 0$.

***   If $L = \dfrac{1}{h_1} - \dfrac{1}{h_2} > 0$, then $\lambda_0 = 0$ is an eigenvalue, with $\phi_0 = x - \dfrac{1}{h_1}$.

Table 4.2: Eigenvalue Problem in Polar Coordinates

Eq (4.37), subject to the boundary conditions shown below, with ν and h real constants, $\nu \geq -\frac{1}{2}$.

	Boundary conditions at $r = a$	$\phi_n(\lambda_n r)$	$\|\phi_n\|^2$	λ_n are the roots of
1.	Dirichlet $\left[y(a) = 0 \right]$	$J_\nu(\lambda_n r)$	$\dfrac{2}{a^2 J_\nu'(\lambda_n a)}$	$J_\nu(\lambda a) = 0$
2.	Neumann $\left[\dfrac{dy}{dx}(a) = 0 \right]$	$J_\nu(\lambda_n r)$	$\dfrac{2\lambda_n^2}{(a^2\lambda_n^2 - \nu^2)J_\nu^2(\lambda_n a)}$	$J_\nu'(\lambda_n a) = 0$ †
3.	Robin $\left[\dfrac{dy}{dx}(a) + hy(a) = 0 \right]$	$J_\nu(\lambda_n r)$	$\dfrac{2\lambda_n^2}{[a^2(h^2 + \lambda_n^2) - \nu^2]J_\nu^2(\lambda_n a)}$	$\lambda J_\nu'(\lambda a) + h J_\nu(\lambda a) = 0$

† $\lambda_0 = 0$ is also an eigenvalue for $\nu = 0$; then $\phi_0 = 1$ and $\|\phi_0\|^2 = \dfrac{a^2}{2}$.

5

Separation of Variables

The method of separation of variables is a well-established technique for solving ordinary differential equations. This method is easily adaptable to almost all linear homogeneous partial differential equations with constant coefficients in canonical form, and exhibits the power of the superposition principle to construct the general solution of such equations. Since linear first-order partial differential equations can always be solved by the method of characteristics, the method of separation of variables is usually applied to solve higher-order partial differential equations. The basic idea of this method is to transform a partial differential equation into as many ordinary differential equations as the number of independent variables in the partial differential equation by representing the solution as a product of functions of each independent variable. After these ordinary differential equations are solved, the method reduces to solving eigenvalue problems and constructing the general solution as an eigenfunction expansion, where the coefficients are evaluated by using the boundary conditions and the initial conditions. In most cases the solution is written in terms of a series of orthogonal functions.

5.1. Introduction

Consider the partial differential equation with constant coefficients

$$a\, u_{xx} + 2b\, u_{xy} + c\, u_{yy} + e\, u_x + f\, u_y + g\, u = 0. \qquad (5.1)$$

The first step in the general technique is to eliminate the term with mixed partial derivatives by introducing a new set of coordinates ξ, η (called the characteristic

121

coordinates, see §2.7). Thus, we have

$$a_1 u_{\xi\xi} + c_1 u_{\eta\eta} + e_1 u_\xi + f_1 u_\eta + g_1 u = 0. \tag{5.2}$$

Now, we assume a solution of the form $u = X(\xi)Y(\eta)$ in (5.2), and obtain

$$a_1 X'' Y + C X Y'' + e_1 X' Y + f_1 X Y' + g_1 X Y = 0,$$

or, formally,

$$\frac{L_1(D_\xi) X}{X} + \frac{L_2(D_\eta) Y}{Y} + g_1 = 0, \tag{5.3}$$

where $L_1(D_\xi)$ is a linear differential operator in ξ, and $L_2(D_\eta)$ a linear differential operator in η. Since the first term in (5.3) is a function of ξ only and the second term is a function of η only, while the third term is a constant, the only way Eq (5.3) can be solved is if each of the first two terms is also constant, thus

$$\frac{L_1 X}{X} = \lambda, \quad \frac{L_2 Y}{Y} = \mu, \quad \text{such that } \lambda + \mu + g_1 = 0.$$

We will explain this method by some examples.

5.2. Hyperbolic Equations

EXAMPLE 5.1. The problem of a vibrating string is governed by the one-dimensional wave equation (§1.4). Consider the boundary value problem

$$\frac{\partial^2 u}{\partial t^2} = c^2 \frac{\partial^2 u}{\partial x^2}, \quad 0 < x < l, \tag{5.4}$$

$$u(0,t) = 0 = u(l,t), \quad t > 0, \tag{5.5}$$

$$u(x,0) = f(x), \quad u_t(x,0) = h(x), \quad 0 < x < l, \tag{5.6}$$

where $f \in C^1(0,l)$ is a given function. We seek a solution of the form

$$u(x,t) = X(x)T(t), \tag{5.7}$$

where X is a function of x only and T a function of t only. We assume that a solution of the form (5.7) exists. Sometimes this method requires certain modifications, as in Example 5.3. We will, however, carry out the details to see if the method works for this problem. Note that x

$$\frac{\partial^2 u}{\partial t^2} = X T'', \quad \text{and} \quad \frac{\partial^2 u}{\partial x^2} = X'' T,$$

where the primes denote the derivative with respect to its corresponding independent variable. Then Eq (5.4) reduces to

$$X\,T'' = c^2 X'' T,$$

or, after separating the variables, it becomes

$$\frac{T''}{T} = c^2 \frac{X''}{X}. \tag{5.8}$$

In Eq (5.8) we have been able to separate the variables. It is only at this stage in the development of this method that we may either continue or abandon the method, depending on whether or not we are successful in separating the variables.

Since the left side of Eq (5.8) is a function of t only and the right side a function of x only, the only situation where $X(x)$ and $T(t)$ have solutions for all x and all t is when $\dfrac{c^2 X''(x)}{X} = \dfrac{T''(t)}{T} = \text{const}.$ Hence, from (5.8) we can write

$$\frac{1}{c^2}\frac{T''}{T} = \frac{X''}{X} = k, \quad k = \text{const.} \tag{5.9}$$

The set of equations (5.9) is equivalent to two ordinary differential equations:

$$T'' - k\,c^2\,T = 0, \tag{5.10}$$
$$X'' - k\,X = 0. \tag{5.11}$$

It is apparent from Eq (5.9) that the constant k in Eqs (5.10) and (5.11) has the same value. The general solution of Eq (5.10) is

$$T(t) = \begin{cases} c_1 e^{c\sqrt{k}t} + c_2 e^{-c\sqrt{k}t} & \text{for } k > 0 \\ c_1 t + c_2 & \text{for } k = 0 \\ c_1 \cos c\sqrt{-k}t + c_2 \sin c\sqrt{-k}t & \text{for } k < 0, \end{cases} \tag{5.12}$$

and of Eq (5.11) is

$$X(x) = \begin{cases} d_1 e^{\sqrt{k}x} + d_2 e^{-\sqrt{k}x} & \text{for } k > 0 \\ d_1 x + d_2 & \text{for } k = 0 \\ d_1 \cos \sqrt{-k}x + d_2 \sin \sqrt{-k}x & \text{for } k < 0. \end{cases} \tag{5.13}$$

In view of the boundary conditions (5.5), we must have

$$X(0)T(t) = 0 = X(l)T(t) \quad \text{for all } t \ge 0. \tag{5.14}$$

Using these conditions in (5.13) for $k > 0$, we get the system of equations

$$X(0) = d_1 + d_2 = 0,$$
$$X(l) = d_1 e^{\sqrt{k}l} + d_2 e^{-\sqrt{k}l} = 0. \tag{5.15}$$

The system (5.15) has a nontrivial solution iff the determinant of its coefficients vanishes. But since

$$\begin{vmatrix} 1 & 1 \\ e^{\sqrt{k}l} & e^{-\sqrt{k}l} \end{vmatrix} = e^{-\sqrt{k}l} - e^{\sqrt{k}l} \neq 0,$$

a nonzero solution for $X(x)$ in (5.13) for $k > 0$ is not possible. Next, for $k = 0$, the boundary conditions (5.14) imply that $d_1 = 0$ and $d_2 = 0$. Hence, there is no nonzero solution for $k = 0$. Finally, for $k < 0$, let us set $k = -\lambda^2$. Then the general solution (5.13) in this case becomes

$$X(x) = d_1 \cos \lambda x + d_2 \sin \lambda x,$$

which, under the boundary conditions (5.14), yields

$$X(0) = d_1 = 0, \quad \text{and} \quad X(l) = d_2 \sin \lambda l = 0.$$

To avoid a trivial solution in this case, we choose λ such that λl is a positive multiple of π, i.e., $\lambda l = n\pi$, or $\lambda = \dfrac{n\pi}{l}$. The positive values of λ are chosen because the negative multiples give the same eigenfunctions as the positive ones. This result leads to an infinite set of solutions which are denoted by

$$X_n(x) = d_{2,n} \sin \frac{n\pi x}{l},$$

where each eigenfunction $\sin \dfrac{n\pi x}{l}$ corresponds to the eigenvalue

$$k = -\frac{n^2\pi^2}{l^2}. \tag{5.16}$$

The solutions for $T(t)$ for the choice of $k < 0$, as in (5.16), are then obtained from (5.12) as

$$T_n(t) = c_{1,n} \cos \frac{n\pi ct}{l} + c_{2,n} \sin \frac{n\pi ct}{l}.$$

Then the infinite set of solutions is

$$u_n(x,t) = X_n(x)T_n(t) = \left[A_n \cos \frac{n\pi ct}{l} + B_n \sin \frac{n\pi ct}{l} \right] \sin \frac{n\pi x}{l}, \tag{5.17}$$

where the constants A_n and B_n are determined from the initial conditions. The eigenfunctions are contained in the solution (5.17), whereas the eigenvalues for this boundary value problem are given by (5.16).

The next step is to obtain the particular solution which satisfies the initial conditions (5.6). At this point it may so happen that none of the solutions (5.17) will satisfy (5.6). In view of the superposition principle (see §1.5), any finite sum of the solutions

(5.17) is also a solution of this boundary value problem. We should, therefore, find a linear combination of those solutions, which also satisfies the initial conditions (5.6). Even if this technique fails, we can always try an infinite series of solutions (5.17), i.e.,

$$u(x,t) = \sum_{n=1}^{\infty} X_n(x)T_n(t) = \sum_{n=1}^{\infty} \left[A_n \cos \frac{n\pi ct}{l} + B_n \sin \frac{n\pi ct}{l} \right] \sin \frac{n\pi x}{l}. \quad (5.18)$$

For convergence of the series (5.18), see Chapter 4. Now, we can take this series expansion formally and verify that the boundary conditions (5.5) are still satisfied. We will use the initial conditions to find the constants A_n and B_n. Using the first of the initial conditions (5.6), we get

$$u(x,0) = \sum_{n=1}^{\infty} A_n \sin \frac{n\pi x}{l} = f(x). \quad (5.19)$$

The infinite series (5.19) is a Fourier sine series. Hence, $f(x)$ can be regarded as an odd function with period $2l$. Thus, we expand this function $f(x)$ on the interval $0 \leq x \leq l$ such that $f(-x) = -f(x)$ on the interval $-l \leq x \leq 0$, and $f(x+2l) = f(x)$ for all x. Then the coefficients A_n for $n = 1, 2, \ldots$ are given by

$$A_n = \frac{1}{l} \int_{-l}^{l} f(x) \sin \frac{n\pi x}{l} \, dx = \frac{2}{l} \int_{0}^{l} f(x) \sin \frac{n\pi x}{l} \, dx. \quad (5.20)$$

Taking the derivative of (5.18), we get

$$u_t = \frac{\pi c}{l} \sum_{n=1}^{\infty} n \left[B_n \cos \frac{n\pi ct}{l} - A_n \sin \frac{n\pi ct}{l} \right] \sin \frac{n\pi x}{l}, \quad (5.21)$$

which, in view of the second of the initial conditions (5.6), gives

$$u_t(x,0) = \frac{\pi c}{l} \sum_{n=1}^{\infty} n B_n \sin \frac{n\pi x}{l} = h(x),$$

where

$$B_n = \frac{2}{n\pi c} \int_{0}^{l} h(x) \sin \frac{n\pi x}{l} \, dx, \quad n = 1, 2, \ldots . \quad (5.22)$$

Hence, the solution (5.18) is completely determined.

We will now derive the d'Alembert's solution for this problem. From (5.18) we have

$$u = \frac{1}{2} \sum_{n=1}^{\infty} A_n \left\{ \sin \frac{n\pi(x+ct)}{l} + \sin \frac{n\pi(x-ct)}{l} \right\}$$

$$+ \frac{1}{2} \sum_{n=1}^{\infty} B_n \left\{ \cos \frac{n\pi(x-ct)}{l} - \cos \frac{n\pi(x+ct)}{l} \right\} \quad (5.23)$$

$$= \frac{1}{2} \left[f(x+ct) + f(x-ct) \right] + \frac{1}{2} \left[-g(x+ct) + g(x-ct) \right],$$

where

$$f(z) = \sum_{n=1}^{\infty} A_n \sin \frac{n\pi z}{l},$$

as in (5.19). Let

$$g(z) = \sum_{n=1}^{\infty} B_n \cos \frac{n\pi z}{l}.$$

Then

$$g'(z) = -\frac{1}{c}\frac{\pi c}{l} \sum_{n=1}^{\infty} n B_n \sin \frac{n\pi z}{l} = -\frac{1}{c} h(z),$$

and from (5.23) we obtain the formal solution, known as the d'Alembert's solution for this problem, as

$$u(x,t) = \phi(x - ct) + \psi(x + ct), \tag{5.24}$$

where c is the wave velocity, and

$$\phi(x - ct) = \frac{1}{2}\left[f(x - ct) + g(x - ct) \right]$$

$$\psi(x + ct) = \frac{1}{2}\left[f(x + ct) - g(x + ct) \right].$$

An interpretation of the solution of this problem is as follows: At each point x of the string, we have

$$u(x,t) = \sum_{n=1}^{\infty} \alpha_n \cos \frac{n\pi c}{l}(t + \delta_n) \sin \frac{n\pi x}{l}.$$

This equation describes a harmonic motion with amplitudes $\alpha_n \sin \dfrac{n\pi x}{l}$. Each such motion of the string is called a standing wave, which has its nodes at the points where $\sin(n\pi x/l) = 0$; these points remain fixed during the entire process of vibration. But the string vibrates with maximum amplitudes α_n at the points where $\sin(n\pi x/l) = \pm 1$. For any t the structure of the standing wave is described by

$$u(x,t) = \sum_{n=1}^{\infty} C_n(t) \sin \frac{n\pi x}{l},$$

where

$$C_n(t) = \alpha_n \cos \omega_n(t + \delta_n), \quad \omega_n = \frac{n\pi c}{l}.$$

At times t when $\cos \omega_n(t + \delta_n) = \pm 1$, the velocity becomes zero and the displacement reaches its maximum value. ∎

5.3. Parabolic Equations

EXAMPLE 5.2. Consider the one-dimensional heat conduction equation

$$\frac{\partial u}{\partial t} = k \frac{\partial^2 u}{\partial x^2}, \quad 0 < x < l, \tag{5.25}$$

subject to the boundary conditions

$$u(0,t) = 0 = u(l,t), \quad t > 0, \tag{5.26}$$

and the initial condition

$$u(x,0) = f(x), \quad 0 < x < l, \tag{5.27}$$

where $f \in C^1$ is a prescribed function. In physical terms, this problem represents the heat conduction in a rod when its ends are maintained at zero temperature while the initial temperature u at any point of the rod is prescribed as $f(x)$. Let us assume the solution in the form

$$u(x,t) = X(x)T(t),$$

which after substitution into Eq (5.25) yields the set of equations

$$\frac{1}{k}\frac{T'}{T} = \frac{X''}{X}. \tag{5.28}$$

As in Example 5.1, the only situation where these two expressions can be equal is for each of them to be constant, say each equal to c. Then Eq (5.28) yields two ordinary differential equations

$$T' - ck\,T = 0, \tag{5.29}$$
$$X'' - c\,X = 0, \tag{5.30}$$

where the boundary conditions (5.26) reduce to

$$X(0)\,T(t) = 0 = X(l)\,T(t), \quad \text{or} \quad X(0) = 0 = X(l), \tag{5.31}$$

except for the case when the rod has zero initial temperature at every point. This situation, being uninteresting, can be neglected. As in the case of Example 5.1, we notice that for a nonzero solution of the problem (5.30)–(5.31) we must choose negative values of c. Hence, we set $c = -\lambda^2$, and find that the eigenvalues $c = -n^2\pi^2/l^2$ have the corresponding eigenfunctions

$$X_n(x) = A_n \sin \frac{n\pi x}{l}.$$

Eq (5.29) then becomes

$$T' + \frac{kn^2\pi^2}{l^2}T = 0$$

whose general solution for each n is given by

$$T_n(t) = B_n e^{-kn^2\pi^2 t/l^2}.$$

Hence, we consider an infinite series of the form

$$u(x,t) = \sum_{n=1}^{\infty} X_n(x)T_n(t) = \sum_{n=1}^{\infty} C_n \sin\frac{n\pi x}{l} e^{-kn^2\pi^2 t/l^2}. \tag{5.32}$$

Now, we use the initial condition (5.27) in (5.32) and obtain

$$u(x,0) = \sum_{n=1}^{\infty} C_n \sin\frac{n\pi x}{l} = f(x), \tag{5.33}$$

which shows that $f(x)$ can be represented as a Fourier sine series, by extending f as an odd, piecewise continuous function of period $2l$ with piecewise continuous derivatives. Equation (5.33) gives the coefficients C_n as

$$C_n = \frac{2}{l}\int_0^l f(x)\sin\frac{n\pi x}{l}\,dx, \ n = 1, 2, \ldots. \tag{5.34}$$

Hence, the solution (5.32) is completely determined for this problem. Note that the series in (5.33) converges since $u(x,0)$ does, and the exponential expression in (5.32) is less than 1 for each n and all $t > 0$ and approaches zero as $t \to \infty$. ∎

An interesting situation, considered in the next example, arises if the function $f(x)$ is zero in the initial condition (5.27), but the boundary conditions are nonhomogeneous.

EXAMPLE 5.3. Consider the dimensionless partial differential equation governing the transient heat conduction for a plane wall

$$u_t = u_{xx}, \quad 0 < x < 1, \tag{5.35}$$

with the boundary conditions

$$u(0,t) = 1, \quad u(1,t) = 0, \quad t > 0, \tag{5.36}$$

and the initial condition

$$u(x,0) = 0, \quad 0 < x < 1. \tag{5.37}$$

Since the nonhomogeneous boundary condition in (5.36) does not allow us to compute the eigenfunctions, as in Example 5.2, we proceed as follows: First, we find a particular solution of the problem, which satisfies only the boundary conditions. Although there is more than one way to determine the particular solution, we can, for example, take the steady-state case, where the equation becomes $\tilde{u}_{xx} = 0$, which, after integrating twice, has the general solution

$$\tilde{u}(x) = c_1 x + c_2,$$

with the boundary conditions $\tilde{u}(0) = 1$, $\tilde{u}(1) = 0$. Thus, $c_1 = -1$, $c_2 = 1$, and the steady-state solution is

$$\tilde{u}(x) = 1 - x.$$

Next, we formulate a homogeneous problem by writing $u(x, t)$ as a sum of the steady-state solution $\tilde{u}(x)$ and a transient term $v(x, t)$, i.e.,

$$u(x, t) = \tilde{u}(x) + v(x, t),$$

or

$$v(x, t) = u(x, t) - \tilde{u}(x). \tag{5.38}$$

Hence, the problem reduces to finding $v(x, t)$. If we substitute v from (5.38) into (5.35), we get

$$v_t = v_{xx}, \tag{5.39}$$

where the boundary conditions (5.36) and the initial condition (5.37) reduce to

$$\begin{aligned} v(0, t) &= u(0, t) - \tilde{u}(0) = 0, \\ v(1, t) &= u(1, t) - \tilde{u}(1) = 0, \end{aligned} \tag{5.40}$$

and

$$v(x, 0) = u(x, 0) - \tilde{u}(x) = x - 1. \tag{5.41}$$

Notice that the problem (5.39)–(5.41) is the same as in Example 5.2 with $k = 1$, $l = 1$, $f(x) = x - 1$, and u replaced by v. Hence, its general solution from (5.32) is given by

$$v(x, t) = \sum_{n=1}^{\infty} C_n e^{-n^2 \pi^2 t} \sin n\pi x, \tag{5.42}$$

and the coefficients C_n are determined from (5.34) as

$$C_n = 2 \int_0^1 (x - 1) \sin n\pi x \, dx = -\frac{2}{n\pi}.$$

Thus,

$$v(x,t) = -\frac{2}{\pi} \sum_{n=1}^{\infty} \frac{1}{n} e^{-n^2\pi^2 t} \sin n\pi x, \tag{5.43}$$

and finally from (5.38)

$$u(x,t) = 1 - x - \frac{2}{\pi} \sum_{n=1}^{\infty} \frac{1}{n} e^{-n^2\pi^2 t} \sin n\pi x. \tag{5.44}$$

In general, if the thickness of the plate is l, the solution is

$$u(x,t) = 1 - \frac{x}{l} - \frac{2}{\pi} \sum_{n=1}^{\infty} \frac{1}{n} e^{-n^2\pi^2 t/l^2} \sin \frac{n\pi x}{l}.$$

The solution for the half-space is derived by letting $l \to \infty$. Since

$$\lim_{l\to\infty} u(x,t) = 1 - \frac{2}{\pi} \sum_{n=1}^{\infty} \frac{l}{n\pi} e^{-n^2\pi^2 t/l^2} \sin \frac{n\pi x}{l} \cdot \frac{\pi}{l},$$

let $n\pi/l = \xi$ and $\pi/l = d\xi$. Then

$$\lim_{l\to\infty} u(x,t) = 1 - \frac{2}{\pi} \int_0^{\infty} \frac{1}{\xi} e^{-\xi^2 t} \sin \xi t \, d\xi = 1 - \mathrm{erf}\left(\frac{x}{2\sqrt{t}}\right) = \mathrm{erfc}\left(\frac{x}{2\sqrt{t}}\right).$$

For the evaluation of the above integral, see Example 6.11. ∎

5.4. Elliptic Equations

EXAMPLE 5.4. We consider the steady-state heat conduction (or potential) problem for the rectangle R $\{0 < x < a, \ 0 < y < b\}$:

$$u_{xx} + u_{yy} = 0, \quad x, y \in R, \tag{5.45}$$

subject to the Dirichlet boundary conditions

$$u(0,y) = 0 = u(a,y), \ u(x,0) = 0, \ u(x,b) = f(x). \tag{5.46}$$

Physically, this problem arises if three edges of a thin isotropic rectangular plate are insulated and maintained at zero temperature, while the fourth edge is subjected to a variable temperature $f(x)$ until the steady-state conditions are attained throughout R. Then the steady-state value of $u(x,y)$ represents the distribution of temperature in the interior of the plate. As before, we seek a solution of the form $u(x,y) = X(x)Y(y)$, which, after substitution into Eq (5.45) leads to the set of two ordinary differential equations:

$$X'' - cX = 0, \tag{5.47}$$
$$Y'' + cY = 0, \tag{5.48}$$

where c is a constant, as in Example 5.2. Since the first three boundary conditions in (5.46) are homogeneous, they become

$$X(0) = 0, \quad X(a) = 0, \quad Y(0) = 0, \tag{5.49}$$

but the fourth boundary condition which is nonhomogeneous must be used separately. Now, taking $c = -\lambda^2$, as before, the solution of (5.47) subject to the first two boundary conditions in (5.49) leads to the eigenvalues and the corresponding eigenfunctions as

$$\lambda_n^2 = \frac{n^2\pi^2}{a^2}, \quad X_n(x) = \sin\frac{n\pi x}{a}, \quad n = 1, 2, \ldots,$$

while for these eigenvalues the solutions of (5.45) satisfying the third boundary condition in (5.49) are

$$Y_n(y) = \sinh\frac{n\pi y}{a}, \quad n = 1, 2, \ldots. \tag{5.50}$$

Hence, for arbitrary constants C_n, $n = 1, 2, \ldots$, we get

$$u(x, y) = \sum_{n=1}^{\infty} C_n \sin\frac{n\pi x}{a} \sinh\frac{n\pi y}{a}. \tag{5.51}$$

The coefficients C_n are then determined by using the fourth boundary condition in (5.46). Thus,

$$u(x, b) = f(x) = \sum_{n=1}^{\infty} C_n \sin\frac{n\pi x}{a} \sinh\frac{n\pi b}{a}, \quad 0 < x < a,$$

which, in view of formula (4.23), yields

$$C_n \sinh\frac{n\pi b}{a} = \frac{2}{a} \int_0^a f(x) \sin\frac{n\pi x}{a}\, dx, \quad n = 1, 2, \ldots. \tag{5.52}$$

This solves the problem completely. In particular, if $f(x) = f_0 = $ const, then

$$C_n \sinh\frac{n\pi b}{a} = \frac{2f_0[1 - (-1)^n]}{n\pi}.$$

Thus, from (5.51), we have

$$u(x,y) = \frac{2f_0}{\pi} \sum_{n=1}^{\infty} \frac{1-(-1)^n}{n} \frac{\sin(n\pi x/a)\sinh(n\pi y/a)}{\sinh(n\pi b/a)}. \ \blacksquare \tag{5.53}$$

EXAMPLE 5.5. Consider the steady-state heat conduction or potential problem

$$u_{xx} + u_{yy} = 0, \quad 0 < x < \pi, \quad 0 < y < 1, \tag{5.54}$$

subject to the mixed boundary conditions

$$u(x,0) = u_0 \cos x, \quad u(x,1) = u_0 \sin^2 x, \tag{5.55}$$
$$u_x(0,y) = 0 = u_x(\pi,y).$$

The separation of variables technique leads to the same set of ordinary differential equations as in (5.47)–(5.48), i.e.,

$$X'' + \lambda^2 X = 0, \quad X'(0) = 0 = X'(\pi), \tag{5.56}$$

and

$$Y'' - \lambda^2 Y = 0. \tag{5.57}$$

The eigenvalues and the corresponding eigenfunctions for (5.56) are

$$\lambda_0 = 0, \quad X_0(x) = 1,$$
$$\lambda_n = n^2, \quad X_n(x) = \cos nx, \quad n = 1, 2, \ldots,$$

and subsequently the solutions of (5.57) are

$$Y_n(y) = \begin{cases} A_0 + B_0 y, & n = 0, \\ A_n \cosh ny + B_n \sinh ny, & n = 1, 2, \ldots. \end{cases} \tag{5.58}$$

Hence, using the superposition principle, we get

$$u(x,y) = A_0 + B_0 y + \sum_{n=1}^{\infty} [A_n \cosh ny + B_n \sinh ny] \cos nx. \tag{5.59}$$

Now, the first boundary condition in (5.55) leads to

$$u(x,0) = A_0 + \sum_{n=1}^{\infty} A_n \cos nx = u_0 \cos x. \tag{5.60}$$

By matching the coefficients of similar terms on both sides of (5.60), we find that $A_0 = 0$, $A_1 = u_0$, and $A_n = 0$ for $n \geq 2$. Hence, the solution becomes

$$u(x,y) = B_0 y + u_0 \cosh y \cos x + \sum_{n=1}^{\infty} B_n \sinh ny \cos nx. \tag{5.61}$$

Similarly, using the second boundary condition in (5.55) we find from (5.61) that

$$u(x, 1) = B_0 + u_0 \cosh 1 \cos x + \sum_{n=1}^{\infty} B_n \sinh n \, \cos nx$$

$$= u_0 \sin^2 x = u_0 \frac{1 - \cos 2x}{2},$$

from which, after comparing the coefficients of similar terms on both sides, we get

$$B_0 = \frac{u_0}{2}, \qquad\qquad B_1 = -\frac{u_0 \cosh 1}{\sinh 1}$$

$$B_2 = -\frac{u_0}{2 \sinh 2}, \qquad B_n = 0 \quad \text{for } n \geq 3.$$

Hence, from (5.61) the general solution is given by

$$u(x, y) = \frac{u_0}{2} y + u_0 \left[\cosh y - \frac{\cosh 1 \sinh y}{\sinh 1} \right] \cos x - u_0 \frac{\sinh 2y}{2 \sinh 2} \cos 2x$$

$$= u_0 \left[\frac{1}{2} y + \frac{\sinh(1 - y)}{\sinh 1} \cos x - \frac{\sinh 2y}{2 \sinh 2} \cos 2x \right]. \; \blacksquare \tag{5.62}$$

5.5. Cylindrical Polar Coordinates

The three-dimensional Laplacian in the cylindrical polar coordinates is

$$\nabla^2 \equiv \frac{\partial^2}{\partial r^2} + \frac{1}{r} \frac{\partial}{\partial r} + \frac{1}{r^2} \frac{\partial^2}{\partial \theta^2} + \frac{\partial^2}{\partial z^2}. \tag{5.63}$$

EXAMPLE 5.6. (*Circular drum*) If a circular drum is struck in the center, its vibrations are radially symmetric. We solve the boundary value problem

$$\frac{\partial^2 u}{\partial t^2} = \nabla^2 u \equiv \frac{\partial^2 u}{\partial r^2} + \frac{1}{r} \frac{\partial u}{\partial r}, \quad r < 1, \tag{5.64}$$

subject to the boundary conditions

$$u(r, 0) = f(r), \quad r < 1,$$
$$u_t(r, 0) = 0, \quad u(1, t) = 0, \quad t \geq 0, \tag{5.65}$$
$$|u(r, t)| < \infty \quad \text{as } r \to 0.$$

If we take $u(r, t) = T(t) R(r)$, then Eq (5.64) reduces to the system of ordinary differential equations

$$\frac{T''}{T} = \frac{rR'' + R'}{rR} = k. \tag{5.66}$$

Here, again, $k = -\lambda^2$ yields nontrivial solutions. Then the system (5.66) gives two uncoupled ordinary differential equations:

$$T'' + \lambda^2 T = 0, \tag{5.67}$$

$$\frac{d^2 R}{dr^2} + \frac{1}{r}\frac{dR}{dr} + \lambda^2 R = 0. \tag{5.68}$$

Eq (5.68) is the Bessel equation. The eigenvalues λ_n are the positive zeros of $J_0(\lambda)$, with the corresponding eigenfunctions $J_0(\lambda_n r)$. The solutions of Eq (5.67) are $T_n = \cos \lambda_n t$. Hence, the solution of the vibrating circular drum struck at the center is given by

$$u(x,t) = \sum_{n=1}^{\infty} C_n \cos \lambda_n t J_0(\lambda_n r), \tag{5.69}$$

where the coefficients C_n are given by (see Example 4.16)

$$C_n = \frac{\int_0^1 f(r) J_0(\lambda_n r) r \, dr}{\int_0^1 [J_0(\lambda_n r)]^2 r \, dr} = \frac{2 \int_0^1 f(r) J_0(\lambda_n r) r \, dr}{J_1^2(\lambda_n)}. \tag{5.70}$$

Marc Kac (1966) asked the question: "Can one hear the shape of a drum?" This means one should answer the question whether two drums of different shapes and struck in their centers have the same eigenvalues (Protter, 1987). This question has been resolved negatively by Gordon, Webb, and Wolpert (1992). Also, see the Mathematica Notebook drum.nb. ∎

5.6. Spherical Coordinates

Using the transformation $x = \rho \sin\phi \cos\theta$, $y = \rho \sin\phi \sin\theta$, $z = \rho \cos\phi$, where $\rho \geq 0$, $0 \leq \theta \leq 2\pi$ and $0 \leq \phi \leq \pi$, the Laplacian in the spherical coordinate system becomes

$$\nabla^2 \equiv \frac{\partial^2}{\partial \rho^2} + \frac{2}{\rho}\frac{\partial}{\partial \rho} + \frac{1}{\rho^2 \sin^2\phi}\frac{\partial^2}{\partial \theta^2} + \frac{1}{\rho^2}\frac{\partial^2}{\partial \phi^2} + \frac{\cot\phi}{\rho^2}\frac{\partial}{\partial \phi}.$$

EXAMPLE 5.7. (*Cooling ball*) Consider the boundary value problem

$$u_t = \nabla^2 u, \quad 0 \leq \rho < 1, \quad 0 \leq \phi \leq \pi, \quad t > 0,$$

subject to the conditions $u(\rho, \phi, 0) = f(\rho, \phi)$ and $u(1, \phi, t) = 0$, where the function $u = u(\rho, \phi, t)$ is independent of θ. The problem describes the temperature distribution in the interior of the unit ball dropped in cold water. The first condition implies that the temperature u is not uniform but depends only on ρ, ϕ, and t. Thus, the solution is assumed formally as

$$u(\rho, \phi, t) = R(\rho)\,\Phi(\phi)\,T(t), \tag{5.71}$$

which, after separating the variables, gives

$$\frac{R'' + \dfrac{2}{\rho}R'}{R} + \frac{\Phi'' + \cot\phi\,\Phi'}{\rho^2\,\Phi} = -\alpha^2 = \frac{T'}{T} = \lambda. \tag{5.72a}$$

The left side of Eq (5.72a) can be expressed as

$$\left(\frac{R'' + \dfrac{2}{\rho}R'}{R} + \alpha^2\right)\rho^2 = -\frac{\Phi'' + \cot\phi\,\Phi'}{\Phi}.$$

In order that this equation be satisfied, the terms on each side must be constant. Thus,

$$\left(\frac{R'' + \dfrac{2}{\rho}R'}{R} + \alpha^2\right)\rho^2 = \mu = -\frac{\Phi'' + \cot\phi\,\Phi'}{\Phi}. \tag{5.72b}$$

It is known that $\mu = n(n+1)$ (Courant and Hilbert, 1963; also, the file mu.pdf on the CRC website). Then Eq (5.72b) yields

$$\begin{aligned}
\Phi'' + \cot\phi\,\Phi' + n(n+1)\,\Phi &= 0, \\
\rho^2\,R'' + 2\rho\,R' + \alpha^2\rho^2\,R - n(n+1)\,R &= 0.
\end{aligned} \tag{5.73}$$

If we set $x = \alpha\rho$ in the second equation, it becomes

$$x^2 R'' + 2x\,R' + \left(x^2 - n(n+1)\right)R = 0,$$

which under the transformation $w = \sqrt{x}\,R$ reduces to the Bessel equation

$$x^2 w'' + x\,w' + \left(x^2 - (n + \tfrac{1}{2})^2\right)w = 0, \tag{5.74}$$

and has a bounded solution $w = J_{n+1/2}(x)$. Hence,

$$R(\rho) = \frac{J_{n+1/2}(\alpha\rho)}{\sqrt{\alpha\rho}}.$$

The eigenfunctions $\psi_{mn} = \dfrac{J_{n+1/2}(\alpha_{mn}\rho)}{\sqrt{\alpha_{mn}\rho}}\,P_n(\cos\phi)$ form an orthogonal set. Using the boundary condition at $\rho = 1$, we get $J_{n+1/2}(\alpha) = 0$. Let α_{mn} denote the positive zeros of $J_{n+1/2}(\alpha)$. Then the solution for the temperature distribution in the unit ball is given by

$$u(\rho, \phi, t) = \sum_{\substack{m=1 \\ n=0}}^{\infty} C_{mn} \frac{e^{-\alpha_{mn}^2 t} J_{n+1/2}(\alpha_{mn}\rho)}{\sqrt{\alpha_{mn}\rho}} P_n(\cos\phi), \qquad (5.75)$$

where, after using $\int_{-1}^{1} P_n^2(x)\, dx = \dfrac{2}{2n+1}$,

$$C_{mn} = \frac{(2n+1)\sqrt{\alpha_{mn}}}{\beta(\alpha_{mn})} \int_0^\pi \int_0^1 \rho^{3/2} J_{n+1/2}(\alpha_{mn}\rho) P_n(\cos\theta) f(\rho,\phi) \, \sin\phi \, d\rho \, d\phi,$$

and

$$\beta(\alpha_{mn}) = 2 \int_0^1 \rho\, J_{n+1/2}^2(\alpha_{mn}\rho)\, d\rho = \left[J'_{n+1/2}(\alpha_{mn}) \right]^2.$$

The factor $\sqrt{\alpha_{mn}}$ can be absorbed in C_{mn}. ∎

5.7. Nonhomogeneous Problems

In the above examples we have seen that the method of separation of variables is applicable to steady-state linear problems with homogeneous governing equations and three homogeneous and one nonhomogeneous boundary conditions. Nonhomogeneity, however, occurs from other conditions as well. For example, there may be more than one nonhomogeneous boundary condition, or the governing equation may be nonhomogeneous. To use the method of separation of variables, a linear nonhomogeneous problem can be divided into finitely many simple problems with homogeneous equations and/or homogeneous boundary conditions. Then the solution of the given problem is obtained from the superposition of the solutions of all these simple problems.

EXAMPLE 5.8. (*with four nonhomogeneous boundary conditions*) Consider the steady-state temperature distributions governed by Eq (5.45) in the region R, with more than one nonhomogeneous boundary condition, viz.,

$$\begin{aligned} u(x,0) = f_1(x), \qquad u(x,b) = f_2(x), \quad 0 < x < a, \\ u(0,y) = f_3(y), \qquad u(a,y) = f_4(y), \quad 0 < y < b. \end{aligned} \qquad (5.76)$$

This problem can be resolved as a superposition of the four problems, shown in Fig. 5.1. Hence,

$$u(x,y) = u_1(x,y) + u_2(x,y) + u_3(x,y) + u_4(x,y), \qquad (5.77)$$

where the solution of each simple problem is obtained as in Example 5.4. ∎

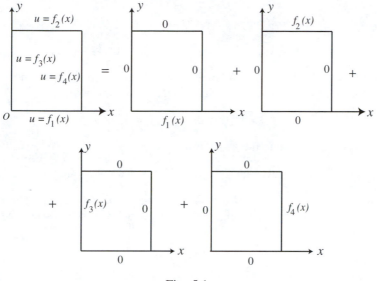

Fig. 5.1.

In certain cases with mixed boundary conditions, the method of separation of variables can be readily used by translating the function $u(x, y)$, which depends on the geometry and material symmetry of the problem.

EXAMPLE 5.9. Consider the Laplace equation (5.45) in a half-strip (see Fig. 5.2) subject to the boundary conditions

$$u(0, y) = f(y), \qquad \lim_{x \to +\infty} u(x, y) = u_\infty,$$

$$u_y(x, 0) = 0, \qquad u_y(x, b) + \beta[u(x, b) - u_\infty] = 0, \qquad (5.78)$$

where β is known as the film coefficient.

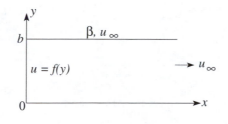

Fig. 5.2.

This problem has more than one nonhomogeneous boundary conditions. By using the translation $U(x, y) = u(x, y) - u_\infty$, the problem (5.45) subject to the boundary conditions (5.78) reduces to

$$U_{xx} + U_{yy} = 0,$$

$$U(0, y) = f(y) - u_\infty = F(y), \quad \lim_{x \to +\infty} U(x, y) = 0, \tag{5.79}$$

$$U(y) = 0, \quad U_y(x, b) = -\beta U(x, b).$$

Now, we assume that $U(x, y) = X(x)Y(y)$, which, after taking $c = \lambda^2$ in Eqs (5.47) and (5.48), yields a set of the two ordinary differential equations:

$$X'' - \lambda^2 X = 0, \quad Y'' + \lambda^2 Y = 0.$$

These two equations lead to the general solution

$$U(x, y) = \left(A_1 e^{-\lambda x} + A_2 e^{\lambda x} \right) \left(B_1 \cos \lambda y + B_2 \sin \lambda y \right). \tag{5.80}$$

Now, since $Y_y(0) = 0$ and $Y_y(b) + \beta Y(b) = 0$, we obtain the eigenfunctions as $\cos \lambda_n y$ with the corresponding eigenvalues λ_n, which are the consecutive positive roots of the equation

$$\lambda \tan \lambda b = \beta. \tag{5.81}$$

Using the boundary condition $\lim_{x \to \infty} X(x) = 0$, we find that

$$U(x, y) = \sum_{n=1}^{\infty} C_n e^{-\lambda_n x} \cos \lambda_n y. \tag{5.82}$$

Then, in view of the nonhomogeneous boundary condition $U(0, y) = F(y)$, we have

$$F(y) = f(y) - u_\infty = \sum_{n=1}^{\infty} C_n \cos \lambda_n y,$$

where the coefficients C_n are given by

$$C_n = \frac{2\lambda_n}{\lambda_n b + \sin \lambda_n b \cos \lambda_n b} \int_0^b [f(y) - u_\infty] \cos \lambda_n y \, dy,$$

(see Table 4.1, # 6). Hence, the temperature distribution is given by

$$U(x, y) = u(x, y) - u_\infty$$

$$= 2 \sum_{n=1}^{\infty} \frac{\lambda_n e^{-\lambda_n x} \cos \lambda_n y}{\lambda_n b + \sin \lambda_n b \cos \lambda_n b} \int_0^b [f(\eta) - u_\infty] \cos \lambda_n \eta \, d\eta. \tag{5.83}$$

In particular, if $f(y) = u_0 = \text{const}$, the temperature distribution reduces to

$$\frac{u(x, y) - u_\infty}{u_0 - u_\infty} = 2 \sum_{n=1}^{\infty} \frac{\sin \lambda_n b}{\lambda_n b + \sin \lambda_n b \cos \lambda_n b} e^{-\lambda_n x} \cos \lambda_n y \, dy. \tag{5.84}$$

In view of (5.81), let ξ_n denote the consecutive positive roots of $\tan \xi - \dfrac{Bi}{\xi} = 0$, where $\xi = \lambda b$, and $Bi = \beta b$ is the Biot number. Three of these roots, denoted by ξ_1, ξ_2, and ξ_3, are shown in Fig. 5.3. ∎

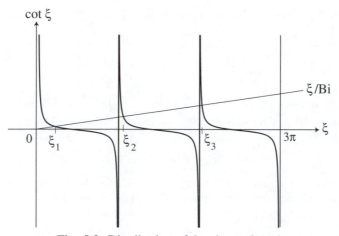

Fig. 5.3. Distribution of the eigenvalues ξ_n.

EXAMPLE 5.10. We consider a problem, whose governing equation is linear and nonhomogeneous, with linear and homogeneous boundary conditions. Assume that heat is generated in a rectangular bar at a constant rate q per unit volume, that there is no temperature gradient in the z-direction, and that the thermal conductivity k of the bar is constant. Then the steady-state temperature distribution is governed by the Poisson's equation

$$k\,(u_{xx} + u_{yy}) + q = 0. \tag{5.85}$$

Let the boundary conditions be given by

$$\begin{aligned}
u_x(0, y) &= 0, & u(a, y) &= 0, \\
u_y(x, 0) &= 0, & u(x, b) &= 0.
\end{aligned} \tag{5.86}$$

Equation (5.85), being nonhomogeneous, is not separable. But, if we assume the solution as

$$u(x, y) = V(x, y) + \phi(x), \tag{5.87}$$

then problem (5.85)–(5.86) reduces to the following two problems:

$$\frac{d^2\phi}{dx^2} + \frac{q}{k} = 0, \quad \frac{d\phi(0)}{dx} = 0, \quad \phi(a) = 0, \tag{5.88}$$

and

$$\begin{aligned}
&V_{xx} + V_{yy} = 0, \\
&V_x(0, y) = 0 = V(a, y), \quad V_y(x, 0) = 0, \quad V(x, b) = -\phi(x).
\end{aligned} \tag{5.89}$$

Solution of problem (5.88) is readily obtained as

$$\phi(x) = \frac{qa^2}{2k}\left(1 - \frac{x^2}{a^2}\right). \tag{5.90}$$

The problem (5.89) is separable and its solution is given by

$$V(x, y) = -\frac{2q}{ka}\sum_{n=0}^{\infty}\frac{(-1)^n}{\lambda_n^3}\frac{\cos\lambda_n x \cosh\lambda_n y}{\cosh\lambda_n b}, \tag{5.91}$$

where the eigenvalues are $\lambda_n = \dfrac{(2n+1)\pi}{2a}$, $n = 0, 1, \ldots$. The solution of this problem is then obtained by adding (5.90) and (5.91). Note that this problem can also be solved by taking

$$u(x, y) = V(x, y) + \psi(y). \tag{5.92}$$

In the case when the rate is variable, say, $q = q(x)$, we should use the substitution (5.87); if $q = q(y)$, then the substitution (5.92) will make the equation in V separable. If q is not constant, then a particular solution can be found. ■

EXAMPLE 5.11. Consider the nonhomogeneous wave equation

$$u_{tt} = c^2 u_{xx} + f(x, t), \quad 0 < x < l, \tag{5.93}$$

with the homogeneous (Dirichlet) boundary conditions $u(0, t) = 0 = u(l, t)$, $t > 0$, and the initial conditions $u(x, 0) = g(x)$, $u_t(x, 0) = h(x)$, $0 \le x \le l$. In the method of separation of variables, the Dirichlet boundary conditions suggest that we assume a Fourier series solution of the form

$$u(x, t) = \sum_{n=1}^{\infty} u_n(t) \sin\frac{n\pi x}{l},$$

where t is regarded as a parameter. The functions f, g, h are written in terms of the Fourier series as

$$f(x, t) = \sum_{n=1}^{\infty} f_n(t) \sin\frac{n\pi x}{l}, \quad \text{where} \quad f_n(t) = \frac{2}{l}\int_0^l f(\xi, t) \sin\frac{n\pi\xi}{l}\,d\xi;$$

$$g(x) = \sum_{n=1}^{\infty} g_n \sin\frac{n\pi x}{l}, \quad \text{where} \quad g_n = \frac{2}{l}\int_0^l g(\xi) \sin\frac{n\pi\xi}{l}\,d\xi; \tag{5.94}$$

$$h(x) = \sum_{n=1}^{\infty} h_n \sin\frac{n\pi x}{l}, \quad \text{where} \quad h_n = \frac{2}{l}\int_0^l h(\xi) \sin\frac{n\pi\xi}{l}\,d\xi.$$

After substituting (5.94) into (5.93), we get

$$\sum_{n=1}^{\infty}\left\{\ddot{u}_n(t) + \frac{n^2\pi^2 c^2}{l^2}u_n(t) - f_n(t)\right\}\sin\frac{n\pi x}{l} = 0,$$

where $\ddot{u} = d^2u/dt^2$. This relation is satisfied if all the coefficients of the series are zero, i.e., if

$$\ddot{u}_n(t) + \frac{n^2\pi^2c^2}{l^2}\, u_n(t) = f_n(t). \tag{5.95}$$

The solution $u_n(t)$ of this ordinary differential equation with constant coefficients is easily obtained under the initial conditions

$$u(x,0) = g(x) = \sum_{n=1}^{\infty} u_n(0) \sin \frac{n\pi x}{l} = \sum_{n=1}^{\infty} g_n \sin \frac{n\pi x}{l},$$

$$u_t(x,0) = h(x) = \sum_{n=1}^{\infty} \dot{u}_n(0) \sin \frac{n\pi x}{l} = \sum_{n=1}^{\infty} h_n \sin \frac{n\pi x}{l}.$$

Thus, $u_n(0) = g_n$, and $\dot{u}_n(0) = h_n$. Now, we define the solutions $u_n(t)$ in the form

$$u_n(t) = u_n^{(1)}(t) + u_n^{(2)}(t),$$

where $u_n^{(1)}(t)$ is a particular solution of Eq (5.95), which, using the variation of parameters method, is given by

$$u_n^{(1)}(t) = \frac{1}{n\pi c} \int_0^t \sin \frac{n\pi c(t-\tau)}{l}\, f_n(\tau)\, d\tau,$$

and represents the solution of the nonhomogeneous equation with the homogeneous initial conditions, and

$$u_n^{(2)}(t) = g_n \cos \frac{n\pi ct}{l} + \frac{lh_n}{n\pi c} \sin \frac{n\pi ct}{l}$$

is the solution of the homogeneous equation with the prescribed initial conditions. Hence,

$$u(x,t) = \sum_{n=1}^{\infty} \left[u_n^{(1)}(t) + u_n^{(2)}(t) \right]$$

$$= \sum_{n=1}^{\infty} \frac{1}{n\pi c} \int_0^t \sin \frac{n\pi c(t-\tau)}{l} \sin \frac{n\pi x}{l}\, f_n(\tau)\, d\tau \tag{5.96}$$

$$+ \sum_{n=1}^{\infty} \left(g_n \cos \frac{n\pi ct}{l} + \frac{lh_n}{n\pi c} \sin \frac{n\pi ct)}{l} \right) \sin \frac{n\pi x}{l}.$$

Note that the second term is the solution of the corresponding problem with $f = 0$ (representing a freely vibrating string with prescribed initial conditions; see Exercise 5.16). The first term represents the forced vibrations of the string under the influence of an external force. ∎

5.8. Mathematica Projects

PROJECT 5.1. We solve Example 5.1 by Mathematica with the following data: $l = 1, c = 1, f(x) = \sin x$, and $g(x) = x + 1$.

In[1]:   L=1
```
c:= 1
f[x_]:= Sin[x]
g[x_]:= x+1
A[n_]:= A[n] = 2/(n Pi L) NIntegrate[g[x] Sin[n Pi x/L],
   {x, 0, L}]//N//Chop
B[n_]:= B[n] = 2/L NIntegrate[f[x] Sin[n Pi x/L],{x,0,L}]
   //Chop
Table[{n, A[n], B[n]}, {n,1,8}]//TableForm
```

Out[7]:
```
{ 1,   0.607927,   0.596094}
{ 2,   -0.0506606,  -0.27481}
{ 3,   0.0675475,   0.180599}
{ 4,   -0.0126651,  -0.134778}
{ 5,   0.0243171,   0.107575}
{ 6,   -0.00562895,  -0.0895348}
{ 7,   0.0124067,   0.0766867}
{ 8,   -0.00316629,  -0.0670683}
```

In[8]:   u[x_,t_,n_] := (A[n] Sin[n Pi c t/L]
```
      + B[n] Cos[n Pi c t/L]) Sin[n Pi x/L];

uapprox[x_,t_] := Sum[u[x,t,k],{k,8}]
```

In[10]:   graphs =
```
   Table[Plot[uapprox[x,t],{x,0,1},
   PlotRange->{-2,2},
   Ticks->{{0,1},{-2,2}},
   DisplayFunction->Identity],
   {t,0,2,1/3}];
```

In[11]: graphsarray = Partition[graphs,2];

In[12]: Show[GraphicsArray[graphsarray],

```
DisplayFunction->$DisplayFunction]
```

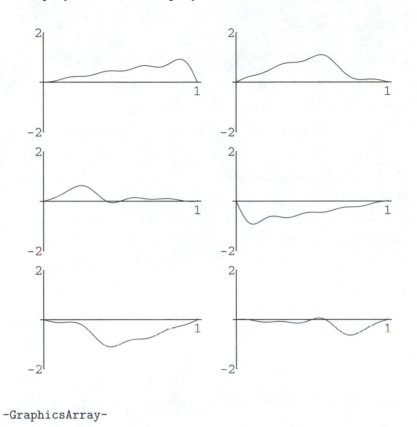

```
-GraphicsArray-
```

PROJECT 5.2. We present Example 5.2 by Mathematica with the following data:
$l = 1$, $k = 1$, and $f(x) = x^2$.

In[13]: L:= 1
    k:= 1
  f[x_]:= x^2
  C[n_]:= C[n] = 2/L NIntegrate[f[x] Sin[n Pi x/L],
    {x, 0, L}]//N//Chop;

  Table[{n, C[n]}, {n,1,8}]//TableForm

Out[17]:
  { 1,0.378607}
  { 2,-0.31831}
  { 3,0.202651}

```
{ 4,-0.159155}
{ 5,0.12526}
{ 6,-0.106103}
{ 7,0.0901935}
{ 8,-0.0795775}
```

In[18]: `u[x_,t_,n_]:= C[n] Sin[n Pi x/L] Exp[-k t (n Pi/L)^2]`

`uapprox[x_,t_]:= Sum[u[x,t,j],{j,5}]`

In[19]:   `graphs = Table[Plot[uapprox[x,t],{x,0,1},`
`PlotRange->{0,1},`
`Ticks->{{0,1},{0,1}},`
`DisplayFunction->Identity],`
`{t,0,1/3,1/24}];`

In[20]: `graphsarray = Partition[graphs,2];`

`Show[GraphicsArray[graphsarray],`
`DisplayFunction->$DisplayFunction]`

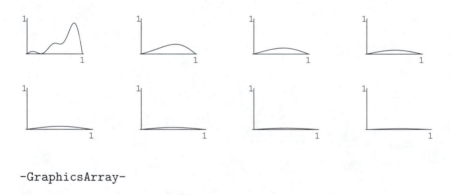

```
-GraphicsArray-
```

PROJECT 5.3. We shall now present a Mathematica session with the nonhomogeneous initial condition

$$u(x,0) = \begin{cases} x^2 & 0 < x < 1/2 \\ x+1 & 1/2 < x < 1. \end{cases}$$

In[21]:   `L:= 1`
`  k:= 1`
`  f[x_]:= x^2`
`  g[x_]:= x+1`

In[25]: `C[n_]:= C[n] = 2/L (NIntegrate[f[x] Sin[n Pi x/L],`
`{x, 0, 1/2}] +`
`NIntegrate[g[x] Sin[n Pi x/L],{x, 1/2, 1}])//N//Chop;`

`Table[{n, C[n]}, {n,1,8}]//TableForm`

Out[26]:
```
   { 1,1.14423}
   { 2,-1.06676}
   { 3,0.419635}
   { 4,-0.119366}
   { 5,0.253616}
   { 6,-0.34603}
   { 7,0.181515}
   { 8,-0.0596831}
```

In[27]: `u[x_,t_,n_] := C[n] Sin[n Pi x/L] Exp[-k t (n Pi/L)^2]`

`uapprox[x_,t_] := Sum[u[x,t,j],{j,5}]`

In[29]: `graphs =`
`Table[Plot[uapprox[x,t],{x,0,1},`
`PlotRange->{0,2.5},`
`Ticks->{{0,1},{0,1,2,2.5}},`
`DisplayFunction->Identity],`
`{t,0,1/3,1/24}];`

In[30]: `graphsarray = Partition[graphs,4];`

In[31]: `Show[GraphicsArray[graphsarray],`
`DisplayFunction->$DisplayFunction]`

```
-GraphicsArray-
```

PROJECT 5.4. We present a Mathematica session for the following more general boundary value problem. Consider $u_{xx} + u_{yy} = 0$, $0 < x < a$, $0 < y < b$, subject to the boundary conditions

$$u(x,0) = f_1(x), \quad u(x,b) = f_2(x) \quad \text{for } 0 < x < a,$$

and the initial conditions

$$u(0, y) = g_1(y), \quad u(a, y) = g_2(y) \quad \text{for } 0 < y < b.$$

The solution is $u(x, y) = u_1(x, y) + u_2(x, y)$, where

$$u_1(x, y) = \sum_{n=1}^{\infty} [A_n \cosh(\lambda_n y) + B_n \sinh(\lambda_n y)] \sin \lambda_n x,$$

$$A_n = \frac{2}{a} \int_0^a f_1(x) \sin \lambda_n x \, dx,$$

$$B_n = \frac{1}{\sinh \lambda_n b} \left[\frac{2}{a} \int_0^a f_2(x) \sin \lambda_n x \, dx - A_n \cosh \lambda_n b \right],$$

with $\lambda_n = \pm n\pi/a$, and

$$u_2(x, y) = \sum_{n=1}^{\infty} [a_n \cosh(\mu_n x) + b_n \sinh(\mu_n x)] \sin \mu_n y,$$

$$a_n = \frac{2}{b} \int_0^b g_1(y) \sin \mu_n y \, dy,$$

$$b_n = \frac{1}{\sinh \mu_n a} \left[\frac{2}{b} \int_0^b g_2(y) \sin \mu_n y \, dy - a_n \cosh \mu_n a \right],$$

with $\mu a_n = \pm n\pi/b$. We shall take $a = 1$, $b = 2$, $f_1(x) = x^2$, $f_2(x) = x + 2$, $g_1(y) = y$, and $g_2(y) = y + 1$. The Mathematica session follows.

```
In[32]:  L:= 1
   M:= 2
   l[n_]:= n Pi/L //N
   m[n_]:= n Pi/M //N

In[36]:  f1[x_]:= x^2
   f2[x_]:= x+2
   g1[y_]:= y
   g2[y_]:= y+1

In[40]:      A[n_]:=
   A[n] =2/L NIntegrate[f1[x] Sin[l[n] x], {x, 0, L}]//Chop;
   B[n_]:=
      B[n] = 1/Sinh[l[n] M]
      (2/L NIntegrate[f2[x] Sin[l[n] x], {x, 0, L}]
      -A[n] Cosh[l[n] M])//Chop;

   Table[{n, A[n],B[n]}, {n,1,8}]//ColumnForm
```

Out[43]:
```
{ 1,0.378607,-0.366722}
{ 2,-0.31831,0.318308}
{ 3,0.202651,-0.202651}
{ 4,-0.159155,0.159155}
{ 5,0.12526,-0.12526}
{ 6,-0.106103,0.106103}
{ 7,0.0901935,-0.0901935}
{ 8,-0.0795775,0.0795775}
```

In[41]: `u1[x_,y_,n_]:= (A[n] Cosh[l[n] y] + B[n] Sinh[l[n] y])`
`Sin[l[n] x]`
`u1approx[x_,y_]:= Sum[u1[x,y,n],{n,8}]`

In[43]: `threeDplot1=Plot3D[u1approx[x,y],{x,0,L},{y,0,M},`
`DisplayFunction->Identity];`

`Show[threeDplot1,DisplayFunction->$DisplayFunction]`

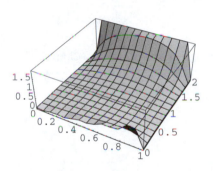 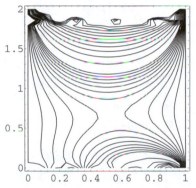

`-GraphicsArray-`

In[45]: `cvals1:= Table[i,{i,0,1/4,1/44}]`
`cvals2:= Table[i,{i,1/4,3/2,5/44}]`
`contourvals:= Union[cvals1,cvals2]`

In[48]: `u1[x_,y_,n_]:= (A[n] Cosh[l[n] y]+B[n] Sinh[l[n] y])`
`Sin[l[n] x]`

In[49]: `u1approx[x_,y_]:= Sum[u1[x,y,n],{n,8}]`

In[50]: `contourgraphs1= ContourPlot[u1approx[x,y],`
`{x,0,1},{y,0,2},`
`PlotPoints->40,`
`Contours->contourvals,`
`ContourShading -> False,`
`DisplayFunction-> Identity];`

In[51]: Show[GraphicsArray[{threeDplot1,contourgraphs1}],
    DisplayFunction -> $DisplayFunction]

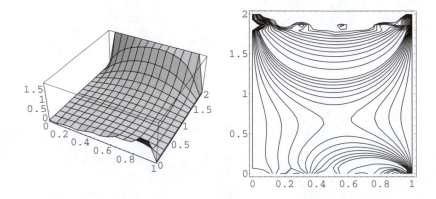

    -GraphicsArray-

In[52]: a[n_]:= a[n] =
    2/M NIntegrate[g1[y] Sin[m[n] y], {y, 0, M}]//Chop
    b[n_]:= b[n] = 1/Sinh[m[n] L]
    (2/M NIntegrate[g2[y] Sin[m[n] y], {y, 0, M}]-
    a[n] Cosh[m[n] L])//Chop

In[54]: u2[x_,y_,n_]:= (a[n] Cosh[m[n] x]+
    b[n] Sinh[m[n] x])* Sin[m[n] y]

    u2approx[x_,y_]:= Sum[u2[x,y,n],{n,1,8}]

    threeDplot2= Plot3D[u2approx[x,y],{x,0,L},{y,0,M},
    DisplayFunction->Identity];

In[56]: contourgraphs2=
    ContourPlot[u2approx[x,y],
        {x,0,1},{y,0,2},
    PlotPoints->40,
    Contours->contourvals,
    ContourShading->False,
    DisplayFunction->Identity];

In[57]: Show[GraphicsArray[{threeDplot2,contourgraphs2}],

```
DisplayFunction->$DisplayFunction]
```

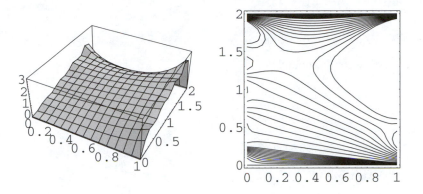

```
-GraphicsArray-
```

In[58]: `uapprox[x_,y_]:= u1approx[x,y] + u2approx[x,y];`

```
threeDplot=
Plot3D[uapprox[x,y],{x,0,L},{y,0,M},
DisplayFunction->Identity];
```

In[60]:   `contourgraphsu=`
```
ContourPlot[uapprox[x,y],{x,0,1},{y,0,2},
PlotPoints->40,Contours->contourvals,
ContourShading->False,
DisplayFunction->Identity];
```

```
Show[GraphicsArray[ {threeDplot,contourgraphsu}],
    DisplayFunction->$DisplayFunction]
```

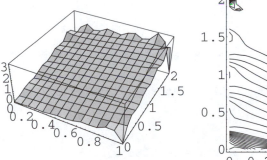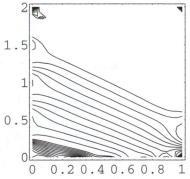

```
-GraphicsArray-
```

```
(* Superpose the SurfaceGraphics plot over the
ContourGraphics plot *)
```

In[62]: myCP=

```
ContourPlot[uapprox[x,y],
{x,0,L},{y,0,M},
Contours->Union[Table[i,{i,0,1/4,1/44}],
Table[i,{i,1/4,3/2,1/8}]],
PlotPoints->30,
AspectRatio->Automatic,
PlotRange->All,
ContourLines->True,
ContourShading->True,
ContourStyle->
(Map[{Hue[#,1,Random[]],Thickness[.006]}&,
Range[0,1,1/12]]),
ColorFunction->Hue,
Ticks->{Range[0,1],Range[0,2],{-1.5,1.5}},
DisplayFunction->Identity]

-GraphicsArray-
```

In[63]: myContourGP= First@Graphics@myCP;

```
myContourGP= N@myContourGP/.{x_AtomQ,y_AtomQ}->{x,y,-20};
```

In[65]: Show[

```
{SurfaceGraphics@myCP,Graphics3D@myContourGP},
Axes->True,
BoxRatios->{1,1,1},
```

`DisplayFunction->$DisplayFunction]`

Surface graph overlay of contour graph.

5.9. Exercises

5.1. Solve $u_{rr} + \dfrac{1}{r}u_r = u_{tt}$, subject to the conditions $u(r,0) = u_0 r$, $u_t(r,0) = 0$, and $u(a,t) = au_0$, $\lim\limits_{r \to 0} |u(r,t)| < \infty$, where u_0 is a constant.

ANS. $u = \displaystyle\sum_{i=1}^{\infty} A_i\, J_0(\alpha_i r)\, \cos \alpha_i t + u_0 a$, where α_i are the consecutive positive roots of $J_0(\alpha a) = 0$, and

$$A_i = u_0\, \frac{\int_0^a r(r-a)\, J_0(\alpha_i r)\, dr}{\int_0^a r\, J_0^2(\alpha_i r)\, dr} = u_0\, \frac{\int_0^a r(r-a)\, J_0(\alpha_i r)\, dr}{a^2\, J_1^2(\alpha_i a)}.$$

5.2. Solve $x^2 u_{xy} + 3y^2 u = 0$, such that $u(x,0) = e^{1/x}$.
ANS. $u = e^{y^3 + 1/x}$.

5.3. Solve $u_{xx} - u_t = A\, e^{-\alpha x}$, $A \ge 0, \alpha > 0$, where $u(0,t) = 0 = u(L,t)$ for $t > 0$, and $u(x,0) = f(x)$ for $0 < x < L$.

ANS. $u = v - \dfrac{A}{\alpha^2} - \dfrac{A}{\alpha^2}\left(e^{-\alpha L} - 1\right)\dfrac{x}{L} + \dfrac{A}{\alpha^2}e^{-\alpha x}$, where

$$v = \sum_{n=1}^{\infty} A_n \sin \frac{n\pi x}{L} e^{-n^2\pi^2 t/L^2},$$

$$A_n = \frac{2}{L}\int_0^L g(x) \sin \frac{n\pi x}{L}\, dx,$$

$$g(x) = f(x) + \frac{A}{\alpha^2}\left[1 + \left(e^{-\alpha L} - 1\right)\frac{x}{L} - e^{-\alpha x}\right].$$

5.4. Solve $u_t = a^2\, u_{xx} + f(x,t)$, $0 < x < l$, subject to the boundary conditions $u(0,t) = 0 = u(l,t)$ for $0 \le x \le l$, and the initial condition $u(x,0) = 0$ for $t > 0$.

ANS. $u(x,t) = \displaystyle\sum_{n=1}^{\infty} u_n(t) \sin \frac{n\pi x}{l}$, where

$$u_n(t) = \int_0^t e^{n^2\pi^2 a^2(\tau-t)/l^2}\, f_n(\tau)\, d\tau, \quad f_n(t) = \frac{2}{l}\int_0^l f(\xi,t) \sin \frac{n\pi\xi}{l}\, d\xi.$$

5.5. Solve $u_{tt} = c^2\, u_{xx}$, $0 < x < l$, subject to the boundary conditions $u(0,t) = u_1$, $u(l,t) = u_2$ for $0 \le x \le l$, where u_1, u_2 are prescribed quantities, and the initial conditions $u(x,0) = g(x)$, $u_t(x,0) = h(x)$ for $t > 0$.

ANS. $u(x,t) = U(x) + v(x,t)$, where $U(x) = u_1 + (u_2 - u_1)\dfrac{x}{l}$, describes the steady-state solution (static deflection), and $v(x,t)$ is the solution of the problem in Example 5.1.

5.6. Find the interior temperature of the cooling ball of Example 5.7, if

$$f(\rho, \phi) = \begin{cases} 1, & 0 \le \phi < \pi/2 \\ 0, & \pi/2 \le \phi < \pi. \end{cases}$$

ANS. $u(\rho, \phi, t) = \displaystyle\sum_{m,n=1}^{\infty} C_{mn} \frac{J_{n+1/2}(\alpha_{mn}\rho)P_n(\cos\phi)}{\sqrt{\alpha_{mn}\rho}} e^{-\alpha_{mn}^2 t}$,

$$\begin{aligned} C_{mn} &= \frac{(2n+1)P_n'(0)\sqrt{\alpha_{mn}}\int_0^1 \rho^{3/2} J_{n+1/2}\left(\alpha_{mn}\rho\right)d\rho}{2n(n+1)\int_0^1 \rho J_{n+1/2}^2\left(\alpha_{mn}\rho\right)d\rho} \\[2mm] &= \frac{(2n+1)P_n'(0)\sqrt{\alpha_{mn}}\int_0^1 \rho^{3/2} J_{n+1/2}\left(\alpha_{mn}\rho\right)d\rho}{n(n+1)J_{n-1/2}^2\left(\alpha_{mn}\right)}, \end{aligned}$$

where α_{mn} are the consecutive positive roots of $J_{n+1/2}(\alpha) = 0$.

5.7. Determine the steady-state temperature inside a solid hemisphere $0 \leq \rho \leq 1$, $0 \leq \phi \leq \pi/2$ (a) when the base $\phi = \pi/2$ is at $0°$ and the curved surface $\rho = 1$, $0 \leq \phi < \pi/2$, is at $1°$; and (b) when the base $\phi = \pi/2$ is insulated, but the temperature on the curved surface is $f(\phi)$.

HINT: $\dfrac{\partial u}{\partial z} = \cos\phi \, \dfrac{\partial u}{\partial \rho} - \dfrac{\sin\phi}{\rho}\dfrac{\partial u}{\partial \phi} = 0$ on the base.

Also use

$$\int_0^1 P_n(x)\,dx = \frac{1}{n(n+1)}\,P_n'(0), \text{ and}$$

$$(x^2 - 1)\,P_n'(x) = \frac{n(n+1)}{2n+1}\,\big[P_{n+1}(x) - P_{n-1}(x)\big].$$

For the Legendre polynomials, see Example 4.4 and Appendix B.

ANS. (a) $u(\rho, \phi) = \displaystyle\sum_{n=0}^{\infty} \rho^{2n+1}[P_{2n}(0) - P_{2n+2}(0)]\,P_{2n+1}(\cos\phi)$.

   (b) $u(\rho, \phi) = \displaystyle\sum_{n=0}^{\infty} c_n\rho^{2n}\,P_{2n}(\cos\phi)$, where

$$c_n = (4n+1)\int_0^{\pi/2} f(\phi)\,P_{2n}(\cos\phi)\,\sin\phi\,d\phi.$$

5.8. Solve $u_t = u_{xx}$, $-\pi < x < \pi$, subject to the conditions $u(x,0) = f(x)$, $u(-\pi, t) = u(\pi, t)$, and $u_x(-\pi, t) = u_x(\pi, t)$, where $f(x)$ is a periodic function of period 2π. This problem describes the heat flow inside a rod of length 2π, which is shaped in the form of a closed circular ring.

HINT: Assume $X(x) = A\cos wx + B\sin wx$.

ANS. $w_n = n$; $u(x,t) = \displaystyle\sum_{n=0}^{\infty} e^{-n^2 t}\,(a_n \cos nx + b_n \sin nx)$, where

$a_n = \dfrac{1}{\pi}\displaystyle\int_{-\pi}^{\pi} f(x)\cos nx\,dx$, $b_n = \dfrac{1}{\pi}\int_{-\pi}^{\pi} f(x)\sin nx\,dx$, $a_0 = \dfrac{1}{2\pi}\int_{-\pi}^{\pi} f(x)\,dx$.

5.9. Solve the problem $u_t = \nabla^2 u$, $r < 1$, $0 < z < 1$, such that $u(r, z, 0) = 1$, $u(1, z, t) = 0$, and $u(r, 0, t) = 0 = u(r, 1, t)$. This problem describes the temperature distribution inside a homogeneous isotropic solid circular cylinder. ANS.

$$u(r, z, t) = \sum_{m,n=1}^{\infty} C_{mn}\, e^{-(\lambda_m^2 + n^2\pi^2)t}\, J_0(\lambda_m r)\,\sin n\pi z,$$

where λ_m are the zeros of J_0, and

$$C_{mn} = \frac{4\,(1 - (-1)^n)}{n\pi\lambda_m\,J_1(\lambda_m)}.$$

5.10. Find the steady-state temperature in a solid circular cylinder of radius 1 and height 1 under the conditions that the flat faces are kept at $0°$ and the curved surface at $1°$.

ANS. $u(r, z) = 4 \sum\limits_{\substack{n=1 \\ n \text{ odd}}}^{\infty} \dfrac{I_0(n\pi r)}{I_0(n\pi)} \dfrac{\sin n\pi z}{mn}.$

5.11. Solve the steady-state problem of temperature distribution in a half-cylinder $0 \le r \le 1, 0 \le \theta \le \pi, 0 \le z \le 1$, where the flat faces are kept at $0°$ and the curved surface at $1°$.

ANS. $u(r, \theta, z) = \dfrac{16}{\pi^2} \sum\limits_{\substack{m,n=1 \\ m,n \text{ odd}}}^{\infty} \dfrac{I_m(n\pi r)}{I_m(n\pi)} \dfrac{\sin n\pi z}{mn} \sin m\theta.$

5.12. Solve $\dfrac{\partial u}{\partial t} = \dfrac{\partial}{\partial x}\left(x\dfrac{\partial u}{\partial x}\right)$, $0 < x < 1$, $t > 0$, subject to the conditions $u(x, 0) = f(x)$ and $u(1, t) = 0$.

HINT: Set $4x = r^2$.

ANS. $u(x, t) = \sum\limits_{n=1}^{\infty} C_n e^{-\lambda_n^2 t/4} J_0(\lambda_n \sqrt{x})$, where λ_n are the zeros of $J_0(\lambda)$, and

$$C_n = \frac{\int_0^1 f(x) J_0(\lambda_n \sqrt{x}) \, dx}{\int_0^1 [J_0(\lambda_n \sqrt{x})]^2 \, dx}.$$

5.13. Solve $u_{tt} = c^2 (u_{xx} + u_{yy})$ in the rectangle $R = \{(x, y) : 0 < x < a, 0 < y < b\}$, subject to the condition $u = 0$ on the boundary of R for $t > 0$, and the initial conditions $u(x, y, 0) = f(x, y)$, $u_t(x, y, 0) = g(x, y)$. This problem describes a vibrating rectangular membrane. Interpret the solutions $u_{11}, u_{12}, u_{21}, u_{22}, u_{13}$, and u_{31} for a square membrane $a = b = 1$.

ANS.

$$u(x, y, t) = \sum\limits_{m,n=1}^{\infty} (A_{mn} \cos \lambda_{mn} t + B_{mn} \sin \lambda_{mn}) \sin \frac{m\pi x}{a} \sin \frac{n\pi y}{b},$$

where

$$A_{mn} = \frac{4}{ab} \int_0^b \int_0^a f(x, y) \sin \frac{m\pi x}{a} \sin \frac{n\pi y}{b} \, dx \, dy,$$

$$B_{mn} = \frac{4}{ab\lambda_{mn}} \int_0^b \int_0^a g(x, y) \sin \frac{m\pi x}{a} \sin \frac{n\pi y}{b} \, dx \, dy,$$

for $m, n = 1, 2, \ldots$; the eigenvalues are

$$\lambda_{mn} = c\pi \sqrt{\frac{m^2}{a^2} + \frac{n^2}{b^2}}.$$

The solutions u_{11}, u_{12}, u_{21}, u_{22}, u_{13}, and u_{31} are represented in Fig. 5.4.

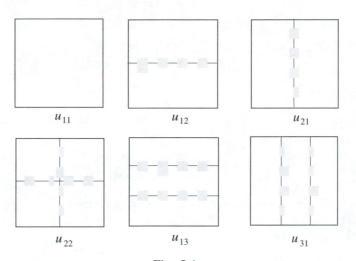

Fig. 5.4.

5.14. Solve $u_{xy} - 4xyu = 0$, such that $u(0, y) = e^{y^2}$.

ANS. $u = e^{x^2 + y^2}$.

5.15. Solve $u_{xx} + u_{yy} = 0$, under the conditions $u(0, y) = 0 = u(\pi, y)$, $u(x, 0) = \sin x$, $\lim\limits_{y \to \infty} u(x, y) < +\infty$.

ANS. $u = e^{-y} \sin x$.

5.16. Solve $u_{xx} - u_{tt} = e^{-a^2 \pi^2 t} \sin a\pi x$, subject to the conditions $u(x, 0) = 0$, $u_t(x, 0) = 0$, and $u(0, t) = u(1, t) = 0$, where a is an integer.

ANS. $\dfrac{1}{a^2 \pi^2 (1 + a^2 \pi^2)} \Big[\cos a\pi t - e^{-a^2 \pi^2 t} - a\pi \sin a\pi t \Big] \sin a\pi x$.

5.17. Solve $r^2 u_{rr} + r u_r + u_{\theta\theta} = 0$, such that $u(b, \theta) = f(\theta)$, $u(r, \theta + 2\pi) = u(r, \theta)$, and $\lim\limits_{r \to 0} u(r, \theta) < +\infty$ (circular disk problem).

HINT: Separate the variables and show that the only relevant part of the solution reduces to

$$u(r, \theta) = c_0 + \sum_\alpha r^\alpha (A(\alpha) \cos \alpha\theta + B(\alpha) \sin \alpha\theta).$$

Note that under the given conditions $u(r, \theta)$ must have a Fourier series representation in θ and, therefore, $\alpha = n$ is a positive integer.

5.18. Solve $u_{xx} + u_{yy} = 0$, under the conditions $u(x, 0) = 0 = u(x, \pi)$, $u(0, y) = 0$, and $u(\pi, y) = \cos^2 y$.

ANS.

$$u = \sum_{n=1}^{\infty} C_n \sinh nx \sin ny,$$

where

$$C_n = \frac{2}{\pi \sinh n\pi} \int_0^\pi \cos^2 y \sin n y \, dy$$

$$= \frac{1}{\pi \sinh n\pi} \left[1 - (-1)^n\right] \left[\frac{1}{n} - \frac{n}{n^2 - 4}\right], \quad n \neq 2,$$

and $C_2 = 0$.

5.19. Solve $u_{rr} + \dfrac{1}{r} u_r + \dfrac{1}{r^2} u_{\theta\theta} = 0$, subject to the conditions $u = 0$ for $\theta = 0$ or $\pi/2$, and $u_r = \sin\theta$ at $r = a$.

ANS. $u = \displaystyle\sum_{n=1}^\infty C_n r^{2n} \sin 2n\theta$, where $C_n = \dfrac{4(-1)^{n+1}}{\pi(4n^2 - 1)a^{2n}}$.

5.20. Solve the problem of transverse vibrations of a beam: $u_{tt} + a^2 u_{xxxx} = 0$, subject to the conditions $u(0,t) = u(L,t) = u_{xx}(0,t) = u_{xx}(L,t) = u_t(x,0) = 0$, and $u(x,0) = f(x)$.

SOLUTION. Let $u = X(x)T(t)$, then we have $\dfrac{X^{(4)}}{X} = -\dfrac{T''}{a^2 T} = \lambda$, where λ is a parameter. By standard arguments it can be shown that the relevant values of λ are positive. Let $\lambda = \alpha^4$. Then the solutions for X and T are given by

$$X = A \cos \alpha x + B \sin \alpha x + C \cosh \alpha x + D \sin \alpha x,$$

and

$$T = E \cos \alpha^2 t + F \sin \alpha^2 t.$$

The condition $X(0) = 0$ means $A + C = 0$, and $X(L) = 0$ yields

$$A \cos \alpha L + B \sin \alpha L + C \cosh \alpha L + D \sin \alpha L = 0,$$

and $X_{xx}(0) = 0$ implies $2\alpha^2 A = 0$, which gives $A = C = 0$, and $X_{xx}(L) = 0$, which yields $\alpha^2(B \sinh \alpha L - D \sin \alpha L) = 0$. Thus, we have a pair of two homogeneous equations:

$$B \sinh \alpha L - D \sin \alpha L = 0, \quad B \sinh \alpha L + D \sin \alpha L = 0.$$

For B and D to have nontrivial values, we must have

$$\sinh \alpha L \sin \alpha L = 0,$$

i.e., $\alpha L = n\pi$, and $B = 0$ and $T(0) = 0$ are equivalent to $F = 0$. Absorbing E in D and using the initial condition, we get

$$u = \sum_{n=1}^\infty D_n \sin \frac{n\pi x}{L} \cos \frac{a n^2 \pi^2 t}{L^2},$$

where $D_n = \dfrac{2}{L} \displaystyle\int_0^L f(x) \sin \dfrac{n\pi x}{L} dx.$

5.21. Solve $r u_{rr} + u_r + r u_{zz} = 0$, $u(a, z) = u_0$, under the conditions $u(a, 0) = u_0$, $u(r, 0) = 0 = u(r, h)$, and $\lim\limits_{r \to 0} u(r, z) < +\infty$ (steady-state temperature in a finite cylinder).

ANS.

$$u = \frac{4u_0}{\pi} \sum_{n=0}^{\infty} \frac{I_0\left((2n+1)\pi r/h\right)}{(2n+1)\, I_0\left((2n+1)\pi a/h\right)} \sin \frac{(2n+1)\pi z}{h}.$$

5.22. Solve $u_{xx} - u_{tt} = e^{-\pi^2 t} \sin \pi x$, such that $u(x, 0) = 0 = u_t(x, 0) = u(0, t) = u_x(1, t)$.

SOLUTION. A particular solution is given by

$$u_p = -\frac{e^{-\pi^2 t} \sin \pi x}{\pi^2 (1 + \pi^2)}.$$

Now, define $u = v + u_p$. Then the problem becomes

$$v_{xx} - v_{tt} = 0, \quad v(x, 0) = \frac{\sin \pi x}{\pi^2 (1 + \pi^2)}, \quad v_t(x, 0) = -\frac{\sin \pi x}{(1 + \pi^2)},$$

$$v(0, t) = 0, \quad v_x(1, t) = -\frac{e^{-\pi^2 t}}{\pi (1 + \pi^2)}.$$

Assume $v = X(x)T(t)$. Then we have

$$\frac{X''}{X} = \frac{T''}{T} = \text{const.}$$

In the case when the constant is zero, the solution will not make any contribution to v. So we consider two cases: (i) const $= \lambda^2$, and (ii) const $= -\alpha^2$. In the first case the solution is

$$v_1 = \sum_\lambda e^{-\lambda t} (A \cosh \lambda x + B \sinh \lambda x),$$

and in the second case

$$v_2 = \sum_\alpha (A_1 \cos \alpha x \cos \alpha t + A_2 \cos \alpha x \sin \alpha t$$

$$+ A_3 \sin \alpha x \cos \alpha t + A_4 \sin \alpha x \sin \alpha t).$$

Applying the boundary conditions to $v + v_1 + v_2$ we get $A = 0$, $B = \dfrac{-1}{\pi^3(1 + \pi^2)}$, $\lambda = \pi^2$, $A_1 = A_2 = 0$, $\alpha = (n + \dfrac{1}{2})\pi$, and A_3, A_4 are to be determined from the initial conditions. Thus, we have

$$u = -\frac{e^{-\pi^2 t}\sin \pi x}{\pi^2(1 + \pi^2)} - \frac{e^{-\pi^2 t}\sinh \pi^2 x}{\pi^3(1 + \pi^2)}$$
$$+ \sum_\alpha \left(A_3 \sin \alpha x \cos \alpha t + A_4 \sin \alpha x \sin \alpha t\right).$$

Now, we apply the initial conditions

$$u(x, 0) = -\frac{\sin \pi x}{\pi^2(1 + \pi^2)} - \frac{\sinh \pi^2 x}{\pi^3(1 + \pi^2)} + \sum_\alpha A_3 \sin \alpha x = 0,$$

$$\sum_\alpha A_3 \sin \alpha x = \frac{\sin \pi x}{\pi^2(1 + \pi^2)} + \frac{\sinh \pi^2 x}{\pi^3(1 + \pi^2)} = g_1(x),$$

and

$$u_t(x, 0) = \frac{\sin \pi x}{(1 + \pi^2)} + \frac{\sinh \pi^2 x}{\pi(1 + \pi^2)} + \sum_\alpha \alpha A_4 \sin \alpha x = 0,$$

$$\sum_\alpha \alpha A_4 \sin \alpha x = -\frac{\sin \pi x}{(1 + \pi^2)} - \frac{\sinh \pi^2 x}{\pi(1 + \pi^2)} = g_2(x).$$

Define

$$I_1 = \int_0^1 \sin \pi x \sin \frac{(2n + 1)\pi x}{2}\, dx = \frac{-4(-1)^n}{(2n + 3)(2n - 1)},$$
$$I_2 = \int_0^1 \sinh \pi^2 x \sin \frac{(2n + 1)\pi x}{2}\, dx = \frac{4(-1)^n \pi^2 \cosh \pi^2}{2\pi^2 + (2n + 1)^2}.$$

Then

$$A_3 = 2\left[\frac{I_1}{\pi^2(1 + \pi^2)} + \frac{I_4}{\pi^3(1 + \pi^2)}\right],$$

$$A_4 = -\frac{4}{(2n + 1)\pi}\left[\frac{I_1}{(1 + \pi^2)} + \frac{I_4}{\pi(1 + \pi^2)}\right].$$

5.23. Solve the Poisson's equation $u_{xx} + u_{yy} = -1$, $0 < x, y < 1$, subject to the Dirichlet boundary conditions $u(0, y) = 0 = u(1, y) = u(x, 0) = u(x, 1)$.
ANS.

$$u(x, y) = \frac{16}{\pi^4} \sum_{\substack{j,k=1 \\ j,k\,\text{odd}}}^{\infty} \frac{\sin j\pi x \sin k\pi y}{jk\,(j^2 + k^2)}.$$

6

Integral Transforms

The technique of integral transforms is a powerful tool for the solution of linear ordinary or partial differential equations. A function $f(x)$ may be transformed by the formula

$$F(s) = \int_a^b f(x)K(s,x)\,dx,$$

provided $F(s)$ exists, where s denotes the variable of the transform, $F(s)$ is the integral transform of $f(x)$, and $K(s,x)$ is known as the *kernel* of the transform. An integral transform is a linear transformation, which, when applied to a linear initial or boundary value problem, reduces the number of independent variables by one for each application of the integral transform. Thus, a partial differential equation can be reduced to an algebraic equation by repeated application of integral transforms. The algebraic problem is generally easy to solve for the function $F(s)$, and the solution of the problem is obtained if we can obtain the function $f(x)$ from $F(s)$ by some inversion formula. The integral transform methods are used to solve a diverse range of initial and boundary value problems, such as problems of circuits, current flow, heat flow, fluid flow, elastic deformation, and control theory.

6.1. Integral Transform Pairs

Some well-known integral transforms and their inversion formulas, known as the transform pairs, are given below.

1. The Fourier cosine transform $\tilde{f}_c(\alpha)$ of $f(x)$ is defined as

$$\mathcal{F}_c\{f(x)\} \equiv \tilde{f}_c(\alpha) = \sqrt{\frac{2}{\pi}} \int_0^\infty f(x) \cos(x\alpha) \, dx, \qquad (6.1a)$$

and its inverse is

$$\mathcal{F}^{-1}\{\tilde{f}_c(\alpha)\} \equiv f(x) = \sqrt{\frac{2}{\pi}} \int_0^\infty \tilde{f}_c(\alpha) \cos(x\alpha) \, d\alpha. \qquad (6.1b)$$

2. The Fourier sine transform $\tilde{f}_s(\alpha)$ of $f(x)$ is defined as

$$\mathcal{F}_s\{f(x)\} \equiv \tilde{f}_s(\alpha) = \sqrt{\frac{2}{\pi}} \int_0^\infty f(x) \sin(x\alpha) \, dx, \qquad (6.2a)$$

and its inverse is

$$\mathcal{F}^{-1}\{\tilde{f}_s(\alpha)\} \equiv f(x) = \sqrt{\frac{2}{\pi}} \int_0^\infty \tilde{f}_s(\alpha) \sin(x\alpha) \, d\alpha. \qquad (6.2b)$$

3. The Fourier complex transform $\mathcal{F}f(x) = \tilde{f}(\alpha)$ of $f(x)$ is defined as

$$\mathcal{F}\{f(x)\} = \tilde{f}(\alpha) = \frac{1}{\sqrt{2\pi}} \int_{-\infty}^\infty f(x) e^{ix\alpha} \, dx, \qquad (6.3a)$$

and its inverse is

$$\mathcal{F}^{-1}\{\tilde{f}(\alpha)\} \equiv f(x) = \frac{1}{\sqrt{2\pi}} \int_{-\infty}^\infty \tilde{f}(\alpha) e^{-ix\alpha} \, d\alpha. \qquad (6.3b)$$

4. The Laplace transform is defined as

$$\mathcal{L}\{f(t)\} \equiv F(s) = \bar{f}(s) = \int_0^\infty f(t) e^{-st} \, dt, \qquad (6.4a)$$

and its inverse is

$$\mathcal{L}^{-1}\{F(s)\} \equiv f(t) = \frac{1}{2\pi i} \int_{c-i\infty}^{c+i\infty} F(s) e^{st} \, ds. \qquad (6.4b)$$

5. The Mellin transform is defined as

$$\mathcal{M}\{f(x)\} \equiv F_M(s) = \int_0^\infty f(x) \, x^{s-1} \, dx, \qquad (6.5a)$$

and its inverse is

$$\mathcal{M}^{-1}\{F_M(s)\} \equiv f(x) = \frac{1}{2\pi i} \int_{c-i\infty}^{c+i\infty} F_M(s) x^{-s} \, ds. \qquad (6.5b)$$

6. The Hankel transform of order n is defined as

$$\mathcal{H}\{f(x)\} \equiv F_n(s) = \int_0^\infty x f(x) J_n(sx)\, dx, \qquad (6.6a)$$

where its inverse is

$$\mathcal{H}^{-1}\{F_n(s)\} \equiv f(x) = \int_0^\infty s F_n(s) J_n(sx)\, ds. \qquad (6.6b)$$

These definitions are not unique, particularly in the case of Fourier and Hankel transforms, which are sometimes defined in a different manner. In fact, we can develop an integral transform corresponding to any orthogonal set of functions. However, the six transforms defined above are most frequently used. Some of the other better-known transforms are Meijer, Kontorowich-Lebedev, Mehler-Foch, Hilbert, and Laguerre. We will discuss only the Laplace and Fourier transforms. Once the use of one transform is understood, it is a simple matter to extend this understanding to another transform.

In most cases, sufficient conditions for the existence of an integral transform of a function are known as *Dirichlet's conditions*, which are defined on an interval (a,b) as follows: (i) the function has only a finite number of extremum points in the interval (a,b), and (ii) the function has only a finite number of finite discontinuities in the interval (a,b). Unless otherwise stated, it will be assumed that all functions in the sequel satisfy these conditions. It is also important to maintain the distinction between these Dirichlet's conditions and the Dirichlet boundary conditions that are defined in §1.2.

The student who lacks the knowledge of contour integration technique may omit in this chapter all material where this technique is used.

I. Laplace Transforms

We will first discuss the Laplace transforms and their applications in detail.

6.2. Notation

It is expected that the reader is familiar with the elementary theory of the Laplace transforms. The following notation is used:

$$\mathcal{L}\{f(t)\} = F(s) = \bar{f}(s) = \int_0^\infty f(t) e^{-st}\, dt,$$

and

$$\mathcal{L}^{-1}\{F(s)\} = \mathcal{L}^{-1}\{\bar{f}(s)\} = f(t),$$

where s is the variable of the transform, which is, in general, a complex variable. Note that the Laplace transform $F(s)$ exists for $s > \alpha$, if the function $f(t)$ is piecewise continuous in every finite closed interval $0 \le t \le b\,(b > 0)$, and $f(t)$ is of exponential order α, i.e., there exist α, M, and $t_0 > 0$ such that $e^{-\alpha t}|f(t)| < M$ for $t > t_0$.

We state some basic properties of the Laplace transforms.

(i) $\mathcal{L}\left\{e^{at}f(t)\right\} = F(s-a)$,
 and $\mathcal{L}^{-1}\left\{F(s-a)\right\} = e^{at}f(t)$.

(ii) $\mathcal{L}\left\{H(t-a)f(t-a)\right\} = e^{-as}F(s)$,
 and $\mathcal{L}^{-1}\left\{e^{-as}F(s)\right\} = H(t-a)f(t-a)$.

(iii) Convolution theorem:

$$\mathcal{L}^{-1}\left\{G(s)F(s)\right\} = \int_0^t f(t-u)g(u)du = \int_0^t f(u)g(t-u)du.$$

(iv) $\mathcal{L}\left\{\dfrac{d^n f(t)}{dt^n}\right\} = s^n F(s) - s^{n-1}f(0) - s^{n-2}f'(0) - \cdots - sf^{(n-2)}(0) - f^{(n-1)}(0)$,

 and $\mathcal{L}^{-1}\left\{s^n F(s)\right\} = \dfrac{d^n}{dt^n}f(t)$.

(v) $\mathcal{L}^{-1}\left\{\dfrac{1}{s}F(s)\right\} = \displaystyle\int_0^t f(u)du.$

(vi) $\mathcal{L}\left\{t^n f(t)\right\} = (-1)^n \dfrac{d^n F}{ds^n}$,

 and $\mathcal{L}^{-1}\left\{(-1)^n \dfrac{d^n F}{ds^n}\right\} = t^n f(t)$.

(vii) If $\mathcal{L}\left\{f(x,t)\right\} = F(x,s)$, then

$$\mathcal{L}\left\{\frac{\partial f(x,t)}{\partial x}\right\} = \frac{\partial F(x,s)}{\partial x}, \quad \text{and} \quad \mathcal{L}^{-1}\left\{\frac{\partial F(x,s)}{\partial x}\right\} = \frac{\partial f(x,t)}{\partial x}.$$

The last two results are based on the Leibniz rule and are extremely effective. The Leibniz rule states that if $g(x,t)$ is an integrable function of t for each value of x, and the partial derivative $\dfrac{\partial g(x,t)}{\partial x}$ exists and is continuous in the region under consideration, and if

$$f(x) = \int_a^b g(x,t)dt,$$

then

$$f'(x) = \int_a^b \frac{\partial g(x,t)}{\partial x}dt. \tag{6.7}$$

6.3. Basic Laplace Transforms

A table of basic Laplace transform pairs is given in Appendix C. We will show the effectiveness of the above properties in the derivation of certain Laplace transforms.

We start with the easily established result that

$$\mathcal{L}\left\{e^{at}\right\} = \frac{1}{s-a}. \tag{6.8}$$

Differentiating both sides with respect to a, we get

$$\mathcal{L}\left\{te^{at}\right\} = \frac{1}{(s-a)^2}, \tag{6.9}$$

and repeating this differentiation n times, we find that

$$\mathcal{L}\left\{t^n e^{at}\right\} = \frac{n!}{(s-a)^{n+1}}. \tag{6.10}$$

After replacing a by ib, choosing an appropriate n, and comparing the real and imaginary parts on both sides, we get the Laplace transforms of functions $t^n \cos bt$ and $t^n \sin bt$, and then combining with Property (i), we obtain the Laplace transforms of functions $t^n e^{at} \cos bt$ and $t^n e^{at} \sin bt$. For example, if we choose $n = 2$, then we have

$$\mathcal{L}\left\{t^2 e^{at}\right\} = \frac{2!}{(s-a)^3}. \tag{6.11}$$

Now letting $a = ib$, we get

$$\mathcal{L}\left\{t^2 e^{ibt}\right\} = \frac{2!}{(s-ib)^3}, \tag{6.12}$$

which yields

$$\mathcal{L}\left\{t^2(\cos bt + i\sin bt)\right\} = \frac{2(s+ib)^3}{(s^2+b^2)^3}. \tag{6.13}$$

Expanding the numerator on the right side of (6.13), we get

$$\mathcal{L}\left\{t^2(\cos bt + i\sin bt)\right\} = \frac{2(s^3 + 3is^2 b - 3sb^2 - ib^3)}{(s^2+b^2)^3}. \tag{6.14}$$

Then, equating the real and imaginary parts in (6.14), we obtain

$$\mathcal{L}\left\{t^2 \cos bt\right\} = \frac{2(s^3 - 3sb^2)}{(s^2+b^2)^3}, \tag{6.15}$$

and

$$\mathcal{L}\left\{t^2 \sin bt\right\} = \frac{2(3s^2 b - b^3)}{(s^2 + b^2)^3}. \tag{6.16}$$

The Laplace transforms of $\mathcal{L}\left\{e^{at} t^2 \cos bt\right\}$ and $\mathcal{L}e^{at}\left\{t^2 \sin bt\right\}$ can now be easily obtained.

An important Laplace inverse is

$$\mathcal{L}^{-1}\left\{\frac{e^{-a\sqrt{s}}}{s}\right\} = \operatorname{erfc}\frac{a}{2\sqrt{t}}, \tag{6.17}$$

where

$$\operatorname{erf}(x) = \frac{2}{\sqrt{\pi}} \int_0^x e^{-u^2}\,du, \quad \text{and} \quad \operatorname{erfc}(x) = 1 - \operatorname{erf}(x) = \frac{2}{\sqrt{\pi}} \int_x^\infty e^{-u^2}\,du. \tag{6.18}$$

We can derive several Laplace inverses by using Properties (i)–(vii).

EXAMPLE 6.1. $\mathcal{L}^{-1}\left\{\dfrac{e^{-a\sqrt{s}}}{\sqrt{s}}\right\}$ is obtained by differentiating formula (6.17) with respect to a. Thus,

$$\mathcal{L}^{-1}\left\{\frac{e^{-a\sqrt{s}}}{\sqrt{s}}\right\} = \frac{1}{\sqrt{\pi t}} e^{-a^2/(4t)}$$

is obtained after differentiating (6.17) with respect to a and canceling out the negative sign on both sides. Although the usual method of deriving the Laplace inverse of $\dfrac{e^{-a\sqrt{s}}}{\sqrt{s}}$ is by contour integration, or by using the Laplace inverse of $\dfrac{e^{-a\sqrt{s}}}{s}$, an interesting method is as follows (Churchill, 1972): Define $\dfrac{e^{-a\sqrt{s}}}{\sqrt{s}} = y$ and $e^{-a\sqrt{s}} = z$. Then

$$y' = \frac{dy}{ds} = -\frac{1}{2s^{3/2}} e^{-a\sqrt{s}} - \frac{a}{2s} e^{-a\sqrt{s}},$$

which yields

$$2sy' + y + az = 0.$$

Similarly, $z' = -\dfrac{a}{2\sqrt{s}} e^{-a\sqrt{s}}$ yields

$$2z' + ay = 0.$$

Taking the inverse transform of these equations, we get

$$aG - F - 2tF' = 0, \quad \text{and} \quad aF - 2tG = 0,$$

where $\mathcal{L}^{-1}\{y\} = F(t)$ and $\mathcal{L}^{-1}\{z\} = G(t)$. From these two equations in F and G, we get

$$F' = \frac{1}{2t}\left(\frac{a^2 F}{2t} - F\right). \tag{6.19}$$

The solution of (6.19) is

$$F = \frac{A}{\sqrt{t}}\, e^{-a^2/(4t)},$$

which gives

$$G = \frac{aA}{2\sqrt{t^3}}\, e^{-a^2/(4t)}.$$

Note that if $a = 0$, then $y = \dfrac{1}{\sqrt{s}}$, and $F(t) = \dfrac{1}{\sqrt{\pi t}}$ implies that $A = \dfrac{1}{\sqrt{\pi}}$. Hence,

$$F(t) = \frac{1}{\sqrt{\pi t}}\, e^{-a^2/(4t)}, \quad G = \frac{a}{\sqrt{\pi t^3}}\, e^{-a^2/(4t)}. \tag{6.20}$$

Then we integrate $\mathcal{L}^{-1}\left\{\dfrac{e^{-a\sqrt{s}}}{\sqrt{s}}\right\} = \dfrac{1}{\sqrt{\pi t}} e^{-a^2/(4t)}$ with respect to a from 0 to a and

obtain $\mathcal{L}^{-1}\left\{\dfrac{e^{-a\sqrt{s}}}{s}\right\}$. In this problem we have assumed that $\mathcal{L}\left\{\dfrac{1}{\sqrt{t}}\right\} = \sqrt{\dfrac{\pi}{s}}$ (see Exercise 6.12). ∎

EXAMPLE 6.2. $\mathcal{L}^{-1}\left\{e^{-a\sqrt{s}}\right\} = \dfrac{a}{2\sqrt{\pi t^3}} e^{-a^2/(4t)}$ is obtained by differentiat-

ing with respect to a the formula for $\mathcal{L}^{-1}\left\{\dfrac{e^{-a\sqrt{s}}}{\sqrt{s}}\right\}$ in the previous example and

canceling out the negative sign. ∎

EXAMPLE 6.3. If we integrate the formula $\mathcal{L}^{-1}\left\{\dfrac{e^{-a\sqrt{s}}}{s}\right\} = \operatorname{erfc}\dfrac{a}{2\sqrt{t}}$ with

respect to a from 0 to a, we get

$$\int_0^a \mathcal{L}^{-1}\left\{\frac{e^{-x\sqrt{s}}}{s}\right\} dx = \int_0^a \operatorname{erfc}\frac{x}{2\sqrt{t}}\,dx.$$

Now, after changing the order of integration and the Laplace inversion and carrying out the integration on the left side, we get

$$\int_0^a \mathcal{L}^{-1}\left\{\frac{e^{-x\sqrt{s}}}{s}\right\} dx = \mathcal{L}^{-1}(s^{-3/2} - s^{-3/2}e^{-a\sqrt{s}}), \tag{6.21}$$

and the right side yields

$$\int_0^a \mathrm{erfc}\, \frac{x}{2\sqrt{t}}\, dx = \left[x\, \mathrm{erfc}\, \frac{x}{2\sqrt{t}} \right]_0^a + \frac{1}{\sqrt{\pi t}} \int_0^a x\, e^{-x^2/(4t)}\, dx$$

$$= a\, \mathrm{erfc}\, \frac{a}{2\sqrt{t}} - 2\sqrt{\frac{t}{\pi}} e^{-a^2/(4t)} + 2\sqrt{\frac{t}{\pi}}.$$

Since $\mathcal{L}^{-1}\left\{ s^{-3/2} \right\} = 2\sqrt{\dfrac{t}{\pi}}$, we get

$$\mathcal{L}^{-1}\left\{ s^{-3/2} e^{-a\sqrt{s}} \right\} = 2\sqrt{\frac{t}{\pi}} e^{-a^2/(4t)} - a\, \mathrm{erfc}\, \frac{a}{2\sqrt{t}}. \quad \blacksquare \qquad (6.22)$$

EXAMPLE 6.4. Evaluate $\mathcal{L}^{-1}\left\{ \dfrac{e^{-a\sqrt{s+c}}}{s} \right\}$. We know from (6.20) that

$$\mathcal{L}^{-1}\left\{ e^{-a\sqrt{s}} \right\} = \frac{a}{2\sqrt{\pi t^3}} e^{-a^2/(4t)}. \qquad (6.23)$$

Hence, using Property (i),

$$\mathcal{L}^{-1}\left\{ e^{-a\sqrt{s+c}} \right\} = \frac{a}{2\sqrt{\pi t^3}} e^{-ct - a^2/(4t)}. \qquad (6.24)$$

Using the convolution theorem (Property (iii)) with $F(s) = \dfrac{1}{s}$ and $G(s) = e^{-a\sqrt{s+c}}$, we get

$$\mathcal{L}^{-1}\left\{ \frac{e^{-a\sqrt{s+c}}}{s} \right\} = \int_0^t \frac{a}{2\sqrt{\pi u^3}} e^{-cu - a^2/(4u)}\, du. \qquad (6.25)$$

Note that

$$\frac{a}{2\sqrt{u^3}} = \frac{a}{4\sqrt{u^3}} + \frac{1}{2}\sqrt{\frac{c}{u}} + \frac{a}{4\sqrt{u^3}} - \frac{1}{2}\sqrt{\frac{c}{u}},$$

and

$$cu + \frac{a^2}{4u} = \left(\sqrt{cu} + \frac{a}{2\sqrt{u}} \right)^2 - a\sqrt{c} = \left(\sqrt{cu} - \frac{a}{2\sqrt{u}} \right)^2 + a\sqrt{c}.$$

Define $x = \dfrac{a}{2\sqrt{u}} + \sqrt{cu}$ and $y = \dfrac{a}{2\sqrt{u}} - \sqrt{cu}$, and use the notation

$$\frac{a}{2\sqrt{t}} + \sqrt{ct} = x_1, \quad \text{and} \quad \frac{a}{2\sqrt{t}} - \sqrt{ct} = y_1.$$

Then the integral on the right side of (6.25) becomes

$$\frac{1}{\sqrt{\pi}} \left[e^{a\sqrt{c}} \int_{x_1}^{\infty} e^{-x^2} \, dx + e^{-a\sqrt{c}} \int_{y_1}^{\infty} e^{-y^2} \, dy \right]$$

$$= \frac{1}{2} \left[e^{a\sqrt{c}} \, \mathrm{erfc} \left(\frac{a}{2\sqrt{t}} + \sqrt{ct} \right) + e^{-a\sqrt{c}} \, \mathrm{erfc} \left(\frac{a}{2\sqrt{t}} - \sqrt{ct} \right) \right].$$

Hence,

$$\mathcal{L}^{-1} \left\{ \frac{e^{-a\sqrt{s+c}}}{s} \right\} = \frac{1}{2} \left[e^{a\sqrt{c}} \, \mathrm{erfc} \left(\frac{a}{2\sqrt{t}} + \sqrt{ct} \right) \right.$$

$$\left. + e^{-a\sqrt{c}} \, \mathrm{erfc} \left(\frac{a}{2\sqrt{t}} - \sqrt{ct} \right) \right]. \ \blacksquare \tag{6.26}$$

We state a very useful theorem without proof.

Theorem 6.1. *If* $G(s) = \sum_{1}^{\infty} G_k(s)$ *is uniformly convergent, then*

$$\mathcal{L}^{-1} G(s) = g(t) = \sum_{k=1}^{\infty} g_k(t), \tag{6.27}$$

where $\mathcal{L}^{-1} G_k(s) = g_k(t)$.

EXAMPLE 6.5. Since

$$\mathcal{L}^{-1} \left\{ s^{-3/2} e^{-1/s} \right\}$$

$$= \mathcal{L}^{-1} \left\{ \frac{1}{s^{3/2}} \left[1 - \frac{1}{s} + \frac{1}{2! s^2} - \frac{1}{3! s^3} + \cdots + (-1)^n \frac{1}{n! s^n} + \cdots \right] \right\} \tag{6.28}$$

$$= \mathcal{L}^{-1} \sum_{0}^{\infty} \frac{(-1)^n}{n! \, s^{n+3/2}} = \frac{1}{\sqrt{\pi}} \sum_{0}^{\infty} \frac{(-1)^n (2\sqrt{t})^{2n+1}}{(2n+1)!} = \frac{1}{\sqrt{\pi}} \sin(2\sqrt{t}),$$

we find that this result and second part of Property (iv) give

$$\mathcal{L}^{-1} \left\{ s^{-1/2} e^{-1/s} \right\} = \frac{1}{\sqrt{\pi t}} \cos(2\sqrt{t}). \ \blacksquare \tag{6.29}$$

EXAMPLE 6.6. Consider a semi-infinite medium bounded by $0 \leq x \leq \infty$, $-\infty < y, z < \infty$, which has an initial zero temperature, while its face $x = 0$ is maintained at a time-dependent temperature $f(t)$. The problem is to find the temperature for $t > 0$. By applying the Laplace transform to the heat conduction equation $k T_{xx} = T_t$, we get $\overline{T}_{xx} = \frac{s}{k} \overline{T}$, where $\overline{T} = \mathcal{L}\{T\}$. The solution of this equation is

$$\overline{T} = A e^{mx} + B e^{-mx}, \tag{6.30}$$

where $m = \sqrt{s/k}$.

Since $\overline{T}$ remains bounded as $x \rightarrow \infty$, we find that $A = 0$. The boundary condition at $x = 0$ in the transform domain yields $B = \bar{f}(s)$, where $\bar{f}(s)$ is the Laplace transform of $f(t)$. Thus, the solution in the transform domain is

$$\overline{T} = \bar{f}(s)\,e^{-mx}.$$

To carry out the inversion, we use the convolution property and Example 6.2 and get

$$T = \int_0^t \frac{x\,e^{-x^2/(4k\tau)}}{2\tau\sqrt{\pi k\tau}}\,f(t-\tau)\,d\tau.$$

If $\bar{f}(s) = 1$, then the solution for T reduces to

$$T = \frac{x\,e^{-x^2/(4kt)}}{2t\sqrt{\pi k t}}.$$

This solution is the fundamental solution for the heat conduction equation for the half-space. In the special case when $f(t) = T_0$, the solution is given by

$$T = T_0\,\operatorname{erfc}\left(\frac{x}{2\sqrt{kt}}\right). \blacksquare$$

In the above example, we have assumed a function whose Laplace transform is 1. The question arises: Is there such a function? We will try to answer this question in a heuristic manner. Consider the Heaviside unit step function $H(t)$ which is defined by

$$H(t) = \begin{cases} 0 & \text{for } t < 0, \\ 1 & \text{for } t \geq 0. \end{cases}$$

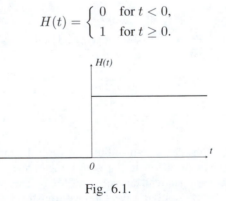

Fig. 6.1.

The Laplace transform of $H(t)$ is $\overline{H}(s) = \dfrac{1}{s}$. Then, by Property (iv) of the Laplace transforms, $\mathcal{L}H'(t) = s\overline{H}(s) = 1$. Let us examine $H'(t)$ closely. Obviously, it vanishes for $|t| > 0$ and does not exist for $t = 0$. From the graph of $H(t)$, it is clear that there is a vertical jump at $t = 0$ (see Fig. 6.1). Therefore, it is reasonable to assume that $\lim_{t \to 0} H'(t) \to \infty$. But since $\int_{-\varepsilon}^{\varepsilon} H'(t)\, dt = 1$, it is obvious that a function like $H'(t)$ does not exist in the classical sense. Such a function is called a *generalized function* or a *distribution* (see §7.1). The function $H'(t)$ is generally denoted by $\delta(t)$ and is known as the Dirac delta function. This function is defined such that

$$\delta(t) = \begin{cases} 0 & \text{for } |t| > 0, \\ \infty & \text{for } t = 0, \end{cases} \quad \text{and} \quad \int_{-\varepsilon}^{\varepsilon} \delta(t)\, dt = 1. \tag{6.31}$$

To make this function acceptable in the classical sense, we modify its definition as follows:

$$\delta(t) = \begin{cases} 0 & \text{for } |t| > \varepsilon, \\ \dfrac{1}{2\varepsilon} & \text{for } |t| < \varepsilon \end{cases} \quad \text{as } \varepsilon \to 0.$$

This definition is consistent with the classical definition of a function and automatically satisfies the second definition in (6.31). An important consequence of Eq (6.31) is that if $f(t)$ is any continuous function, then

$$\int_{-\infty}^{\infty} \delta(t) f(t)\, dt = \lim_{\varepsilon \to 0} \int_{-\varepsilon}^{\varepsilon} \delta(t) f(t)\, dt = f(0).$$

To prove this assertion, we note that by definition

$$\int_{-\infty}^{\infty} \delta(t) f(t)\, dt = \lim_{\varepsilon \to 0} \int_{-\varepsilon}^{\varepsilon} \frac{1}{2\varepsilon} f(t)\, dt = \lim_{\varepsilon \to 0} \frac{1}{2\varepsilon} 2\varepsilon f(t') = f(0),$$

where t' is a point at which $f(t)$ takes its average value in $(-\varepsilon, \varepsilon)$, such that $t' \in (-\varepsilon, \varepsilon)$, and therefore, $t' \to 0$ as $\varepsilon \to 0$. For more details on the Dirac delta function, see §7.1.1.

EXAMPLE 6.7. Consider an infinite slab bounded by $0 \le x \le l$, $-\infty < y, z < \infty$, with initial zero temperature. The face $x = 0$ is maintained at a constant temperature T_0, and the face $x = l$ is maintained at zero temperature. The problem is to find the temperature inside the slab for $t > 0$. Proceeding as in the above example, the solution in the transform domain is given by Equation (6.30). Applying the boundary conditions in the transform domain, we get

$$A + B = \frac{T_0}{s}, \quad \text{and} \quad A e^{ml} + B e^{-ml} = 0.$$

These two equations yield

$$B = \frac{T_0\, e^{ml}}{2s \sinh ml}, \quad \text{and} \quad A = \frac{T_0}{s} - B.$$

Substituting these values into $\overline{T}$, given by (6.30), and simplifying, we find that

$$\overline{T} = \frac{T_0 \sinh m(l - x)}{s \sinh ml}.$$

Rewriting this solution as

$$\overline{T} = \frac{T_0}{s} e^{-ml} \left(e^{m(l-x)} - e^{-m(l-x)} \right) \left(1 - e^{-2ml} \right)^{-1},$$

and expanding the last factor by the binomial theorem, we get

$$\overline{T} = \frac{T_0}{s} e^{-ml} \left(e^{m(l-x)} - e^{-m(l-x)} \right) \sum_{0}^{\infty} e^{-2nml}$$

$$= \frac{T_0}{s} \sum_{0}^{\infty} \left(e^{-m(2nl+x)} - e^{-m[(2n+2)l-x]} \right),$$

which, on inversion, yields

$$T = T_0 \sum_{0}^{\infty} \left[\operatorname{erf}\left(\frac{2(n+1)l - x}{2\sqrt{kt}} \right) - \operatorname{erf}\left(\frac{2nl + x}{2\sqrt{kt}} \right) \right].$$

Alternatively, we can use the Cauchy residue theorem and obtain the solution in terms of the Fourier series. Thus,

$$T = \sum \text{residues of } \frac{T_0 e^{st} \sinh m(l - x)}{s \sinh ml}$$

$$= T_0 \left[1 - \frac{x}{l} - \sum_{1}^{\infty} \frac{2}{n\pi} e^{-n^2 \pi^2 kt/l^2} \sin\left(n\pi x/l \right) \right]. \blacksquare \tag{6.32}$$

EXAMPLE 6.8. Consider a solid sphere of radius a. Suppose that its initial temperature is zero and its surface is maintained at a temperature T_0 for $t \geq 0$. The problem is to determine the temperature inside the sphere at any subsequent time. In this case the heat conduction equation in the spherical coordinates is

$$T_{\rho\rho} + \frac{2}{\rho} T_\rho = \frac{1}{k} T_t. \tag{6.33}$$

If we introduce a new independent variable u, such that $u = \rho T$, then the heat conduction equation reduces to

$$u_{\rho\rho} = \frac{1}{k} u_t, \tag{6.34}$$

which can be solved as in Example 6.6. $\blacksquare$

EXAMPLE 6.9. Solve the wave equation

$$u_{tt} = c^2 u_{xx}, \tag{6.35}$$

subject to the initial conditions

$$u = u_t = 0 \quad \text{for } t \leq 0,$$

and the boundary conditions

$$u = 0 \quad \text{at } x = 0, \quad \text{and} \quad u_x = T \quad \text{at } x = l.$$

If we apply the Laplace transform to the wave equation (6.35), we get

$$\bar{u}_{xx} = c^{-2} s^2 \bar{u}.$$

Its solution is

$$\bar{u} = A\, e^{-sx/c} + B\, e^{sx/c}.$$

Applying the boundary conditions in the transform domain, we get

$$A + B = 0, \quad \text{and} \quad -A\, e^{-sl/c} + B\, e^{sl/c} = \frac{cT}{s^2}.$$

Solving for A and B and substituting their values in the solution for $\bar{u}$, we get

$$\bar{u} = \frac{cT}{s^2} \frac{\sinh(sx/c)}{\cosh(sl/c)}. \tag{6.36}$$

This equation can be expressed as

$$\bar{u} = \frac{cT}{s^2} \left(\frac{e^{sy} - e^{-sy}}{e^{sL} + e^{-sL}} \right),$$

where $y = x/c$, and $L = l/c$. After some simplifications similar to those in Example 6.7, it becomes

$$\bar{u} = \frac{T}{s^2} \sum_0^\infty (-1)^n \left(e^{-s[(2n+1)L-y]} - e^{-s[(2n+1)L+y]} \right),$$

which, after inversion, gives

$$u = T \sum_0^\infty (-1)^n \big[(t - (2n+1)L + y) H(t - (2n+1)L + y) -$$

$$(t - (2n+1)L - y) H(t - (2n+1)L - y) \big]. \tag{6.37}$$

Alternatively,

$$u = \sum \text{residues of } \{\bar{u}\,e^{st}\}$$

$$= T\left[x - \frac{8l}{\pi^2}\sum_0^\infty \frac{(-1)^n}{(2n+1)^2}\sin\frac{(2n+1)\pi x}{2l}\cos\frac{(2n+1)\pi ct}{2l}\right]. \quad\blacksquare \qquad (6.38)$$

EXAMPLE 6.10. The problem of the propagation of sound waves produced by the motion of a sphere of radius a in an infinite expanse of fluid is governed by

$$\rho^2\frac{\partial^2 u}{\partial t^2} = c^2\frac{\partial}{\partial\rho}\left(\rho^2\frac{\partial u}{\partial\rho}\right), \qquad (6.39)$$

subject to the initial conditions

$$u(\rho,0) = u_t(\rho,0) = 0, \qquad (6.40)$$

and the boundary conditions

$$\left.\frac{\partial u(\rho,t)}{\partial\rho}\right|_{\rho=a} = f(t), \quad\text{and}\quad u \to 0 \quad\text{as } \rho \to \infty. \qquad (6.41)$$

First, we introduce a new independent variable $v = \rho u$. This substitution reduces the partial differential equation to the standard wave equation

$$v_{tt} = c^2\,v_{\rho\rho}. \qquad (6.42)$$

Applying the Laplace transform and using the second boundary condition, the solution in the transform domain is given by

$$\bar{v} = A\,e^{-s\rho/c},$$

or

$$\bar{u} = \frac{A}{\rho}\,e^{-s\rho/c}.$$

Applying the first boundary condition in the transform domain, we get

$$\bar{u} = -\frac{ac}{s+\dfrac{c}{a}}\frac{F(s)}{\rho}\,e^{-s(\rho-a)/c}.$$

By the convolution theorem, we have

$$\mathcal{L}^{-1}\left\{\frac{F(s)}{s+k}\right\} = \int_0^t e^{-k(t-x)}f(x)\,dx \equiv \phi(t), \quad k = \frac{c}{a}. \qquad (6.43)$$

Thus, the solution is

$$u = \mathcal{L}^{-1} \left\{ -\frac{ac}{s+c/a} \frac{F(s)}{\rho} e^{-s(\rho-a)/c} \right\}, \tag{6.44}$$

which, by Property (ii) and Eq (6.43), yields

$$u = -\frac{ca}{\rho} H\left(t - \frac{\rho-a}{c}\right) \phi\left(t - \frac{\rho-a}{c}\right). \tag{6.45}$$

If $f(t) = \delta(t)$, then $\phi(t) = e^{-ct/a}$, and the solution becomes

$$u = -\frac{ac}{\rho} H\left(t - \frac{\rho-a}{c}\right) \exp\left\{ -\frac{c}{a}\left(t - \frac{\rho-a}{c}\right) \right\}. \tag{6.46}$$

We can derive solutions for other values of $f(t)$ by evaluating the appropriate function ϕ. ∎

6.4. Inversion Theorem

We will establish the inversion theorem:

Theorem 6.2. *If $F(s)$ is the Laplace transform of $f(t)$, then*

$$f(t) = \frac{1}{2\pi i} \int_{c-i\infty}^{c+i\infty} F(s) e^{st} \, ds, \tag{6.47}$$

where $F(s)$ is of order $O\left(s^{-k}\right)$.

To prove this theorem, we first state and prove a lemma.

Lemma 6.1. *If $f(z)$ is analytic and of order $O(z^{-k})$ in the half-plane $\Re\{z\} > \gamma$, where γ and $k > 0$ are real constants, then*

$$f(z_0) = \frac{1}{2\pi i} \lim_{\beta \to \infty} \int_{\gamma-i\beta}^{\gamma+i\beta} \frac{f(z)}{z_0 - z} \, dz, \quad \Re\{z_0\} > \gamma. \tag{6.48}$$

PROOF. Consider the rectangle in Fig. 6.2. Choose $\beta > |\gamma|$ and such that z_0 lies inside this rectangle. By the Cauchy integral formula, we have

$$\int_\Gamma \frac{f(z)}{z - z_0} \, dz = 2\pi i f(z_0), \tag{6.49}$$

where Γ is the contour ABCDA. Let S denote the contour ABCD, then

$$\int_\Gamma \frac{f(z)}{z - z_0}\, dz = \int_{DA} \frac{f(z)}{z - z_0}\, dz + \int_S \frac{f(z)}{z - z_0}\, dz.$$

Since

$$\int_{DA} \frac{f(z)}{z - z_0}\, dz = -\int_{AD} \frac{f(z)}{z - z_0}\, dz,$$

we get from (6.49)

$$-\int_{\gamma - i\beta}^{\gamma + i\beta} \frac{f(z)}{z - z_0}\, dz + \int_S \frac{f(z)}{z - z_0}\, dz = 2\pi i f(z_0). \qquad (6.50)$$

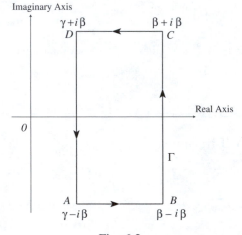

Fig. 6.2.

Now, consider $\displaystyle\int_S \frac{f(z)}{z - z_0}\, dz$ as $\beta \to \infty$. Obviously, $\beta \to \infty$ implies that $|z| \to \infty$ on S. Thus, $|z| \geq \beta$ for points on S. If we take β large enough so that $\beta > 2|z_0|$, then $|z_0| < \frac{1}{2}\beta \leq \frac{1}{2}|z|$, or $\left|\frac{z_0}{z}\right| < \frac{1}{2}$ implies that $\left|1 - \frac{z_0}{z}\right| \geq 1 - \left|\frac{z_0}{z}\right| > \frac{1}{2}$. Noting that $|f(z)| < M|z|^{-k}$ for large z, we get

$$\left|\frac{f(z)}{z - z_0}\right| = \left|\frac{f(z)}{z} \frac{1}{\left(1 - \frac{z_0}{z}\right)}\right| \leq \frac{M}{\left|z^{k+1}\left(1 - \frac{z_0}{z}\right)\right|} \leq \frac{2M}{\beta^{k+1}}.$$

It now follows that

$$\left|\int_S \frac{f(z)}{z - z_0}\, dz\right| < \frac{2M}{\beta^{k+1}} \int_S |dz| = \frac{2M}{\beta^{k+1}}\,(\text{length of } S)$$

$$= \frac{2M}{\beta^k}\left(\frac{4\beta - 2\gamma}{\beta}\right) = \frac{2M}{\beta^k}\left(4 - \frac{2\gamma}{\beta}\right).$$

Thus, $\lim\limits_{\beta \to \infty} \int_S \dfrac{f(z)}{z - z_0}\, dz = 0$. Hence, from (6.50),

$$-\int_{\gamma - i\infty}^{\gamma + i\infty} \frac{f(z)}{z - z_0}\, dz = 2\pi i f(z_0),$$

or

$$F(s) = \frac{1}{2\pi i} \int_{\gamma - i\infty}^{\gamma + i\infty} \frac{F(z)}{s - z}\, dz. \blacksquare \qquad (6.51)$$

The proof of Theorem 6.2 for the Laplace transform now becomes elementary. By taking the Laplace inverse of both sides of Eq (6.51), we have

$$f(t) = L^{-1}\{F(s)\} = \frac{1}{2\pi i} \int_{\gamma - i\infty}^{\gamma + i\infty} L^{-1}\left\{\frac{F(z)}{s - z}\right\} dz = \frac{1}{2\pi i} \int_{\gamma - i\infty}^{\gamma + i\infty} F(z) e^{zt}\, dz. \blacksquare$$

$$(6.52)$$

Lemma 6.2. *If* $|f(z)| < CR^{-k}$, $z = R e^{i\theta}$, $-\pi \le \theta \le \pi$, $R > R_0$, *where* R_0, C, *and* k *are constants, then* $\int_\Gamma e^{zt} f(z)\, dz \to 0$ *as* $R \to \infty$, *provided* $t > 0$, *where* Γ *is the arc* $BB'CA'A$, *and* R *is the radius of the circular arc with chord* AB *(Fig. 6.3).*

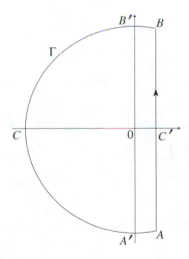

Fig. 6.3.

PROOF. Consider the integral over the arc BB'. Let the angle BOC' be denoted by α. On BB' we have $z = R e^{i\theta}$, where θ varies from α to $\pi/2$, $\alpha = \cos^{-1}(\gamma/R)$, and $\gamma = OC'$. Then we get

$$\left| \int_{BB'} e^{zt} f(z)\, dz \right| < \int_\alpha^{\pi/2} \left| CR^{-k} e^{Rt e^{i\theta}} Ri e^{i\theta} \right| d\theta$$

$$= CR^{-k+1} \int_\alpha^{\pi/2} \left| e^{Rt\cos\theta} \right| d\theta \le CR^{-k+1} \int_\alpha^{\pi/2} \left| e^{\gamma t} \right| d\theta$$

$$= CR^{-k+1}(\pi/2 - \alpha)e^{\gamma t}$$

$$= CR^{-k+1} e^{\gamma t} \sin^{-1} \frac{\gamma}{R} \to 0 \quad \text{as } R \to \infty.$$

Similarly, $\int_{A'A} e^{zt} f(z)\, dz \to 0$ as $R \to \infty$.

Let us now consider the integral over the arc $B'CA'$. By following the above procedure, we get

$$\left| \int_{B'CA'} e^{zt} f(z)\, dz \right| < CR^{-k+1} \int_{\pi/2}^{3\pi/2} \left| e^{Rt\cos\theta} \right| d\theta$$

$$= CR^{-k+1} \int_0^\pi e^{-Rt\sin\phi} d\phi \quad \text{where } \theta = \pi/2 + \phi$$

$$= 2CR^{-k+1} \int_0^{\pi/2} e^{-Rt\sin\phi} d\phi$$

$$\le 2CR^{-k+1} \int_0^{\pi/2} e^{-2Rt\phi/\pi}\, d\phi \; {}^\dagger$$

$$= \frac{\pi CR^{-k}}{t}\left(1 - e^{-Rt}\right) \to 0 \quad \text{as } R \to \infty.$$

Hence, $\int_\Gamma e^{zt} f(z)\, dz \to 0$ as $R \to \infty$, provided $t > 0$. ■

This result enables us to convert the integral $\dfrac{1}{2\pi i} \displaystyle\int_{\gamma-i\beta}^{\gamma+i\beta} F(z)\, e^{zt}\, dz$ into an integral over the contour $-\Gamma$.

EXAMPLE 6.11. Evaluate $\mathcal{L}^{-1}\left\{ \dfrac{e^{-a\sqrt{s}}}{s} \right\}$ by contour integration. If $f(t) = \mathcal{L}^{-1}\left\{ \dfrac{e^{-a\sqrt{s}}}{s} \right\}$, then by the Laplace inversion theorem, we have

$$f(t) = \frac{1}{2\pi i} \int_{c-i\infty}^{c+i\infty} \frac{e^{-a\sqrt{s}}}{s}\, e^{st}\, ds. \tag{6.53}$$

Consider the Bromwich contour $MABC_1CDL$ (Fig. 6.4). Then after using the

---

† To show that $g(\phi) = \sin\phi - 2\phi/\pi \ge 0$ for $0 \le \phi \le \pi/2$, note that $g(0) = 0 = g(\pi/2)$, and there is only one critical point of g at $\phi = \cos^{-1}(2/\pi)$, which gives a maximum. Hence, $e^{-RT\sin\phi} \le e^{-2RT\phi/\pi}$.

Cauchy's theorem

$$I = \int_{c-i\infty}^{c+i\infty} \frac{e^{-a\sqrt{s}}}{s} e^{st}\, ds = \int_{LD} F(s)\, ds + \int_{DC} F(s)\, ds +$$

$$\int_{C_1} F(s)\, ds + \int_{BA} F(s)\, ds + \int_{AM} F(s)\, ds.$$

As established in Lemma 6.2, we have

$$\int_{LD} F(s)\, ds + \int_{AM} F(s)\, ds = 0,$$

where

$$F(s) = \frac{e^{-a\sqrt{s}}}{s} e^{st}.$$

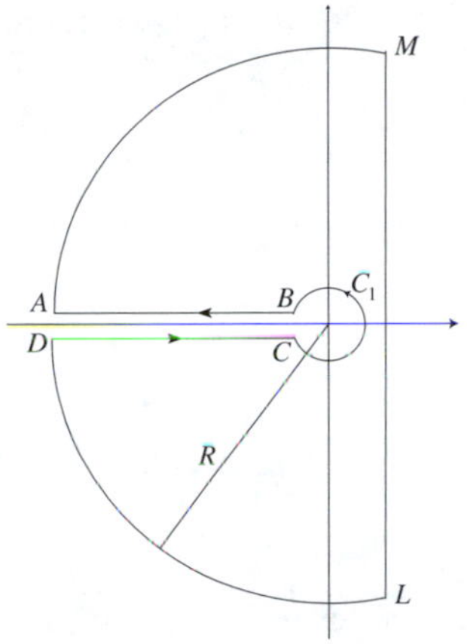

Fig. 6.4.

The integral over the circle C_1 is easily shown to be equal to $2\pi i$. This is done by taking the radius to be ε and substituting $s = \varepsilon e^{i\theta}$. On BA, $s = u\, e^{i\pi}$, and

$$I_{BA} = \int_{\varepsilon \to 0}^{R \to \infty} \frac{1}{u\, e^{i\pi}} e^{-a\sqrt{u}e^{i\pi/2} + ut e^{i\pi}}\, e^{i\pi}\, du$$

$$= \int_0^\infty \frac{1}{u} e^{-ia\sqrt{u} - ut}\, du = \int_0^\infty \frac{1}{u} e^{-ut}(\cos a\sqrt{u} - i\sin a\sqrt{u})\, du$$

$$= 2\int_0^\infty \frac{1}{v} e^{-v^2 t}(\cos av - i\sin av)\, dv,$$

where $u = v^2$. Similarly,

$$\int_{CD} = -2\int_0^\infty \frac{1}{v} e^{-v^2 t}(\cos av + i\sin av)\, dv.$$

Hence,

$$\int_{CD} + \int_{BA} = -4i \int_0^\infty \frac{1}{v} e^{-v^2 t} \sin av \, dv.$$

In order to evaluate the integral $\int_0^\infty \frac{1}{v} e^{-v^2 t} \sin av \, dv$, we consider the integral $\int_0^\infty e^{-v^2 t} \cos av \, dv$. Then

$$\int_0^\infty e^{-v^2 t} \cos av \, dv = \Re \left\{ \int_0^\infty e^{-v^2 t + iav} \, dv \right\}$$

$$= \Re \left\{ e^{-a^2/(4t)} \int_0^\infty e^{-(v\sqrt{t} - ia/(2\sqrt{t}))^2} \, dv \right\}$$

$$= \Re \left\{ \frac{e^{-a^2/(4t)}}{\sqrt{t}} \int_{-ia/(2\sqrt{t})}^\infty e^{-w^2} \, dw \right\}, \quad \text{where } w = v\sqrt{t} - ia/(2\sqrt{t})$$

$$= \Re \left\{ \frac{e^{-a^2/(4t)}}{\sqrt{t}} \left[\int_0^\infty e^{-w^2} \, dw + \int_{-ia/(2\sqrt{t})}^0 e^{-w^2} \, dw \right] \right\}.$$

Hence,

$$\int_0^\infty e^{-v^2 t} \cos av \, dv = \frac{\sqrt{\pi} \, e^{-a^2/(4t)}}{2\sqrt{t}}. \tag{6.54}$$

Integrating both sides of this equation with respect to a from 0 to a, we get

$$\int_0^\infty \frac{1}{v} e^{-v^2 t} \sin av \, dv = \sqrt{\frac{\pi}{4t}} \int_0^a e^{-x^2/(4t)} \, dx = \frac{\pi}{2} \operatorname{erf} \frac{a}{2\sqrt{t}}.$$

Thus,

$$\mathcal{L}^{-1} \left\{ \frac{e^{-a\sqrt{s}}}{s} \right\} = \frac{1}{2\pi i} \left[2\pi i - 4i \frac{\pi}{2} \operatorname{erf} \frac{a}{2\sqrt{t}} \right] = \operatorname{erfc} \frac{a}{2\sqrt{t}}. \quad \blacksquare \tag{6.55}$$

6.5. Mathematica Projects

PROJECT 6.1. Use Mathematica to compute the Laplace transform for the following functions: (i) $t^n e^{at}$, (ii) $t^2 e^{at}$, (iii) $t^2 E^{ibt}$, (iv) $t^2 \cos bt$, and $t^2 \sin bt$.

In[1]:   <<Calculus`LaplaceTransform`

In[2]: `LaplaceTransform[t^n Exp[a t],t,s]`

Out[2]: $(-a + s)^{-1-n} \text{Gamma}[1 + n]$

In[3]: `LaplaceTransform[t^2 Exp[a t],t,s]`

Out[3]: $\dfrac{2}{(-a + s)^3}$

In[4]: `LaplaceTransform[t^2 Exp[I b t],t,s]`

Out[4]: $\dfrac{2}{(-I b + s)^3}$

In[5]: `LaplaceTransform[t^2 Cos[b t],t,s] // Simplify`

Out[6]: $\dfrac{2s(-3b^2 + s^2)}{(b^2 + s^2)^3}$

In[6]: `LaplaceTransform[t^2 Sin[b t],t,s] // Simplify`

Out[6]: $\dfrac{2b(-b^2 + 3s^2)}{(b^2 + s^2)}$

PROJECT 6.2. This example computes an important formula that will be found useful.

In[5]: `(* An important formula *)`

`InverseLaplaceTransform[Exp[-a Sqrt[s]]/s,s,t]`

Out[5]: $1 - \text{Erf}\left[\dfrac{a}{2\,\text{Sqrt}[t]}\right]$

PROJECT 6.3. This example computes an important Laplace inverse that will be found useful.

In[8]: `Needs["Calculus`LaplaceTransform`"]`

`(* An important Laplace Inverse *)`

In[9]: `f[s_]:= Exp[-a Sqrt[s]]/s`

`InverseLaplaceTransform[f[x],s,t] == Erfc[a/(2 Sqrt[t])]`

Out[11]: $1 - \text{Erf}\left[\dfrac{a}{2\,\text{Sqrt}[t]}\right] == \text{Erfc}\left[\dfrac{a}{2\,\text{Sqrt}[t]}\right]$

In[11]: RHS:= Erfc[a/(2 Sqrt[t])]

   (* Differentiate RHS with respect to a *)
In[12]: D[RHS,a]

Out[14]: $-\dfrac{E^{-a^2/(4\,t)}}{\text{Sqrt[Pi] Sqrt[t]}}$

In[15]: InverseLaplaceTransform[-D[f[s],a],s,t] == -D[RHS,a]

Out[15]: True

PROJECT 6.4. Let $F(s)$ be defined as in Project 6.1. Then we compute $\mathcal{L}^{-1}\left\{e^{-a\sqrt{s}}\right\}$, for which see Project 6.2.

In[1]: Needs["Calculus`LaplaceTransform`"];

   (* f[s] is defined as in Project 6.1 *)

   f[s] Sqrt[s]

Out[3]: $-\dfrac{1}{E^{a\ \text{Sqrt[s]}}\ \text{Sqrt[s]}}$

In[3]: D[f[s] Sqrt[s],a]

Out[4]: $-E^{-(a\ \text{Sqrt[s]})}$

In[4]: InverseLaplaceTransform[-D[f[s] Sqrt[s],a],s,t]

Out[5]: $\dfrac{a}{2\ E^{a^2/(4\,t)}\ \text{Sqrt}[\text{Pi}\ t^3]}$

PROJECT 6.5. Let $F(s)$ be defined as in Example 6.1. Then we compute $\mathcal{L}^{-1}\left\{\dfrac{e^{-a\sqrt{s}}}{s}\right\}$, which is given in Example 6.3.

In[1]: Needs["Calculus`LaplaceTransform`"];

(* f[s] is defined as in Project 6.1 *)

f[s]/.a -> x Sqrt[s]

Out[2]: $\dfrac{1}{\mathrm{E}^{\,\mathrm{Sqrt[s]x}}\,s}$

(* Change the order of operations *)

In[3]: LHS = Integrate[f[s]/.a -> x ,{x,0,a}]

Out[3]: $\dfrac{1 - \mathrm{E}^{-a\,\mathrm{Sqrt[s]}}}{s^{3/2}}$

In[4]: Simplify[InverseLaplaceTransform[f[s],s,t,a>0]]
         == Erfc[a/(2 Sqrt[t])]

Out[4]:
$1 - \mathrm{Erf}\left[\dfrac{a}{2\ \mathrm{Sqrt[t]}}\right] == \mathrm{Erfc}\left[\dfrac{a}{2\,\mathrm{Sqrt[t]}}\right]$

(* Define a substitution *)

In[5]:

         trans =
             Simplify[InverseLaplaceTransform[f[s],s,t, a>0]]
             -> Erfc[a/(2 Sqrt[t])]

Out[5]:
$1 - \mathrm{Erf}\left[\dfrac{a}{2\ \mathrm{Sqrt[t]}}\right] -> \mathrm{Erfc}\left[\dfrac{a}{2\,\mathrm{Sqrt[t]}}\right]$

(*   Collect the terms involving a,
in order to apply the transformation *)

In[6]: RHS=
    Collect[Integrate[Erfc[x/(2 Sqrt[t])],{x,0,a}],a]/.trans

Out[6]:
$\dfrac{2\ \mathrm{Sqrt[t]}}{\mathrm{Sqrt[Pi]}} - \dfrac{2\ \mathrm{Sqrt[t]}}{\mathrm{E}^{\,a^2/(4\,t)}\ \mathrm{Sqrt[Pi]}} + a\ \mathrm{Erfc}\left[\dfrac{a}{2\ \mathrm{Sqrt[t]}}\right]$

(* Prevent evaluation of the InverseLaplaceTransform *)

In[8]: ILT[X_]:= Hold[InverseLaplaceTransform[X,s,t]]

    sexpr:=
    s^(-3/2) Exp[-a Sqrt[s]]]

In[10]:
    Solve[
    InverseLaplaceTransform[
    s^(-3/2),s,t] - ILT[sexpr] == RHS,ILT[sexpr]]

Out[10]:

{{Hold[InverseLaplaceTransform[$\dfrac{1}{E^a \text{ Sqrt[s] } s3/2}$,s,t]]->

$\dfrac{2 \text{ Sqrt}[t]}{E^{a^2/(4\,t)} \text{ Sqrt[Pi]}}$ − a Erfc[$\dfrac{a}{2 \text{ Sqrt}[t]}$]}}

PROJECT 6.6. In Example 6.5 we have computed $\mathcal{L}^{-1}\left\{s^{-3/2}\,e^{-1/s}\right\}$. For the Mathematica session, see `Example6.5.nb`.

For some of the above examples and projects, see the Mathematica Notebook `Example6.nb`.

6.6. Exercises

6.1. Using the properties of §6.2 and $\mathcal{L}\left\{e^{at}\right\} = \dfrac{1}{s-a}$, derive the Laplace transform of $\sin at$, $\cos at$, $e^{bt}\sin at$, $e^{bt}\cos at$, $t^n\,e^{at}$, $\sinh bt$, and $\cosh bt$.

6.2. Show that $\mathcal{L}^{-1}\left\{e^{-a\sqrt{s}}\right\} = \dfrac{a}{2\sqrt{\pi t^3}}\,e^{-a^2/(4t)}$.

6.3. Find (a) $\mathcal{L}^{-1}\left\{\dfrac{\cosh a\sqrt{s}}{s\cosh b\sqrt{s}}\right\}$, and (b) $\mathcal{L}^{-1}\left\{\dfrac{\sinh a\sqrt{s}}{\sinh b\sqrt{s}}\right\}$, $b > a > 0$.

HINT. Use $\cosh x = \left(e^x + e^{-x}\right)/2$, $\sinh x = \left(e^x - e^{-x}\right)/2$, and $(1+z)^{-1} = \sum_{n=0}^{\infty}(-1)^n\,z^n$.

ANS. (a) $\displaystyle\sum_{n=0}^{\infty}(-1)^n\left\{\text{erfc}\left(\dfrac{(2n+1)b-a}{2\sqrt{t}}\right) + \text{erfc}\left(\dfrac{(2n+1)b+a}{2\sqrt{t}}\right)\right\}$,

(b) $\displaystyle-\sum_{n=-\infty}^{\infty}\dfrac{(2n+1)b+a}{\sqrt{4\pi t^3}}\,e^{-[(2n+1)b+a]^2/(4t)}$.

6.4. Using the Laplace transform method, solve the partial differential equation $u_t = u_{zz} - cu$, where c is a constant, given that $u(z,0) = 0$, and $u(0,t) = u_0$, $\lim_{z\to\infty} u(z,t) = 0$ for $t > 0$. (This problem corresponds to the flow of a viscous fluid on an infinite moving plate under the influence of a constant magnetic field applied perpendicular to the plate.) Derive the solution when $c = 0$.
ANS.

$$u = \dfrac{u_0}{2}\left\{e^{z\sqrt{c}}\,\text{erfc}\left(\dfrac{z}{2\sqrt{t}} + \sqrt{ct}\right) + e^{-z\sqrt{c}}\,\text{erfc}\left(\dfrac{z}{2\sqrt{t}} - \sqrt{ct}\right)\right\}.$$

For more details, see Puri and Kulshrestha (1976).

6.5. Using the Laplace transform method, solve in the transform domain the partial differential equation $u_t = u_{zz} + k u_{tzz}$, given that $u(z,0) = 0$, and $u(0,t) = u_0$, $\lim_{z \to \infty} u(z,t) = 0$ for $t > 0$. Expand the solution in the transform domain in the form $\bar{u} = \dfrac{u_0}{s} e^{-z\sqrt{s}}(1 + \text{powers of } k)$. Invert the first two terms of this expansion.

ANS. $\dfrac{u}{u_0} = \text{erfc}\,\dfrac{z}{2\sqrt{t}} + \dfrac{kz}{4t\sqrt{\pi t}}\left(\dfrac{z^2}{2t} - 1\right)e^{-z^2/(4t)}.$

This problem corresponds to the flow of a viscoelastic fluid on an infinite moving plate. Obtain the exact solution in terms of definite integrals by using the contour integration.

ANS. $\dfrac{u(z,t)}{u_0} = 1 - \dfrac{1}{\pi}\int_0^\lambda \dfrac{1}{x} e^{-xt} \sin\sqrt{\dfrac{\lambda x}{\lambda - x}}\, dx$, where $\dfrac{1}{\lambda} = k.$

6.6. Using the Laplace transform method, solve the partial differential equation $u_t = u_{xx}$, with the initial condition $u(x,0) = 0$ and the boundary conditions $u_x(0,t) = 0$, and $u_x(1,t) = 1$.

HINT. The solution in the transform domain is $\bar{u} = \dfrac{\cosh x\sqrt{s}}{s^{3/2}\sinh\sqrt{s}}.$ Find two different inverses of this solution by expanding the solution in a series of the type shown in Example 6.7 and by the residue theorem.

ANS. $u = \displaystyle\sum_{n=0}^{\infty}\Big\{2\sqrt{t/\pi}\left(e^{-(2n+1-x)^2/(4t)} + e^{-(2n+1+x)^2/(4t)}\right)$

$-(2n+1-x)\,\text{erfc}\,\dfrac{2n+1-x}{2\sqrt{t}} - (2n+1+x)\,\text{erfc}\,\dfrac{2n+1+x}{2\sqrt{t}}\Big\},$

and

$$u = \frac{x^2}{2} + t - \frac{1}{6} - \sum_{n=1}^{\infty}\frac{2(-1)^n}{n^2\pi^2}e^{-n^2\pi^2 t}\cos n\pi x.$$

6.7. Using the Laplace transform method, solve the partial differential equation $u_{tt} = u_{xx}$, with the initial conditions $u(x,0) = -\dfrac{(1-x)^2}{2}$, $u_t(x,0) = 0$, and the boundary conditions $u_x(0,t) = 1$ and $u_x(1,t) = 0$.

ANS. $u = -\dfrac{1}{2}t^2 - \dfrac{(1-x)^2}{2}.$

6.8. Using the Laplace transform method, solve the partial differential equation $u_t = u_{xx}$, with the initial condition $u(x,0) = 0$ and the boundary conditions $u_x(0,t) = 0$ and $u(1,t) = 1$.

HINT. The solution in the transform domain is $\bar{u} = \dfrac{\cosh x\sqrt{s}}{s\cosh\sqrt{s}}.$ Find two different inverses of this solution by expanding the solution in a series of the type shown in Example 6.7 and by the residue theorem.

ANS. $u = \displaystyle\sum_0^{\infty}(-1)^n\left[\text{erfc}\,\dfrac{2n+1-x}{2\sqrt{t}} + \text{erfc}\,\dfrac{2n+1+x}{2\sqrt{t}}\right]$, or

$$u = 1 - \sum_0^\infty (-1)^n \frac{4\cos(2n+1)\pi x/2}{(2n+1)\pi} e^{-(2n+1)^2\pi^2 t/4}.$$

6.9. Using the Laplace transform method, solve the partial differential equation $u_{tt} = u_{xx}$, with the initial conditions $u(x,0) = -\frac{(1-x)^2}{2}$, $u_t(x,0) = 0$ and the boundary conditions $u(0,t) = 1$, $u_x(1,t) = 0$.

ANS. $u = \sum_0^\infty \frac{3(-1)^n}{2}\{H(t-2n-2+x) + H(t-2n-x)\}$

$$+ \sum_0^\infty \frac{(-1)^n}{2}\big[H(t-2n-2+x)\,(t-2n-2+x)^2$$

$$+ H(t-2n-x)\,(t-2n-x)^2\big] - \frac{1}{2}H(t)\,(1-x)^2 - H(t)\frac{t^2}{2}.$$

6.10. Using the Laplace transform method, solve the partial differential equation $u_{xx} - u_{tt} = e^{-\pi^2 t}\sin\pi x$, subject to the conditions $u(x,0) = 0$, $u_t(x,0) = 0$, $u(0,t) = u_x(1,t) = 0$.

SOLUTION. Applying Laplace transform to the partial differential equation, we get $(D^2 - s^2)\bar{u} = \frac{1}{s+\pi^2}\sin\pi x$. Its solution is given by

$$\bar{u} = Ae^{sx} + Be^{-sx} - \frac{1}{(s+\pi^2)(s^2+\pi^2)}\sin\pi x.$$

Applying the boundary conditions, we get $A + B = 0$, and

$$s(Ae^s - Be^{-s}) + \frac{\pi}{(s+\pi^2)(s^2+\pi^2)}\cos\pi = 0.$$

These two equations yield

$$A = -B = -\frac{\pi}{2s(s+\pi^2)(s^2+\pi^2)\sinh s}.$$

Thus,

$$\bar{u} = -\frac{\pi\sinh sx}{s(s+\pi^2)(s^2+\pi^2)\sinh s} - \frac{1}{(s+\pi^2)(s^2+\pi^2)}\sin\pi x$$

$$= \frac{1}{\pi}\Big[\frac{1}{\pi^2(1+\pi^2)(s+\pi^2)} + \frac{1}{(1+\pi^2)(s^2+\pi^2)}(s+1) - \frac{1}{\pi^2 s}\Big]\frac{\sinh sx}{\sinh s}$$

$$- \frac{1}{\pi^2(1+\pi^2)}\Big[\frac{1}{s+\pi^2} - \frac{s}{s^2+\pi^2} + \frac{\pi^2}{s^2+\pi^2}\Big]\sin\pi x$$

$$= \frac{1}{\pi}\Big[\frac{1}{\pi^2(1+\pi^2)(s+\pi^2)} + \frac{s+1}{(1+\pi^2)(s^2+\pi^2)} - \frac{1}{\pi^2 s}\Big]$$

$$\times \Big[\sum_0^\infty \big(e^{(x-2k-1)s} - e^{-(2k+1+x)s}\big)\Big]$$

$$- \frac{1}{\pi^2(1+\pi^2)}\Big[\frac{1}{s+\pi^2} - \frac{s}{s^2+\pi^2} + \frac{\pi^2}{s^2+\pi^2}\Big]\sin\pi x.$$

On inversion we find

$$u = \frac{1}{\pi} \sum_0^\infty H(t + x - 2k - 1) \left\{ \frac{e^{-\pi^2(t+x-2k-1)}}{\pi^2(1+\pi^2)} \right.$$

$$+ \frac{\cos(t + x - 2k - 1) + \sin(t + x - 2k - 1)}{(1 + \pi^2)} - \left. \frac{1}{\pi^2} \right\}$$

$$- \frac{1}{\pi} \sum_0^\infty H(t - x - 2k - 1) \left\{ \frac{e^{-\pi^2(t-x-2k-1)}}{\pi^2(1+\pi^2)} \right.$$

$$+ \frac{\cos(t - x - 2k - 1) + \sin(t - x - 2k - 1)}{(1 + \pi^2)} - \left. \frac{1}{\pi^2} \right\}$$

$$- \frac{1}{\pi^2(1+\pi^2)} [e^{-\pi^2 t} - \cos \pi t + \pi \sin \pi t] \sin \pi x.$$

6.11. Show that

$$\mathcal{L}\{t^p\} = \frac{\Gamma(p+1)}{s^{p+1}},$$

where $\Gamma(x)$ is the gamma function defined by $\Gamma(x) = \displaystyle\int_0^\infty t^{x-1} e^{-t}\, dt$, $x > 0$.

ANS. Set $st = x$ in $\mathcal{L}\{t^p\} = \displaystyle\int_0^\infty t^p e^{-st}\, dt$. Then

$$\int_0^\infty t^p e^{-st}\, dt = \frac{1}{s^{p+1}} \int_0^\infty x^p e^{-x}\, dx = \frac{\Gamma(p+1)}{s^{p+1}}.$$

6.12. Solve the diffusion equation $u_t = k\, u_{xx}$, $\quad 0 < x < \pi$, $\quad t > 0$, subject to the boundary conditions $u(0, t) = 1 - e^{-t}$ and $u(\pi, t) = 0$ for $t \geq 0$, and the initial condition $u(x, 0) = 0$ for $0 < x < \pi$.

SOLUTION. By the Laplace transform method, we get

$$\frac{d^2\bar{u}}{dx^2} = \frac{s}{k}\,\bar{u},$$

with $\bar{u}(0, s) = \dfrac{1}{s(s+1)}$, and $\bar{u}(\pi, s) = 0$, which has the solution

$$\bar{u}(x, s) = \frac{1}{s(s+1)} \frac{\sinh \sqrt{s/k}\,(\pi - x)}{\sinh \sqrt{s/k}\,\pi}.$$

Then the inversion formula gives

$$u(x, t) = \frac{1}{2\pi i} \int_{c-i\infty}^{c+i\infty} \frac{e^{st} \sinh \sqrt{s/k}(\pi - x)}{s(s+1) \sinh \sqrt{s/k}\,\pi}\, ds,$$

where c is any positive constant. Assuming that k is not of the form n^{-2}, the integrand has simple poles at $s = 0, -1, -kn^2, n = 1, 2, \cdots$. The contour is shown in Fig. 6.5, where the left semicircle, with $\Re\{s\} = c$, is defined as the limit of a sequence of semicircles Γ_n that cross the negative s-axis between the poles at $-kn^2$ and $-k(n+1)^2$. The limit of the integrand around Γ_n is zero as $n \to \infty$. The residue at the pole $s = 0$ is $(\pi - x)/\pi$, and at the pole $s = -1$ is

$$- e^{-t} \frac{\sin[(\pi - x)/\sqrt{k}]}{\sin(\pi/\sqrt{k})}.$$

The residue at $s = -kn^2$ is given by

$$\lim_{s \to kn^2} \frac{s+n^2}{s(s+1)} e^{st} \frac{\sinh\sqrt{s/k}(\pi - x)}{\sinh\sqrt{s/k}\,\pi} = \frac{2\sin nx}{n\pi(kn^2 - 1)} e^{-kn^2 t}.$$

Hence,

$$u(x,t) = \frac{\pi - x}{\pi} - e^{-t} \frac{\sin[(\pi - x)/\sqrt{k}]}{\sin(\pi/\sqrt{k})} + \frac{2}{\pi} \sum_{n=1}^{\infty} \frac{\sin nx}{n(kn^2 - 1)} e^{-kn^2 t}.$$

Note that $u \to \dfrac{\pi - x}{\pi}$ as $t \to \infty$, which gives the steady-state temperature in the interval $0 \le x \le \pi$.

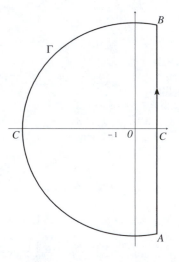

Fig. 6.5.

6.13. Solve the nonhomogeneous Cauchy problem in $R^1 \times R^+$:

$$u_t - u_{xx} = f(x,t), \quad x \in R^1,$$
$$u(0,t) = 0 = u(l,t), \quad u(x,0) = g(x), \quad t > 0,$$

where $g(x)$ is prescribed.

SOLUTION. Using the Laplace transform we get

$$\frac{d^2\bar{u}}{dx^2} - s\bar{u} = -f(x),$$

with the boundary conditions $\bar{u}(0, s) = 0 = \bar{u}(l, s)$. The solution of the above homogeneous equation is given by

$$\bar{u}(x, s) = \frac{1}{\sqrt{s}\sinh(l\sqrt{s})}\left[\sinh((l-x)\sqrt{s})\int_0^x \sinh(y\sqrt{s})f(y)\,dy\right.$$
$$\left. + \sinh(x\sqrt{s})\int_x^l \sinh((l-y)\sqrt{s})f(y)\,dy\right].$$

By inversion, the solution of this problem becomes

$$u(x, t) = \lim_{R\to\infty}\left[\frac{1}{2\pi i}\int_{c-iR}^{c+iR} e^{st}\bar{u}(x, s)\,ds\right].$$

The integrand in this solution has simple poles at $s = 0, -k_n^2$, where $k_n = n\pi/l, n = 1, 2, \ldots$. We choose a contour of integration that avoids these poles and take $u > 0$. Then it can be shown that the residue at the pole $s = 0$ is zero, while at the poles $s = -k_n^2$ it is given by

$$\frac{2}{l}e^{-k_n^2 t}\sin k_n x \int_0^l f(y)\sin k_n y\,dy.$$

Hence, the final formal solution of the problem is given by

$$u(x, t) = \sum_{n=1}^{\infty}\left[\frac{2}{l}\int_0^l f(y)\sin k_n y\,dy\right]e^{-k_n^2 t}\sin k_n x.$$

Alternatively, if we use the series representation

$$\frac{1}{1-x} = \sum_{n=0}^{\infty} x^n, \quad |x| < 1,$$

then we have

$$\sinh(l\sqrt{s}) = 2e^{-l\sqrt{s}}\sum_{n=0}^{\infty} e^{-2nl\sqrt{s}},$$

where $\Re\{s\}$ is chosen such that $e^{-2l\sqrt{s}} < 1$. Then expressing the hyperbolic functions in terms of exponentials, using the formula

$$\mathcal{L}^{-1}\left\{\frac{e^{-z\sqrt{s}}}{\sqrt{s}}\right\} = \frac{e^{-z^2/(4t)}}{\sqrt{\pi t}},$$

where z is independent of s and t (see §C.2), and interchanging the orders of summation and integration when needed, we obtain the required solution.

II. Fourier Transforms

We will not discuss the underlying theory of the Fourier transforms. Instead we will only define and discuss their properties and applications. From the definitions of the transform pairs (6.1a,b) and (6.2a,b) we note that the Fourier cosine and sine transforms and their inverses are symmetric. But the Fourier complex transform and its inverse are related in the following manner: If $\mathcal{F}f(x) = \tilde{f}(\alpha)$, then $\mathcal{F}\tilde{f}(x) = f(-\alpha)$. Various authors have defined the Fourier transform in different ways, but we will follow the notation used by Sneddon (1957). A table of basic Fourier transform pairs is given in Appendix C.

6.7. Fourier Integral Theorems

Theorem 6.3. (Fourier integral theorem) *If $f(x)$ satisfies the Dirichlet's conditions on the entire real line and is absolutely integrable on $(-\infty, \infty)$, then*

$$\frac{1}{2}[f(x+0) + f(x-0)] = \frac{1}{2\pi}\int_{-\infty}^{\infty} e^{-i\alpha x}\, d\alpha \int_{-\infty}^{\infty} f(u)e^{i\alpha u}\, du. \qquad (6.56)$$

Theorem 6.4. (Fourier cosine theorem) *If $f(x)$ satisfies the Dirichlet's conditions on the nonnegative real line and is absolutely integrable on $(0, \infty)$, then*

$$\frac{1}{2}[f(x+0) + f(x-0)] = \frac{2}{\pi}\int_{0}^{\infty} d\alpha \int_{0}^{\infty} f(u)\cos(\alpha u)\cos(\alpha x)\, du. \qquad (6.57)$$

Theorem 6.5. (Fourier sine theorem) *If $f(x)$ satisfies the Dirichlet's conditions on the nonnegative real line and is absolutely integrable on $(0, \infty)$, then*

$$\frac{1}{2}[f(x+0) + f(x-0)] = \frac{2}{\pi}\int_{0}^{\infty} d\alpha \int_{0}^{\infty} f(u)\sin(\alpha u)\sin(\alpha x)\, du. \qquad (6.58)$$

If $f(x)$ is continuous, then

$$\frac{1}{2}[f(x+0) + f(x-0)] = f(x).$$

These three integrals form the basis of the Fourier transforms. For a proof, see Sneddon (1957).

6.8. Properties of Fourier Transforms

We will use the following notation: Let $\mathcal{F}f(x) = \tilde{f}(\alpha)$. Then

(1) $\mathcal{F}f(x-a) = e^{i\alpha a}\,\tilde{f}(\alpha)$.

(2) $\mathcal{F}f(ax) = \dfrac{1}{|a|}\tilde{f}(\alpha/a)$.

(3) $\mathcal{F}e^{iax}f(x) = \tilde{f}(\alpha+a)$.

(4) $\mathcal{F}\tilde{f}(x) = f(-\alpha)$.

(5) $\mathcal{F}x^n f(x) = (-i)^n \dfrac{d^n}{d\alpha^n}\,\tilde{f}(\alpha)$.

(6) $\mathcal{F}f(ax)e^{ibx} = \dfrac{1}{|a|}\tilde{f}\left(\dfrac{\alpha+b}{a}\right)$.

6.8.1. Fourier Transforms of the Derivatives of a Function. We assume that $f(x)$ is differentiable n times and the function and its derivatives approach zero as $|x| \to \infty$. Then it can be easily established that

$$\tilde{f}^{(p)}(\alpha) = (-i\alpha)\tilde{f}^{(p-1)},$$

where $\tilde{f}^{(p)}$ is the Fourier transform of $f^{(p)}(x)$, which is the p-th derivative of $f(x)$ for $0 \le p \le n$.

If $\lim\limits_{x\to\infty} f^{(p)}(x) = 0$, and $\lim\limits_{x\to 0} f^{(p)}(x) = \sqrt{\dfrac{\pi}{2}}\,c_p$, then

$$\tilde{f}_c^{(p)} = -c_{p-1} + \alpha \tilde{f}_s^{(p-1)}, \tag{6.59}$$

and

$$\tilde{f}_s^{(p)} = -\alpha \tilde{f}_c^{(p-1)}. \tag{6.60}$$

6.8.2. Convolution Theorems for Fourier Transform. The convolution (or Faltung) of $f(t)$ and $g(t)$ over $(-\infty, \infty)$ is defined by

$$f \star g = \frac{1}{\sqrt{2\pi}} \int_{-\infty}^{\infty} f(\eta)g(x-\eta)\,d\eta = \frac{1}{\sqrt{2\pi}} \int_{-\infty}^{\infty} f(x-\eta)g(\eta)\,d\eta. \tag{6.61}$$

Theorem 6.6. *Let $\tilde{f}(\alpha)$ and $\tilde{g}(\alpha)$ be the Fourier transforms of $f(x)$ and $g(x)$, respectively. Then the inverse Fourier transform of $\tilde{f}(\alpha)\,\tilde{g}(\alpha)$ is*

$$\mathcal{F}^{-1}\left\{\tilde{f}(\alpha)\tilde{g}(\alpha)\right\} = \frac{1}{\sqrt{2\pi}} \int_{-\infty}^{\infty} f(\eta)g(x-\eta)\,d\eta.$$

PROOF. Consider

$$\int_{-\infty}^{\infty} f(\eta)g(x-\eta)\,d\eta = \frac{1}{\sqrt{2\pi}} \int_{-\infty}^{\infty} f(\eta)\,d\eta \int_{-\infty}^{\infty} \tilde{g}(\alpha)e^{-i\alpha(x-\eta)}\,d\alpha$$

$$= \frac{1}{\sqrt{2\pi}} \int_{-\infty}^{\infty} \tilde{g}(\alpha)e^{-i\alpha x}\,d\alpha \int_{-\infty}^{\infty} f(\eta)e^{i\alpha\eta}\,d\eta$$

$$= \int_{-\infty}^{\infty} \tilde{f}(\alpha)\tilde{g}(\alpha)e^{-i\alpha x}\,d\alpha,$$

which proves the theorem. ∎

Theorem 6.7. *Let $\tilde{f}(\alpha)$ and $\tilde{g}(\alpha)$ be the Fourier transforms of $f(x)$ and $g(x)$, respectively. Then*

$$\int_{-\infty}^{\infty} \tilde{f}(\alpha)\tilde{g}(\alpha)d\alpha = \int_{-\infty}^{\infty} f(-\eta)g(\eta)\,d\eta. \tag{6.62}$$

PROOF. Consider

$$\int_{-\infty}^{\infty} \tilde{f}(\alpha)\tilde{g}(\alpha)\,d\alpha = \int_{-\infty}^{\infty} \tilde{f}(\alpha)\,d\alpha \frac{1}{\sqrt{2\pi}} \int_{-\infty}^{\infty} g(\eta)e^{i\alpha\eta}d\eta$$

$$= \int_{-\infty}^{\infty} g(\eta)\,d\eta \frac{1}{\sqrt{2\pi}} \int_{-\infty}^{\infty} \tilde{f}(\alpha)e^{i\alpha\eta}d\alpha$$

$$= \int_{-\infty}^{\infty} g(\eta)f(-\eta)d\eta, \quad \text{by Property (4), §6.8.} \ \blacksquare$$

6.8.3. Some Fourier Transform Formulas. Using the definition and properties of the Fourier transform we will derive these formulas by the following examples.

EXAMPLE 6.12. Find the Fourier transform of $f(x) = e^{-k|x|}, k > 0$.

$$\mathcal{F}f(x) = \tilde{f}(\alpha) = \frac{1}{\sqrt{2\pi}} \int_{-\infty}^{\infty} e^{-k|x|}e^{i x \alpha}\,dx$$

$$= \frac{1}{\sqrt{2\pi}} \left[\int_{-\infty}^{0} e^{kx}e^{i x \alpha}\,dx + \int_{0}^{\infty} e^{-kx}e^{i x \alpha}\,dx \right]$$

$$= \frac{1}{\sqrt{2\pi}} \left(\frac{1}{k+i\alpha} - \frac{1}{-k+i\alpha} \right) = \frac{k\sqrt{2}}{\sqrt{\pi}(k^2+\alpha^2)}.$$

Now by Property (4), the Fourier transform of $f(x) = \dfrac{k\sqrt{2}}{\sqrt{\pi}(k^2+x^2)}$ should be $\tilde{f}(-\alpha) = e^{-k|\alpha|}$. It is interesting as well as instructive to check if this is indeed the case. Since $f(x) = f(-x)$, $\tilde{f}(\alpha) = \tilde{f}(-\alpha)$, we have

$$\mathcal{F}\frac{k\sqrt{2}}{\sqrt{\pi}(k^2+x^2)} = \frac{1}{\sqrt{2\pi}} \int_{-\infty}^{\infty} \frac{k\sqrt{2}}{\sqrt{\pi}(k^2+x^2)} e^{i x \alpha}\,dx$$

$$= \frac{1}{\pi} \int_{-\infty}^{\infty} \frac{ke^{ix\alpha}}{(k^2 + x^2)} \, dx$$

$$= \frac{1}{\pi} \left[\int_{-\infty}^{0} \frac{ke^{ix\alpha}}{(k^2 + x^2)} \, dx + \int_{0}^{\infty} \frac{ke^{ix\alpha}}{(k^2 + x^2)} \, dx \right]$$

$$= \frac{1}{\pi} \left[\int_{0}^{\infty} \frac{ke^{-ix\alpha}}{(k^2 + x^2)} \, dx + \int_{0}^{\infty} \frac{ke^{ix\alpha}}{(k^2 + x^2)} \, dx \right]$$

$$= \frac{2}{\pi} \int_{0}^{\infty} \frac{k \cos x\alpha}{(k^2 + x^2)} \, dx = \frac{1}{\pi} \int_{-\infty}^{\infty} \frac{k \cos x\alpha}{(k^2 + x^2)} \, dx.$$

The standard method of evaluating this integral is by contour integration. The contour is the upper half-circle with radius R and center at the origin if $\alpha > 0$ and the lower half-circle if $\alpha < 0$. Its value is $e^{-k\alpha}$ if $\alpha > 0$, and $e^{k\alpha}$ if $\alpha < 0$. Thus, the value can be expressed as $e^{-k|\alpha|}$.

A number of other Fourier transforms can be found by differentiating both sides of $\mathcal{F}e^{-k|x|} = \dfrac{k\sqrt{2}}{\sqrt{\pi}(k^2 + \alpha^2)}$ with respect to k. For example, the Fourier transform of $|x|e^{-k|x|}$ is $\sqrt{\dfrac{2}{\pi}} \dfrac{k^2 - \alpha^2}{(k^2 + \alpha^2)^2}$. ∎

EXAMPLE 6.13. Find the Fourier transform of $f(x) = e^{-kx^2}$.

$$\tilde{f}(\alpha) = \frac{1}{\sqrt{2\pi}} \int_{-\infty}^{\infty} e^{-kx^2} e^{ix\alpha} \, dx$$

$$= \frac{1}{\sqrt{2\pi}} \int_{-\infty}^{\infty} e^{-k(x^2 - ix\alpha/k - \alpha^2/(4k^2) + \alpha^2/(4k^2))} \, dx$$

$$= \frac{1}{\sqrt{2\pi}} \int_{-\infty}^{\infty} e^{-k(x - i\alpha/(2k))^2 - \alpha^2/(4k)} \, dx$$

$$= \frac{1}{\sqrt{2\pi k}} e^{-\alpha^2/(4k)} \int_{-\infty}^{\infty} e^{-u^2} \, du$$

$$= \frac{1}{\sqrt{2\pi k}} e^{-\alpha^2/(4k)} \sqrt{\pi} = \frac{1}{\sqrt{2k}} e^{-\alpha^2/(4k)}. \quad ∎$$

EXAMPLE 6.14. Find the Fourier transform of $f(x) = 0$ for $x < 0$ and $f(x) = xe^{-ax}$ for $x > 0$.

$$\mathcal{F}f(x) = \frac{1}{\sqrt{2\pi}} \int_{-\infty}^{\infty} f(x) e^{i\alpha x} \, dx$$

$$= \frac{1}{\sqrt{2\pi}} \int_{0}^{\infty} xe^{-ax} e^{i\alpha x} \, dx$$

$$= \frac{xe^{-ax + i\alpha x}}{\sqrt{2\pi}(i\alpha - a)} \Big|_{0}^{\infty} - \frac{1}{\sqrt{2\pi}(i\alpha - a)} \int_{0}^{\infty} e^{-ax} e^{i\alpha x} \, dx$$

$$= -\frac{1}{\sqrt{2\pi}(i\alpha - a)} \int_{0}^{\infty} e^{-ax} e^{i\alpha x} \, dx = \frac{1}{\sqrt{2\pi}(i\alpha - a)^2}. \quad ∎$$

EXAMPLE 6.15. Find the Fourier transform of $f(x) = 0$ if $x < b$ and $f(x) = e^{-a^2 x^2}$ if $0 < b < x$. The solution is

$$\tilde{f}(\alpha) = \frac{1}{\sqrt{2\pi}} \int_b^\infty e^{-a^2 x^2} e^{ix\alpha} \, dx = \frac{1}{\sqrt{2\pi}} \int_b^\infty e^{-a^2(x - i\alpha/(2a^2))^2 - \alpha^2/(4a^2)} \, dx$$

$$= \frac{e^{-\alpha^2/(4a^2)}}{a\sqrt{2\pi}} \int\limits_{ab - i\alpha/(2a)}^\infty e^{-u^2} \, du = \frac{1}{2a\sqrt{2}} e^{-\alpha^2/(4a^2)} \operatorname{erfc}\left(ab - \frac{i\alpha}{2a}\right). \quad \blacksquare$$

EXAMPLE 6.16. Solve the partial differential equation $u_x + u_y + kyu = f(x)$, in the domain $|x| \geq 0, y > 0$, with the boundary conditions $u(x, 0) = 0$, $\lim\limits_{x \to \pm\infty} u(x, y) = 0$, where $f(x)$ is a function such that $f(x) \to 0$ as $|x| \to \infty$.

The given partial differential equation in the domain of the Fourier transform with respect to x becomes

$$\tilde{u}(\alpha, y)_y + (-i\alpha + ky)\,\tilde{u}(\alpha, y) = \tilde{f}(\alpha), \tag{6.63}$$

where $\tilde{u} = \mathcal{F}u$ and $\tilde{f}(\alpha) = \mathcal{F}f$. The differential equation in $\tilde{u}$ is a linear equation of first order and its solution subject to the given boundary conditions is

$$\tilde{u} = e^{i\alpha y - ky^2/2} \tilde{f}(\alpha) \int_0^y e^{-i\alpha t + kt^2/2} \, dt. \tag{6.64}$$

This solution, on inversion, yields

$$u = e^{-ky^2/2} \int_0^y f(x - y + t) e^{kt^2/2} \, dt. \tag{6.65}$$

Note that by using Property 1, we get $\mathcal{F}^{-1}\tilde{f}(\alpha)e^{-i\alpha(t-y)} = f(x - y + t)$. $\blacksquare$

EXAMPLE 6.17. Find the solution of the Laplace equation $u_{xx} + u_{yy} = 0$ in the domain $|x| < \infty$ and $y \geq 0$, subject to the conditions that $u \to 0$ as $|x| \to \infty$ or as $y \to \infty$, and $u(x, 0) = \delta(x)$. After applying the Fourier transform to the partial differential equation with respect to x, we get

$$\tilde{u}_{yy} - \alpha^2 \tilde{u} = 0,$$

whose solution is

$$\tilde{u} = A e^{-|\alpha| y}.$$

Applying the boundary condition at $y = 0$ in the transform domain, we get $\tilde{u}(\alpha, 0) = A = \frac{1}{\sqrt{2\pi}}$. Hence, $\tilde{u} = \frac{1}{\sqrt{2\pi}} e^{-|\alpha| y}$. On inverting, we obtain

$$u(x, y) = \frac{1}{\pi} \frac{y}{x^2 + y^2}.$$

Now, we use the convolution theorem (§6.8.2) to obtain the solution to the problem with arbitrary condition $u(x,0) = f(x)$. Then the solution is

$$u(x,y) = \frac{1}{\pi} \int_{-\infty}^{\infty} \frac{y f(\eta)}{(x-\eta)^2 + y^2} \, d\eta, \tag{6.66}$$

which is known as the Poisson integral representation for the Dirichlet problem in the half-plane. ∎

EXAMPLE 6.18. Find the general solution of the diffusion equation $u_t = k u_{xx}$ under the nonhomogeneous initial condition $u(x,0) = f(x)$, subject to the conditions that $\lim\limits_{x \to \pm\infty} f(x), u(x,t) \to 0$. Applying Fourier transform to the partial differential equation, we get

$$\tilde{u}_t + k\alpha^2 \tilde{u} = 0,$$

with the initial condition $\tilde{u}(\alpha, 0) = \tilde{f}(\alpha)$. Hence, the solution, after applying the initial condition, is

$$\tilde{u}(\alpha, t) = \tilde{f}(\alpha) e^{-k\alpha^2 t},$$

which, on inversion, yields

$$u(x,t) = \frac{1}{2\sqrt{\pi k t}} \int_{-\infty}^{\infty} f(\eta) e^{-(x-\eta)^2/(4kt)} \, d\eta,$$

where we have used the convolution theorem and Example 6.13. ∎

6.9. Fourier Sine and Cosine Transforms

We will discuss Fourier sine and Fourier cosine transform simultaneously. This is because most of the time they are used simultaneously while solving a problem.

6.9.1. Properties of Fourier Sine and Cosine Transforms.

$$\mathcal{F}_c f(x) = \tilde{f}_c(\alpha), \quad \mathcal{F}_s f(x) = \tilde{f}_s(\alpha), \tag{6.67}$$

$$\mathcal{F}_c \tilde{f}_c(x) = f(\alpha), \quad \mathcal{F}_s \tilde{f}_s(x) = f(\alpha), \tag{6.68}$$

$$\mathcal{F}_c f(kx) = \frac{1}{k} \tilde{f}_c(\frac{\alpha}{k}), \quad k > 0, \tag{6.69}$$

$$\mathcal{F}_s f(kx) = \frac{1}{k} \tilde{f}_s(\frac{\alpha}{k}), \quad k > 0, \tag{6.70}$$

$$\mathcal{F}_c f(kx)\cos bx = \frac{1}{2k}\left[\tilde{f}_c\left(\frac{\alpha+b}{k}\right)+\tilde{f}_c\left(\frac{\alpha-b}{k}\right)\right], \quad k>0, \tag{6.71}$$

$$\mathcal{F}_c f(kx)\sin bx = \frac{1}{2k}\left[\tilde{f}_s\left(\frac{\alpha+b}{k}\right)-\tilde{f}_s\left(\frac{\alpha-b}{k}\right)\right], \quad k>0, \tag{6.72}$$

$$\mathcal{F}_s f(kx)\cos bx = \frac{1}{2k}\left[\tilde{f}_s\left(\frac{\alpha+b}{k}\right)+\tilde{f}_s\left(\frac{\alpha-b}{k}\right)\right], \quad k>0, \tag{6.73}$$

$$\mathcal{F}_s f(kx)\sin bx = \frac{1}{2k}\left[\tilde{f}_c\left(\frac{\alpha-b}{k}\right)-\tilde{f}_c\left(\frac{\alpha+b}{k}\right)\right], \quad k>0, \tag{6.74}$$

$$\mathcal{F}_c x^{2n} f(x) = (-1)^n \frac{d^{2n}\tilde{f}_c(\alpha)}{d\alpha^{2n}}, \tag{6.75}$$

$$\mathcal{F}_c x^{2n+1} f(x) = (-1)^n \frac{d^{2n+1}\tilde{f}_s(\alpha)}{d\alpha^{2n+1}}, \tag{6.76}$$

$$\mathcal{F}_s x^{2n} f(x) = (-1)^n \frac{d^{2n}\tilde{f}_s(\alpha)}{d\alpha^{2n}}, \tag{6.77}$$

$$\mathcal{F}_s x^{2n+1} f(x) = (-1)^{n+1} \frac{d^{2n+1}\tilde{f}_c(\alpha)}{d\alpha^{2n+1}}. \tag{6.78}$$

6.9.2. Convolution Theorems for Fourier Sine and Cosine Transforms. These theorems are as follows:

Theorem 6.8. *Let $\tilde{f}_c(\alpha)$ and $\tilde{g}_c(\alpha)$ be the Fourier cosine transforms of $f(x)$ and $g(x)$, respectively, and let $\tilde{f}_s(\alpha)$ and $\tilde{g}_s(\alpha)$ be the Fourier sine transforms of $f(x)$ and $g(x)$, respectively. Then*

$$\mathcal{F}_c^{-1}[\tilde{f}_c(\alpha)\tilde{g}_c(\alpha)] = \frac{1}{\sqrt{2\pi}}\int_0^\infty g(\eta)[f(|x-\eta|)+f(x+\eta)]\,d\eta. \tag{6.79}$$

PROOF. We have

$$\int_0^\infty \tilde{f}_c(\alpha)\,\tilde{g}_c(\alpha)\cos\alpha x\,d\alpha = \sqrt{\frac{2}{\pi}}\int_0^\infty \tilde{f}_c(\alpha)\cos\alpha x\,d\alpha\int_0^\infty g(\eta)\cos\alpha\eta\,d\eta$$

$$= \sqrt{\frac{2}{\pi}}\int_0^\infty g(\eta)\,d\eta\int_0^\infty \tilde{f}_c(\alpha)\cos\alpha x\cos\alpha\eta\,d\alpha$$

$$= \frac{1}{2}\sqrt{\frac{2}{\pi}}\int_0^\infty g(\eta)\,d\eta\int_0^\infty \tilde{f}_c(\alpha)[\cos\alpha(|x-\eta|)+\cos\alpha(x+\eta)]\,d\alpha$$

$$= \frac{1}{2}\int_0^\infty g(\eta)[f(|x-\eta|)+f(x+\eta)]\,d\eta. \blacksquare$$

Theorem 6.9. *If $\tilde{f}_s(\alpha)$, $\tilde{f}_c(\alpha)$ and $\tilde{g}_s(\alpha)$, $\tilde{g}_s(\alpha)$ are the Fourier sine and cosine transforms of $f(x)$ and $g(x)$, respectively, then the following results*

hold:

(i) $\displaystyle\int_0^\infty \tilde{f}_c(\alpha)\,\tilde{g}_s(\alpha)\sin\alpha x\,d\alpha = \frac{1}{2}\int_0^\infty g(\eta)[f(|x-\eta|)-f(x+\eta)]\,d\eta,$
$$\tag{6.80}$$

(ii) $\displaystyle\int_0^\infty \tilde{f}_s(\alpha)\,\tilde{g}_c(\alpha)\sin\alpha x\,d\alpha = \frac{1}{2}\int_0^\infty f(\eta)[g(|x-\eta|)-g(x+\eta)]\,d\eta.$
$$\tag{6.81}$$

(iii) $\displaystyle\int_0^\infty \tilde{f}_s(\alpha)\,\tilde{g}_s(\alpha)\cos\alpha x\,d\alpha$

$$= \frac{1}{2}\int_0^\infty g(t)\left[H(t+x)\,f(t+x)+H(t-x)\,f(t-x)\right]dt$$

$$= \frac{1}{2}\int_0^\infty f(t)\left[H(t+x)\,g(t+x)+H(t-x)\,g(t-x)\right]dt;$$
$$\tag{6.82}$$

(iv) $\displaystyle\int_0^\infty \tilde{f}_c(\alpha)\,\tilde{g}_c(\alpha)\,d\alpha = \int_0^\infty f(\eta)g(\eta)\,d\eta = \int_0^\infty \tilde{f}_s(\alpha)\tilde{g}_s(\alpha)\,d\alpha.$
$$\tag{6.83}$$

The proofs for these theorems are left as exercises. Now, we derive Fourier sine and cosine transforms of some functions.

EXAMPLE 6.19. Define

$$I_1 = \int_0^\infty e^{-ax}\sin bx\,dx, \quad I_2 = \int_0^\infty e^{-ax}\cos bx\,dx.$$

Then $I_2 = \dfrac{1}{a} - \dfrac{b}{a}I_1$, and $I_1 = \dfrac{b}{a}I_2$. Solving for I_1 and I_2, we get

$$I_1 = \frac{b}{a^2+b^2}, \quad\text{and}\quad I_2 = \frac{a}{a^2+b^2}.$$

If $f(x) = e^{-bx}$, then

$$\tilde{f}_c(\alpha) = \sqrt{\frac{2}{\pi}}\frac{b}{\alpha^2+b^2},$$

and

$$\tilde{f}_s(\alpha) = \sqrt{\frac{2}{\pi}}\frac{\alpha}{\alpha^2+b^2}.$$

These two results yield on inversion

$$\int_0^\infty \frac{\cos\alpha x}{\alpha^2+b^2}\,d\alpha = \frac{\pi}{2b}e^{-bx},$$

and

$$\int_0^\infty \frac{\alpha \sin \alpha x}{\alpha^2 + b^2}\, d\alpha = \frac{\pi}{2} e^{-bx}.$$

An interesting integral is obtained by defining

$$f(x) = \begin{cases} 1 & \text{for } 0 < x < b, \\ 0 & \text{for } b < x. \end{cases}$$

Then

$$\tilde{f}_c(\alpha) = \sqrt{\frac{2}{\pi}} \frac{\sin \alpha b}{\alpha}. \tag{6.84}$$

Also, define $g(x) = e^{-ax}$. Then

$$\tilde{g}_c(\alpha) = \sqrt{\frac{2}{\pi}} \frac{a}{\alpha^2 + a^2}.$$

Thus, we have

$$\int_0^\infty \tilde{f}_c(\alpha)\, \tilde{g}_c(\alpha)\, d\alpha = \frac{2}{\pi} \int_0^\infty \frac{a}{\alpha} \frac{\sin \alpha b}{\alpha^2 + a^2}\, d\alpha = \int_0^\infty f(\eta) g(\eta)\, d\eta$$

$$= \int_0^b e^{-a\eta}\, d\eta = \frac{1 - e^{-ab}}{a}. \ \blacksquare$$

EXAMPLE 6.20. Show that

$$\int_0^\infty \frac{1}{\lambda^2} \sin \lambda a \sin \lambda b\, d\lambda = \frac{\pi}{2} \min(a, b).$$

If we define

$$f(x) = \begin{cases} 1 & \text{for } 0 \le x < a \\ 0 & \text{for } a < x, \end{cases} \quad \text{and} \quad g(x) = \begin{cases} 1 & \text{for } 0 \le x < b \\ 0 & \text{for } b < x, \end{cases}$$

then

$$\int_0^\infty \tilde{f}_c(\lambda)\, \tilde{g}_c(\lambda)\, d\lambda = \int_0^\infty f(x)\, g(x)\, dx.$$

Since, in view of (6.84), $\tilde{f}_c(\lambda) = \sqrt{\dfrac{2}{\pi}} \dfrac{\sin \lambda a}{\lambda}$, $\tilde{g}_c(\lambda) = \sqrt{\dfrac{2}{\pi}} \dfrac{\sin \lambda b}{\lambda}$, we have

$$\frac{2}{\pi} \int_0^\infty \frac{1}{\lambda^2} \sin \lambda a \sin \lambda b\, d\lambda = \int_0^\infty f(x) g(x)\, dx = \min(a, b). \ \blacksquare$$

EXAMPLE 6.21. Find the Fourier cosine transform of $f(x)$, where

$$f(x) = \begin{cases} x & \text{for } 0 < x < 1, \\ 2 - x & \text{for } 1 < x < 2, \\ 0 & \text{for } 2 < x < \infty. \end{cases}$$

Here,

$$\mathcal{F}_c[f(x)] = \sqrt{\frac{2}{\pi}} \left[\int_0^1 x \cos \alpha x \, dx + \int_1^2 (2 - x) \cos \alpha x \, dx \right]$$

$$= \sqrt{\frac{2}{\pi}} \left[\frac{x \sin \alpha x}{\alpha} \Big|_0^1 - \frac{1}{\alpha} \int_0^1 \sin \alpha x \, dx + \frac{(2-x) \sin \alpha x}{\alpha} \Big|_1^2 \right.$$

$$\left. + \frac{1}{\alpha} \int_1^2 \sin \alpha x \, dx \right]$$

$$= \sqrt{\frac{2}{\pi}} \left[\frac{2 \cos \alpha - 1 - \cos 2\alpha}{\alpha^2} \right]. \blacksquare$$

EXAMPLE 6.22. We will use the Fourier transform to solve Example 6.6, which was earlier solved by the Laplace transform. The partial differential equation and the boundary and initial conditions are

$$k \, u_{xx} = u_t,$$

$$u = 0 \quad \text{for } t \leq 0, \quad u \to 0 \text{ as } x \to \infty, \quad \text{and} \quad u = T_0 \text{ at } x = 0 \text{ for } t > 0.$$

Applying the Fourier sine transform to the partial differential equation, we get

$$\frac{\partial \tilde{u}_s}{\partial t} + k \alpha^2 \tilde{u}_s = \sqrt{\frac{2}{\pi}} k \alpha T_0,$$

where $\tilde{u}_s$ is the Fourier sine transform of u. Its solution is given by

$$\tilde{u}_s = A e^{-k\alpha^2 t} + \sqrt{\frac{2}{\pi}} \frac{T_0}{\alpha}.$$

By applying the initial condition at $t = 0$, we get

$$A + \sqrt{\frac{2}{\pi}} \frac{T_0}{\alpha} = 0.$$

Hence,

$$\tilde{u}_s = \sqrt{\frac{2}{\pi}} \frac{T_0}{\alpha} \left(1 - e^{-k\alpha^2 t} \right).$$

Thus, $u(x, t)$ is given by

$$
\begin{aligned}
u(x,t) &= \frac{2}{\pi} T_0 \int_0^\infty \frac{1}{\alpha} \left(1 - e^{-k\alpha^2 t} \right) \sin \alpha x \, d\alpha \\
&= \frac{2}{\pi} T_0 \left[\int_0^\infty \frac{\sin \alpha x}{\alpha} \, d\alpha - \int_0^\infty \frac{\sin \alpha x}{\alpha} e^{-k\alpha^2 t} \, d\alpha \right] \\
&= \frac{2}{\pi} T_0 \left[\frac{\pi}{2} - \frac{\pi}{2} \operatorname{erf} \frac{x}{2\sqrt{kt}} \right] \\
&= T_0 \operatorname{erfc} \frac{x}{2\sqrt{kt}}. \quad \blacksquare
\end{aligned}
$$

6.10. Finite Fourier Transforms

When the domain of the physical problem is finite, it is generally not convenient to use the transforms with an infinite range of integration. In many cases finite Fourier transform can be used with advantage. We define

$$
\tilde{f}_s(n) = \int_0^a f(x) \sin \left(\frac{n\pi x}{a} \right) \, dx \tag{6.85}
$$

as the finite Fourier sine transform of $f(x)$. The function $f(x)$ is then given by

$$
f(x) = \frac{2}{a} \sum_1^\infty \tilde{f}_s(n) \sin \left(\frac{n\pi x}{a} \right). \tag{6.86}
$$

Similarly, the finite Fourier cosine transform is defined by

$$
\tilde{f}_c(n) = \int_0^a f(x) \cos \left(\frac{n\pi x}{a} \right) \, dx, \tag{6.87}
$$

and its inverse by

$$
f(x) = \frac{\tilde{f}_c(0)}{a} + \frac{2}{a} \sum_1^\infty \tilde{f}_c(n) \cos \left(\frac{n\pi x}{a} \right). \tag{6.88}
$$

EXAMPLE 6.23. Consider the Laplace equation in the rectangle $\{0 < x < a, \, 0 < y < b\}$

$$
u_{xx} + u_{yy} = 0, \tag{6.89}
$$

with the boundary conditions

$$
u(0, y) = u(a, y) = u(x, b) = 0, \quad \text{and} \quad u(x, 0) = f(x).
$$

After applying the finite Fourier sine transform to $u(x, y)$ with respect to x from 0 to a, we have

$$(\tilde{u}_s)_{xx}(n) = \int_0^a u_{xx} \sin \frac{n\pi x}{a} \, dx$$

$$= \frac{n\pi}{a} \left[u(0, y) - (-1)^n u(a, y) \right] - \frac{n^2\pi^2}{a^2} \tilde{u}_s(n). \tag{6.90}$$

Then Equation (6.89) becomes

$$\left[\frac{d^2}{dy^2} - \frac{n^2\pi^2}{a^2} \right] \tilde{u}_s(n, y) = 0.$$

Solving for $\tilde{u}_s(n, y)$, we get

$$\tilde{u}_s(n, y) = A e^{n\pi y/a} + B e^{-n\pi y/a}.$$

Since $\tilde{u}_s(n, b) = 0$, we can express $\tilde{u}_s(n, y)$ as

$$\tilde{u}_s(n, y) = A_n \left(e^{n\pi(y-b)/a} - e^{-n\pi(y-b)/a} \right).$$

On applying the boundary condition at $y = 0$, we get

$$A_n \left(e^{-n\pi b/a} - e^{n\pi b/a} \right) = \bar{f}_s(n),$$

which, after solving for A_n and substituting its value into $\tilde{u}_s(n, y)$, yields

$$\tilde{u}_s(n, y) = -\frac{\sinh[n\pi(y-b)/a]}{\sinh(n\pi b/a)} \bar{f}_s(n).$$

Hence,

$$u(x, y) = \frac{2}{a} \sum_1^\infty \frac{\sinh[n\pi(b-y)/a]}{\sinh(n\pi b/a)} \bar{f}_s(n) \sin(n\pi x/a),$$

where

$$\bar{f}_s(n) = \int_0^a f(\xi) \sin(n\pi\xi/a) d\xi. \quad \blacksquare$$

EXAMPLE 6.24. Solve the wave equation

$$u_{tt} = c^2 u_{xx}, \quad 0 < x < l,$$

subject to the conditions

$$u(0, t) = g(t) \quad \text{and} \quad u(l, t) = 0 \quad \text{for } 0 < x < l,$$
$$u(x, 0) = 0 \quad \text{for } t > 0.$$

Taking the finite Fourier sine transform with respect to x and using the boundary conditions at $x = 0, l$, and (6.90), we get

$$\frac{d^2\tilde{u}_s}{dt^2} + \frac{n^2\pi^2c^2}{l^2}\tilde{u}_s = \frac{n\pi c^2}{l}g(t).$$

The general solution of this equation is

$$\tilde{u}_s(n) = A\cos\frac{n\pi ct}{l} + B\sin\frac{n\pi ct}{l} + \tilde{u}_{s,p}(n),$$

where $\tilde{u}_{s,p}(n)$ is the particular solution which can be obtained by the method of variation of parameters as

$$\tilde{u}_{s,p}(n) = c\int_0^t g(\tau)\sin\frac{n\pi c(t-\tau)}{l}\,dt$$

$$= c\int_0^t g(t-\tau)\sin\frac{n\pi c\tau}{l}\,d\tau.$$

With this choice of $\tilde{u}_{s,p}(n)$, the constants A and B become zero because of the initial condition $\tilde{u}_s(n, 0) = 0$. Hence, $\tilde{u}_s(n) = \tilde{u}_{s,p}(n)$. Thus, by (6.86)

$$u(x, t) = \frac{2c}{l}\sum_{n=1}^{\infty}\sin\frac{n\pi x}{l}\int_0^t g(t-\tau)\sin\frac{n\pi c\tau}{l}\,d\tau. \ \blacksquare$$

Note that the finite cosine transforms for the derivatives of a function u can be obtained analogous to (6.90) (see Exercise 6.23).

EXAMPLE 6.25. By using both Laplace and Fourier transforms, we will solve the boundary value problem

$$u_{tt} - c^2\,u_{xx} = e^{-|x|}\sin t, \quad u(x, 0) = 0, \quad u_t(x, 0) = e^{-|x|}.$$

Note that this equation in the Laplace transform domain is

$$s^2\,\overline{u}(x, s) - c^2\,\overline{u}_{xx}(x, s) = \frac{2 + s^2}{1 + s^2}e^{-|x|}.$$

Applying the complex Fourier transform, we get

$$(c^2\alpha^2 + s^2)\,\tilde{\overline{u}}(x, s) = \frac{2 + s^2}{1 + s^2}\sqrt{\frac{2}{\pi}}\frac{1}{1 + \alpha^2}.$$

Thus,

$$\tilde{\overline{u}}(x, s) = \frac{2 + s^2}{1 + s^2}\sqrt{\frac{2}{\pi}}\frac{1}{(1 + \alpha^2)(c^2\alpha^2 + s^2)}$$

$$= \frac{2 + s^2}{(1 + s^2)(s^2 - c^2)}\sqrt{\frac{2}{\pi}}\left[\frac{1}{1 + \alpha^2} - \frac{c^2}{c^2\alpha^2 + s^2}\right].$$

On inverting the Fourier transform, we have

$$\overline{u}(x,s) = \frac{2+s^2}{(1+s^2)(s^2-c^2)}\left(e^{-|x|} - \frac{c}{s}e^{-s|x|/c}\right)$$

$$= \left\{B\left(\frac{1}{s-c} - \frac{1}{s+c}\right) - \frac{1}{(1+c^2)(1+s^2)}\right\}e^{-|x|}$$

$$+ \left\{\frac{2}{cs} - B\left(\frac{1}{s-c} + \frac{1}{s+c}\right) - \frac{cs}{(1+c^2)(1+s^2)}\right\}e^{-s|x|/c},$$

where $B = \dfrac{2+c^2}{2c(1+c^2)}$. After taking the Laplace inverse, we find that

$$u(x,t) = \left[B\left(e^{ct} - e^{-ct}\right) - \frac{1}{1+c^2}\sin t\right]e^{-|x|}$$

$$+ H(ct-|x|)\left[\frac{2}{c} - B\left(e^{ct-|x|} + e^{-ct+|x|}\right) - \frac{c}{1+c^2}\cos(t-|x|/c)\right]. \blacksquare$$

6.11. Mathematica Projects

We present some projects involving Fourier transforms as well as Fourier sine and cosine transforms.

PROJECT 6.7. First we set the Fourier overall constant as $1/2\sqrt{\pi}$, and then compute $\mathcal{F}\{e^{-k|x|}\}$.

```
(* Load the package *)
```
$In[1]:$ Needs["Calculus`FourierTransform"];

$In[2]:$
?$FourierOverallConstant

$Out[2]:$
   $FourierOverallConstant is the default setting for the
   option FourierOverallConstant (an option to FourierTransform and
   related functions).

$In[3]:$ $FourierOverallConstant

$Out[3]:$ 1

```
(* Reset the constant *)
```
In[4]:
```
$FourierOverallConstant = 1/Sqrt[2 Pi]
```

Out[4]: $\dfrac{1}{2\ \text{Sqrt}[\text{Pi}]}$

In[5]: `FourierTransform[Exp[-k Abs[x]],x,w]`

Out[5]: $\dfrac{k\ \text{Sqrt}[\frac{2}{\text{Pi}}]}{k^2 + w^2}$

```
(* Alternate calculation; takes time *)
```
In[6]:
```
  int1:= Integrate[Exp[k x] Exp[I x a],x,-Infinity,0]
  int2:= Integrate[Exp[-k x] Exp[I x a],x,0,Infinity]
```
In[8]:
```
  result = 1/Sqrt[2 Pi] (int1 + int2)//Simplify
```

```
  General::intinit:  Loading integration packages -- please wait.
```
Out[7]:

$\dfrac{k\ \text{Sqrt}[\frac{2}{\text{Pi}}]}{k^2 + a^2}$

PROJECT 6.8. Compute $\mathcal{F}\{e^{-kx^2}\}$.

In[1]: `Needs["Calculus'FourierTransform"];`

In[2]: `FourierTransform[Exp[-k t^2],t,w]`

Out[2]: $\dfrac{1}{\text{Sqrt}[2]\ \text{E}^{w^2/(4k)}\ \text{Sqrt}[k]}$

PROJECT 6.9. Mathematica is used to solve Example 6.15.

In[1]: `<<Declare.m`

Out[1]: `{Declare, NewDeclare, NonPositive, RealQ}`

In[2]:
```
  Declare[a,Positive];
```

```
f[x_]:= Exp[-a^2 x^2]
int:= Integrate[f[x] Exp[I x alpha],{x,b,Infinity}]
sub[X_]:= 1 - Erf[X] -> Erfc[X//Together]
Simplify[1/Sqrt[2 Pi] int]/.sub[a*((-I/2*alpha)/ a^2 + b)]
```

Out[6]:

$$\frac{\text{Erfc}[\dfrac{-I\ alpha + 2\ a^2\ b}{2\ a}]}{2\ \text{Sqrt}[2]\ a\ E^{alpha^2/(4a^2)}}$$

PROJECT 6.10. Compute the Fourier cosine transform of $f(t)$, where

$$f(x) = \begin{cases} x & \text{if } 0 < x < 1, \\ 2-x & \text{if } 1 < x < 2, \\ 0 & \text{if } x \geq 2. \end{cases}$$

In[1]: Needs[
"Calculus`FourierTransform`","Algebra`Trigonometry`"];

In[2]: Clear[f];
```
    f[x_]:= If[0< x <1, x ,If[1< x < 2, 2-x ,0]]
```

In[4]:
```
    FourierCosTransform[f[t],t,w]
```

```
On::none:  Message SeriesData::csa not found.
On::none:  Message SeriesData::csa not found.
On::none:  Message SeriesData::csa not found.
General::stop:  Further output of On::none
will be suppressed during this calculation.
```

Out[4]: FourierCosTransform[f[t],t,w]

In[5]:
```
    int1:= Integrate[x Cos[alpha x],{x,0,1}]
    int2:= Integrate[(2-x) Cos[alpha x],{x,1,2}]
```

```
    result= Sqrt[2/Pi] (int1 + int2)//Simplify
```

Out[7]:

$$\frac{4\ \text{Sqrt}[\dfrac{2}{\text{Pi}}]\ \text{Cos[alpha]}\ \text{Sin}[\dfrac{alpha}{2}]^2}{alpha^2}$$

6.12. Exercises

6.14. Find the complex Fourier transform of of the following functions:

(a) $f(x) = \begin{cases} 0 & \text{if } x < 0 \\ e^{-ax} & \text{if } x > 0. \end{cases}$

ANS. $\dfrac{1}{\sqrt{2\pi}(a - i\alpha)}$.

(b) $f(x) = \begin{cases} 0 & \text{if } x < 0, \\ \dfrac{1}{x}[e^{-ax} - e^{-bx}] & \text{if } x > 0 \text{ and } b > a > 0. \end{cases}$

HINT. Use part (a) and integrate with respect to a.

ANS. $\dfrac{1}{\sqrt{2\pi}} \ln \dfrac{b - i\alpha}{a - i\alpha}$.

(c) $f(x) = \begin{cases} 0 & \text{if } |x| > a, \\ 1 - \dfrac{|x|}{a} & \text{if } |x| < a. \end{cases}$

ANS. $\sqrt{\dfrac{2}{\pi}} \dfrac{1}{a\alpha^2}(1 - \cos \alpha a)$.

(d) $f(x) = \cos ax^2$ and $f(x) = \sin ax^2$.

HINT: Use Example 6.13 and define k as ia.

ANS. $\dfrac{1}{2\sqrt{a}}\left(\sin \dfrac{\alpha^2}{4a} + \cos \dfrac{\alpha^2}{4a}\right)$ and $\dfrac{1}{2\sqrt{a}}\left(\sin \dfrac{\alpha^2}{4a} - \cos \dfrac{\alpha^2}{4a}\right)$.

(e) $f(x) = \begin{cases} 1 & \text{if } |x| < 1 \\ 0 & \text{if } |x| > 1. \end{cases}$

ANS. $\dfrac{\sqrt{2}}{\alpha\sqrt{\pi}} \sin \alpha$.

(f) $f(x) = \begin{cases} \sin kx & \text{if } |x| < 1 \\ 0 & \text{if } |x| > 1. \end{cases}$

ANS. $\dfrac{i}{\sqrt{2\pi}}\left[\dfrac{\sin(k - \alpha)}{k - \alpha} - \dfrac{\sin(k + \alpha)}{k + \alpha}\right]$.

(g) $f(x) = \begin{cases} \dfrac{1}{\sqrt{x}} & \text{if } x > 0 \\ = 0 & \text{if } x < 0 \end{cases}$.

HINT: Substitute $\alpha x = v^2 e^{i\pi/2}$.

ANS. $\dfrac{e^{i\pi/4}}{\sqrt{2\alpha}}$, $\alpha > 0$.

(h) $f(x) = \begin{cases} \dfrac{1}{\sqrt{|x|}} & \text{if } x < 0 \\ 0 & \text{if } x > 0. \end{cases}$

ANS. $\dfrac{e^{-i\pi/4}}{\sqrt{2\alpha}}$, $\alpha > 0$.

(i) $f(x) = \dfrac{1}{\sqrt{|x|}}$.

Ans. $\dfrac{1}{\sqrt{|\alpha|}}$.

6.15. Solve the partial differential equation $u_{tt} - c^2\, u_{xx} = 0$, subject to the conditions $u(x,0) = f(x) + g(x)$ and $u_t(x,0) = c\,(f'(x) - g'(x))$, where $u(x,t)$, $f(x)$, $g(x)$, $u'(x,t)$, $f'(x)$, and $g'(x)$ all go to zero as $|x| \to \infty$.
Ans. $u = f(x+ct) + g(x-ct)$.

6.16. Solve the following system for $h(x,t)$: $u_t - v = -h_x$, $v_t + u = 0$, $h_t + c^2\, u_x = 0$, subject to the conditions $u(x,0) = v(x,0) = 0$ and $h(x,0) = k$ if $|x| < a$ and $h(x,0) = 0$ if $|x| > a$; in addition $u(x,t)$, $v(x,t)$ and $h(x,t)$ approach 0 as $|x| \to \infty$, where $a > 0$ is a real number. This problem is connected with the shallow water waves.
Ans.

$$h(x,t) = \frac{2k}{\pi} \int_0^\infty \frac{\sin(a\alpha)}{\alpha} \left[\frac{1 + \alpha^2 c^2 \cos(t\sqrt{1 + \alpha^2 c^2})}{1 + \alpha^2 c^2}\right] \cos(\alpha x)\, d\alpha.$$

6.17. Solve the partial differential equation $\rho^2 \dfrac{\partial^2 u}{\partial \rho^2} = c^2 \dfrac{\partial}{\partial \rho}\left(\rho^2 \dfrac{\partial u}{\partial \rho}\right)$ for $\rho > a$, with initial conditions $u(\rho,t) = u_t(\rho,t) = 0$ for $t \leq 0$ and the boundary condition $u_\rho(a,t) = H(t)f(t)$, by using Fourier transform with respect to t. Note that $\rho \geq a$, and the equation satisfies Sommerfeld's radiation conditions $\lim\limits_{\rho\to\infty} \left[u(\rho,t), u_\rho(\rho,t)\right] = 0$.

Hint. Note that $\mathcal{F}^{-1}\left\{\dfrac{1}{i\alpha - b}\right\} = -\sqrt{2\pi}\, H(t)\, e^{-bt}$, and $\mathcal{F}^{-1}\left\{\dfrac{e^{i\alpha k(\rho - a)}}{i\alpha - b}\right\} = -\sqrt{2\pi}\, H[t - k(\rho - a)]e^{-b[t - k(\rho - a)]}$. Use the convolution theorem.
Ans. See Example 6.10.

6.18. Find the Fourier sine and cosine transforms of $f(x) = \dfrac{1}{\sqrt{x}} e^{-ax}$.

Ans. $\mathcal{F}_c f(x) = \dfrac{\sqrt{\sqrt{a^2 + \alpha^2} + a}}{\sqrt{a^2 + \alpha^2}}$, $\mathcal{F}_s f(x) = \dfrac{\sqrt{\sqrt{a^2 + \alpha^2} - a}}{\sqrt{a^2 + \alpha^2}}$.

6.19. Find the Fourier cosine transform of $\dfrac{1}{x}\left[e^{-ax} - e^{-bx}\right]$, $\Re\{a\}, \Re\{b\} > 0$.

Ans. $\dfrac{1}{2\pi} \ln\left(\dfrac{a^2 + \alpha^2}{b^2 + \alpha^2}\right)$.

6.20. Find the Fourier sine transform of $\dfrac{1}{x} e^{-ax}$, $\Re\{a\} > 0$.

Ans. $\sqrt{\dfrac{2}{\pi}} \tan^{-1} \dfrac{\alpha}{a}$.

6.21. Derive additional formulas for Fourier sine and cosine transforms from Example 6.19 by differentiating or integrating with respect to a.

6.22. Find Fourier sine and cosine transforms of $f(x) = \sqrt{x}e^{-ax}$, $a > 0$.

HINT: $\tilde{f}_c(\alpha) + i\tilde{f}_s(\alpha) = \dfrac{1}{\sqrt{2(a - i\alpha)^3}}$; then express $(a - i\alpha)$ in the polar form.

ANS. $\tilde{f}_c(\alpha) + i\tilde{f}_s(\alpha) = \dfrac{e^{(3i/2)\,\tan^{-1}(\alpha/2)}}{\sqrt{2}(a^2 + \alpha^2)^{3/4}}$.

6.23. Derive the formulas for the finite sine and cosine transforms for the first, second, and third derivatives of a function $u(x)$ in the interval $[0, l]$.

ANS. For the finite sine transform

$$\left(\frac{du}{dx}\right)_s = -\frac{n\pi}{l}\,\tilde{u}_c(n),$$

$$\left(\frac{d^2u}{dx^2}\right)_s \text{ is given by (6.90),}$$

$$\left(\frac{d^3u}{dx^3}\right)_s = -\frac{n\pi}{l}\left[(-1)^n u'(l) - u'(0)\right] - \frac{n^3\pi^3}{l^3}\,\tilde{u}_c(n).$$

For the finite cosine transform

$$\left(\frac{du}{dx}\right)_c = \left[(-1)^n u(l) - u(0)\right] + \frac{n\pi}{l}\,\tilde{u}_s(n),$$

$$\left(\frac{d^2u}{dx^2}\right)_c = \left[(-1)^n u'(l) - u'(0)\right] - \frac{n^2\pi^2}{l^2}\,\tilde{u}_c(n),$$

$$\left(\frac{d^3u}{dx^3}\right)_c = \left[(-1)^n u''(l) - u''(0)\right] - \frac{n^2\pi^2}{l^2}\left[(-1)^n u(l) - u(0)\right]$$

$$- \frac{n^3\pi^3}{l^3}\,\tilde{u}_s(n).$$

7

Green's Functions

A Green's function for a partial differential equation is the solution of its adjoint equation, where the forcing term is the Dirac delta function due to a unit point source in a given domain Ω. This solution enables us to generate solutions of partial differential equations subject to a range of boundary conditions and internal sources. This technique is important in a variety of physical problems. For the derivation of Green's functions, we can assume the presence of an internal source or a certain boundary condition which results in the same effect as the point source. If L is a linear differential operator, and $Lu(\mathbf{x}) = f(\mathbf{x})$ is a linear differential equation, where $\mathbf{x}$ denotes a point in the domain $\Omega \in R^n$, and $f(\mathbf{x})$ is the nonhomogeneous term in the differential equation, then the solution of this equation subject to homogeneous boundary conditions can be written in the form $u(\mathbf{x}) = \int \cdots \int_\Omega G(\mathbf{x}, \mathbf{y}) f(\mathbf{y}) \, d\mathbf{y}$, where $G(\mathbf{x}, \mathbf{y})$ is the Green's function, which depends only on the adjoint operator L^* of L and on the geometry of the problem for homogeneous adjoint initial and boundary conditions.

7.1. Generalized Functions

To derive Green's functions, it is important to understand the mathematical behavior of generalized functions. Therefore, we will study certain elementary aspects of the distribution theory, followed by the concept of Green's functions and their construction for different types of boundary value problems.

The point source is represented by the Dirac delta function, which belongs to a class of functions known as generalized functions or distributions. There exist relations which are not functions in the traditional sense. We start our discussion with an example. Consider a spring vibrating system, and suppose that an external force $F(t)$ is applied from $t_0 - t_1 \leq t \leq t_0 + t_1$, where $F(t)$ is defined as

$$F(t) = \begin{cases} 0, & t_0 < |t - t_1|, \\ f(t), & t_0 \geq |t - t_1|. \end{cases}$$

The measure $I(t)$ of the strength of this force at the point t_1 can be represented as

$$I(t_1) = \int_{-\infty}^{\infty} F(t)\, dt = \int_{t_1 - t_0}^{t_1 + t_0} f(t)\, dt.$$

Now, suppose that we increase $f(t)$ and decrease t_0 in such a way that $I(t_1)$ remains constant. Then we have what is known as an impulsive force. Let us assume that $I(t_1) = 1$. Then

$$\lim_{t_0 \to 0} \int_{t_1 - t_0}^{t_1 + t_0} f(t)\, dt = 1. \tag{7.1}$$

While there are many functions $f(t)$ that satisfy the requirement (7.1), one such function is $f(t) = \dfrac{1}{2t_0}$. Thus, we define $F(t)$ as

$$F(t) = \begin{cases} 0 & \text{if } t_0 < |t - t_1|, \\ \dfrac{1}{2t_0} & \text{if } t_0 \geq |t - t_1|. \end{cases} \tag{7.2}$$

A relationship of the type expressed by (7.2) is called a *generalized function* or a *distribution*. This function is denoted by $\delta(t - t_1)$ or $\delta(t, t_1)$, and is known as the *Dirac delta function*. There are other descriptions of this function but they all lead to the same properties, which are

$$\delta(t - t_1) = \begin{cases} 0 & \text{if } t \neq t_1, \\ \infty & \text{if } t \to t_1, \end{cases} \quad \text{and} \quad \int_{t_1 - \varepsilon}^{t_1 + \varepsilon} \delta(t - t_1)\, dt = 1, \tag{7.3}$$

where $\varepsilon > 0$ is an arbitrarily small real quantity. A consequence of the property (7.3) is

$$\int_{t_1 - \varepsilon}^{t_1 + \varepsilon} f(t)\, \delta(t - t_1)\, dt = f(t_1). \tag{7.4}$$

Note that the Dirac delta function, introduced in §6.3, was developed there from a different point of view. It satisfies the properties (7.3) and (7.4), which are proved there.

7.1.1. Dirac Delta Function in Curvilinear Coordinates.
The Dirac delta function in the polar coordinates (r, θ) in R^2 is given by

$$\delta(x)\, \delta(y) = \frac{\delta(r)}{2\pi r}, \quad r^2 = x^2 + y^2, \tag{7.5}$$

and in the spherical coordinates (ρ, θ, ϕ) in R^3 by

$$\delta(x)\,\delta(y)\,\delta(z) = \frac{\delta(\rho)}{4\pi\rho^2}, \quad \rho^2 = x^2 + y^2 + z^2. \tag{7.6}$$

In general, if the Cartesian coordinates $\mathbf{x} = (x, y)$ are transformed into general coordinates $\mathbf{u} = (u, v)$, then

$$\delta(\mathbf{x} - \mathbf{x}') = \delta(x - x')\,\delta(y - y') = \delta(\mathbf{u} - \mathbf{u}')\left[J\left(\frac{(x', y')}{(u', v')}\right)\right]^{-1}, \tag{7.7}$$

provided $J\left(\dfrac{(x', y')}{(u', v')}\right) \neq 0$, where $\mathbf{x}' = (x', y')$. For the three-dimensional case, where $\mathbf{x} = (x, y, z)$, $\mathbf{x}' = (x', y', z')$, $\mathbf{u} = (u, v, w)$, and $\mathbf{u}' = (u', v', w')$, the transformation is similar and is given by

$$\begin{aligned}
\delta(\mathbf{x} - \mathbf{x}') &= \delta(x - x')\,\delta(y - y')\,\delta(z - z') \\
&= \delta(\mathbf{u} - \mathbf{u}')\left[J\left(\frac{(x', y', z')}{(u', v', w')}\right)\right]^{-1},
\end{aligned} \tag{7.8}$$

provided $J\left(\dfrac{(x', y', z')}{(u', v', w')}\right) \neq 0$. For the proofs and the singular cases when the Jacobian J is zero, see Stakgold (1979) or Trim (1990). Some useful results in the polar and the spherical coordinates are:

$$\delta(x - x')\,\delta(y - y') = \frac{\delta(r - r')\,\delta(\theta - \theta')}{r} \quad \text{in } R^2,$$

$$\delta(x - x')\,\delta(y - y')\,\delta(z - z') = \frac{\delta(\rho - \rho')\,\delta(\theta - \theta')\,\delta(\phi - \phi')}{\rho^2 \sin\phi} \quad \text{in } R^3.$$

The Dirac delta function in all dimensions has the following properties:

$$\int_\Omega f(\mathbf{x})\,\delta(\mathbf{x} - \mathbf{x}')\,d\mathbf{x} = \begin{cases} f(\mathbf{x}') & \text{if } \mathbf{x}' \in \Omega, \\ 0 & \text{if } \mathbf{x}' \notin \Omega, \end{cases} \tag{7.9}$$

and

$$\int_\Omega \delta(k(\mathbf{x} - \mathbf{x}'))\,d\mathbf{x} = \int_\Omega \frac{1}{k}\delta(\mathbf{x} - \mathbf{x}')\,d\mathbf{x}, \tag{7.10}$$

where the integral is single, double, or triple, depending on the domain Ω.

Note that $\delta(x - x')$ is also written as $\delta(x, x')$, and has the units $[L^{-1}]$ if x denotes length, and the units $[T^{-1}]$ if x represents time. Also, we often write $\delta(\mathbf{x}, \mathbf{x}')$ for $\delta(\mathbf{x} - \mathbf{x}')$, $\delta(\mathbf{x}, t)$ for $\delta(\mathbf{x})\,\delta(t)$, and $\delta(\mathbf{x})$ for $\delta(\mathbf{x}, 0)$.

EXAMPLE 7.1. In view of Example 5.2, the spatial orthogonal eigenfunctions $f_n(x)$ for the one-dimensional problem $u_t = k\,u_{xx}$, $-a < x < a$, subject to the initial and boundary conditions $u(x,0) = F(x)$ for $-a < x < a$, and $u(-a,t) = 0 = u(a,t)$ for $t > 0$, are given by $\dfrac{1}{2a} \sin \dfrac{n\pi x}{a}$, which in complex form are

$$f_n(x) = \frac{1}{2a} \Im\{e^{in\pi x/a}\}.$$

Hence, the Dirac delta function in the region $-a < x < a$ for the steady-state (as $t \to \infty$) one-dimensional Laplace equation is represented by

$$\delta(x, x') = \begin{cases} \frac{1}{4a^2} \sum_{-\infty}^{\infty} e^{in\pi(x-x')/a}, & \text{for } |x| < a, \\ 0, & \text{for } |x| > a. \; \blacksquare \end{cases}$$

EXAMPLE 7.2. To determine the Fourier transform of the Dirac delta function $\delta(x)$, consider the function

$$f_\kappa(x) = \begin{cases} \dfrac{1}{2\kappa}, & |x| < \kappa, \\ 0, & |x| > \kappa. \end{cases}$$

Note that $\lim\limits_{\kappa \to 0} f_\kappa(x) = \delta(x)$. Also, $e^{-ax} f_\kappa(x) \in L_1(-\infty, \infty)$ for all real a, which implies that $\mathcal{F} f_\kappa(x) = F_\kappa(\alpha)$ is analytic in the entire α-plane of the transform domain. Then

$$F_\kappa(\alpha) = \frac{1}{\sqrt{2\pi}} \int_{-\infty}^{\infty} f_\kappa(x) e^{i\alpha x} \, dx$$

$$= \frac{1}{\sqrt{2\pi}} \int_{-\kappa}^{\kappa} \frac{1}{2\kappa} e^{i\alpha x} \, dx = \frac{1}{\sqrt{2\pi}} \left(\frac{e^{ik\alpha} - e^{-ik\alpha}}{2ik\alpha} \right).$$

Hence, $\lim\limits_{\kappa \to 0} F_\kappa(\alpha) = \dfrac{1}{\sqrt{2\pi}}$, which, in view of $\delta(-\alpha) = \delta(\alpha)$, implies that

$$\mathcal{F}\delta(x) = \frac{1}{\sqrt{2\pi}}, \quad \text{and} \quad \mathcal{F}\{1\} = \sqrt{2\pi}\, \delta(\alpha). \; \blacksquare$$

7.2. Green's Functions and Adjoint Operators

A Green's function $G(\mathbf{x}, \mathbf{z})$ for a linear partial differential operator L satisfies the following system: $L^* G(\mathbf{x}, \mathbf{z}) = \delta(\mathbf{x} - \mathbf{z})$ in Ω and $MG(\mathbf{x}, \mathbf{z}) = 0$ on $\partial\Omega$, where M is a linear partial differential operator whose order is less than the order of L or L^*. If the domain Ω is the whole space, then the solution to the system is known as the *fundamental* or *singularity solution* for the self-adjoint operator L. For the definition of the adjoint and self-adjoint operators, and the development of the concept of Green's functions, see §7.2.2.

7.2.1. The Concept of a Green's Function. To understand the concept of Green's functions, we will first discuss Green's functions for ordinary differential equations. Consider the following problem:

$$Ly = y'' - k^2 y = f(x), \quad a < x < b, \quad y(a) = y(b) = 0. \tag{7.11}$$

The solution for this problem, using the method of variation of parameters, is

$$y = \frac{1}{2k} \int_a^x \frac{\cosh k(b - x - z + a) - \cosh k(b - x + z - a)}{\sinh k(b - a)} f(z)\, dz$$

$$+ \frac{1}{2k} \int_x^b \frac{\cosh k(b + x - z - a) - \cosh k(b + x - z - a)}{\sinh k(b - a)} f(z)\, dz.$$

This can be simplified as

$$y = -\frac{1}{k} \left[\int_a^x \frac{\sinh k(b - x)\sinh k(z - a)}{\sinh k(b - a)} f(z)\, dz \right.$$

$$\left. + \int_x^b \frac{\sinh k(x - a)\sinh k(b - z)}{\sinh k(b - a)} f(z)\, dz \right]. \tag{7.12}$$

Now, if we define

$$G(x, z) = \begin{cases} -\dfrac{\sinh k(x - a)\sinh k(b - z)}{k \sinh k(b - a)} & \text{if } x < z, \\[3mm] -\dfrac{\sinh k(b - x)\sinh k(z - a)}{k \sinh k(b - a)} & \text{if } z < x, \end{cases}$$

we can express the solution (7.12) as

$$y = \int_a^b G(x, z) f(z)\, dz.$$

The function $G(x, z)$ is known as the *Green's function* for the problem (7.11). Obviously, $G(x, z) = G(z, x)$. We can also introduce the inverse operator L^{-1} and express the solution as $y = L^{-1} f(z) = \int_a^b G(x, z) f(z)\, dz$, i.e., $L^{-1}(\cdot) = \int_a^b G(x, z)\,(\cdot)\, dz$.

We will show that $G(x, z)$ is the solution of the boundary value problem

$$y'' - k^2 y = \delta(x - z), \quad a < x < b, \quad y(a) = y(b) = 0. \tag{7.13}$$

Before we solve this differential equation, we will state and prove the following result:

Theorem 7.1. *The solution of the differential equation*

$$P(x)\, y'' + Q(x)\, y' + R(x)\, y = \delta(x - z), \quad a < x < b, \quad y(a) = y(b) = 0, \quad (7.14)$$

satisfies the condition $y'(z_+) - y'(z_-) = \dfrac{1}{P(z)}$, *provided z is not a singular point of $P(x)$.*

PROOF. Define $F(x) = e^{\int Q/P\, dx}$. Then Eq (7.14) can be written as

$$F\, y'' + \frac{Q}{P}\, F\, y' + \frac{R}{P}\, F\, y = \frac{F}{P}\, \delta(x - z), \qquad (7.15)$$

where we have dropped the arguments in the functions F, P, Q, and R. Since Eq (7.15) can be expressed as

$$\frac{d}{dx}\, (F\, y') + \frac{R}{P}\, F\, y = \frac{F}{P}\, \delta(x - z),$$

integrating both sides of this equation from $z - \varepsilon$ to $z + \varepsilon$, we get

$$\int_{z-\varepsilon}^{z+\varepsilon} \frac{d}{dx}(F\, y')\, dx + \int_{z-\varepsilon}^{z+\varepsilon} \frac{R}{P}\, F\, y\, dx = \int_{z-\varepsilon}^{z+\varepsilon} \frac{F(x)}{P(x)}\, \delta(x - z)\, dx = \frac{F(z)}{P(z)}. \quad (7.16)$$

Now, if we take the limit as $\varepsilon \to 0$, the second term in (7.16) vanishes and we have

$$\lim_{\varepsilon \to 0} \left[F(z + \varepsilon) y'(z + \varepsilon) - F(z - \varepsilon) y'(z - \varepsilon) \right] = \frac{F(z)}{P(z)},$$

or

$$F(z) \left[y'(z_+) - y'(z_-) \right] = \frac{F(z)}{P(z)},$$

which yields the required result. ∎

Now, the solution of problem (7.13) is given by

$$y(x) = \begin{cases} A \sinh k(x - a) & \text{if } x < z, \\ B \sinh k(b - x) & \text{if } x > z, \end{cases} \qquad (7.17)$$

where the solution (7.17) satisfies the boundary conditions and the constants A and B satisfy the condition of continuity for $y(x)$ and the jump condition for $y'(x)$ (see Theorem 7.1). Thus,

$$\begin{aligned} A \sinh k(z - a) &= B \sinh k(b - z), \\ -Bk \cosh k(b - z) - Ak \cosh k(z - a) &= 1. \end{aligned} \qquad (7.18)$$

Solving the set of equations (7.18) for A and B and substituting them into Eq (7.17), we get

$$y(x) = G(x, z) = \begin{cases} -\dfrac{\sinh k(x - a) \sinh k(b - z)}{k \sinh k(b - a)} & \text{if } x < z, \\[2ex] -\dfrac{\sinh k(z - a) \sinh k(b - x)}{k \sinh k(b - a)} & \text{if } x > z. \end{cases}$$

In this case the Green's function $G(x, z)$ has the following properties:

(1) $\left(\dfrac{d^2}{dx^2} - k^2\right) G(x, z) = 0$ for $a < x < z$ and for $z < x < b$;

(2) $G(x, z)$ is continuous at $x = z$;

(3) $\dfrac{dG}{dx}$ is discontinuous at $x = z$, and $\dfrac{dG(x, z)}{dx}\bigg|_{x=z+} - \dfrac{dG(x, z)}{dx}\bigg|_{x=z-} = 1$;

(4) $G(x, z) = G(z, x)$;

(5) $G(a, z) = G(b, z) = G(x, a) = G(x, b) = 0$; and

(6) Green's function $G(x, z)$ is singular at the source point $z \in \Omega$, such that

$$LG(x, z) = \delta(x, z),$$

where L is a differential operator.

In some textbooks, Green's function for this case is defined as the solution of the problem

$$y'' - k^2 y = -\delta(x - z), \quad a < x < b, \quad y(a) = y(b) = 0.$$

7.2.2. Adjoint Operator. The Green's function $G(\mathbf{x}, \mathbf{z})$ for the boundary value problem $L u = 0$ in Ω such that $B u = 0$ on $\partial\Omega$, satisfies the following system: $L^* G(\mathbf{x}, \mathbf{z}) = \delta(\mathbf{x}, \mathbf{z})$, where L^* is the adjoint operator of L in Ω, and $B^* G(\mathbf{x}, \mathbf{z}) = 0$ on $\partial\Omega$; here B and B^* denote linear partial differential operators whose order is less than that of L and L^*, respectively, and the boundary condition $B^* G(\mathbf{x}, \mathbf{z}) = 0$ is the adjoint boundary condition of $B u = 0$ on $\partial\Omega$. If $L = L^*$, then the operator L is said to be a self-adjoint operator.

DEFINITION 7.1. If L is a linear differential operator with dependent variable u and independent variable x, then an operator L^* that satisfies the relation

$$\int_a^b v \, Lu \, dx = \left[M\left(u, v, u', v', x\right) \right]_a^b + \int_a^b u \, L^* v \, dx, \tag{7.19}$$

where $M\left(u, v, u', v', x\right)$ represents the boundary terms obtained after integration by parts, is called the *adjoint operator* of L.

We will limit our discussion to second-order operators. The following problems clarify the concept of Green's functions. First, consider the problem:

$$Lu = f(x) \quad \text{in } (a, b), \text{ such that } B_1 u \Big|_{x=a} = 0, \, B_2 u \Big|_{x=b} = 0,$$

where $Lu = A(x)u'' + B(x)u' + C(x)u$, and the order of B_1 and B_2 is ≤ 1. Then integration by parts yields

$$\int_a^b v \, Lu \, dx = \Big[A(x)vu' - \big(A(x)v\big)'u + B(x)vu\Big]_a^b + \int_a^b uL^*v \, dx,$$

where $L^*v = A(x)v'' + [2A'(x) - B(x)]\, v' + [A''(x) - B'(x) + C(x)]\, v$. If we require v to satisfy the boundary conditions $B_1^* v \Big|_{x=a} = 0$ and $B_2^* v \Big|_{x=b} = 0$, where B_1^* and B_2^* are chosen such that $\Big[A(x)vu' - \big(A(x)v\big)' + B(x)vu\Big]_a^b = 0$, then the pair of the boundary conditions $B_1^* v \Big|_{x=a} = 0$ and $B_2^* v \Big|_{x=b} = 0$ become the adjoint boundary conditions of the given pair of the boundary conditions $B_1 u \Big|_{x=a} = 0$ and $B_2 u \Big|_{x=b} = 0$. Next, consider the following problem:

$$L^*v = \delta(x - x') \quad \text{in } (a, b), \text{ such that } B_1^* v \Big|_{x=a} = 0, \, B_2^* v \Big|_{x=b} = 0.$$

Let $G(x, x')$ denote its solution. Then if we replace v by $G(x, x')$ in Eq (7.19), it reduces to

$$\int_a^b G(x, x') \, Lu \, dx = \int_a^b uL^* G(x, x') \, dx = \int_a^b u\delta(x - x') \, dx = u(x').$$

Since $Lu = f(x)$ in (a, b), the above equation reduces to

$$\int_a^b G(x, x') \, f(x) \, dx = u(x'). \tag{7.20}$$

Moreover, since x' is an arbitrary point, we interchange x and x' in Eq (7.20) and get

$$u(x) = \int_a^b G(x', x) \, f(x') \, dx'. \tag{7.21}$$

Now, we consider the second-order partial differential operators. If L denotes such an operator, let $u(\mathbf{x})$ and $v(\mathbf{x})$ be two C^2-functions. Then, integrating by parts, we get

$$\iiint_\Omega v(\mathbf{x}) \, Lu(\mathbf{x}) \, d\Omega = \iint_{\partial\Omega} M(u, v) \, dS + \iiint_\Omega u(\mathbf{x}) L^* v(\mathbf{x}) \, d\Omega, \tag{7.22}$$

where $M(u, v)$ is a differential operator of the first order, $\partial\Omega$ is the boundary of the region Ω, and L^* is the adjoint operator of L. Consider the following problem:

$$Lu = f(\mathbf{x}) \quad \text{in } \Omega, \text{ such that } Bu = 0 \text{ on } \partial\Omega,$$

where B is a linear differential operator of the first order. To find the solution of this problem, we use Eq (7.22) and the solution of the following problem:

$$L^*v = \delta(\mathbf{x}, \mathbf{x}') \quad \text{in } \Omega, \text{ such that } B^*v = 0 \text{ on } \partial\Omega, \tag{7.23}$$

where B^* is a linear differential operator of the first order such that $M(u, v) = 0$. The boundary condition $B^*v = 0$ on $\partial\Omega$ is known as the adjoint boundary condition of $Bu = 0$ on $\partial\Omega$. The solution $v(\mathbf{x}, \mathbf{x}')$ of the problem (7.23) is denoted by $G(\mathbf{x}, \mathbf{x}')$ and is known as Green's function for the problem (7.23). Using (7.21), we have

$$\iiint_\Omega G(\mathbf{x}, \mathbf{x}')\, Lu(\mathbf{x})\, d\Omega = \iint_{\partial\Omega} M(u, G)\, dS + \iiint_\Omega u(\mathbf{x})\, L^*G(\mathbf{x}, \mathbf{x}')\, d\Omega, \tag{7.24}$$

which gives

$$\iiint_\Omega G(\mathbf{x}, \mathbf{x}')\, f(\mathbf{x})\, d\Omega = \iiint_\Omega u(\mathbf{x})\, \delta(\mathbf{x} - \mathbf{x}')\, d\Omega = u(\mathbf{x}').$$

Since $\mathbf{x}'$ is arbitrary, we interchange $\mathbf{x}$ and $\mathbf{x}'$ and obtain

$$u(\mathbf{x}) = \iiint_\Omega G(\mathbf{x}', \mathbf{x})\, f(\mathbf{x}')\, d\Omega.$$

It turns out that even when the boundary condition on u is nonhomogeneous, we can find the solution from Eq (7.24). We will prove this assertion for special cases in later sections.

Theorem 7.2. *Green's function is symmetric for the self-adjoint operators with respect to $\mathbf{x}$ and $\mathbf{x}'$, i.e., $G(\mathbf{x}, \mathbf{x}') = G(\mathbf{x}', \mathbf{x})$.*

PROOF. Consider the equations $LG(\mathbf{x}, \mathbf{y}) = \delta(\mathbf{x}, \mathbf{y})$ and $LG(\mathbf{x}, \mathbf{z}) = \delta(\mathbf{x}, \mathbf{z})$. Multiplying the first of these equations by $G(\mathbf{x}, \mathbf{z})$ and the second by $G(\mathbf{x}, \mathbf{y})$ and subtracting, we get

$$G(\mathbf{x}, \mathbf{z})\, LG(\mathbf{x}, \mathbf{y}) - G(\mathbf{x}, \mathbf{y})\, LG(\mathbf{x}, \mathbf{z}) = \delta(\mathbf{x}, \mathbf{y})G(\mathbf{x}, \mathbf{z}) - \delta(\mathbf{x}, \mathbf{z})G(\mathbf{x}, \mathbf{y}). \tag{7.25}$$

Since for self-adjoint operators

$$\iiint_\Omega G(\mathbf{x}, \mathbf{z})LG(\mathbf{x}, \mathbf{y})\, d\Omega = \iiint_\Omega G(\mathbf{x}, \mathbf{y})LG(\mathbf{x}, \mathbf{z})\, d\Omega,$$

the boundary terms in (7.25) disappear, and after integrating Eq (7.25) over the region Ω, we obtain

$$0 = \iiint_{\Omega} \left[\delta(\mathbf{x}, \mathbf{y}) G(\mathbf{x}, \mathbf{z}) - \delta(\mathbf{x}, \mathbf{z}) G(\mathbf{x}, \mathbf{y}) \right] d\Omega = G(\mathbf{y}, \mathbf{z}) - G(\mathbf{z}, \mathbf{y}). \ \blacksquare$$

7.3. Elliptic Equations

We consider the Laplace and Helmholtz operators and derive Green's functions for them. In general, we will develop Green's function $G(\mathbf{x})$ for the source of strength -1 at the origin, and then obtain the general Green's function $G(\mathbf{x} - \mathbf{x}') = G(\mathbf{x}; \mathbf{x}')$ for the source at a point $\mathbf{x}' \neq \mathbf{0}$ by simply replacing $\mathbf{x}$ by $\mathbf{x} - \mathbf{x}'$.

7.3.1. Green's Function for the Laplacian. Those functions u that satisfy the Laplace equation $\nabla^2 u(\mathbf{x}) = \delta(\mathbf{x})$ in a domain D subject to homogeneous boundary conditions are known as Green's functions for the Laplace equation.* We will give some basic methods for finding Green's functions.

EXAMPLE 7.3. BY DIRECT METHOD. Assuming that the source point $\mathbf{x}'$ is at the origin, the Green's function $G(\mathbf{x}) = G(\mathbf{x}; \mathbf{0})$ for the Laplacian ∇^2 in R^3 satisfies the equation

$$\frac{\partial^2 u}{\partial x^2} + \frac{\partial^2 u}{\partial y^2} + \frac{\partial^2 u}{\partial z^2} = \delta(x)\,\delta(y)\,\delta(z). \tag{7.26}$$

Since the problem has radial symmetry, we transform Eq (7.26) into the spherical coordinates, which gives

$$\frac{\partial^2 u}{\partial \rho^2} + \frac{2}{\rho}\frac{\partial u}{\partial \rho} = \frac{\delta(\rho)}{4\pi \rho^2}.$$

Multiplying both sides of this equation by ρ^2, we get

$$\frac{\partial}{\partial \rho}\left(\rho^2 \frac{\partial u}{\partial \rho} \right) = \frac{\delta(\rho)}{4\pi}.$$

Integrating both sides of this equation from 0 to ρ, we find that

$$\rho^2 \frac{\partial u}{\partial \rho} = \frac{1}{4\pi}, \quad \text{or} \quad u = -\frac{1}{4\pi\rho},$$

---

*In some books the equation is taken as $-\nabla^2 u(\mathbf{x}) = \delta(\mathbf{x})$ to signify the source of strength $+1$ at the origin. In such cases Green's function is the negative of the one derived in this chapter. This practice is prevalent in physics.

where the arbitrary constant due to integration is taken to be zero, in order to satisfy the requirement that u must vanish as $\rho \to \infty$. Hence, Green's function in this case is given by

$$G(\mathbf{x}) = -\frac{1}{4\pi\rho} = -\frac{1}{4\pi|\mathbf{x}|}, \tag{7.27}$$

and for an arbitrary singular point $\mathbf{x}'$ by

$$G(\mathbf{x} - \mathbf{x}') = -\frac{1}{4\pi \log \rho_1} = -\frac{1}{4\pi |\mathbf{x} - \mathbf{x}'|}. \;\blacksquare \tag{7.28}$$

For R^2, we use the polar coordinates and have

$$\frac{1}{r}\frac{\partial}{\partial r}\left(r\frac{\partial u}{\partial r}\right) = \frac{\delta(r)}{2\pi r}, \quad r^2 = x^2 + y^2.$$

Multiplying both sides of this equation by r and integrating from 0 to r, we get $\dfrac{\partial u}{\partial r} = \dfrac{1}{2\pi r}$, which gives

$$u = \frac{1}{2\pi}\log r + B.$$

Since it is not possible to normalize u by requiring that it must vanish as $r \to \infty$, we normalize it such that $u = 0$ at $r = 1$, which results in $B = 0$. Hence, Green's function in this case is given by

$$G(\mathbf{x}) = \frac{1}{2\pi}\log r = \frac{1}{2\pi}\log|\mathbf{x}|, \tag{7.29}$$

and for an arbitrary singular point $\mathbf{x}'$ by

$$G(\mathbf{x} - \mathbf{x}') = \frac{1}{2\pi}\log r_1 = \frac{1}{2\pi}\log|\mathbf{x} - \mathbf{x}'|. \;\blacksquare \tag{7.30}$$

BY A METHOD BASED ON THE PHYSICAL CONCEPTS. In view of the property (7.9), Eq (7.26) yields

$$\iiint_\Omega \nabla^2 G(\mathbf{x})\, dV = \iiint_{S_\varepsilon} \delta(\mathbf{x})\, dV = 1, \tag{7.31}$$

where S_ε is a sphere of radius ε with center at the origin. Since the operator ∇^2 is invariant under a rotation of the coordinate axes, we seek a solution that depends only on $\rho = |\mathbf{x}|$. For $\rho > 0$, the function $G(\rho)$ satisfies the homogeneous equation $\nabla^2 G = 0$, which in the spherical coordinates is

$$\frac{1}{\rho^2}\frac{\partial}{\partial \rho}\left(\rho^2\frac{\partial u}{\partial \rho}\right) = 0.$$

This equation has a solution $G(\rho) = A/\rho + B$. If we require this solution to vanish at infinity, then $B = 0$, and thus, $G(\rho) = A/\rho$. To determine A, we consider the strength of the source at $x = 0$ (in this case it is -1), which, on using (7.31) and considering the flux across a sphere of radius ε, gives

$$-\iint_{\partial S_\varepsilon} \frac{\partial G}{\partial \rho}\, dS = \int_0^{2\pi} \int_0^\pi \frac{A}{\varepsilon^2}\, \varepsilon^2 \sin\phi\, d\phi\, d\theta = 4\pi A = -1, \qquad (7.32)$$

where ∂S_ε is the surface of the sphere S_ε. Physically, Eq (7.32) represents the conservation of the charge, i.e., the flux of the electric field through the closed surface ∂S_ε (of area $4\pi\varepsilon^2$) is equal to the charge in the interior of S_ε. Using (7.32), we find that $A = -1/(4\pi)$, and thus, Green's function of the three-dimensional Laplacian is given by

$$G(\mathbf{x}) = -\frac{1}{4\pi\rho} = -\frac{1}{4\pi|\mathbf{x}|},$$

as in (7.27). If the source is at $\mathbf{x}'$, then Green's function is the same as in (7.28).

For R^2, we have

$$\frac{1}{r}\frac{\partial}{\partial r}\left(r\frac{\partial G}{\partial r}\right) = 0 \quad \text{for } r \neq 0, \text{ where } r = |\mathbf{x}| = \sqrt{x^2 + y^2},$$

which has a solution $G(r) = A\ln r + B$. To evaluate A, we consider the flux across a circle C_ε of radius ε with center at the origin. Thus,

$$-\int_{C_\varepsilon}\left[\frac{\partial G}{\partial r}\right]_{r=\varepsilon} ds = -2\pi A = -1,$$

which gives $A = 1/(2\pi)$. If we normalize as before, we have $B = 0$, and Green's function is the same as (7.29). If the source is at $\mathbf{x}'$, then Green's function is the same as given by (7.30). ∎

EXAMPLE 7.4. (METHOD OF IMAGES) The problem of finding Green's function $G(\mathbf{x}; \mathbf{x}') = G(x, y; x', y')$ inside some region Ω bounded by a closed curve Γ with the homogeneous boundary condition $G = 0$ on Γ amounts to that of finding the electrostatic potential due to a point charge at the point (x', y') inside a grounded conductor in the shape of the boundary Γ. Consider, for example, the region Ω as the half-plane $y > 0$. Then Green's function is given by the point charge of strength -1 at (x', y') together with an equal but opposite charge (of strength $+1$) at the point $(x', -y')$ which is the image of the point (x', y') in the x-axis (Fig.7.1). Thus, Green's function for the half-plane such that $G = 0$ on the x-axis and $G \to 0$ as $r \to \infty$, where $r = |\mathbf{x} - \mathbf{x}'| = \sqrt{(x - x')^2 + (y - y')^2}$, is given by

$$G(x, y; x', y') = \frac{1}{4\pi} \ln\left\{\frac{(x - x')^2 + (y - y')^2}{(x - x')^2 + (y + y')^2}\right\}. \qquad (7.33)$$

Since the algebraic sum of the charges over the entire (x, y)-plane is zero, the condition $G \to 0$ as $r \to \infty$ becomes possible because any nonzero residual charge will make Green's function behave like $\ln r$. This means that by this method we cannot determine a Green's function for the half-plane $y > 0$ subject to the conditions $G_y = 0$ on the x-axis and $G \to 0$ as $r \to \infty$. However, we can determine Green's function for the quarter-plane $x > 0$, $y > 0$, subject to the conditions $G = 0$ on $y = 0$ and $G_y = 0$ on $x = 0$. As seen in Fig. 7.1, we have the charge of strength $+1$ at each $(x', -y')$ and $(-x', y')$, and the charge of strength -1 at each (x', y') and $(-x', -y')$. Thus, Green's function in this case is

$$G(x, y; x', y') = \frac{1}{4\pi} \ln \left\{ \frac{[(x-x')^2 + (y-y')^2][(x+x')^2 + (y+y')^2]}{[(x-x')^2 + (y+y')^2][(x+x')^2 + (y-y')^2]} \right\}. \tag{7.34}$$

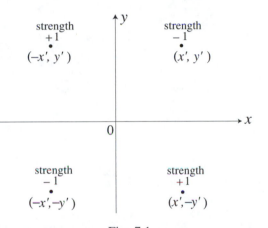

Fig. 7.1.

EXAMPLE 7.5. (SERIES EXPANSION METHOD) Consider $\nabla^2 u = 0$ in the rectangle $R = \{0 < x < a, 0 < y < b\}$. Then Green's function $G(x, y; x', y')$ for this rectangle satisfies the equation

$$\nabla^2 G(x, y; x', y') = \delta(x - x') \, \delta(y - y'), \tag{7.35}$$

subject to the boundary conditions

$$G(0, y; x', y') = G(a, y; x', y') = G(x, 0; x', y') = G(x, b; x', y') = 0. \tag{7.36}$$

A double Fourier series representation for $G(x, y; x', y')$ and $\delta(x - x') \, \delta(y - y')$, which is of the form

$$G(x, y; x', y') = \sum_{m=0}^{\infty} \sum_{n=0}^{\infty} A_{mn} \sin \frac{m\pi x}{a} \sin \frac{n\pi y}{b}, \tag{7.37}$$

$$\delta\left(x-x'\right)\delta\left(y-y'\right)=\sum_{m=0}^{\infty}\sum_{n=0}^{\infty}B_{mn}\sin\frac{m\pi x}{a}\sin\frac{n\pi y}{b}, \tag{7.38}$$

satisfies the conditions (7.36). To determine A_{mn}, we first find B_{mn} by using formula (4.23), which gives

$$B_{mn}=\frac{4}{ab}\sin\frac{m\pi x'}{a}\sin\frac{n\pi y'}{b}.$$

Then substituting (7.37) and (7.38) into Eq (7.35), we get

$$-\sum_{m=0}^{\infty}\sum_{n=0}^{\infty}A_{mn}\left(\frac{m^2\pi^2}{a^2}+\frac{n^2\pi^2}{b^2}\right)\sin\frac{m\pi x}{a}\sin\frac{n\pi y}{b}$$

$$=\frac{4}{ab}\sum_{m=0}^{\infty}\sum_{n=0}^{\infty}\sin\frac{m\pi x'}{a}\sin\frac{n\pi y'}{b}\sin\frac{m\pi x}{a}\sin\frac{n\pi y}{b}.$$

Hence,

$$A_{mn}=-\frac{4ab}{m^2\pi^2 b^2+n^2\pi^2 a^2}\sin\frac{m\pi x'}{a}\sin\frac{n\pi y'}{b},$$

and

$$G\left(x,y;x',y'\right)$$

$$=-\sum_{m=0}^{\infty}\sum_{n=0}^{\infty}\frac{4ab}{m^2\pi^2 b^2+n^2\pi^2 a^2}\sin\frac{m\pi x'}{a}\sin\frac{n\pi y'}{b}\sin\frac{m\pi x}{a}\sin\frac{n\pi y}{b}. \ \blacksquare \tag{7.39}$$

7.3.2. Harmonic Functions. A real-valued function $u(x,y)\in C^2(D)$ is said to be *harmonic* in a region D if $\nabla^2 u=0$ in D. Some properties of harmonic functions in R^2 are as follows:

(i) The function

$$\frac{1}{r}=\frac{1}{\sqrt{(x-x_0)^2+(y-y_0)^2}}$$

is harmonic in a region that does not contain the point (x_0,y_0).

(ii) If $u(x,y)$ is a harmonic function in a simply connected region D, then there exists a conjugate harmonic function $v(x,y)$ in D such that $u(x,y)+i\,v(x,y)$ is an analytic function of $z=x+iy=(x,y)$ in D. In view of the Cauchy-Riemann equations*,

$$v(x,y)-v(x_0,y_0)=\int_{x_0,y_0}^{x,y}\left(-u_y\,dx+u_x\,dy\right), \tag{7.40}$$

---

* The Cauchy-Riemann equations for a function $w=f(z)=u(x,y)+iv(x,y)$, $z=x+iy$, which is an analytic on a region D are $u_x=v_y$, $u_y=-v_x$.

where $(x_0, y_0) = z_0$ is a given point in D. This property is also true if D is multiply connected. However, in that case the conjugate function $v(x, y)$ can be multiple-valued, as we see by considering $u(x, y) = \log r = \log \sqrt{x^2 + y^2}$ defined on a region D containing the origin which has been indented by a small circle centered at the origin. Then, in view of (7.40),

$$v(x, y) - v(x_0, y_0) = \tan^{-1} \frac{y}{x} \pm 2n\pi + \text{const}, \quad n = 1, 2, \dots,$$

which is multiple-valued.

(iii) Since derivatives of all orders of an analytic function exist and are themselves analytic, any harmonic function will have continuous partial derivatives of all orders, i.e., a harmonic function belongs to the class $C^\infty(D)$, and a partial derivative of any order is again harmonic.

(iv) In view of the maximum modulus theorem[†], the maximum (and also the minimum) of a harmonic function u in a region D occurs only on ∂D.

Theorem 7.3. (Maximum Principle) *A nonconstant function that is harmonic inside a bounded region D with boundary Γ and continuous in the closed region $\bar{D} = D \cup \Gamma$ attains its maximum and minimum values only on the boundary of the region.*

Thus, u has a maximum (or minimum) at $z_0 \in D$, i.e., if $u(z) \leq u(z_0)$ (or $u(z) \geq u(z_0)$) for z in a neighborhood $B(z_0, \varepsilon)$ of z_0, then $u = \text{const}$ in $B(z_0, \varepsilon)$.

(v) The value of a harmonic function u at an interior point in terms of the boundary values u and $\dfrac{\partial u}{\partial n}$ is given by Green's third identity (A.10).

(vi) A harmonic function satisfies the mean-value theorem, where the mean value at a point is evaluated for the circumference or the area of the circle around that point.

Proofs of these results are available in textbooks on complex analysis, e.g., Carrier, Krook, and Pearson (1966) and Levinson and Redheffer (1970).

7.3.3. Symmetry of Green's Functions. The functions u that satisfy the Laplace equation $\nabla^2 u = 0$ are known as harmonic functions. We have the following result which is a particular case of Theorem 7.2.

Theorem 7.4. Green's function for the Laplacian is symmetric, i.e.,

$$G(\mathbf{x}, \mathbf{y}) = G(\mathbf{y}, \mathbf{x}).$$

PROOF. Consider two functions $G(\mathbf{x}, \mathbf{y})$ and $G(\mathbf{x}, \mathbf{z})$; these functions are solutions of

$$\nabla^2 G(\mathbf{x}, \mathbf{y}) = \delta(\mathbf{x} - \mathbf{y}) \quad \text{in } \Omega \text{ and } G(\mathbf{x}, \mathbf{y}) = 0 \text{ on } \partial\Omega,$$
$$\nabla^2 G(\mathbf{x}, \mathbf{z}) = \delta(\mathbf{x} - \mathbf{z}) \quad \text{in } \Omega \text{ and } G(\mathbf{x}, \mathbf{z}) = 0 \text{ on } \partial\Omega.$$

[†] This theorem states that if f is a nonconstant analytic function on a region D with a simple boundary Γ, then $|f|$ cannot have a local maximum anywhere on the interior of Γ.

Green's identity (A.7) states that

$$\iiint_{\Omega} (u\nabla^2 v - v\nabla^2 u)\, d\Omega = \iint_{\partial\Omega} \left(u\frac{\partial v}{\partial n} - v\frac{\partial v}{\partial n}\right) dS,$$

where n is the outward unit normal at $\partial\Omega$. Taking $u = G(\mathbf{x}, \mathbf{y})$ and $v = G(\mathbf{x}, \mathbf{z})$, we get

$$\iiint_{\Omega} \left(G(\mathbf{x}, \mathbf{y})\nabla^2 G(\mathbf{x}, \mathbf{z}) - G(\mathbf{x}, \mathbf{z})\nabla^2 G(\mathbf{x}, \mathbf{y})\right) d\Omega$$

$$= \iint_{\partial\Omega} \left(G(\mathbf{x}, \mathbf{y})\frac{\partial G(\mathbf{x}, \mathbf{z})}{\partial n} - G(\mathbf{x}, \mathbf{z})\frac{\partial G(\mathbf{x}, \mathbf{y})}{\partial n}\right) dS,$$

where the right side of the above equation vanishes because of the boundary conditions and the left side reduces to

$$\iiint_{\Omega} (G(\mathbf{x}, \mathbf{y})\delta(\mathbf{x} - \mathbf{z}) - G(\mathbf{x}, \mathbf{z})\delta(\mathbf{x} - \mathbf{y})\, d\Omega = 0,$$

or, since Ω is arbitrary,

$$G(\mathbf{z}, \mathbf{y}) - G(\mathbf{y}, \mathbf{z}) = 0.$$

Since the point $\mathbf{z}$ is arbitrary, we can replace it by $\mathbf{x}$ and we have

$$G(\mathbf{x}, \mathbf{y}) - G(\mathbf{y}, \mathbf{x}) = 0.$$

7.3.4. Green's Functions for the Helmholtz Operator. The Helmholtz operator is defined as $(\nabla^2 + k^2)$, where $k > 0$ is real. Green's function for this operator satisfies the equation

$$(\nabla^2 + k^2)G(\mathbf{x} - \mathbf{x}') = \delta(\mathbf{x} - \mathbf{x}'). \tag{7.41}$$

In some books this operator is taken as $-\left(\nabla^2 + k^2\right)$ to signify the source of strength $+1$ (see the footnote on page 216).

EXAMPLE 7.6. For the operator $\nabla^2 + k^2$ in R^2 we solve the equation

$$\left(\nabla^2 + k^2\right) u = \delta(x)\, \delta(y).$$

Since the right side is zero everywhere except at the origin, where a negative source (sink) of unit strength exists, we find a solution with singularity at the origin such that

$$-\lim_{\varepsilon\to 0} \int_{r=\varepsilon} \frac{\partial G}{\partial n}\, ds = -1, \tag{7.42}$$

where r is the distance from the origin. The solution is symmetric with respect to the origin, and therefore, dependent only on r. Thus, the equation $\left(\nabla^2 + k^2\right) u = 0$ reduces to

$$\frac{\partial^2 G}{\partial r^2} + \frac{1}{r} \frac{\partial G}{\partial r} + k^2 G = 0 \quad \text{for } r \neq 0.$$

Its solution is $G(r) = A H_0^{(1)}(kr) + B H_0^{(2)}(kr)$, where $H_0^{(1)}(kr) = J_0(kr) + i Y_0(kr)$ and $H_0^{(2)}(kr) = J_0(kr) - i Y_0(kr)$ are the Hankel functions of the first and second kind, respectively. Applying the condition (7.42), we find that

$$
\begin{aligned}
-\lim_{\varepsilon \to 0} \int_{r=\varepsilon} \frac{\partial G}{\partial n}\, ds &= -\lim_{\varepsilon \to 0} 2\pi\varepsilon \left[\frac{\partial}{\partial r} \left\{ A H_0^{(1)}(kr) + B H_0^{(2)}(kr) \right\} \right]_{r=\varepsilon} \\
&= \lim_{\varepsilon \to 0} 2\pi\varepsilon \left[k \left\{ (A + B) J_1(kr) + (A - B) i Y_1(kr) \right\} \right]_{r=\varepsilon} \qquad (7.43) \\
&= 2\pi \left(-\frac{2i}{\pi} \right)(A - B) = -1,
\end{aligned}
$$

which gives $A - B = -i/4$. Since the contribution from the coefficient $(A + B)$ vanishes, we can assign it any value. Traditionally, B is taken zero. Hence, the Green's function is

$$G(r) = -\frac{i\, H_0^{(1)}(kr)}{4}.$$

In R^3, it is obvious from physical considerations that Green's function for the Helmholtz operator must be spherically symmetric. Then Green's function $G(\mathbf{x})$ for $\mathbf{x}' = 0$ in the spherical coordinates satisfies the equation

$$\frac{1}{\rho^2} \frac{d}{d\rho} \left(\rho^2 \frac{dG}{d\rho} \right) + k^2 G = \frac{\delta(\rho)}{4\pi\rho^2}. \qquad (7.44)$$

If we substitute $G = w/\rho$, then Eq (7.44) reduces to

$$\frac{d^2 w}{d\rho^2} + k^2 w = \frac{\delta(\rho)}{4\pi\rho}. \qquad (7.45)$$

Applying the Fourier sine transform to Eq (7.45), we get

$$
\begin{aligned}
\alpha c_0 + \left(k^2 - \alpha^2 \right) \tilde{w}_s &= \sqrt{\frac{2}{\pi}} \int_0^\infty \frac{\delta(\rho)}{4\pi\rho} \sin(\alpha\rho)\, d\rho \\
&= \sqrt{\frac{2}{\pi}} \frac{1}{4\pi} \lim_{\rho \to 0} \frac{\sin(\alpha\rho)}{\rho} = \frac{\alpha}{4\pi} \sqrt{\frac{2}{\pi}},
\end{aligned}
$$

where $c_0 \sqrt{\pi} = \sqrt{2} \lim_{\rho \to 0} w$. Thus,

$$\tilde{w}_s = \left(c_0 - \frac{1}{4\pi} \sqrt{\frac{2}{\pi}} \right) \frac{\alpha}{\alpha^2 - k^2},$$

which, on inverting (using formula 5 in §C.3 with $a = ik$), and assuming $c_0 = 0$, yields $w = -e^{-ik\rho}/(4\pi)$. Hence,

$$G(\rho) = -\frac{e^{-ik\rho}}{4\pi\rho}. \tag{7.46}$$

The physical interpretation of this Green's function is as follows: If we consider the flux of the quantity $G(\rho)$ across a sphere of radius ρ and take the limit as $\varepsilon \to 0$, we obtain the limiting value -1. This implies the presence of a sink of unit strength at the origin. If we were to find the solution of the equation

$$\frac{1}{\rho^2}\frac{d}{d\rho}\left(\rho^2\frac{dG}{d\rho}\right) + k^2 G = -\frac{\delta(\rho)}{4\pi\rho^2}, \tag{7.47}$$

the solution $G(\rho) = \dfrac{e^{-ik\rho}}{4\pi\rho}$ would then represent Green's function due to a source of unit strength. We point out that the flux of a quantity u away from the source is always given by $-\dfrac{\partial u}{\partial n}$, where n is the outward unit normal to the surface. ∎

Theorem 7.5. *Green's function for the Helmholtz Operator is symmetric, i.e.,*

$$G(\mathbf{x}, \mathbf{y}) = G(\mathbf{y}, \mathbf{x}).$$

The proof is similar to that of Theorem 7.4 and is left as an exercise (see Exercise 7.19).

7.4. Parabolic Equations

In this section we will derive Green's functions for some parabolic equations.

EXAMPLE 7.7. Find Green's function for the operator $\dfrac{\partial}{\partial t} - k\dfrac{\partial^2}{\partial x^2}$ in R^1, i.e., for the region $(-\infty, \infty) \times (0, \infty)$. Note that the diffusion operator is not self-adjoint. However, in the Laplace transform domain it is self-adjoint. The equation and the boundary and initial conditions to be satisfied by the Green's function $G(x, x'; t, t')$ are

$$\frac{\partial G}{\partial t} - k\frac{\partial^2 G}{\partial x^2} = \delta(x - x')\,\delta(t - t'), \tag{7.48}$$

$$G(x, x'; 0, t') = 0, \qquad \lim_{|x|\to\infty} G(x, x'; t, t') = 0. \tag{7.49}$$

Applying the Laplace transform to Eq (7.48), we have

$$s\overline{G} - k\frac{d^2\overline{G}}{dx^2} = \delta(x - x')\,e^{-st'},$$

where the boundary condition becomes

$$\lim_{|x| \to \infty} \overline{G}(x, x'; s, t') = 0. \tag{7.50}$$

The discontinuity condition at x' is (see Theorem 7.1),

$$\frac{d\overline{G}(x'_+, x'; s, t')}{dx} - \frac{d\overline{G}(x'_-, x'; s, t')}{dx} = -\frac{1}{k} e^{-st'}. \tag{7.51}$$

The complementary function is then given by

$$\overline{G}_c = A e^{\alpha x} + B e^{-\alpha x}, \quad \alpha = \sqrt{s/k}. \tag{7.52}$$

The particular solution G_p is found by variation of parameters as

$$G_p = u e^{\alpha x} + v e^{-\alpha x},$$

where u and v satisfy the following conditions:

$$u' e^{\alpha x} + v' e^{-\alpha x} = 0, \tag{7.53}$$

$$\alpha(u' e^{\alpha x} - v' e^{-\alpha x}) = -\frac{1}{k} e^{-st'} \delta(x - x'), \tag{7.54}$$

where the prime over u and v denotes their derivative with respect to x. Solving Eqs (7.53) and (7.54), we have

$$u' = -\frac{1}{2k\alpha} e^{-\alpha x} e^{-st'} \delta(x - x'), \quad v' = \frac{1}{2k\alpha} e^{\alpha x} e^{-st'} \delta(x - x'). \tag{7.55}$$

From Eq (7.55), we find that

$$u = -\frac{H(x - x')}{2k\alpha} e^{-\alpha x' - st'}, \quad v = \frac{H(x - x')}{2k\alpha} e^{\alpha x' - st'}.$$

Combining Eqs (7.52) and (7.55), we get

$$\overline{G}(x, x', s, t') = \begin{cases} A_1 e^{\alpha x} + B_1 e^{-\alpha x} & \text{if } x < x', \\ A_2 e^{\alpha x} + B_2 e^{-\alpha x} + \frac{H(x-x')}{2k\alpha}[e^{-\alpha(x-x')} - e^{\alpha(x-x')}] e^{-st'} \\ \qquad\qquad\qquad \text{if } x > x'. \end{cases} \tag{7.56}$$

Now, we apply the boundary conditions, the continuity of $\overline{G}$ at $x = x'$, and the jump condition (7.51). The condition $\lim_{x \to -\infty} \overline{G} = 0$ yields $B_1 = 0$, and the continuity of $\overline{G}$ at $x = x'$ gives

$$A_1 e^{\alpha x'} = A_2 e^{\alpha x'} + B_2 e^{-\alpha x'}. \tag{7.57}$$

Also, the condition $\lim_{x \to \infty} \overline{G} = 0$ yields

$$A_2 + \frac{1}{2k\alpha} e^{-\alpha x' - st'} = 0, \tag{7.58}$$

and from the condition (7.51), we have

$$A_2\, e^{\alpha x'} + B_2\, e^{-\alpha x'} - A_1\, e^{\alpha x'} = \frac{1}{k\alpha}\, e^{-st'}. \tag{7.59}$$

Solving Eqs (7.57), (7.58) and (7.59), we find that

$$A_1 = -\frac{1}{2k\alpha} e^{-\alpha x' - st'} = A_2, \quad B_2 = \frac{1}{k\alpha}\, e^{\alpha x' - st'}.$$

Substituting the values of A_1, B_1, A_2, B_2 into Eq (7.56), we get

$$\overline{G} = -\frac{1}{2k\alpha} e^{-\alpha|x - x'| - st'} = -\frac{1}{2}\sqrt{\frac{k}{s}}\, e^{r\sqrt{s/a} - st'}, \quad r = |x - x'|,$$

which, on inversion, gives

$$G(x, x'; t, t') = \frac{H(t - t')}{2\sqrt{\pi k(t - t')}}\, e^{-(x - x')^2/(4a(t - t'))}. \tag{7.60}$$

This solution is also the fundamental solution for the diffusion operator.

Another method is to apply the Fourier transform followed by the Laplace transform to Eq (7.48). This gives

$$s\, \overline{\widetilde{G}} + k\, \alpha^2\, \overline{\widetilde{G}} = -\frac{1}{2\pi}\, e^{-st' + i\alpha x'},$$

where the tilde and the bar over G denote its Fourier and Laplace transforms, respectively. The solution for $\overline{\widetilde{G}}$ is

$$\overline{\widetilde{G}} = \frac{1}{\sqrt{2\pi}\,(s + k\alpha^2)}\, e^{-st' + i\alpha x'}.$$

Inversion of the Laplace transform yields

$$\widetilde{G}(x, \alpha) = \frac{H(t - t')}{\sqrt{2\pi}}\, e^{-k\alpha^2(t - t') + i\alpha x'},$$

which, after inverting the Fourier transform, gives

$$G(x, x'; t, t') = \frac{H(t - t')}{\sqrt{2\pi k(t - t')}}\, e^{-(x - x')^2/(4k(t - t'))}. \tag{7.61}$$

The graphs for $G(x,t)$ for $0 < t_1 < t_2 < t_3$ are shown in Fig. 7.2.

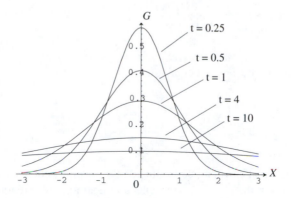

Fig. 7.2. Graphs of $G(x,t)$.

EXAMPLE 7.8. Find Green's function for the operator $\dfrac{\partial}{\partial t} - k\,\nabla^2$ in R^3, $t > 0$, assuming that the singularity is at the origin at $t = 0$. The Green's function $G(\mathbf{x}; t)$ satisfies the following equation and the boundary conditions:

$$\frac{\partial G}{\partial t} - k\,\nabla^2 G = \delta(\mathbf{x})\,\delta(t),$$

$$G(\mathbf{x}; t) \to 0 \quad \text{as } |\mathbf{x}| \to \infty \text{ and for } t \le 0. \tag{7.62}$$

To Eq (7.62) we apply the triple Fourier transform with respect to the space variables and the Laplace transform with respect to t and obtain

$$s\,\overline{\tilde{\tilde{\tilde{G}}}}(\alpha,\beta,\gamma,s) - k\,(\alpha^2 + \beta^2 + \gamma^2)\,\overline{\tilde{\tilde{\tilde{G}}}}(\alpha,\beta,\gamma,s) = \frac{1}{(2\pi)^{3/2}},$$

where α, β, and γ are the variables of the Fourier transform with respect to x, y, and z, respectively. Hence,

$$\overline{\tilde{\tilde{\tilde{G}}}}(\alpha,\beta,\gamma,s) = \frac{1}{(2\pi)^{3/2}\left[s - k\,(\alpha^2 + \beta^2 + \gamma^2)\right]}.$$

Inverting with respect to the Laplace transform, we have

$$\tilde{\tilde{\tilde{G}}}(\alpha,\beta,\gamma,s) = \frac{H(t)}{(2\pi)^{3/2}}\,e^{-k(\alpha^2 + \beta^2 + \gamma^2)t}.$$

Inverting the triple Fourier transform, we get

$$G(\mathbf{x}; t) = \frac{H(t)}{(2\pi)^{3/2}\,(2kt)^{3/2}}\,e^{-(x^2 + y^2 + z^2)/(4kt)} = \frac{H(t)}{8(\pi kt)^{3/2}}\,e^{-|\mathbf{x}|^2/(4kt)}.$$

By translating the singularity to $\mathbf{x}'$ and t', the Green's function is

$$G(\mathbf{x} - \mathbf{x}'; t - t') = \frac{H(t - t')}{8[\pi k(t - t')]^{3/2}} \, e^{- |\mathbf{x} - \mathbf{x}'|^2 / (4k(t - t'))}. \quad \blacksquare \qquad (7.63)$$

EXAMPLE 7.9. (SCHRÖDINGER EQUATION) If we consider the nonhomogeneous case, then the Fourier heat equation

$$\left(\frac{1}{k} \frac{\partial}{\partial t} - \nabla^2 \right) u(x, t) = f(x, t) \qquad (7.64)$$

has two interpretations:

(i) If $k > 0$ is a real constant, depending on the specific heat and thermal conductivity of the medium, then $u(x, t)$ represents a temperature distribution. The source function $f(x, t)$ on the right side describes local heat production minus absorption.

(ii) The function $u(x, t)$ represents a particle density and k is the diffusion coefficient. If k is purely imaginary such that $k = \dfrac{i\hbar}{2m}$, where m is the mass of the quantum particle, and $\hbar = 1.054 \times 10^{-34}$ joules-sec is the Planck's constant, then Eq (7.64) defines the Schrödinger equation

$$i\hbar \frac{\partial u(x, t)}{\partial t} + \frac{\hbar^2}{2m} \nabla^2 u(x, t) = f(x, t).$$

In view of (7.61), we have $K = \hbar/(2im)$, and Green's function for Eq (7.63) in R^1 is given by

$$G(x, t) = H(t) \frac{1 + i}{2} \sqrt{\frac{m}{\pi \hbar t}} \, e^{imx^2/(2\hbar t)}.$$

Sometimes the notation h is used for the Plank's constant; it is related to $\hbar$ by $h = 2\pi \hbar$. $\blacksquare$

The notation $G(\mathbf{x}, t; \mathbf{x}', t')$ is composed of two parts in its argument. The first part $\mathbf{x}, t$ denotes the field point where the *effect* of the impulsive heat source located at the source point signifies the temperature at the point $\mathbf{x}$ at time t. The second part $\mathbf{x}', t'$ denotes the *cause* which is the impulsive heat source situated at the point $\mathbf{x}'$ generating an instantaneous (impulsive) heat at an earlier time t'. The combined notation has the physical significance of an entire space-time process which can be visualized as $G(\text{effect}; \text{cause}) \equiv G(\mathbf{x}, t; \mathbf{x}', t')$.

Theorem 7.6. (Reciprocity Principle) *For the Green's function* $G(\mathbf{x}, t; \mathbf{x}', t')$, *we have*

$$G(\mathbf{x}, t; \mathbf{x}', t') = G(\mathbf{x}', -t'; \mathbf{x}, -t).$$

The physical significance of this relation is that the *effect* at $\mathbf{x}, t$ due to a *cause* at $\mathbf{x}', t'$ for $t' < t$ is the same as the *effect* at $\mathbf{x}', -t'$ due to a *cause* at $\mathbf{x}, -t$. For details about this principle, see Stakgold (1968, p. 303).

7.5. Hyperbolic Equations

We will denote Green's function for the wave operator $\Box_c = \dfrac{\partial^2}{\partial t^2} - c^2 \nabla^2$ in R^n by $G_n(\mathbf{x}, t)$. It satisfies the equation

$$\Box_c \, G_n(\mathbf{x}, t) = \delta(\mathbf{x}, t), \tag{7.65}$$

where ∇^2 is the Laplacian in the space variable $\mathbf{x}$, and $t > 0$. We will use the transform method and derive Green's functions for the wave operator in R^1, R^2, and R^3, respectively.

EXAMPLE 7.10. We assume that the singularity is at the origin and at $t = 0$. In R^1, the Green's function $G_1(x; t)$ satisfies the equation

$$\frac{\partial^2 G_1}{\partial t^2} - c^2 \frac{\partial^2 G_1}{\partial x^2} = \delta(x)\, \delta(t),$$

such that $G_1(x; t) \rightarrow 0$ as $|x| \rightarrow \infty$, $t > 0$. Applying the Laplace and Fourier transforms to this equation, we get

$$\left(s^2 + \alpha^2 c^2\right) \overline{\tilde{G}}_1 = \frac{1}{\sqrt{2\pi}},$$

or

$$\overline{\tilde{G}}_1 = \frac{1}{\sqrt{2\pi}\,(s^2 + \alpha^2 c^2)},$$

which on the Laplace inversion gives

$$\tilde{G}_1 = \frac{1}{\alpha c \sqrt{2\pi}} \sin \alpha c t.$$

After inverting the Fourier transform, we have

$$G_1(x; t) = \frac{1}{2c} \left[H(x + ct) - H(x - ct) \right].$$

However, if we reverse the process by carrying out the Fourier inversion prior to the Laplace inversion, we get

$$G_1(x, t) = \frac{1}{2c} H\left(t - |x|/c\right) = \frac{1}{2c} H(ct - |x|).$$

It is easy to check that the two solutions are equivalent. Thus, if the source is at $(x'; t')$, the solution becomes

$$G_1(x, t; x', t') = \frac{1}{2c} H\big[c(t - t') - (|x - x'|)\big]. \blacksquare \qquad (7.66)$$

EXAMPLE 7.11. In R^2, the Green's function G_2 satisfies the equation

$$\frac{\partial^2 G_2}{\partial t^2} - c^2 \left(\frac{\partial^2 G_2}{\partial x^2} + \frac{\partial^2 G_2}{\partial y^2} \right) = \delta(x - x')\, \delta(y - y')\, \delta(t - t').$$

Applying the Laplace transform, we get

$$s^2 \overline{G}_2 - c^2 \left(\frac{\partial^2 \overline{G}_2}{\partial x^2} + \frac{\partial^2 \overline{G}_2}{\partial y^2} \right) = e^{-st'} \delta(x - x')\, \delta(y - y'),$$

or using the axial symmetry in the polar coordinates with $r^2 = (x - x')^2 + (y - y')^2$,

$$s^2 \overline{G}_2 - c^2 \left(\frac{d^2 \overline{G}_2}{dr^2} + \frac{1}{r} \frac{\partial \overline{G}_2}{\partial r} \right) = \frac{e^{-st'} \delta(r)}{2\pi r}. \qquad (7.67)$$

The forcing term $\dfrac{e^{-st'} \delta(r)}{2\pi r}$ in Eq (7.67) can be considered as representing a source of strength $e^{-st'}$ at the point $r = 0$. The solution of the homogeneous part of Eq (7.67) is $A\, I_0(kr) + B\, K_0(kr)$, where I_0 and K_0 are the modified Bessel functions of the first and the second kind of order zero, respectively. Since $I_0 \to \infty$ as $r \to \infty$, we take $A = 0$, and then B is obtained by equating the flux across a circle C_ε of radius ε to $e^{-st'}$. Thus,

$$- \lim_{\varepsilon \to 0} \int_{\partial C_\varepsilon} B\, \frac{\partial K_0(sr/c)}{\partial r}\, ds = \lim_{\varepsilon \to 0} 2\pi\varepsilon\, B\, (s/c)\, K_1(s\varepsilon/c) = 2\pi\, B = e^{-st'}.$$

Hence, $B = \dfrac{e^{-st'}}{2\pi}$, and

$$\overline{G}_2(r, s; t') = \frac{e^{-st'}}{2\pi} K_0(sr/c).$$

Since

$$\mathcal{L}^{-1}\{K_0(\alpha s)\} = \frac{H(t - \alpha)}{\sqrt{t^2 - \alpha^2}}$$

(see Erdélyi et al., 1954 or Abramowitz and Stegun, 1965), on inversion we get

$$G_2(r, t; t') = \begin{cases} -\dfrac{1}{2\pi c \sqrt{c^2(t - t')^2 - r^2}} & \text{for } r < c(t - t') \\[2mm] 0 & \text{for } r > c(t - t') \end{cases} \qquad (7.68)$$

$$= -\frac{H(c(t - t') - r)}{2\pi c \sqrt{c^2(t - t')^2 - r^2}}. \blacksquare$$

EXAMPLE 7.12. In R^3, the Green's function G_3 satisfies the equation

$$\frac{\partial^2 G_3}{\partial t^2} - c^2 \left(\frac{\partial^2 G_3}{\partial x^2} + \frac{\partial^2 G_3}{\partial y^2} + \frac{\partial^2 G_3}{\partial z^2} \right) = \delta(x - x')\delta(y - y')\delta(z - z')\delta(t - t').$$

Applying the Laplace transform, we get

$$s^2 \overline{G}_3 - c^2 \left(\frac{\partial^2 \overline{G}_3}{\partial x^2} + \frac{\partial^2 \overline{G}_3}{\partial y^2} + \frac{\partial^2 \overline{G}_3}{\partial z^2} \right) = e^{-st'} \delta(x - x')\delta(y - y')\delta(z - z'),$$

or using the axial symmetry with $\rho^2 = (x - x')^2 + (y - y')^2 + (z - z')^2$,

$$s^2 \overline{G}_3 - c^2 \left(\frac{d^2 \overline{G}_3}{d\rho^2} + \frac{2}{\rho} \frac{\partial \overline{G}_3}{\partial \rho} \right) = \frac{e^{-st'} \delta(\rho)}{4\pi \rho^2}. \tag{7.69}$$

Set $\rho \overline{G}_3 = V$ in Eq (7.69). Then it reduces to

$$s^2 \overline{V} - c^2 \frac{\partial^2 \overline{V}}{\partial \rho^2} = \frac{e^{-st'} \delta(\rho)}{4\pi \rho}, \quad \rho = |\mathbf{x} - \mathbf{x}'|.$$

The solution of this equation is $\overline{V} = A e^{s\rho/c} + B e^{-s\rho/c}$. Since $e^{s\rho/c}$ becomes unbounded as $\rho \to \infty$, we take $A = 0$. Thus, $\overline{G}_3 = B e^{-(s\rho/c)}/\rho$. To evaluate B, the forcing term, which is $\dfrac{e^{-st'} \delta(\rho)}{4\pi \rho^2}$ in Eq (7.69), can be considered as representing a source of strength $e^{-st'}$ at the point $\rho = 0$. Then by equating $e^{-st'}$ to the flux across a sphere S_ε of radius ε and letting $\varepsilon \to 0$, we have

$$-\lim_{\varepsilon \to 0} \iint_{\partial S_\varepsilon} B \frac{\partial}{\partial \rho} \left(\frac{e^{-s\rho/c}}{\rho} \right) dS = e^{-st'},$$

or

$$-4\pi B \lim_{\varepsilon \to 0} \varepsilon^2 \left[-\frac{s}{c} \frac{e^{-s\rho/c}}{\rho} - \frac{e^{-s\rho/c}}{\rho^2} \right]_{\rho=\varepsilon} = e^{-st'},$$

which gives $4\pi B = e^{-st'}$, or $\overline{G}_3 = \dfrac{e^{-st'-s\rho/c}}{4\pi \rho}$. On inversion, this yields

$$G_3(\rho; t, t') = \frac{1}{4\pi \rho} \delta \left(t - t' - \frac{\rho}{c} \right),$$

or

$$G_3(x, y, z, t; x', y', z', t')$$
$$= \frac{1}{4\pi \rho} \delta \left(t - t' - \frac{1}{c} \sqrt{(x - x')^2 + (y - y')^2 + (z - z')^2} \right). \ \blacksquare \tag{7.70}$$

The graphs of Green's functions G_1, G_2, and G_3 are presented in Figs. 7.3, 7.4, and 7.5.

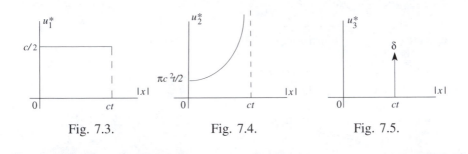

Fig. 7.3.   Fig. 7.4.   Fig. 7.5.

7.6. Applications of Green's Functions

In this section we will develop formal solutions of some well-known initial and boundary value problems in terms of Green's functions.

7.6.1. Dirichlet Problem. The solution of the boundary value problem $\nabla^2 u = f(\mathbf{x})$ in Ω such that $u = g(\mathbf{x}_s)$ on $\partial\Omega$, where $\mathbf{x}_s$ is an arbitrary point on the boundary, is given by

$$u(\mathbf{x}') = \iiint_\Omega f(\mathbf{x}) G(\mathbf{x}, \mathbf{x}')\, d\Omega + \iint_{\partial\Omega} g(\mathbf{x}_s) \frac{\partial G(\mathbf{x}, \mathbf{x}')}{\partial n}\, dS. \qquad (7.71)$$

PROOF. Let $\mathbf{x}'$ denote an arbitrary point in Ω; then by (A.7) we have

$$\iiint_\Omega \left(u\nabla^2 v - v\nabla^2 u \right) d\Omega = \iint_{\partial\Omega} \left(u\frac{\partial v}{\partial n} - v\frac{\partial u}{\partial n} \right) dS.$$

If we replace v by $G(\mathbf{x}, \mathbf{x}')$, we get

$$\iiint_\Omega \left(u\nabla^2 G(\mathbf{x}, \mathbf{x}') - G(\mathbf{x}, \mathbf{x}')\nabla^2 u \right) d\Omega$$
$$= \iint_{\partial\Omega} \left(u\frac{\partial G(\mathbf{x}, \mathbf{x}')}{\partial n} - G(\mathbf{x}, \mathbf{x}')\frac{\partial u}{\partial n} \right) dS.$$

which yields

$$\iiint_\Omega \left(u\, \delta(\mathbf{x}, \mathbf{x}') - f(\mathbf{x}) G(\mathbf{x}, \mathbf{x}') \right) d\Omega = \iint_{\partial\Omega} g(\mathbf{x}_s) \frac{\partial G(\mathbf{x}, \mathbf{x}')}{\partial n}\, dS,$$

and the result follows. ∎

7.6.2. Neumann Problem. Find the solution of $\nabla^2 u = F(\mathbf{x})$ in Ω such that $\dfrac{\partial u}{\partial n} = P(\mathbf{x}_s)$ on $\partial\Omega$, where $\mathbf{x}_s$ is an arbitrary point on the boundary. First, we note that the solution to the problem stated here is not always possible, for if we apply the divergence theorem to ∇u, we have

$$\iiint_\Omega \nabla \cdot \nabla u \, d\Omega = \iiint_\Omega \nabla^2 u \, d\Omega = \iint_{\partial\Omega} \frac{\partial u}{\partial n} dS,$$

which implies that

$$\iiint_\Omega F(\mathbf{x}) \, d\Omega = \iint_{\partial\Omega} P(\mathbf{x}_s) dS. \tag{7.72}$$

Thus, the consistency condition is necessary for the existence of the solution to the Neumann problem. We note that the homogeneous problem, i.e., when $F(x)$ and $P(x)$ are both zero, always has a nontrivial solution $u = C = $ (const). Hence, an arbitrary constant can always be added to any solution of the Neumann problem. The Green's function in this case needs to be modified such that it satisfies the condition (7.72). If $F(\mathbf{x}) = \delta(\mathbf{x} - \mathbf{x}')$ and $P(\mathbf{x}_s) = 0$, the condition (7.72) is not satisfied. However, (7.72) is satisfied if $F(\mathbf{x}) = \delta(\mathbf{x} - \mathbf{x}') - 1/V$, where V is the volume of Ω. This modified Green's function, also known as the *Neumann function*, satisfies

$$\nabla^2 G(\mathbf{x}, \mathbf{x}') = \delta(\mathbf{x} - \mathbf{x}') - \frac{1}{V}, \quad \frac{\partial G(\mathbf{x}, \mathbf{x}')}{\partial n} = 0.$$

Proceeding as before, Eq (7.71) in the present case becomes

$$\iiint_\Omega \left[u \left\{ \delta\left(\mathbf{x} - \mathbf{x}' \right) - \frac{1}{V} \right\} - F(\mathbf{x}) G(\mathbf{x}, \mathbf{x}') \right] d\Omega$$
$$= \iint_{\partial\Omega} (u(\mathbf{x}_s) V - P(\mathbf{x}_s) G(\mathbf{x}, \mathbf{x}')) \, dS,$$

or

$$u(\mathbf{x}') = \iiint_\Omega \left[\frac{u}{V} + F(\mathbf{x}) G(\mathbf{x}, \mathbf{x}') \right] d\Omega - \iint_{\partial\Omega} P(\mathbf{x}_s) G(\mathbf{x}, \mathbf{x}') \, dS. \tag{7.73}$$

In the solution (7.73) the only unknown integral is $\iiint_\Omega \frac{u}{V} \, d\Omega$ which is equal to the average value of u in Ω and is a constant. Since we can add an arbitrary constant to any solution of the Neumann problem, we can express the solution (7.73) as

$$u(\mathbf{x}') = \iiint_\Omega F(\mathbf{x}) G(\mathbf{x}, \mathbf{x}') \, d\Omega - \iint_{\partial\Omega} P(\mathbf{x}_s) G(\mathbf{x}, \mathbf{x}') \, dS + C.$$

This solution is valid only when the consistency condition (7.72) is satisfied.

In the one-dimensional case, the corresponding problem is

$$y'' = f(x), \quad y'(0) = \alpha, \quad y'(l) = \beta.$$

If we integrate this equation from 0 to l, we get

$$y'\big|_0^l = \int_0^l f(x)\, dx$$

or

$$\beta - \alpha = \int_0^l f(x)\, dx. \tag{7.74}$$

Thus, the solution to the one-dimensional Neumann problem does not always exist. The condition (7.74) is a consistency condition required for the solution to exist.

7.6.3. Robin Problem. Find the solution of $\nabla^2 u = f(\mathbf{x})$ in Ω such that $u + \alpha \dfrac{\partial u}{\partial n} = g(\mathbf{x}_s)$ on $\partial\Omega$, where $\mathbf{x}_s$ is an arbitrary point on the boundary. The solution for the Robin problem is given by

$$u(\mathbf{x}') = \iiint_\Omega f(\mathbf{x}) G(\mathbf{x}, \mathbf{x}')\, d\Omega - \frac{1}{\alpha} \iint_{\partial\Omega} g(\mathbf{x}_s) G(\mathbf{x}, \mathbf{x}')\, dS$$

$$= \iiint_\Omega f(\mathbf{x}) G(\mathbf{x}, \mathbf{x}')\, d\Omega + \iint_{\partial\Omega} g(\mathbf{x}_s) \frac{\partial G(\mathbf{x}, \mathbf{x}')}{\partial n}\, dS.$$

Green's function in this case is the solution of the following problem:

$$\nabla^2 G(\mathbf{x}, \mathbf{x}') = \delta(\mathbf{x} - \mathbf{x}') \text{ in } \Omega \text{ such that } G(\mathbf{x}, \mathbf{x}') + \alpha \frac{\partial G(\mathbf{x}, \mathbf{x}')}{\partial n} = 0 \text{ on } \partial\Omega.$$

The proof is similar to the previous two cases and is left as an exercise (see Exercise 7.20).

7.6.4. Solution of Helmholtz Equation in Terms of Green's Function. We will solve the Dirichlet problem, which in this case is

$$(\nabla^2 + k^2) u(\mathbf{x}) = F(\mathbf{x}) \quad \text{in } \Omega, \quad u(\mathbf{x}_s) = K(\mathbf{x}_s) \quad \text{on } \partial\Omega. \tag{7.75}$$

Green's function satisfies the following conditions:

$$(\nabla^2 + k^2) G(\mathbf{x}, \mathbf{y}) = \delta(\mathbf{x} - \mathbf{y}) \quad \text{in } \Omega, \quad G(\mathbf{x}_s, \mathbf{y}) = 0 \quad \text{on } \partial\Omega, \tag{7.76}$$

where $\mathbf{x}_s$ is any point on the boundary $\partial\Omega$. Multiplying the first equation in (7.75) by $G(\mathbf{x}, \mathbf{y})$ and the first equation in (7.76) by $u(\mathbf{x})$ and subtracting, we get

$$G(\mathbf{x}, \mathbf{y}) \nabla^2 u(\mathbf{x}) - u(\mathbf{x}) \nabla^2 G(\mathbf{x}, \mathbf{y}) = F(\mathbf{x}) G(\mathbf{x}, \mathbf{y}) - u(\mathbf{x}) \delta(\mathbf{x} - \mathbf{y}). \tag{7.77}$$

Integrating both sides of Eq (7.77) over Ω and using Green's identity (A.7), we have

$$\iiint_\Omega \left[G(\mathbf{x}, \mathbf{y}) \nabla^2 u(\mathbf{x}) - u(\mathbf{x}) \nabla^2 (G(\mathbf{x}, \mathbf{y}) \right]\, d\Omega$$

$$= \iint_{\partial\Omega} \left(G(\mathbf{x}, \mathbf{y}) \frac{\partial u(\mathbf{x})}{\partial n} - u(\mathbf{x}) \frac{\partial G(\mathbf{x}, \mathbf{y})}{\partial n} \right)\, dS$$

$$= \iiint_\Omega [F(\mathbf{x}) G(\mathbf{x}, \mathbf{y}) - u(\mathbf{x}) \delta(\mathbf{x} - \mathbf{y})]\, d\Omega$$

$$= \iiint_\Omega F(\mathbf{x}) G(\mathbf{x}, \mathbf{y})\, d\Omega - u(\mathbf{y}).$$

Thus,

$$u(\mathbf{y}) = \iiint_\Omega F(\mathbf{x}) G(\mathbf{x}, \mathbf{y}) \, d\Omega + \iint_{\partial\Omega} K(\mathbf{x}_s) \frac{\partial G(\mathbf{x}_s, \mathbf{y})}{\partial n} \, dS, \qquad (7.78)$$

since $G(\mathbf{x}_s, \mathbf{y}) = 0$, and $u(\mathbf{x}_s) = K(\mathbf{x}_s)$. The solution (7.78) implies that if both $F(\mathbf{x}) = K(\mathbf{x}_s) = 0$, then only a trivial solution exists. But we note that if Ω is the square $R: \{0 < x, y < a\}$, then $u = \phi(x,y) = \dfrac{1}{2}\sin(n\pi x/a)\sin(n\pi y/a)$ is a nontrivial solution of $(\nabla^2 + n^2\pi^2/a^2)u(x,y) = 0$, with $u = 0$ on the boundary. A consequence of this result is:

Theorem 7.7. *If $u(x,y)$ is a solution of*

$$\left(\nabla^2 + 2n^2\pi^2/a^2\right) u(x,y) = F(x,y) \quad \text{in } R, \text{ and } u = 0 \text{ on } \partial R, \qquad (7.79)$$

then $\phi(x,y) = \dfrac{1}{2}\sin(n\pi x/a)\sin(n\pi y/a)$ and $F(x,y)$ satisfy the condition

$$\iint_R F(x,y)\phi(x,y)\,dx\,dy = 0.$$

Proof. Let $u(x,y)$ be a solution of Eq (7.79). Multiplying both sides of Eq (7.79) by $\phi(x,y)$ and noting that $(\nabla^2 + n^2\pi^2/a^2)\,\phi(x,y) = 0$, we get

$$\phi(x,y)(\nabla^2 + n^2\pi^2/a^2)u(x,y) - u(x,y)(\nabla^2 + n^2\pi^2/a^2)\phi(x,y) = \phi(x,y)F(x,y),$$

or

$$\phi(x,y)\nabla^2 u(x,y) - u(x,y)\nabla^2\phi(x,y) = \phi(x,y)F(x,y).$$

Integrating both sides over R and applying Green's identity (A.7) to the left side, we find that

$$\int_{\partial R} \left[(\phi(x,y)\frac{\partial u(x,y)}{\partial n} - u(x,y)\frac{\partial(\phi(x,y))}{\partial n}\right] ds = \iint_R \phi(x,y)F(x,y)\,dx\,dy.$$

However, the left side is zero because both $u(x,y)$ and $\phi(x,y)$ vanish on ∂R. Obviously, the solution is not unique, since $u(x,y) + C\phi(x,y)$ is also a solution for any arbitrary constant C.

The solution of the Neumann or the Robin boundary value problem can be derived in the same manner as that for the Laplace equation.

7.6.5. Solution of Hyperbolic or Parabolic Equations. To solve the wave equation in terms of Green's function, consider

$$u_{tt} - c^2\,\nabla^2 u = f(\mathbf{x}, t) \quad \text{in } \Omega, \qquad (7.80)$$

subject to the conditions $u(\mathbf{x}, 0) = 0 = u_t(\mathbf{x}, 0)$ and $u(\mathbf{x}, t) = 0$ on the boundary $\partial\Omega$. The Green's function $G(\mathbf{x}, t; \mathbf{x}', t')$ for this problem satisfies the equation

$$G_{tt} - c^2 \nabla^2 G = \delta(\mathbf{x} - \mathbf{x}')\,\delta(t - t') \quad \text{in } \Omega, \tag{7.81}$$

and the conditions $G(\mathbf{x}, \mathbf{x}'; 0) = 0 = G_t(\mathbf{x}, \mathbf{x}'; 0)$ and $G(\mathbf{x}, \mathbf{x}'; t, t') = 0$ on $\partial\Omega$. Multiplying Eq (7.80) by G and Eq (7.81) by u and subtracting, we get

$$G\,u_{tt} - c^2\,G\,\nabla^2 u - u\,G_{tt} + c^2\,u\,\nabla^2 G = G\,f(\mathbf{x}, t) - u\,\delta(\mathbf{x} - \mathbf{x}')\,\delta(t - t').$$

Integrating this equation with respect to t from 0 to t', we have

$$\int_0^{t'} \left[(G\,u_{tt} - u\,G_{tt}) + c^2\left(u\,\nabla^2 G - G\,\nabla^2 u\right) \right] dt$$

$$= \int_0^{t'} \left[G\,f(\mathbf{x}, t) - u\,\delta(\mathbf{x} - \mathbf{x}')\,\delta(t - t') \right] dt,$$

which gives

$$\left. (G\,u_t - u\,G_t) \right|_0^{t'} - \int_0^{t'} \left[(G_t\,u_t - u_t\,G_t) + c^2\left(u\,\nabla^2 G - G\,\nabla^2 u\right) \right] dt$$

$$= \int_0^{t'} \left[G\,f(\mathbf{x}, t) - u\,\delta(\mathbf{x} - \mathbf{x}')\,\delta(t - t') \right] dt.$$

Since, in view of the initial conditions, $u(\mathbf{x}, 0) = 0 = u_t(\mathbf{x}, 0)$, and $G(\mathbf{x}, \mathbf{x}'; 0) = G(\mathbf{x}, \mathbf{x}'; t', t') = 0 = G_t(\mathbf{x}, \mathbf{x}'; 0) = G_t(\mathbf{x}, \mathbf{x}'; t, t')$, we find that

$$c^2 \int_0^{t'} \left(u\nabla^2 G - G\nabla^2 u \right) dt = \int_0^{t'} \left[G\,f(\mathbf{x}, t) - u(\mathbf{x}, t')\,\delta(\mathbf{x} - \mathbf{x}') \right] dt.$$

Integrating this equation over the region Ω, we obtain

$$c^2 \iiint_\Omega \int_0^{t'} \left(u\nabla^2 G - G\nabla^2 u \right) dt\, d\Omega$$

$$= \iiint_\Omega \int_0^{t'} G\,f(\mathbf{x}, t)\, dt - \iiint_\Omega u(\mathbf{x}, t')\,\delta(\mathbf{x} - \mathbf{x}')\, d\Omega.$$

Applying Green's second identity (A.7) after interchanging the time integral with the space integral, we get

$$c^2 \int_0^{t'} \iint_{\partial\Omega} \left(u\frac{\partial G}{\partial n} - G\frac{\partial u}{\partial n} \right) dS\, dt = \int_0^{t'} \iiint_\Omega G\,f(\mathbf{x}, t)\, d\Omega\, dt - u(\mathbf{x}', t').$$

Since u and G both vanish on the boundary $\partial\Omega$, we have

$$u\left(\mathbf{x}',t'\right) = \int_0^{t'} \iiint_\Omega G\left(\mathbf{x},\mathbf{x}';t,t'\right)\, f(\mathbf{x},t)\, d\Omega\, dt.$$

Interchanging $\mathbf{x}$ and t with $\mathbf{x}'$ and t', respectively, and noting the symmetry of $G\left(\mathbf{x},\mathbf{x}';t,t'\right)$ with respect to $\mathbf{x}$ and $\mathbf{x}'$ and with respect to t and t', we obtain

$$u\left(\mathbf{x},t\right) = \int_0^{t'} \iiint_\Omega G\left(\mathbf{x},\mathbf{x}';t,t'\right)\, f(\mathbf{x}',t')\, d\Omega\, dt'. \qquad (7.82)$$

For the general problem of Eq (7.80), subject to the conditions

$$u(\mathbf{x},0) = \phi(\mathbf{x}), \quad u_t(\mathbf{x},0) = \psi(x), \quad \alpha\,\frac{\partial u}{\partial n} + \beta\,u(\mathbf{x},t) = g(\mathbf{x},t) \quad \text{on } \partial\Omega,$$

the solution is given by

$$
\begin{aligned}
u\left(\mathbf{x},t\right) ={}& \int_0^{t'} \iiint_\Omega G\left(\mathbf{x},\mathbf{x}';t,t'\right)\, f(\mathbf{x}',t')\, d\Omega\, dt \\
&+ \iiint_\Omega \left[\psi(\mathbf{x})\, G\left(\mathbf{x},\mathbf{x}';t,0\right) - \phi(\mathbf{x})\,\frac{\partial\left(\mathbf{x},\mathbf{x}';t,0\right)}{\partial t'}\right] d\Omega \\
&+ \frac{c^2}{\alpha} \int_0^{t'} \iint_{\partial\Omega} G\left(\mathbf{x},\mathbf{x}';t,t'\right)\, g\left(\mathbf{x}',t'\right)\, dS\, dt,
\end{aligned}
$$

or

$$
\begin{aligned}
u\left(\mathbf{x},t\right) ={}& \int_0^{t} \iiint_\Omega G\left(\mathbf{x},\mathbf{x}';t,t'\right)\, f(\mathbf{x}',t')\, d\Omega\, dt' \\
&+ \iiint_\Omega \left[\psi(\mathbf{x})\, G\left(\mathbf{x},\mathbf{x}';t,0\right) - \phi(\mathbf{x})\,\frac{\partial\left(\mathbf{x},\mathbf{x}';t,0\right)}{\partial t'}\right] d\Omega \qquad (7.83) \\
&+ \frac{c^2}{\beta} \int_0^{t'} \iint_{\partial\Omega} \frac{\partial G\left(\mathbf{x},\mathbf{x}';t,t'\right)}{\partial n}\, g\left(\mathbf{x}',t'\right)\, dS\, dt'.
\end{aligned}
$$

Similarly, for the diffusion problem $u_t - k\,\nabla^2 u = f(\mathbf{x},t)$ in Ω, subject to the conditions $u(\mathbf{x},0) = \phi(\mathbf{x})$, $\alpha\,\dfrac{\partial u}{\partial n} + \beta\,u(\mathbf{x},t) = g(\mathbf{x},t)$ on $\partial\Omega$, the solution in terms of its Green's function is given by

$$
\begin{aligned}
u\left(\mathbf{x},t\right) ={}& \int_0^{t} \iiint_\Omega G\left(\mathbf{x},\mathbf{x}';t,t'\right)\, f(\mathbf{x}',t')\, d\Omega\, dt' \\
&+ \iiint_\Omega \phi(\mathbf{x}\, G\left(\mathbf{x},\mathbf{x}';t,0\right)\, d\Omega \\
&+ \frac{k}{\alpha} \int_0^{t} \iint_{\partial\Omega} G\left(\mathbf{x},\mathbf{x}';t,t'\right)\, g\left(\mathbf{x}',t'\right)\, dS\, dt',
\end{aligned}
$$

or

$$u(\mathbf{x}, t) = \int_0^t \iiint_\Omega G(\mathbf{x}, \mathbf{x}'; t, t') \, f(\mathbf{x}', t') \, d\Omega \, dt'$$

$$+ \iiint_\Omega \psi(\mathbf{x} \, G(\mathbf{x}, \mathbf{x}'; t, 0) \, d\Omega \tag{7.84}$$

$$- \frac{k}{\beta} \int_0^t \iint_{\partial\Omega} \frac{\partial G(\mathbf{x}, \mathbf{x}'; t, t')}{\partial n} g(\mathbf{x}', t') \, dS \, dt'. \ \blacksquare$$

Recall that the diffusion operator is not self-adjoint. The proof is complicated and will be omitted. In the Laplace transform domain the diffusion operator becomes self-adjoint, and we will establish (7.84) by the Laplace transform technique in §7.6.6.

We remark that the singularity solutions are synonymous to fundamental solutions that are the free space Green's functions, or Green's functions for the whole space. On the other hand, the solutions of the equation $L\left(\frac{\partial}{\partial t}, \frac{\partial}{\partial x}\right) = \delta(\mathbf{x} - \mathbf{x}') \, \delta(t - t')$ are known as the causal Green's functions, or simply Green's functions. We will not discuss the fundamental solutions for partial differential operators; details about these solutions are available in Friedlander (1982), Kythe (1996), and Vladimirov (1984).

EXAMPLE 7.13. Let Ω be the half-plane $y > 0$. Then Green's function associated with the Dirichlet boundary condition $u = 0$ on the boundary $y = 0$ is given by (7.33). Hence, for $u(x, 0) = f(x)$ we find from (7.82) that

$$u(x', y') = -\frac{1}{4\pi} \int_{-\infty}^\infty f(x) \frac{\partial}{\partial y} \left[\ln \frac{(x - x')^2 + (y - y')^2}{(x - x')^2 + (y + y')^2} \right]_{y=0} dx$$

$$= -\frac{1}{4\pi} \int_{-\infty}^\infty f(x) \frac{-4y'}{(x - x')^2 + y'^2} \, dx$$

$$= \frac{y'}{\pi} \int_{-\infty}^\infty \frac{f(x)}{(x - x')^2 + y'^2} \, dx.$$

Alternatively, the problem is to solve

$$\frac{\partial^2 u}{\partial x^2} + \frac{\partial^2 u}{\partial y^2} = 0, \ y > 0, \ u(x, 0) = f(x).$$

Applying the Fourier transform with respect to x, we have

$$\frac{d^2 \tilde{u}}{dy^2} - \alpha^2 \tilde{u} = 0.$$

Its solution which remains bounded at $y \to \infty$ is given by

$$\tilde{u} = A e^{-|\alpha| y}.$$

At $y = 0$, we have $\tilde{u}(\alpha, 0) = \tilde{f}(\alpha)$, which gives $A = \tilde{f}(\alpha)$; thus, $\tilde{u} = \tilde{f}(\alpha)e^{-|\alpha|y}$. The Fourier inverse of $e^{-|\alpha|y}$ is $\sqrt{\dfrac{2}{\pi}} \dfrac{y}{x^2 + y^2}$. Using the convolution theorem, we have

$$u(x, y) = \frac{y}{\pi} \int_{-\infty}^{\infty} \frac{f(\eta)}{(x - \eta)^2 + y^2}.$$

EXAMPLE 7.14. Let Γ be the circle $r = a$, and let $u(a, \theta) = f(\theta)$. In this case Green's function is given in Exercise 7.15. Since $\dfrac{\partial}{\partial n} = \dfrac{\partial}{\partial r}$ on the circle Γ, we find from (7.82) that

$$u(r', \theta') = \frac{1}{4\pi} \int_0^{2\pi} f(\theta) \frac{\partial}{\partial r} \left[\ln \frac{a^2[r^2 - 2rr'\cos(\theta - \theta') + r'^2]}{r^2 r'^2 - 2rr'a^2 \cos(\theta - \theta') + a^4} \right]_{r=a} a\, d\theta$$

$$= \frac{1}{4\pi} \int_0^{2\pi} f(\theta) \left[\frac{2r - 2r'\cos(\theta - \theta')}{r^2 - 2rr'\cos(\theta - \theta') + r'^2} \right.$$

$$\left. - \frac{2rr'^2 - 2a^2r'\cos(\theta - \theta')}{r^2 r'^2 - 2rr'a^2 \cos(\theta - \theta') + a^4} \right]_{r=a} a\, d\theta$$

$$= \frac{a^2 - r'^2}{2\pi} \int_0^{2\pi} \frac{f(\theta)}{a^2[r^2 - 2rr'\cos(\theta - \theta') + r'^2]} d\theta. \blacksquare$$

EXAMPLE 7.15. To find the harmonic function ϕ in the quarter-plane $x > 0$, $y > 0$, subject to the boundary conditions

$$\phi(x, 0) = f(x), \quad 0 < x < \infty, \quad \text{and} \quad \frac{\partial \phi}{\partial x}(0, y) = g(y), \quad 0 < y < \infty,$$

note that in view of (7.81), the solution is given by

$$\phi(x', y') = \int_\Gamma \left(\phi \frac{\partial G}{\partial n} - g \frac{\partial \phi}{\partial n} \right) ds,$$

where G is defined by Eq (7.67) (Example 7.3), and Γ is boundary of the quarter-plane $\{x > 0, y > 0\}$. Then

$$\phi(x', y') = -\int_0^\infty f(x) \left[\frac{\partial G}{\partial y} \right]_{y=0} dx + \int_0^\infty [G]_{x=0}\, g(y)\, dy$$

$$= \frac{y'}{\pi} \int_0^\infty f(x) \left[\frac{1}{(x - x')^2 + y'^2} + \frac{1}{(x + x')^2 + y'^2} \right] dx$$

$$+ \frac{1}{2\pi} \int_0^\infty \ln \frac{x'^2 + (y - y')^2}{x'^2 + (y + y')^2}\, g(y)\, dy.$$

Note that the signs in the two integrals above result from the fact that $\dfrac{\partial}{\partial n} = -\dfrac{\partial}{\partial y}$ and $\dfrac{\partial}{\partial n} = -\dfrac{\partial}{\partial x}$ for the quarter-region. $\blacksquare$

7.6.6. Integral Transform Method. The Laplace transform can be used effectively to solve parabolic or hyperbolic equations. For example, consider the equation $\left[L\left(\dfrac{\partial}{\partial t}\right) - \nabla^2\right] u = f(\mathbf{x}, t)$. Applying the Laplace transform with respect to t to this equation and assuming homogeneous initial condition, we get

$$\left[L(s) - \nabla^2\right] \overline{u} = F(\mathbf{x}, s). \tag{7.85}$$

Since the operator is now self-adjoint, Green's function satisfies the equation

$$\left[L(s) - \nabla^2\right] \overline{G} = e^{-st'} \delta(\mathbf{x} - \mathbf{x}'). \tag{7.86}$$

Multiplying (7.85) by $\overline{G}$ and (7.86) by $\overline{u}$ we obtain

$$\overline{u} \nabla^2 \overline{G} - \overline{G} \nabla^2 \overline{u} = \overline{G} F(\mathbf{x}, s) - \overline{u} e^{-st'} \delta(\mathbf{x} - \mathbf{x}'). \tag{7.87}$$

Integrating both sides of Eq (7.87) over the region under consideration and using Green's identity (A.7), we have

$$\iiint_\Omega \left[\overline{u} \nabla^2 \overline{G} - \overline{G} \nabla^2 \overline{u}\right] d\Omega = \iint_{\partial\Omega} \left(\overline{u} \frac{\partial \overline{G}}{\partial n} - \overline{G} \frac{\partial \overline{u}}{\partial n}\right) dS$$

$$= \iiint_\Omega \left[\overline{G} F(\mathbf{x}, s) - \overline{u} e^{-st'} \delta(\mathbf{x} - \mathbf{x}')\right] d\Omega,$$

which yields

$$\overline{u} = e^{st'} \left[\iiint_\Omega \overline{G} F(\mathbf{x}, s)\, d\Omega - \iint_{\partial\Omega} \left(\overline{u} \frac{\partial \overline{G}}{\partial n} - \overline{G} \frac{\partial \overline{u}}{\partial n}\right) dS\right]. \tag{7.88}$$

The integrand in the surface integral in (7.88) consists of either known functions or functions that vanish on the boundary. For the Dirichlet problem $\overline{G} = 0$ and $\overline{u}$ is known on $\partial\Omega$. For the Neumann problem, $\dfrac{\partial \overline{G}}{\partial n} = 0$ and $\dfrac{\partial \overline{u}}{\partial n}$ is known on $\partial\Omega$. Hence, the solution is known for the transform domain, and the solution to the problem can be found by carrying out the Laplace inversion.

In the case of the Robin problem, let $u + \lambda \dfrac{\partial u}{\partial n} = \phi(\mathbf{x}_B, t)$ be satisfied on the boundary, where $\mathbf{x}_B$ is any point on the boundary. For the corresponding Green's function, the function $\phi(\mathbf{x}_B, t)$ is replaced by zero. Hence,

$$\iint_{\partial\Omega} \left(\overline{u} \frac{\partial \overline{G}}{\partial n} - \overline{G} \frac{\partial \overline{u}}{\partial n}\right) dS$$

$$= \iint_{\partial\Omega} \left[\frac{\partial \overline{G}}{\partial n} \left(\overline{\phi}(\mathbf{x}_B, s) - \lambda \frac{\partial \overline{u}}{\partial n}\right) - \overline{G} \frac{\partial \overline{u}}{\partial n}\right] dS$$

$$= \iint_{\partial\Omega} \left[\frac{\partial \overline{G}}{\partial n} \left(\overline{\phi}(\mathbf{x}_B, s) - \lambda \frac{\partial \overline{u}}{\partial n}\right) + \lambda \frac{\partial \overline{G}}{\partial n} \frac{\partial \overline{u}}{\partial n}\right] dS \tag{7.89}$$

$$= \iint_{\partial\Omega} \overline{\phi}(\mathbf{x}_B, s) \frac{\partial \overline{G}}{\partial n}\, dS.$$

Once again the right side of Eq (7.89) is in terms of known functions. The final solution can be found by inverting the Laplace transform (see §7.6.4).

EXAMPLE 7.16. Solve the boundary value problem

$$u_{tt} - c^2\, u_{xx} = e^{-|x|}\,\sin t, \quad u(x,0) = 0, \quad u_t(x,0) = e^{-|x|}.$$

The given equation in the Laplace transform domain is (see Example 6.25)

$$\bar{u}_{xx}(x,s) - \frac{s^2}{c^2}\,\bar{u}(x,s) = -\frac{2+s^2}{c^2\,(1+s^2)}\,e^{-|x|}.$$

Green's function for this problem is

$$G(x,x';s) = -\frac{c}{2s}\,e^{-s|x-x'|/c}.$$

The solution is then given by

$$\bar{u}(x,s) = \frac{2+s^2}{2cs\,(1+s^2)} \int_{-\infty}^{\infty} e^{-|x'|-s|x-x'|/c}\, dx'.$$

Hence, for $x < 0$, we have

$$\bar{u}(x,s) = \frac{2+s^2}{2cs\,(1+s^2)} \left\{ \int_{-\infty}^{x} e^{-x'-s(x-x')/c}\, dx' + \int_{x}^{0} e^{x'+s(x-x')/c}\, dx' \right.$$
$$\left. + \int_{0}^{\infty} e^{-x'+s(x-x')/c}\, dx' \right\}$$

$$= \frac{2+s^2}{s\,(1+s^2)\,(s^2-c^2)} \left[s\,e^x - c\,e^{sx/c} \right]$$

$$= B\left(\frac{1}{s-c} - \frac{1}{s+c}\right) e^x - \frac{e^x}{(1+c^2)\,(1+s^2)}$$
$$+ \left[\frac{2}{cs} - B\left(\frac{1}{s-c} + \frac{1}{s+c}\right) - \frac{cs}{(1+c^2)\,(1+s^2)} \right] e^{s|x|/c},$$

where $B = \dfrac{2+c^2}{2c\,(1+c^2)}$, and

$$u(x,t) = B\left(e^{ct+x} - e^{-ct+x}\right) - \frac{e^x\,\sin t}{1+c^2}$$
$$+ H(ct+x)\left[\frac{2}{c} - B\left(e^{ct+x} + e^{-ct-x}\right) - \frac{c}{1+c^2}\,\cos\left(t - x/c\right) \right].$$

For $x > 0$, the solution in the transform domain becomes

$$\bar{u}(x, s) = \frac{2 + s^2}{2cs \left(1 + s^2\right)} \left\{ \int_{-\infty}^{x} e^{x' - s(x-x')/c}\, dx' + \int_{x}^{0} e^{-x' - s(x-x')/c}\, dx' \right.$$

$$\left. + \int_{0}^{\infty} e^{-x' + s(x-x')/c}\, dx' \right\}$$

$$= \frac{2 + s^2}{s \left(1 + s^2\right)} \left[s\,e^{-x} - c\,e^{-sx/c} \right]$$

$$= B \left(\frac{1}{s - c} - \frac{1}{s + c} \right) e^{-x} - \frac{e^{-x}}{\left(1 + c^2\right)\left(1 + s^2\right)}$$

$$+ \left[\frac{2}{cs} - B \left(\frac{1}{s - c} + \frac{1}{s + c} \right) - \frac{cs}{\left(1 + c^2\right)\left(1 + s^2\right)} \right] e^{-sx/c}.$$

Hence, for $x > 0$,

$$u(x, t) = B \left(e^{ct - x} - e^{-ct - x} \right) - \frac{e^{-x} \sin t}{1 + c^2}$$

$$+ H(ct + x) \left[\frac{2}{c} - B \left(e^{ct - x} + e^{-ct + x} \right) - \frac{c}{1 + c^2} \cos\left(t - x/c\right) \right].$$

This solution matches with that of Example 6.25. ∎

7.7. Computation of Green's Functions

We will develop a method to numerically compute Green's function for a simply connected domain $D \in R^2$ that has the origin as an interior point. If the origin is outside D, we carry out a suitable translation to bring the origin inside D. We will use the notation z and z' instead of $\mathbf{x} = (x, y)$ and $\mathbf{x}' = (x', y')$, respectively. The method is based on the following results from the theory of conformal mapping: If the Green's function $G(z; z')$ with a pole (singularity) at a point $z' \in D$ and an analytic function $f(z)$ each map the domain D conformally onto the unit disk, then $G(z; z')$ and $f(z)$ are related by

$$G(z; z') = \frac{1}{2\pi} \log |f(z)|, \quad z = x + iy,$$

and the mapping function $f(z)$ is given by

$$f(z) = (z - z')\, e^{-2\pi(g + ih)},$$

where g and h are harmonic functions in D. Since $\log |f(z)| = \log |z - z'| - 2\pi g(z)$, the construction of Green's function $G(z; z')$ involves determining the harmonic function $g(z) \equiv g(x, y)$ such that $G(z; z') = 0$ on the boundary ∂D. Thus, we solve the Dirichlet problem for the domain D with the boundary condition $g(x, y)\Big|_{\partial D} = \dfrac{1}{2\pi} \log r$, $r = |z - z'|$ (see §7.3.1). Hence, Green's function for the domain D is constructed by the formula*

$$G(z; z') = \frac{1}{2\pi} \log r - g(x, y). \tag{7.90}$$

To determine the function $g(x, y)$, we use an interpolation method, developed by Kantorovich (see Kantorovich and Krylov, 1958), as follows: Assuming the series representation for $g(z)$ as $g(z) = \sum_{k=0}^{\infty} c_k (z - z')^k$, $c_k = a_k + i\, b_k$, we approximate it by the harmonic polynomial $p_n(r, \theta) = \Re\left\{ \sum_{k=0}^{n} c_k (z - z')^k \right\}$. Thus, in the polar coordinates we have

$$g(x, y) \approx p_n(r, \theta) = a_0 + \sum_{k=1}^{n} r^k \left(a_k \cos k\theta - b_k \sin k\theta \right),$$

where $z - z' = r\, e^{i\theta}$. Since this polynomial has $(2n+1)$ coefficients, we take $(2n+1)$ arbitrary points $z_1, \ldots, z_{2n+1}$ on the boundary ∂D and then choose the coefficients a_k and b_k such that at each of the points z_j, $j = 1, \ldots, 2n+1$, the polynomial $p_n(r, \theta)$ takes the same value as $g(z_j)$. Since $g(z)$ has the boundary value $g(x, y) = \dfrac{1}{2\pi} \log r$, the coefficients a_k, b_k are determined by solving the system of equations

$$a_0 + \sum_{k=1}^{n} r_1^k \left(a_k \cos k\theta_1 - b_k \sin k\theta_1 \right) = \frac{1}{2\pi} \log r_1,$$

$$a_0 + \sum_{k=1}^{n} r_2^k \left(a_k \cos k\theta_2 - b_k \sin k\theta_2 \right) = \frac{1}{2\pi} \log r_2, \tag{7.91}$$

$$\cdots \quad \cdots \quad \cdots \quad \cdots \quad \cdots \quad \cdots \quad \cdots$$

$$a_0 + \sum_{k=1}^{n} r_{2n+1}^k \left(a_k \cos k\theta_{2n+1} - b_k \sin k\theta_{2n+1} \right) = \frac{1}{2\pi} \log r_{2n+1},$$

where $z_j - z' = r_j\, e^{i\theta_j}$, $j = 1, \ldots, 2n + 1$. The determinant of the system (7.91) depends on the choice of the points z_j. Let us assume that these points lie on an equipotential line of a harmonic polynomial $Q_n(r, \theta)$ of degree at most n. This line has the equation

$$Q_n(r, \theta) = a_0 + \sum_{k=1}^{n} r^k \left(a_k \cos k\theta - b_k \sin k\theta \right) = 0,$$

---

* This formula is valid in view of §7.3.1, where we have assumed a source of strength -1 at z'. If this strength is taken as $+1$, formula (7.90) becomes $G(z, z') = \dfrac{1}{2\pi} \log \dfrac{1}{r} - g(x, y)$, which is the same as in Kantorovich and Krylov (1958) and Kythe (1998).

where the coefficients a_k and b_k are all not zero at the same time. Then this line exists iff the homogeneous system of $(2n+1)$ algebraic equations

$$a_0 + \sum_{k=1}^{n} r_j^k \left(a_k \cos k\theta_j - b_k \sin k\theta_j \right) = 0, \quad j = 1, \ldots, 2n+1, \qquad (7.92)$$

has a nonzero solution for $a_0, a_1, b_1, \ldots$. Hence, Green's function has the approximate representation

$$G(z; z') \approx \frac{1}{2\pi} \log r - a_0 - \sum_{k=1}^{n} r^k \left(a_k \cos k\theta - b_k \sin k\theta \right). \qquad (7.93)$$

It is assumed here that the difference between $p_n(r, \theta)$ and $g(x, y)$ decreases as n increases and the above interpolation method becomes justified, provided that $\lim_{n \to \infty} p_n(r, \theta) = g(z)$. In view of the Riemann mapping theorem, the function $f(z)$ maps the domain D uniquely onto the unit disk iff $f(0) = 0$ and $f'(0) = 1$. The latter condition is used to compute the distortion in the mapping, and hence, the relative error in the computation of $G(z; z')$.

EXAMPLE 7.17. We will approximate the Green's function for the square $\{(x, y) : -1 \le x, y \le 1\}$ in the z-plane with the pole z', which is taken, without loss of generality, at the origin (see Fig. 7.6). In view of the symmetry about both coordinate axes, the values of $g(x, y)$ are arranged symmetric to the y-axis; thus, all $b_k = 0$ in (7.92). Also, since the values of $g(x, y)$ are symmetric to x-axis and bisectors of coordinate angles, the series (7.92) will contain cosine terms of angles $4n\theta$, $n = 0, 1, 2, \ldots$. Hence, the series expansion for $g(x, y)$ becomes

$$g(x, y) \approx a_0 + a_4 \, r^4 \cos 4\theta + a_8 \, r^8 \cos 8\theta + \cdots.$$

We take the points z_j on the boundary as follows: $z_1 = 1$, $z_2 = \dfrac{2}{\sqrt{3}} e^{i\pi/6}$, $z_3 = \sqrt{2} \, e^{i\pi/4}$. Then the coefficients a_0, a_4, and a_8 are obtained by solving the system (7.91), i.e.,

$$\begin{bmatrix} 1 & 1 & 1 \\ 1 & -\frac{8}{9} & -\frac{128}{81} \\ 1 & -4 & 16 \end{bmatrix} \begin{Bmatrix} a_0 \\ a_4 \\ a_8 \end{Bmatrix} = \frac{1}{2\pi} \begin{Bmatrix} 0 \\ \ln \frac{2}{\sqrt{3}} \\ \ln \sqrt{2} \end{Bmatrix}.$$

Thus, $a_0 = 0.120286 = 0.075578/(2\pi)$, $a_4 = -0.0117794 = -0.0740122/(2\pi)$, and $a_8 = -0.000249209 = -0.00156583/(2\pi)$. Note that additional points $z_4 = \dfrac{2}{\sqrt{3}} e^{i\pi/3}$ and $z_4 = e^{i\pi/2}$ add nothing to the solution. The approximate Green's function is given by

$$\begin{aligned} G(z; 0) &= \frac{1}{2\pi} \log r - g(z) \\ &\approx \frac{1}{2\pi} \left[\log r - 0.075578 + 0.0740122 \, r^4 \cos 4\theta + 0.00156583 \, r^8 \cos 8\theta \right] \\ &= \frac{1}{2\pi} \, \Re \left\{ \log z - 0.075578 + 0.0740122 \, z^4 + 0.00156583 \, z^8 \right\}. \end{aligned}$$

Then, the function $f(z)$ that maps the square onto the unit disk is found from
$$\log |f(z)| = \Re \left\{ \log z - 0.075578 + 0.0740122\, z^4 + 0.00156583\, z^8 \right\},$$
up to a purely imaginary additive constant which we will ignore. Hence,
$$f(z) = z\, e^{-0.075578 + 0.0740122\, z^4 + 0.00156583\, z^8}.$$
Note that $f'(0) = e^{-2\pi a_0} = e^{-0.075578} = 0.927207$. The value is $f'(0) =$
$\displaystyle \int_0^1 \frac{d\zeta}{\sqrt{1 + \zeta^4}} \approx 0.927037$, which shows the error of about 0.017%. ∎

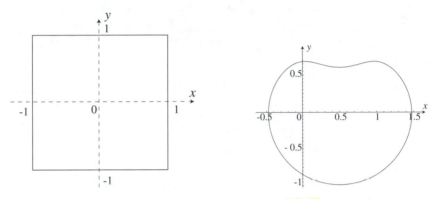

Fig. 7.6. The unit square.   Fig. 7.7. The curve for $\alpha = 1$.

EXAMPLE 7.18. To approximate Green's function for the ellipse
$$x = \left(1 + \lambda^2\right) \cos t, \quad y = \left(1 - \lambda^2\right) \sin t, \quad 0 \le t < 2\pi,$$
or, in complex notation, $z(t) = e^{it}\left(1 + \lambda^2\, e^{-2it}\right), 0 \le t < 2\pi$, we consider the first quadrant because of the axial symmetry and take the points z_j for $t = 0, \pi/8, \pi/4, 3\pi/8$, and $\pi/2$. Thus,
$$g(x, y) \approx a_0 + a_1 r_1 \cos 2t + a_2 r_2 \cos 4t + a_3 r_3 \cos 6t + a_4 r_4 \cos 8t.$$
Then, for example, for $\lambda = 0.1$ we find that
$$g(z) = -0.0000159163 + 0.00159187\, z^2 - 0.000023828\, z^4$$
$$+ 5.31236 \times 10^{-7}\, z^6 - 1.39292 \times 10^{-8}\, z^8,$$
which gives
$$G(z; 0) = \frac{1}{2\pi} \Re\left\{ \log z + 0.0001 - 0.010002\, z^2 + 0.00015\, z^4 \right.$$
$$\left. - 3.3 \times 10^{-6}\, z^6 + 8.75 \times 10^{-8}\, z^8 \right\}.$$
Note that $f'(0) = e^{0.0001} \approx 1.0001$. For computational details for $\lambda = 0.1$ and also for $\lambda = 0.5$, see Example7.18.nb. Note that if $\lambda = 0$, then $g(x, y) = 0$, and the Green's function $G(z; 0)$ is the same as (7.29). ∎

7.8. Mathematica Projects

PROJECT 7.1. The curve

$$F(x, y, \alpha) = \left[(x - 0.5)^2 + (y - \alpha)^2 \right] \left[1 - y^2 - (x - 0.5)^2 \right] = 0.1$$

for $\alpha = \infty$, 1, 0.5, 0.3, and 0.2746687749 is plotted by Mathematica as follows:

```
(* Need graphics package for implicit plots *)
```
In[1]: <<Graphics`ImplicitPlot`
```
(* define the function as f(x, y, α) *)
```
In[2]: f[x_, y_,α_]:=
         ((x-0.5)² + (y−α)²) ∗ (1 − y² − (−0.5)² − 0.1)
```
(* plot at α = ∞ *)
```
In[3]: p1= ImplicitPlot[1-y² − (x − 0.5)² == 0, {x, −1, 2}];
```
(* plot for α=1, 0.5, 0.3, and 0.2746687749 *)
```
         p2= ImplicitPlot[f[x, y, 1] == 0, {x, -1, 2}];
         p3= ImplicitPlot[f[x, y, 0.5] == 0, {x, -1, 2}];
         p4= ImplicitPlot[f[x, y, 0.3] == 0, {x, -1, 2}];
         p5= ImplicitPlot[f[x, y, 0.2746687749] == 0, {x, -1, 2}];
         Show[p2, p3, p4, p5]
```

*Out[3]*:

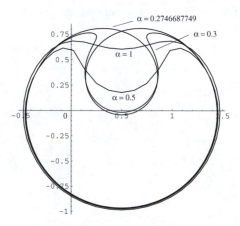

```
-Graphics-
```
```

The plot for $\alpha = 1$ is shown in Fig. 7.7. For $\alpha = \infty$ this curve represents the unit circle with center at $(0.5, 0)$. The region does not remain simply connected for $\alpha = 0.2746687749$.

PROJECT 7.2. Cassini's ovals, defined by

$$F(x, y, \alpha) = \left[(x + \alpha)^2 + y^2\right]\left[(x - \alpha)^2 + y^2\right] = 1,$$

are plotted by Mathematica as follows:

```
(* Need graphics package for implicit plots *)
```
$In[1]:$ `<<Graphics'ImplicitPlot'`

```
(* define the function as f(a) *)
```
$In[2]:$ $\text{f[a\_]} := (((x+a)^2 + (y^2)) * (((x - a)^2 + (y^2)) == 1$

$In[3]:$

```
g1=ImplicitPlot[f[1]];
g2=ImplicitPlot[f[0.99]];
g3=ImplicitPlot[f[0.9]];
g4=ImplicitPlot[f[0.5]];
g5=ImplicitPlot[f[0]];
Show[g1, g2, g3, g4, g5]
```

$Out[3]:$ `(* The plots are shown in Fig.7.8 *)`

`-Graphics-`

The graphs for $\alpha = 0$, 0.5, 0.9, 0.99, and 1 are plotted in Fig. 7.8. For $\alpha = 0$ the curve becomes the unit circle. The region does not remain simply connected for $\alpha = 1$.

PROJECT 7.3. We shall consider the curve of Project 7.1 for $\alpha = 1$ and approximate Green's function for this region which is a nearly circular cardioid (see Fig. 7.7). Since the region is symmetric about the line $x = 0.5$, we shall choose seven boundary points z_j as follows: $z_1 = 0.5 - 0.98725\,i$, $z_2 = 0.5 + 0.60343\,i$, $z_3 = 0.8 - 0.94024\,i$, $z_4 = 0.8 + 0.65588\,i$, $z_5 = 1.2 - 0.69296\,i$, $z_6 = 1.2 + 0.59722\,i$, and $z_7 = 1.474003$. The Green's function is given by

$$G(z; 0.5) \approx \frac{1}{2\pi}\,\log r - 0.0431944 + 0.0404596\,r\,\cos\theta$$

$$+ 0.0107145\,r^2\,\cos 2\theta - 0.0114393\,r^3\,\cos 3\theta$$

$$- 0.0569173\,r\,\sin\theta + 0.039137\,r^2\,\sin 2\theta - 0.00477567\,r^3\,\sin 3\theta,$$

where $r = |z - 0.5|$. Since the origin is at the point $(-0.5, 0)$, the distortion, given by $f'(0.5) = e^{0.0431944} = 1.04415$. The Mathematica session is given in the Notebook `proj7.3.nb`.

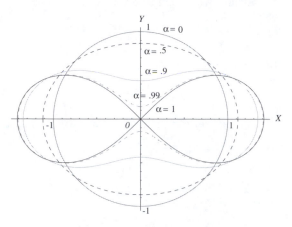

Fig. 7.8. Cassini's ovals.

PROJECT 7.4. We will consider Cassini's ovals of Project 7.2 and approximate Green's function for the region at $\alpha = 0.5$ which is bounded by a nearly circular ellipse.

Because of the symmetry of the region about the coordinate axes, the function $g(z)$ has a series expansion of the form

$$g(z) = a_0 + a_2\, r^2 \cos 2\theta + a_4\, r^4 \cos 4\theta + \cdots .$$

Let the points z_j on the boundary be chosen as $z_1 = 1.11803$, $z_2 = 1.04942\, e^{i\pi/6}$, $z_3 = 0.98399\, e^{i\pi/4}$, $z_4 = 0.92265\, e^{i\pi/3}$, and $z_5 = 0.866\, e^{i\pi/2}$. Then, as in Project 7.3, we can compute $a_0 = -0.00513$, $a_2 = 0.02124$, $a_4 = -0.00282$, $a_6 = 0.0005$, $a_8 = -0.000096$. Thus, Green's function is given by

$$G(z; 0) \approx \frac{1}{2\pi}\, \Re\Big\{ \log z - 0.0322236 + 0.133465\, z^2 - 0.0177425\, z^4$$
$$+ 0.0031616\, z^6 - 0.000605\, z^8 \Big\},$$

with $f'(0) = e^{0.0322236} \approx 1.03276$, which has an error of about 3%. The Mathematica session is given in the Notebook `proj7.4.nb`.

PROJECT 7.5. Green's functions are presented in the Mathematica Notebook `greens.nb`.

7.9. Exercises

7.1. Find Green's function for the Laplace operator in the circle $r \le a$, where $r = \sqrt{x^2 + y^2}$.

ANS. $G\left(x, t; x', t'\right) = \dfrac{1}{2\pi} \ln \dfrac{a\, r_1}{r'\, r_2}$, where $r'^2 = x'^2 + y'^2$, $r_1^2 = (x - x')^2 + (y - y')^2$, and $r_2^2 = \left(x - r^2\, x'/r'^2\right) + \left(y - r^2\, y'/r'^2\right)$.

7.2. Find Green's function for the Laplace operator in the sphere $\rho \le a$, where $\rho = \sqrt{x^2 + y^2 + z^2}$.

ANS. $G\left(\mathbf{x}; \mathbf{x}'\right) = -\dfrac{1}{4\pi} \left\{ \dfrac{1}{|\mathbf{x} - \mathbf{x}'|} - \dfrac{a\,|\mathbf{x}'|}{\big|\,|\mathbf{x}'|^2\, \mathbf{x} - a^2\, \mathbf{x}'\,\big|} \right\}$.

7.3. Find Green's function for the Laplace operator in the semicircle $x^2 + y^2 \le a^2$, $y \ge 0$.

ANS. $G\left(x, y; x', y'\right) = \dfrac{1}{2\pi} \ln \dfrac{r_1\, R_2}{r_2\, R_1}$, where $x = r\cos\theta$, $y = r\sin\theta$, $x' = r'\cos\theta'$, $y' = r'\sin\theta'$,

$r'^2 = x'^2 + y'^2$, $r_1^2 = (x - x')^2 + (y - y')^2 = r^2 - 2rr'\cos(\theta - \theta') + r'^2$,

$r_2^2 = \left(x - \dfrac{a^2 x'}{r'^2}\right)^2 + \left(y - \dfrac{a^2 y'^2}{r'^2}\right) = r^2 - 2\dfrac{a^2 r}{r'}\cos(\theta - \theta') + \dfrac{a^4}{r'^2}$,

$R_1^2 = (x - x')^2 + (y - y')^2 = r^2 - 2rr'\cos(\theta + \theta') + r'^2$,

$R_2^2 = \left(x - \dfrac{a^2 x'}{r'^2}\right)^2 + \left(y + \dfrac{a^2 y'^2}{r'^2}\right) = r^2 - 2\dfrac{a^2 r}{r'}\cos(\theta + \theta') + \dfrac{a^4}{r'^2}$.

7.4. Use Green's function to find the solution of the problem $\nabla^2 u = 0$, in the region $\{x^2 + y^2 \le a^2, y > 0\}$, subject to the boundary conditions $u(x, y) = k$ on $x^2 + y^2 = a^2$ and on $y = 0$ for $x < 0$, and $u(x, y) = 0$ on $y = 0$ for $x > 0$.

ANS. $u(x, y) = k - \dfrac{k}{\pi}\left[\tan^{-1}\left(\dfrac{a - x}{y}\right) - \tan^{-1}\left(\dfrac{x^2 + y^2 - ax}{ay}\right) \right]$.

7.5. Find Green's function for the Laplace operator in the upper hemisphere $\rho = \sqrt{x^2 + y^2 + z^2} \le a$, $z \ge 0$.

ANS. $G\left(\mathbf{x}; \mathbf{x}'\right) = -\dfrac{1}{4\pi\,|\mathbf{x} - \mathbf{x}'|} + \dfrac{a\,|\mathbf{x}'|}{4\pi\,\big|\,|\mathbf{x}'|^2\, \mathbf{x} - a^2\, \mathbf{x}'\,\big|} + \dfrac{1}{4\pi\,|\mathbf{x} - \mathbf{x}'_1|}$

$-\dfrac{a\,|\mathbf{x}'|}{4\pi\,\big|\,|\mathbf{x}'|^2\, \mathbf{x} - a^2\, \mathbf{x}'_1\,\big|}$, where $\mathbf{x}'_1 = (x', -y')$.

7.6. Find Green's function for the operator $L \equiv \dfrac{\partial^2}{\partial x^2} + \dfrac{\partial^2}{\partial y^2} + 2\alpha\dfrac{\partial}{\partial x} + 2\beta\dfrac{\partial}{\partial y}$ in the region $\{0 < x < a, 0 < y < b\}$ for the Dirichlet boundary conditions.

ANS. $G\left(x, t; x', t'\right)$

$$= -4\,e^{\alpha(x'-x)+\beta\,(y'-y)} \sum_{m=1}^{\infty} \sum_{n=1}^{\infty} \frac{ab\sin\dfrac{m\pi x}{a}\sin\dfrac{n\pi y}{b}}{\pi^2\left(m^2 b^2 + n^2 a^2\right) + a^2 b^2\left(\alpha^2 + \beta^2\right)}.$$

7.7. Find Green's function $G\left(x, t; x', t'\right)$ for the operator $L \equiv \dfrac{\partial}{\partial t} - k\dfrac{\partial^2}{\partial x^2}$ in the region $\{0 < x < a, t > 0\}$ subject to the conditions $G\left(0, t; x', t'\right) = G\left(a, t; x', t'\right) = 0$ and $G\left(x, t; x', t'\right) = 0$ for $t < t'$.

ANS. $G\left(x, t; x', t'\right) = \dfrac{2}{a}\,H(t - t')\displaystyle\sum_{n=1}^{\infty} e^{-n^2\pi^2 k(t-t')/a^2}\sin\dfrac{n\pi x}{a}\sin\dfrac{n\pi x'}{a}.$

7.8. Use Green's function to find the solution of the problem $\dfrac{\partial u}{\partial t} - k\dfrac{\partial^2 u}{\partial x^2} = 0$, subject to the boundary conditions $u(x, 0) = 0$, $u(0, t) = T_0$, and $u(a, t) = 0$.

ANS. $u(x, t) = T_0\left(1 - \dfrac{x}{a}\right) - 2T_0\displaystyle\sum_{n=1}^{\infty}\dfrac{1}{n\pi}\,e^{-n^2\pi^2 kt/a^2}\sin\dfrac{n\pi x}{a}.$

7.9. Use Green's function to find the solution of the problem $u_t - u_{xx} = 0$, $x \in R^1$, $0 < t$, subject to the conditions $u(x, 0) = H(x) - H(-x)$.

ANS. $u(x, t) = \mathrm{erf}\left(\dfrac{x}{2\sqrt{kt}}\right).$

7.10. Find Green's function for the Fokker-Planck operator $\dfrac{\partial}{\partial t} - \dfrac{\partial}{\partial x}\left(\dfrac{\partial}{\partial x} + x\right)$ in R^1.

HINT. Use the transformation $X = x\,e^t$ and $2T = e^{2t}$. Then the given operator reduces to the diffusion operator.

ANS. $G\left(x, t; x', t'\right) = \dfrac{H(t - t')}{\sqrt{2\pi\left(e^{2t} - e^{2t'}\right)}}\exp\left\{\dfrac{-\left(x - x'\,e^{-(t-t')}\right)^2}{2\left(1 - e^{-2(t-t')}\right)}\right\}.$

7.11. Find Green's function for the operator $\nabla^2 - k^2$ in R^1, R^2, and R^3.

ANS. $-\dfrac{1}{2k}\,e^{-k|x-x'|}$ in R^1; $-\dfrac{1}{2k}\,K_0(kr)$ in R^2, where $r = |\mathbf{x} - \mathbf{x}'|$; and $-\dfrac{1}{4\pi\rho}\,e^{-k\rho}$ in R^3, where $\rho = |\mathbf{x} - \mathbf{x}'|$.

7.12. Use Green's function for the half-plane (Example 7.4) to solve the following problems:

(a) $\nabla^2 u = 0$, $x \in R$, $0 < y$, such that $u(x, 0) = g(x)$.

(b) $\nabla^2 u = 0$, $x \in R$, $0 < y$, such that $u(x, 0) = k$.

(c) $\nabla^2 u = 0$, $x \in R$, $0 < y$, such that $u(x, 0) = |x|/x$.

ANS. (a) $u(\mathbf{x}') = \int_{-\infty}^{\infty} g(\mathbf{x}_s) \left(-\dfrac{\partial G(\mathbf{x}, \mathbf{x}')}{\partial y} \right) dx$; (b) $u(x, y) = k$; and

(c) $u(x, y) = \dfrac{2}{\pi} \tan^{-1}(x/y) = 1 - \dfrac{2\theta}{\pi}$.

7.13. Use Green's function to solve the following problem (see Example 7.10):
$u_{tt} - c^2 u_{xx} = e^{-x} \sin t$, $u(x, 0) = 0$, $u_t(x, 0) = e^{-x}$ for $x > 0$, $t > 0$, and
$\lim\limits_{x \to \infty} u(x, t) = 0$.

ANS. $u(x, t) = B \left(e^{ct-x} - e^{-ct-x} \right) - \dfrac{1}{1+c^2} e^{-x} \sin t$

$\qquad + H(ct - x) \left[\dfrac{2}{c} - B \left(e^{ct-x} + e^{-ct+x} \right) - \dfrac{c}{1+c^2} \cos(t - x/c) \right]$.

7.14. Find Green's function for the Laplacian in R^1.

ANS. For Green's function of the one-dimensional Laplace equation, we note

that (7.26) becomes $\dfrac{d^2 G}{dx^2} = \delta(x, x')$, whose general solution for a fixed x' is

$G(x) = \dfrac{1}{2} |x - x'| + Ax + B$. If we require the symmetry about x', i.e.,
$G = G(|x - x'|)$, then $A = 0$, and set $B = 0$, then Green's function is given by
$G(x) = \dfrac{1}{2} |x - x'|$.

7.15. Solve the Laplace equation $\dfrac{\partial^2 u}{\partial r^2} + \dfrac{1}{r}\dfrac{\partial u}{\partial r} + \dfrac{1}{r^2}\dfrac{\partial^2 u}{\partial \theta^2} = 0$, subject to the
conditions $u(r, 0) = f(r)$ and $u(r, \pi) = \pi$ for $r > 0$.

ANS. This is the well-known problem of solving the Laplace equation in the
half-plane $y > 0$. Taking the Fourier sine transform defined by

$$\tilde{u}(r, n) = \int_0^\pi u(r, \theta) \sin n\theta \, d\theta,$$

we get

$$r^2 \dfrac{d^2 \tilde{u}}{dr^2} + r \dfrac{d\tilde{u}}{dr} - n^2 \tilde{u} = -nf(r).$$

This equation is solved such that the solution is finite at $r = 0$ and tends to zero
as $r \to \infty$. To do this, first we determine Green's function associated with the
above problem. The solutions for the homogeneous equation are $r^{\pm n}$, and thus,

$$G(r, r') = \begin{cases} A \left(\dfrac{r}{r'} \right)^n, & r \le r', \\[2mm] A \left(\dfrac{r'}{r} \right)^n, & r \ge r', \end{cases}$$

where A is such that $G(r, r')$ has a discontinuity of the derivative of amount $1/r'^2$
at $r = r'$, i.e., $A = -1/(2nr)$. Then

$$\tilde{u} = -n \int_0^\infty G(r, r') f(r') \, dr',$$

and

$$\tilde{u}(r,n) = \frac{1}{2}\left[\int_0^r \left(\frac{r'}{r}\right)^n \frac{f(r')}{r'}\,dr' + \int_r^\infty \left(\frac{r}{r'}\right)^n \frac{f(r')}{r'}\,dr'\right],$$

which gives

$$\begin{aligned}
u(r,\theta) &= \frac{1}{\pi}\left[\int_0^r \frac{1}{r'}\sum_{n=1}^\infty \left(\frac{r'}{r}\right)^n \sin n\theta\, f(r')\,dr'\right.\\
&\quad \left. + \int_r^\infty \frac{1}{r'}\sum_{n=1}^\infty \left(\frac{r}{r'}\right)^n \sin n\theta\, f(r')\,dr'\right]\\
&= \frac{1}{\pi}\left[\int_0^r \frac{1}{r'}\Im\sum_{n=0}^\infty \left(\frac{r'}{r}e^{i\theta}\right)^n f(r')\,dr'\right.\\
&\quad \left. + \int_r^\infty \frac{1}{r'}\Im\sum_{n=1}^\infty \left(\frac{r}{r'}e^{i\theta}\right)^n f(r')\,dr'\right]\\
&= \frac{1}{\pi}\left[\int_0^r \frac{1}{r'}\Im\frac{r}{r-r'e^{i\theta}} f(r')\,dr'\right.\\
&\quad \left. + \int_r^\infty \frac{1}{r'}\Im\frac{r'}{r'-re^{i\theta}} f(r')\,dr'\right]\\
&= \frac{1}{\pi}\left[\int_0^r \frac{r\sin\theta}{r^2-2rr'\cos\theta+r'^2} f(r')\,dr'\right.\\
&\quad \left. + \int_r^\infty \frac{r\sin\theta}{r'^2-2rr'\cos\theta+r^2} f(r')\,dr'\right]\\
&= \frac{1}{\pi}\int_0^\infty \frac{r\sin\theta\, f(r')}{r^2-2rr'\cos\theta+r'^2}\,dr'\\
&= \frac{y}{\pi}\int_0^\infty \frac{f(r')}{(x-r')^2+y^2}\,dr' \quad \text{(in Cartesian coordinates).}
\end{aligned}$$

Note that if we remove the restriction on u being zero on half of the x-axis, and take $u = f(x)$ on the entire x-axis, then we must add the solution for $x < 0$, i.e., the integral $\int_{-\infty}^0$ to the above solution. Hence, the solution under this condition is

$$u(x,y) = \frac{y}{\pi}\int_{-\infty}^\infty \frac{f(r')}{(x-r')^2+y^2}\,dx'.$$

7.16. For the wave operator show that

$$\int_{-\infty}^\infty G_3(x,y,z,t;x',y',z',t')\,dz' = G_2(x,y,t;x',y',t').$$

SOLUTION. Since $\int_{-\infty}^{\infty} \delta(z - z') \, dz' = 1$, we find from (7.82) that

$$\int_{-\infty}^{\infty} G_3(x, y, z, t; x', y', z', t') \, dz'$$

$$= \frac{1}{4\pi} \int_{-\infty}^{\infty} \frac{\delta\left(t - t' - \dfrac{1}{c}\sqrt{(x - x')^2 + (y - y')^2 + (z - z')^2}\right)}{\sqrt{(x - x')^2 + (y - y')^2 + (z - z')^2}} \, dz'$$

$$= -\frac{1}{4\pi} \int_{-\infty}^{\infty} \frac{\delta\left(t - t' - \dfrac{1}{c}\sqrt{(x - x')^2 + (y - y')^2 + u^2}\right)}{\sqrt{(x - x')^2 + (y - y')^2 + u^2}} \, dz'$$

where $z - z' = u$

$$= -\frac{c}{2\pi} - \int_{r/c}^{\infty} \frac{\delta(t - t' - v)}{\sqrt{c^2 v^2 - r^2}} \, dv$$

where $r^2 + u^2 = v^2$, $r^2 = (x - x')^2 + (y - y')^2$

$$= -\frac{c}{2\pi \sqrt{c^2 v^2 - r^2}} \quad \text{for } t - t' < r/c$$

$$= G_2(x, y, t; x', y', t').$$

7.17. For the wave operator show that

$$\int_{-\infty}^{\infty} G_2(x, y, t; x', y', t') \, dz' = G_1(x, t; x', t').$$

HINT. Follow the method in Exercise 7.16.

7.18. Compute numerically Green's function for an equilateral triangle of side 2 units, where the x-axis is parallel to the base of the triangle and $1/2$ units above it and the y-axis passes through the top vertex.

HINT. Use the symmetry and assume $g(x, y) \approx a_0 + a_1 r^2 \cos 2\theta + a_1 r^2 \cos 2\theta + a_2 r^4 \cos 4\theta + \cdots$. Consider the two sets of points z_j:
(i) Four points at $(0, -1/2)$, $(1, -1/2)$, $\left(1/2, (\sqrt{3} - 1)/2\right)$, and $(0, \sqrt{3} - 1/2)$,
and (ii) Seven points at $(0, -1/2)$, $(1/2, -1/2)$, $(1, -1/2)$, $\left(3/4, (\sqrt{3} - 2)/4\right)$, $\left(1/2, (\sqrt{3} - 1)/2\right)$, $\left(1/4, (3\sqrt{3} - 2)/4\right)$, and $(0, \sqrt{3} - 1/2)$.

ANS. (i) $G(z; 0) = \dfrac{1}{2\pi} \Re \{ \log z - 0.012785 + 0.023988 \, z^2 - \cdots \}$, with $f'(0) = e^{0.012785} = 1.0128$;

(ii) $G(z; 0) = \dfrac{1}{2\pi} \Re \{ \log z - 0.042729 - 0.0476354 \, z^2 - \cdots \}$, with $f'(0) = e^{0.042729} = 1.04366$.

For computational details, see `equilateral.nb`, where results for many points on the boundary are discussed, and a technique for capturing an arbitrary number of boundary points is described.

7.19. Prove Theorem 7.5.

7.20. Verify the solution given in §7.6.3 for the Robin problem $\nabla^2 u = f(\mathbf{x})$ in Ω such that $u + \alpha \dfrac{\partial u}{\partial n} = g(\mathbf{x}_s)$ on $\partial \Omega$, where $\mathbf{x}_s$ is an arbitrary point on the boundary.

8

Initial and Boundary Value Problems

In previous chapters we studied analytical methods for solving different types of initial and boundary value problems. They include the characteristics methods for the first- and second-order equations (Chapter 2); inverse operator method (Chapter 3); separation of variables method in Cartesian, cylindrical polar, and spherical coordinates (Chapter 5); integral transform methods, including the Laplace and Fourier transforms, Fourier sine and cosine transforms, and the finite Fourier transform (Chapter 6); and Green's function methods for solving elliptic, parabolic, and hyperbolic equations (Chapter 7). In this chapter we discuss some additional initial and boundary value problems, and analyze the kinematics of wave propagation and dispersion, boundary layer flows, and certain ill-posed problems.

8.1. Initial and Boundary Conditions

Initial conditions, also known as Cauchy conditions, are defined in §1.2 and initial value problems for the first-order equations are studied in Exercises 2.22 through 2.26. Three types of boundary conditions (Dirichlet, Neumann, and Robin) are defined in §1.2. These conditions are also known as boundary conditions of the first, second, and third kind, respectively. The boundary conditions are said to be *linear* if they express a linear relation between the dependent variable u and its partial derivatives up to an appropriate finite order on the boundary ∂D of the domain D, along which the boundary value is prescribed. The *mixed boundary conditions* of the second-order partial differential equations are represented by

$$\left[\alpha\, u + \beta\, \frac{\partial u}{\partial n}\right]_{\partial D} = f\left(\mathbf{x}\right)\Big|_{\partial D}, \tag{8.1}$$

where α and β are constants, $\dfrac{\partial u}{\partial n}$ denotes the normal derivative defined by

$$\frac{\partial u}{\partial n} = \mathbf{n} \cdot \nabla u = n_{x_1}\frac{\partial u}{\partial x_1} + \cdots + n_{x_n}\frac{\partial u}{\partial x_n},$$

where $\mathbf{x} = (x_1, \ldots, x_n)$, and $\mathbf{n}$ is the outward normal to the boundary ∂D. A mixed boundary value problem can have a Dirichlet boundary condition on one part of the boundary, a Neumann boundary condition on another part, and a Robin boundary condition on still a third part of the boundary. Examples of boundary value problems that we have already studied are spread over Chapters 3 through 7. It should be pointed out that Neumann boundary conditions do not lead to a unique solution of a boundary value problem. For example, if u is a solution of the Laplace equation $\nabla^2 u = 0$ on the rectangle of Fig. 8.1, subject to the Neumann boundary conditions

$$\frac{\partial u}{\partial x}(x,0) = f_1(x), \qquad \frac{\partial u}{\partial y}(x,b) = f_2(x),$$

$$\frac{\partial u}{\partial x}(0,y) = g_1(y), \qquad \frac{\partial u}{\partial x}(a,y) = g_2(y),$$

where $f_{1,2}(x)$ and $g_{1,2}(y)$ are prescribed functions, then $w = u + c$, where c is a constant, is also a solution of this boundary value problem.

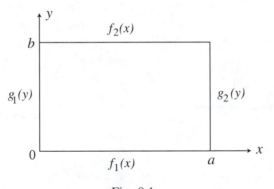

Fig. 8.1.

We will present a few additional examples of different types of boundary value problems.

EXAMPLE 8.1. Consider the Laplace equation $\nabla^2 u = 0$ on the rectangle $\{0 \le x \le a, 0 \le y \le b\}$ (Fig. 8.1, where the indicated boundary values refer to those of u), subject to the linear boundary conditions

$$u(x,0) = f_1(x), \qquad u(x,b) = f_2(x),$$

$$u(0,y) = g_1(y), \qquad u(a,y) = g_2(y).$$

This is an example of a boundary value problem that is solved by the method of separation of variables if we divide it into two parts, each with its own homogeneous boundary conditions, which are zero on a pair of opposite sides. Thus, the above boundary value problem is composed of the following two problems:

$$\nabla^2 u_1 = 0, \qquad\qquad \text{and} \qquad \nabla^2 u_2 = 0,$$
$$u_1(0, y) = 0 = u_1(a, y), \qquad\qquad u_2(x, 0) = 0 = u_2(x, b),$$
$$u_1(x, 0) = f_1(x), \qquad\qquad u_2(0, y) = g_1(y),$$
$$u_1(x, b) = f_2(x), \qquad\qquad u_2(a, y) = g_2(y),$$

which are solved separately for u_1 and u_2, yielding the solution of the original problem as $u = u_1 + u_2$. ∎

Note that a similar boundary value problem that is split into four problems with more than one nonhomogeneous boundary condition is given in Example 5.8, which has the solution $u = u_1 + u_2 + u_3 + u_4$.

8.2. Implicit Conditions

In most cases of boundary value problems, the initial and boundary conditions lead to a unique solution. But in certain cases, especially when the partial differential equation represents a model of some real-life physical situation, certain restrictions are imposed by means of boundary conditions, which are implicit in nature. These restrictions are also known as natural boundary conditions as opposed to essential boundary conditions, which in the case of mixed boundary value problems take the form of initial conditions. These types of implicit conditions are discussed in detail in §9.4 in connection with the weak variational formulation of physical problems. We give some simple examples to show the importance of implicit conditions.

EXAMPLE 8.2. Consider the one-dimensional heat equation

$$\frac{\partial u}{\partial t} = k \frac{\partial^2 u}{\partial x^2}, \quad -\infty < x < \infty, \quad t > 0,$$

where k denotes the thermal diffusivity. This equation has the following sets of solutions:

(i) $u_0(x, t) = A x + B$, where A and B are arbitrary constants; and

(ii) $u_\lambda(x, t) = \cosh(\lambda x) e^{k\lambda^2 t}$, where $\lambda \geq 0$ is a real parameter.

Note that for any fixed t, $\lim\limits_{x \to \pm\infty} u_\lambda(x, t) = +\infty$. Since these solutions are not bounded as $x \to \pm\infty$, they may not be acceptable for physical problems. If we require the solution u to be bounded at infinity in both x and t, we must impose the following conditions (natural conditions):

$$\lim_{x \to \pm\infty} u(x,t) < \infty, \quad \text{and} \quad \lim_{t \to \infty} u(x,t) < \infty,$$

as well as the initial condition $u(x,0) = f(x)$, $-\infty < x < \infty$, where $f(x)$ is a preassigned bounded function for all x, e.g., $f(x) = e^{-x^2}$. ∎

EXAMPLE 8.3. Consider the two-dimensional Laplace equation in polar coordinates, which is defined by

$$\nabla^2 u = \frac{1}{r} \frac{\partial}{\partial r} \left(r \frac{\partial u}{\partial r} \right) + \frac{1}{r^2} \frac{\partial^2 u}{\partial \theta^2} = 0.$$

Let the domain be the disk $0 \leq r \leq a$. This equation has the following two sets of solutions:

(i) $u_n(r, \theta) = r^n \left(A_n \sin n\theta + B_n \cos n\theta \right)$, $n = 1, 2, \ldots$;

(ii) $u_n(r, \theta) = \frac{1}{r^n} \left(C_n \sin n\theta + D_n \cos n\theta \right)$, $n = 1, 2, \ldots$.

Notice that the set (ii) has a singularity at $r = 0$ and, therefore, it is not acceptable for the given domain unless there is a source at the singularity. But the set (i) being bounded on the disk is acceptable. However, if the domain is the exterior $r > a$ of the above disk, then the set (ii) will be bounded and, therefore, acceptable for this domain. ∎

8.3. Periodic Conditions

In Definition 4.10 we have defined periodic boundary conditions. Now, we will present some examples involving such boundary conditions.

EXAMPLE 8.4. We will investigate earth's temperature u averaged annually as well as daily. Assuming that earth's surface is a plane, let $f(t)$ denote an average purely periodic temperature. We will ignore the conditions in the interior of the earth and, therefore, not consider any increase in temperature due to radioactive or nuclear processes. Using rectilinear Cartesian coordinates with z-axis positive downward, let $z = 0$ denote earth's surface, $z = \infty$ a great depth, say, of a cellar (see Fig. 8.2), and suppose $u(0, t) = f(t)$ at $z = 0$, where t denotes time, and $f(t)$ is a periodic function of period l. Then expanding $f(t)$ into a complex Fourier series, we have

$$f(t) = \sum_{n=-\infty}^{+\infty} C_n e^{2i\pi nt/l}, \tag{8.2}$$

where l denotes the length of a year ($l = 365$ days) or that of one day ($l = 1$). We will assume the temperature $u(z, t)$ at a depth $z > 0$ in the interior as

$$u(z, t) = \sum_{n=-\infty}^{+\infty} C_n \, u_n(z) \, e^{2i\pi nt/l}, \tag{8.3}$$

where each term of the series (8.3) individually satisfies the heat conduction equation $u_t = k \, u_{zz}$, where k is the thermal diffusivity. This leads to the following differential equation for u_n:

$$\frac{d^2 u_n}{dz^2} = \alpha_n^2 \, u_n, \quad \alpha_n^2 = \frac{2i\pi n}{kl}. \tag{8.4}$$

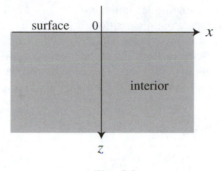

Fig. 8.2.

The functions u_n must satisfy the conditions $u_n(0) = 1$ and $u_n(\infty) < +\infty$. The first condition justifies reduction of (8.3) to (8.2) at $z = 0$, and the second condition ensures that the solution remains bounded at great depths. For n positive or negative, we set $2in = (1 + i)^2 \, |n|$, and $\alpha_n = \pm(1 + i) \, \beta_n$, $\beta_n = \sqrt{\dfrac{|n| \, \pi}{kl}} > 0$. Since the general solution of Eq (8.4) is

$$u_n(z) = A_n \, e^{(1+i)\beta_n z} + B_n \, e^{-(1+i)\beta_n z}, \tag{8.5}$$

so, in view of the second condition, we have $A_n = 0$; then the first condition gives $B_n = 1$. Substituting these values into (8.3), we obtain the solution as

$$u(z, t) = \sum_{n=-\infty}^{+\infty} C_n \, e^{-(1+i)\beta_n z} \, e^{2i\pi nt/l}. \tag{8.6}$$

Let $C_n = |C_n| \, e^{i\gamma}$ for $n > 0$. Then C_n for $n < 0$ has the same absolute value but negative phase. Thus, the real part of the solution (8.6) becomes

$$u(z, t) = C_0 + 2 \sum_{n=1}^{+\infty} |C_n| \, e^{-\beta_n z} \cos\left(\frac{2\pi nt}{l} + \gamma - \beta_n z\right). \tag{8.7}$$

Hence, we conclude that

(i) The amplitude $|C_n| e^{-\beta_n z}$ of the n-th partial wave is damped exponentially with increasing depth z; this damping increases with increasing n.

(ii) The phase shift of the n-th partial wave is given by $(\beta_n z - \gamma) l/(2\pi n)$.

(iii) The situation with the annual averaged temperature distribution is as follows: Take $k = 2 \times 10^{-3}$ cm²/sec. Then for $l = 365 \times 24 \times 60^2 = 31536000$ sec and at $z = 1$ m $= 100$ cm, the first approximation $n = 1$ gives

$$\beta_1 z = \sqrt{\frac{\pi}{2 \times 10^{-3} \times 31536 \times 10^3}} \times 100 = 0.705759 \approx \frac{\pi}{4},$$

so that $e^{-\beta_1 z} = 0.4937335 \approx 1/2$. Thus, at a depth of $z = 4$ m we have a phase lag $\beta_1 z = \pi$ and an amplitude damping of $2^{-4} = 1/16$. Even with the first approximation for the temperature distribution, it means physically that it is winter at a depth of 4 m when it is summer at the surface, and the amplitude at 4 m is only a fraction of the amplitude at the surface. For higher approximations ($n > 1$) the phase lag and amplitude damping in partial waves become correspondingly greater by the ratio $\sqrt{|n|}$ (which is a factor in β_n). Thus, for higher approximations, note that

$$\beta_3 z = \sqrt{\frac{3\pi}{63072}} \times 100 = 1.2224 \approx \frac{\sqrt{3}\,\pi}{4},$$

$$e^{-\beta_3 z} = 0.29452 \approx \frac{1}{3.4},$$

$$\beta_5 z = \sqrt{\frac{5\pi}{63072}} \times 100 = 1.57812 \approx \frac{\sqrt{5}\,\pi}{4},$$

$$e^{-\beta_5 z} = 0.20636 \approx \frac{1}{4.8},$$

and so on. The Fourier sine series for $u(z, t)$ with $\gamma = -\pi/2$ is given by

$$u(z, t) = \frac{4}{\pi} \left[e^{-\beta_1 z} \sin(\tau - \beta_1 z) + \frac{1}{3} e^{-\beta_3 z} \sin(\tau - \beta_3 z) \right.$$
$$\left. + \frac{1}{5} e^{-\beta_5 z} \sin(\tau - \beta_5 z) + \cdots \right], \quad \tau = \frac{2\pi t}{l},$$

so that at $z = 0$

$$u(0, t) = \frac{4}{\pi} \left(\sin\tau + \frac{1}{3} \sin 3\tau + \frac{1}{5} \sin 5\tau + \cdots \right). \tag{8.8}$$

For example, at $z = 100$ cm, we find that

$$u(100, t) \approx \frac{4}{\pi} \left[\frac{1}{2} \sin\left(\tau - \frac{\pi}{4}\right) + \frac{1}{3(3.4)} \sin\left(3\tau - \frac{\sqrt{3}\,\pi}{4}\right) \right.$$
$$\left. + \frac{1}{5(4.8)} \sin\left(5\tau - \frac{\sqrt{5}\,\pi}{4}\right) + \cdots \right], \tag{8.9}$$

and at $z = 400$ cm, we get

$$u(400, t) \approx \frac{4}{\pi} \left[\frac{1}{16} \sin\left(\tau - \pi\right) + \frac{10^{-2}}{3(1.3)} \sin\left(3\tau - \sqrt{3}\,\pi\right) \right.$$
$$\left. + \frac{10^{-2}}{5(5.3)} \sin\left(5\tau - \sqrt{5}\,\pi\right) + \cdots \right], \tag{8.10}$$

In the absence of any air flow in natural or man-made cellars in the interior of earth, a comparison of (8.8), (8.9) and (8.10) brings out the influence of depth on the amplitude and phase of the temperature fluctuations as follows: The larger the depth of the cellars is, the smaller are the temperature fluctuations as compared to those on the surface, and therefore, they are warmer in winter than in summer.

The situation with the daily averaged temperature distribution is as follows: The values of β_n increase by a factor of $\sqrt{365} \approx 19$, which means that the damping and phase lag that belong to a depth z in the above (annual) case will now occur at depth $z/19$. Thus, for example, from (8.10) we find that at a depth of $z = 400/19 \approx 21$ cm the amplitude decreases to $1/16$ for the first approximation ($n = 1$), which corresponds to the principal term in (8.10). Hence, the daily temperature fluctuations are very intense at very small depths, and this process occurs in a thin surface layer (this can be compared to the boundary layer theory in fluid flows). ∎

EXAMPLE 8.5. (Ring-shaped heat conductor) Consider a heat conductor of unit length such that periodic boundary conditions are prescribed at its two ends $x = \pm 1/2$, i.e., the heat pole at $x = 0$ repeats periodically (see Fig. 8.3(a)).*

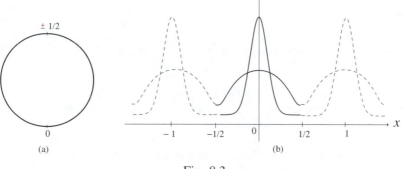

Fig. 8.3.

Because of this periodicity, the temperature u and all its derivatives coincide at the end points $x = \pm 1/2$ and there is no jump discontinuity at these end points. This is achieved by bending the conductor rod into a ring such that its two ends coincide. Although this ring is drawn circular in the above figure, its shape is immaterial for a

---

*If the length of the conductor is l, then introduce a dimensionless coordinate $x' = x/l$ and write x instead of x'.

linear heat conduction problem. However, the lateral surface of the ring is assumed to be adiabetically closed. The initial temperature is taken as $u(x, 0) = f(x)$, where $f(x)$ is an arbitrary function that is symmetric with respect to $x = 0$. Then the Fourier expansion of $f(x)$ is a cosine series satisfying the periodicity condition at the end points, i.e.,

$$f(x) = \sum_{n=0}^{\infty} A_n \cos 2\pi nx,$$

$$A_0 = \int_{-1/2}^{1/2} f(x)\, dx, \quad A_n = 2 \int_{-1/2}^{1/2} f(x) \cos 2\pi nx\, dx. \tag{8.11}$$

The general solution of the heat equation $u_t = k\, u_{xx}$ is given by

$$u(x,t) = \sum_{n=0}^{\infty} A_n\, e^{-4\pi^2 n^2 kt} \cos 2\pi nx. \tag{8.12}$$

Let $f(x)$ denote the unit source, i.e., $f(x) = \delta(x)$, which means that $f(x) = 0$ for $x \neq 0$ and $\int_{-1/2}^{1/2} f(x)\, dx = 1$. Then the coefficients A_n in (8.11) are given by $A_0 = 1$, $A_n = 2$ for $n \geq 1$. In this case the solution $u(x,t)$, given by (8.12), becomes the theta-function $\vartheta(x,t)$. Hence, the solution of this problem is

$$\vartheta(x,t) = 1 + 2 \sum_{n=1}^{\infty} e^{-4\pi^2 n^2 kt} \cos 2\pi nx, \tag{8.13}$$

or, taking $4i\pi kt = \tau$,

$$\vartheta(x,\tau) = 1 + 2 \sum_{n=1}^{\infty} e^{i\pi\tau n^2} \cos 2\pi nx. \tag{8.14}$$

This series converges faster for large kt, and represents the later phases of exponential damping due to the unit source. The curves sketched in Fig. 8.3(b) for the solution (8.13) show the behavior for large kt (where $kt > 1$) by flat curves and the behavior for small kt ($kt < 1$) by steep curves. However, to understand the temperature distribution we go back to the Fourier series (8.12). From the heat source $U_0(x,t)$ given in the ring we find that at points $x = n$, where $n = \pm 1, \pm 2, \ldots$, the identical heat sources are

$$U_n(x,t) = \frac{1}{\sqrt{4\pi kt}}\, e^{-(x-n)^2/(4kt)}. \tag{8.15}$$

Now, consider the series

$$u(x,t) = \sum_{n=-\infty}^{+\infty} U_n(x,t) = \frac{1}{\sqrt{4\pi kt}} \sum_{n=-\infty}^{\infty} e^{-(x-n)^2/(4kt)}, \tag{8.16}$$

which for small values of kt converges rapidly. In this series the dominant terms are U_0 and possibly U_{-1} and U_1; other terms have no effect because of the factor $e^{n^2/(4kt)}$ in (8.15). Therefore, the representation (8.16) is a good complement of (8.13). If we rewrite the series (8.16) as

$$u(x,t) = \frac{1}{\sqrt{4\pi kt}} e^{-x^2/(4kt)} \left[1 + 2 \sum_{n=1}^{+\infty} e^{-n^2/(4kt)} \cos \frac{inx}{2kt} \right],$$

where we have used the identity $\cos ix = \left(e^x + e^{-x} \right)/2$, then in terms of τ the terms in the square brackets become

$$1 + 2 \sum_{n=1}^{+\infty} e^{-i\pi n^2/\tau} \cos \frac{2\pi nx}{\tau} = \vartheta \left(\frac{x}{\tau}, -\frac{1}{\tau} \right).$$

Hence,

$$\vartheta(x, \tau) = \sqrt{\frac{i}{\tau}} \, e^{-i\pi x^2/\tau} \cdot \vartheta \left(\frac{x}{\tau}, -\frac{1}{\tau} \right),$$

or, conversely,

$$\vartheta \left(\frac{x}{\tau}, -\frac{1}{\tau} \right) = \sqrt{\frac{\tau}{i}} \, e^{i\pi x^2/\tau} \cdot \vartheta(x, \tau), \tag{8.17}$$

which is known as the *transformation formula* for the theta-function. ∎

8.4. Wave Propagation and Dispersion

We discuss three important aspects of the hyperbolic equation, which deal with wave propagation, dispersion, and damping.

8.4.1. Wave Propagation. The solution of the problem of a vibrating string, defined by the one-dimensional wave equation (5.4) and subject to the boundary conditions (5.5) and the initial conditions $u(x, 0) = f(x)$, $u_t(x, 0) = g'(x)$, is given in (5.23), which can be represented as $u(x, t) = u_1(x, t) + u_2(x, t)$, where

$$u_1(x, t) = \frac{f(x + ct) + f(x - ct)}{2},$$
$$u_2(x, t) = \frac{g(x + ct) - g(x - ct)}{2c}. \tag{8.18}$$

The function $u_1(x, t)$ represents the path of the propagation of the initial displacement without the initial velocity, i.e., for $g'(x) = 0$. The other function $u_2(x, t)$ contains the initial velocity (initial impulse) with zero initial displacement. Geometrically, the function $u(x,t)$ represents a surface in the (u, x, t)-space (Fig. 8.4(a)). The intersection of this surface by a plane $t = t_0$ is described analytically by $u = u(x, t_0)$, which exhibits the profile of the string at time t_0. However, the intersection of the surface $u(x,t)$ by a plane $x = x_0$ is given by $u = u(x_0, t)$, which represents the path of the motion of the point x_0.

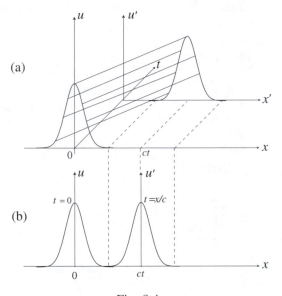

Fig. 8.4.

The function $u = f(x - ct)$ represents a *propagating wave*. The structure of a propagating wave at different times t is described as follows: Assume that an observer moves parallel to the x-axis with velocity c (Fig. 8.4(b)). If the observer is at the initial time $t = 0$ at the position $x = 0$, then he(she) has moved along the path toward the right until some time t. Let a new coordinate system, defined by $x' = x - ct$, $t' = t$, move along with the observer. Then $u(x, t) = f(x - ct)$ is defined in this new coordinate system by

$$u(x', t') = f(x'),$$

which means that the observer sees one and the same profile $f(x')$ during the entire time t. Thus, $f(x - ct)$ represents a fixed profile $f(x')$, which moves to the right with velocity c (hence, a propagating wave). In other words, in the (x, t)-plane the function $u = f(x - ct)$ remains constant on the line $x - ct = $ const.

The other function $f(x + ct)$ similarly represents a wave propagating toward the left with velocity c. For this wave we have a similar explanation. Thus, the initial form of both waves is characterized by the function $f(x')/2$, which is equal to one half of the original displacement.

EXAMPLE 8.6. The one-dimensional wave equation $u_{tt} = c^2 u_{xx}$, subject to the initial conditions $u(x,0) = f(x)$, $u_t(x,0) = g'(x)$, has the solution

$$u(x,t) = \phi(x - ct) + \psi(x + ct), \quad x \in R^1, \qquad (8.19)$$

where ϕ represents a disturbance (wave) traveling in the positive x direction with velocity c, while ψ a disturbance traveling in the negative x direction with the same velocity. Since Eq (8.19) gives $\phi(-ct) + \psi(ct) = 0$ at $x = 0$, we find that

$$u(x,t) = \phi(x - ct) - \phi(-x - ct).$$

The initial conditions give

$$\begin{aligned} f(x) &= \phi(x) + \psi(x) = \phi(x) - \phi(-x), \\ g'(x) &= -c\,\phi'(x) + c\,\psi'(x) = -c\,\phi'(x) + c\,\phi'(-x). \end{aligned} \qquad (8.20)$$

If we integrate the second equation in (8.20), with x_0 as an arbitrary point in R^1, we obtain

$$g(x) = -c\big[\phi(x) + \phi(-x)\big] + 2c\phi\,(x_0) + g\,(x_0),$$

which, when added to the first equation in (8.20), gives

$$2\phi(x) = f(x) - \frac{1}{c}\,g(x) + A, \quad \text{where} \quad A = -2\phi\,(x_0) + \frac{1}{c}\,g\,(x_0). \qquad (8.21)$$

Substituting (8.21) into (8.19), we find the solution as

$$\begin{aligned} u(x,t) &= \frac{1}{2}\left[f(x - ct) + f(x + ct) + \frac{1}{c}\left(\int_{x_0}^{x+ct} - \int_{x_0}^{x-ct} \right) g'(s)\,ds \right] \\ &= \frac{1}{2}\left[f(x - ct) + f(x + ct) + \frac{1}{c} \int_{x-ct}^{x+ct} g'(s)\,ds \right] \\ &= \frac{1}{2}\left[f(x - ct) + f(x + ct) + \frac{1}{c}\,g(x + ct) - \frac{1}{c}\,g(x - ct) \right], \end{aligned} \qquad (8.22)$$

which is valid only for all $x \in R^1$.

A geometrical interpretation of Eq (8.22) is obvious from Fig. 8.5, where we draw two lines PA and PB through the point P in the (x,t)-plane, which satisfy the equation $x \pm ct = $ const. The disturbance at the point P at the time T is caused by those initial data that lie on the segment $AB = (x_1, x_2)$. In optics, the triangle APB is called the *retrograde light cone*. If the initial velocities are zero, i.e., if $g = 0$, then

$$u(x,t) = \frac{1}{2}\left[f(x - ct) + f(x + ct) \right]. \qquad (8.23)$$

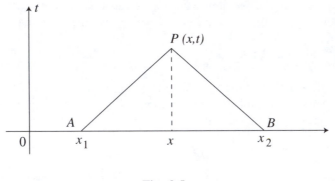

Fig. 8.5.

This represents the *Huygens' principle*, which states that each point of an advancing wavefront becomes a new source of secondary waves such that the envelope tangent to all these secondary waves forms a new wavefront. The secondary waves have the same frequency and speed as the primary advancing waves. Thus, the disturbance at x at time t, originating from the sources at A and B at time zero, needs the time $\tau = (x_2 - x)/c = (x - x_1)/c$ to reach the point x. ∎

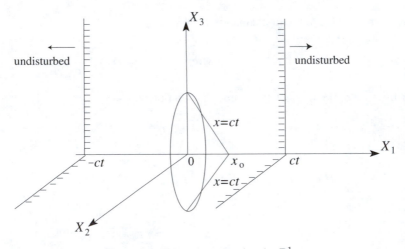

Figs. 8.6. Wave propagation in R^1.

The Green's functions for the wave operator in R^1, R^2, and R^3 are derived in Chapter 7. In R^1, the solution (7.66) shows that the wave originating instantaneously at a point source $\delta(x, t)$ at time $t > 0$ covers the interval $-ct \leq x \leq ct$, where there exist two edges defined by $x = \pm ct$ that move forward with velocity c. This wave is observed behind the front edge and has amplitude $1/(2c)$ (see Fig. 7.3). Hence, the wave diffusion occurs in this case. A three-dimensional representation of Green's function in R^1 is shown in Fig. 8.6. It can be viewed as that of a wave starting at the point source and propagating as a plane wave $|x| \leq ct$ whose front edge $|x| = ct$ moves with the velocity c perpendicular to the plane $x = 0$. In this case there does not exist a rear edge of the wave.

In R^2, Green's function defined by (7.68), with $(\mathbf{x}', t') = (\mathbf{0}, 0)$, shows that the disturbance originates instantaneously at the point source $\delta(\mathbf{x}, t)$, and at time $t > 0$ it occupies the entire circle $|\mathbf{x}| \leq ct$ (see Fig. 8.7). The wavefront at $|\mathbf{x}| = ct$ propagates throughout the plane with velocity c, but wave propagation exists behind the front edge at all subsequent times, and the wave has no rear edge. The wave diffusion occurs in this case, and Huygens' principle does not apply.

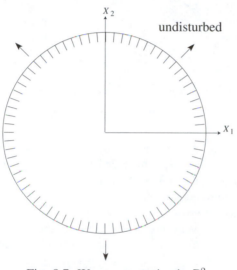

Fig. 8.7. Wave propagation in R^2.

In R^3, the Green's function (7.70), with $(\mathbf{x}', t') = (\mathbf{0}, 0)$, implies that the disturbance originating at a point source $\delta(\mathbf{x}, t)$ at time $t > 0$ occupies a spherical surface of radius ct and center at the origin. The wave propagates as a spherical wave with wavefront at $|\mathbf{x}| = ct$ and velocity c, and after the wave has passed, there is no disturbance (see Fig. 8.8). Huygens' principle applies in this case. The amplitude of the wave decays like r^{-1} as the radius increases. There is a significant difference between the two- and three-dimensional cases. If a stone is dropped in a calm shallow pond, the leading water wave spreads out in a circular form with its radius increasing uniformly with time, but the water contained by this wave continues to move after its passage. This is because of the Heaviside function in the solution (7.68), which leaves a wake behind it. On the other hand, in the three-dimensional case if a shot fired suddenly at time $t = t'$ in still air is heard only on expanding spherical surfaces with center at the firing gun and radius $c|t - t'|$, where c is the velocity of sound. However, the air does not continue to reverberate after the passage of this wave. This is because of the presence of the Dirac delta function in the solution (7.70), which represents a sharp bang and no tail effect. Huygens' principle accounts for the simplicity of communications in our three-dimensional world. If it were two-dimensional, communication would be impossible since utterances could be hardly distinguished from one another.

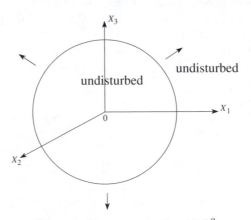

Figs. 8.8. Wave propagation in R^3.

8.4.2. Wave Dispersion. The one-dimensional wave equation $u_{tt} = c^2 u_{xx}$ has solutions of arbitrary form given by (5.24). Consider the general form (1.8) of the homogeneous, linear partial differential equation of the second order with constant coefficients. Then the problem of determining which solutions are of arbitrary form (5.24) reduces to that of constructing relationships among the coefficients in Eq (1.8), which guarantee the solution of this equation to be of the form

$$u(x,t) = f(x - ct), \tag{8.24}$$

where f is an arbitrary function. Substituting (8.24) into (1.8), we get the linear ordinary differential equation

$$\left(a_{11} - 2a_{12}c + a_{22}c^2\right) f''(x - ct) + (b_1 - b_2 c) f'(x - ct) + c_0 f(x - ct) = 0,$$

which must satisfy the wave profile. This equation is solvable when all its coefficients are zero, i.e., when

$$a_{11} - 2a_{12}c + a_{22}c^2 = 0,$$
$$b_1 - b_2 c = 0, \tag{8.25}$$
$$c_0 = 0.$$

If the differential equation (1.8) is hyperbolic, then it is necessary and sufficient that the conditions (8.25) be satisfied. The first equation in (8.25) yields the wave velocity

$$c = \frac{a_{12} \pm \sqrt{a_{12}^2 - a_{11}a_{22}}}{a_{22}}.$$

Thus, there exist two velocities of wave propagation for hyperbolic differential equations for which $a_{12}^2 - a_{11}a_{22} > 0$. Then for both values of c, we find from the remaining two equations in (8.25) that $b_1 = b_2 = c_0 = 0$. Therefore, a solution of the form (8.24) for a propagating wave without decay or growth is possible only for an equation of the form

$$a_{11}u_{xx} + 2a_{12}u_{xt} + a_{22}u_{tt} = 0. \tag{8.26}$$

If $a_{22} \neq 0$, then Eq (8.26) represents a wave equation in a moving coordinate system. To see this, we use the characteristic coordinates ξ, $\eta = x - \beta_{1,2}\,t$, where

$$\beta_{1,2} = \frac{a_{11}}{a_{12} \pm \sqrt{a_{12}^2 - a_{11}a_{22}}} = \frac{a12 \mp \sqrt{a_{12}^2 - a_{11}a_{22}}}{a_{22}} = c_{2,1},$$

provided $a_{12}^2 - a_{11}a_{22} > 0$. Then Eq (8.26) is transformed into

$$a_{11} \left[u_{\xi\xi} + 2u_{\xi\eta} + u_{\eta\eta} \right] - 2a_{12} \left[\beta_1 u_{\xi\xi} + (\beta_1 + \beta_2)\, u_{\xi\eta} + \beta_2 u_{\eta\eta} \right]$$
$$+ a_{22} \left[\beta_1^2 u_{\xi\xi} + 2\beta_1\beta_2 u_{\xi\eta} + \beta_2^2 u_{\eta\eta} \right] = 0.$$

Note that the coefficients of both $u_{\xi\xi}$ and $u_{\eta\eta}$ vanish. Thus, Eq (8.26) reduces to $u_{\xi\eta} = 0$, which is the wave equation with the solution

$$u = f(\xi) + g(\eta) = f(x - \beta_1 t) + g(x - \beta_2 t).$$

This solution represents two waves with speed β_1 and β_2, respectively.

For elliptic equations (i.e., when $a_{12}^2 - a_{11}a_{22} < 0$), we do not get solutions that represent waves with real velocities.

For parabolic equations (i.e., when $a_{12}^2 - a_{11}a_{22} = 0$), Eq (8.26) becomes

$$\left(\alpha \frac{\partial}{\partial x} + \beta \frac{\partial}{\partial t} \right) \left(\alpha \frac{\partial}{\partial x} + \beta \frac{\partial}{\partial t} \right) u = 0, \tag{8.27}$$

where $\alpha = \sqrt{a_{11}}$, $\beta = \sqrt{a_{22}}$, and $\alpha\beta = a_{12}$, and its solution is given by

$$u(x,t) = f(\beta x - \alpha t) + x\, g(\beta x - \alpha t). \tag{8.28}$$

Hence, solutions in the form of propagating waves exist in this case.

In physics, the propagating wave is denoted by $u = f(x - t/c)$, and the concept of wave dispersion is slightly different. Thus, in physics literature, a harmonic wave is represented by $u(x,t) = e^{i(\omega t - \nu x)}$, where ω is the frequency, $\nu = 2\pi/\lambda$ the wave number, and λ the wave length. Then the velocity by which the phase $\delta = \omega t - \nu x$ of the wave moves is known as the phase velocity and is equal to ω/ν. A wave dispersion occurs when the phase velocity of a harmonic wave depends on the frequency, and the harmonic signals are displayed relative to each other, thus creating a distorted signal.

8.4.3. Damped Waves. Eq (1.8) admits solutions in the form of damped waves, which are given by

$$u(x,t) = \mu(t)\, f(x - ct), \quad \text{or} \quad u(x,t) = \mu(x)\, g(x - ct). \tag{8.29}$$

We will consider the first form of the waves; the other form can be analyzed similarly. Substituting (8.29) into (1.8), we get

$$\left(a_{11} - 2a_{12}c + a_{22}c^2\right) \mu f'' + \left[\left(b_1 - b_2 c\right)\mu + 2\left(a_{12} - a_{22}c\right)\mu' \right] f'$$
$$\left(c_0 \mu + b_2 \mu' + a_{22}\mu''\right) f = 0.$$

Now, since f is arbitrary, the coefficients of f, f', and f'' must be equal to zero:

$$a_{11} - 2a_{12}c + a_{22}c^2 = 0,$$
$$\left(b_1 - b_2 c\right)\mu + 2\left(a_{12} - a_{22}c\right)\mu' = 0, \tag{8.30}$$
$$c_0 \mu + b_2 \mu' + a_{22}\mu'' = 0.$$

Since the function $\mu(t)$ satisfies an ordinary differential equation with constant coefficients, it has the form $\mu(t) = e^{\pm \kappa t}$. We consider the case when $\mu(t) = e^{-\kappa t}$. Thus, substituting it into Eqs (8.30), we get

$$a_{11} - 2a_{12}c + a_{22}c^2 = 0,$$
$$\left(b_1 - b_2 c\right) - 2\kappa\left(a_{12} - a_{22}c\right) = 0, \tag{8.31}$$
$$a_{22}\kappa^2 - b_2 \kappa + c_0 = 0.$$

If we eliminate c and κ from the system (8.31), we obtain a condition of compatibility of these three equations. The first equation shows that only the hyperbolic differential equation admits damped waves as solutions. The damping coefficient κ is obtained from the second equation as $\kappa = \dfrac{b_1 - b_2 c}{2\left(a_{12} - a_{22}c\right)}$. If we substitute this value of κ into the third equation and use the first equation to eliminate c, we obtain the following relation for the coefficients:

$$4\left(a_{12}^2 - a_{11}a_{22}\right) c_0 + \left(a_{11}b_2^2 - 2a_{12}b_1 b_2 + a_{22}b_1^2\right) = 0. \tag{8.32}$$

If this relation is satisfied, then the solutions of Eq (1.8) exist in the form of damped waves.

EXAMPLE 8.7. Consider the telegraph equation

$$u_{xx} - CL\, u_{tt} - (CR + LG)\, u_t - GR\, u = 0, \tag{8.33}$$

where C is the capacity, L the induction coefficient, R the resistance, and G the loss coefficient of a conductor through which an electric current passes. This equation is hyperbolic ($a_{12}^2 - a_{11}a_{22} = CL > 0$). Also, it does not have a solution as a propagating wave if G and R are both nonzero, because then the relation (8.32) is satisfied. The velocity c of the damped wave is given by the first equation in (8.31) as $c = 1/\sqrt{CL}$; the damping coefficient κ by the second equation in (8.31) as $\kappa = (CR + LG)/(2CL)$; and the third equation yields $4CLGR - (CR + LG)^2 = -(CR - LG)^2 = 0$, i.e., $CR = LG$, which is the compatibility condition for Eq (8.31). Using these values, the damped wave is given by

$$u(x,t) = e^{-\kappa t} f(x - ct), \quad \kappa = \frac{R}{L} = \frac{G}{C}, \quad c = \sqrt{\frac{1}{CL}}.$$

A physical interpretation of this analysis is as follows: It is important for a telegraphic communication over large distances that no damping of the wave should occur in a cable. This requires that either R or G be small, or L or C be large. Thus, the propagation of an undamped signal should result in an undisturbed reproduction of this signal at the receiving end, because wave dispersion, if it occurs, always impairs the quality of reception independent of the quality of telegraphic equipment and cables. A similar situation also exists in the acoustical effects in telephones. ∎

8.4.4. Thermal Waves. As we have seen above, every sinusoidal wave has its source in some kind of periodic disturbance. The kinematical aspects of wave motion are important. Like other kind of waves, thermal or heat waves possess similar kinematics. We will discuss the problem of propagation of thermal waves in the earth.

EXAMPLE 8.8. As mentioned in Example 8.4, the changes in the earth's surface occur daily (day-night cycle) as well as annually (summer-winter cycle). Disregarding the earth's inhomogeneity, we will assume that the periodic temperature distribution in the earth is homogeneous. Since the effect of the initial temperature becomes small after several temperature variations, we will solve a steady-state boundary value problem without initial conditions. Thus, we find a bounded solution $u(y, t)$ for the following problem:

$$u_t = k\, u_{yy}, \quad 0 \leq y < \infty, \quad -\infty < t, \tag{8.34}$$

$$u(0, t) = A \cos \omega t. \tag{8.35}$$

Note that the range $0 \leq y < \infty$ represents the medium from the earth's surface to its interior. We assume the solution of the form $u(y, t) = A\, e^{\alpha y + \beta t}$, where α and β are constants to be determined, and A is a preassigned constant. Then substituting this solution into Eq (8.34) and using the complex form of the boundary condition $u(0, t) = A\, e^{i\omega t}$, we find that $\beta = i\omega$, and $\alpha^2 = \omega/k = i\omega/k$, which gives $\alpha = \pm(1 + i)\sqrt{\omega/(2k)}$. Hence,

$$u(y, t) = A\, e^{\pm y \sqrt{\omega/(2k)} + i\left[\pm y \sqrt{\omega/(2k)} + \omega t\right]}. \tag{8.36}$$

The bounded solution is obtained by choosing the minus sign. Thus, taking the real part of (8.36), we obtain the required solution as

$$u(y, t) = A\, e^{-y \sqrt{\omega/(2k)}} \cos\left(y \sqrt{\omega/(2k)} - \omega t\right). \tag{8.37}$$

It is known that when the temperature of the earth's surface changes periodically over a long period of time, the temperature fluctuations in its interior develop with the same period. The structure of the thermal waves is as follows:

(i) The amplitude of these waves, given by $A\,e^{-y\,\sqrt{\omega/(2k)}}$, decreases exponentially with the depth y. Thus, an increase in depth results in a decay of the amplitude.

(ii) The temperature fluctuations in the earth occur with a phase lag of $y/\sqrt{2k\omega}$, which denotes the time that lapses between the temperature maximum and minimum inside the earth, and the corresponding time point on the surface is proportional to the depth y.

(iii) The change in the temperature amplitude is given by $e^{-y\,\sqrt{\omega/(2k)}}$, which means that the depth of penetration of the temperature is smaller when the period $2\pi/\omega$ is smaller. Thus, for two distinct temperature distributions at time t_1 and t_2, the corresponding depths y_1 and y_2 at which the relative temperature change is the same are related by

$$y_2 = y_1\,\sqrt{t_2/t_1}.$$

For example, a comparison between the daily ($t_1 = 1$) and the annual variation with $t_2 = 365\,t_1$ yields $y_2 = \sqrt{365}\,y_1 \approx 19.104\,y_1$, i.e., the depth of penetration of the annual temperature distribution with the same amplitude as on the surface is about 19 times larger than the depth of penetration of the daily temperature distribution. ∎

8.5. Boundary Layer Flows

The presence of the convection terms in the momentum equations for the Navier-Stokes fluid flows introduces nonlinearity in the equations and makes it very difficult to solve them. There are, however, certain important classes of fluid flows in which these convection terms do not exist. They are known as the boundary layer flows, and we will present some problems where the boundary of the flow regions are cylindrical surfaces with generators along the x-axis. These flows are of the following two types: (i) Steady-state flows through different pipes of uniform cross section with constant pressure gradient, and (ii) unsteady flows generated by the motion of a solid boundary along the x-axis.

EXAMPLE 8.9. The first type of fluid flows are governed by the Poisson's equation

$$\frac{\partial^2 u}{\partial y^2} + \frac{\partial^2 u}{\partial z^2} = \frac{1}{\mu}\frac{dp}{dx}, \tag{8.38}$$

subject to the boundary condition $u = 0$ at the wall of the pipe, where u is the velocity component of the flow along the x-axis, μ the coefficient of viscosity, and dp/dx the constant pressure gradient. The general analytical solution of Eq (8.38) subject to the given boundary condition is

$$u(y, z) = -f(y, z) \frac{1}{\mu} \frac{dp}{dx}, \tag{8.39}$$

where the function $f(y, z)$ depends on the shape of the cross section of the pipe and is such that $\nabla^2 f(y, z) = -1$. The volume flux across the cross section of the pipe is determined by $-\dfrac{C}{\mu} \dfrac{dp}{dx}$, where C is a constant that depends on the volume flux per unit area of the cross section. Some special cases are as follows:

(a) Two-dimensional channel $-a \leq z \leq a$: In this case

$$f(y, z) = \frac{1}{2} \left(a^2 - z^2\right), \quad C = \frac{2}{3} a^3.$$

(b) Circular pipe of radius a: In this case

$$f(y, z) = \frac{1}{4} \left(a^2 - r^2\right), \quad C = \frac{\pi}{8} a^4, \quad r^2 = y^2 + z^2.$$

(c) Annular cross section $a \leq r \leq b$: In this case

$$f(y, z) = \frac{1}{4} \left[a^2 - r^2 + \frac{b^2 - a^2}{\ln(b/a)} \ln\left(\frac{r}{a}\right)\right],$$

$$C = \frac{\pi}{8} \left[b^4 - a^4 - \frac{(b^2 - a^2)^2}{\ln(b/a)}\right].$$

(d) Elliptic cross section $\dfrac{y^2}{a^2} + \dfrac{z^2}{b^2} = 1$: In this case

$$f(y, z) = \frac{a^2 b^2}{2\left(a^2 + b^2\right)} \left(1 - \frac{y^2}{a^2} - \frac{z^2}{b^2}\right),$$

$$C = \frac{\pi}{4} \frac{a^3 b^3}{a^2 + b^2}.$$

(e) Rectangular cross section $|y| \leq a$, $|z| \leq b$: In this case

$$f(y, z) = \frac{a^2}{2} - \frac{y^2}{2}$$

$$- 2a^2 \left(\frac{2}{\pi}\right)^3 \sum_{n=0}^{\infty} \frac{(-1)^n}{(2n+1)^3} \frac{\cosh(2n+1)(\pi z/(2a))}{\cosh(2n+1)(\pi b/(2a))} \cos \frac{(2n+1)\pi y}{2a},$$

$$C = \frac{4}{3} b a^3 - 8a^4 \left(\frac{2}{\pi}\right)^5 \sum_{n=0}^{\infty} \frac{1}{(2n+1)^5} \tanh \frac{(2n+1)\pi b}{2a}.$$

In particular, if $a = b$, then $C = 0.562308\, a^4$.

If the cross section is none of the above regions, the soap film method is used to experimentally compute f and C. We will not discuss this method here, but the interested readers can see the experimental technique in Taylor (1937). ∎

EXAMPLE 8.10. This is an example of the second type of flows. Consider an infinite plate that is moved in its own plane with a constant velocity U at time $t = 0$ in a fluid which is initially at rest. If the plate lies in the (x, y)-plane, then the velocity distribution satisfies the equation

$$\frac{\partial u}{\partial t} = \nu \frac{\partial^2 u}{\partial z^2}, \tag{8.40}$$

where ν is the kinematic viscosity of the fluid. The initial and boundary conditions are $u(z, 0) = 0$, and $u(0, t) = U$ for $t > 0$. This problem is similar to the heat conduction problems solved in Examples 5.3 and 6.6, and its solution is

$$u = U \ \mathrm{erfc}\left(\frac{z}{2\sqrt{\nu t}}\right). \tag{8.41}$$

A distinction between this boundary layer problem and the similar heat conduction problem is that there is no true propagation velocity in the latter case. But by defining the boundary layer thickness as the distance in which the value of u drops to a certain preassigned fraction of U, we find from (8.41) that the boundary layer thickness is proportional to $\sqrt{\nu t}$, and it grows at a rate proportional to $\sqrt{\nu/t}$.

For periodic oscillations of a plate, see §10.5. ∎

Besides the above examples, there are many situations in which the solutions of the heat conduction equation are applicable to the problems of fluid flows. Some examples are as follows.

EXAMPLE 8.11. In the cylindrical polar coordinates (r, θ, z), let the velocity components of a fluid flow be denoted by u_r, u_θ, and u_z. Consider the case of the flow for which $u_r = 0 = u_z$, $u_\theta = u_\theta(r, t)$, and the pressure $p = \mathrm{const}$. Then u_θ satisfies the equation

$$\frac{\partial u_\theta}{\partial t} = \nu \left(\frac{\partial^2 u_\theta}{\partial r^2} + \frac{1}{r} \frac{\partial u_\theta}{\partial r} - \frac{u_\theta}{r^2} \right), \tag{8.42}$$

and the vorticity $\omega_z = \dfrac{1}{r} \dfrac{\partial}{\partial r} (r u_\theta)$ satisfies the diffusion equation

$$\frac{\partial \omega_z}{\partial t} = \nu \left(\frac{\partial^2 \omega_z}{\partial r^2} + \frac{1}{r} \frac{\partial \omega_z}{\partial r} \right). \tag{8.43}$$

A solution of Eq (8.43) is

$$\omega_z = \frac{\Gamma}{4\pi \nu t} e^{-r^2/(4\nu t)}, \tag{8.44}$$

where Γ is the initial value of the circulation about the origin. This solution describes the dissolution of a vortex filament concentrated at the origin at time $t = 0$. ∎

EXAMPLE 8.12. The diffusion equation (8.42) also describes the motion of a fluid contained inside or outside an infinite cylinder that is rotating about its axis. The vorticity ω_z is initially concentrated at the surface of the cylinder, but it spreads outside into the fluid such that $\omega_z \to 0$ as $t \to \infty$, and then $u_\theta = A/r$, where A is a constant. However, inside the cylinder the vorticity w_z approaches a constant value which is equal to twice the angular velocity of the cylinder, and the fluid tends to rotate like a rigid body. ∎

EXAMPLE 8.13. Consider the case of a steady flow parallel to an infinite plate on which the nonzero component of velocity assumes a prescribed nonzero value. This represents the steady-state flow far downstream of the leading edge of a semi-infinite plate. If there is no suction in the plate, the boundary layer grows indefinitely downstream such that at a finite distance the velocity eventually becomes zero. But if there is suction in the plate, the boundary layer eventually stops growing and we obtain the "asymptotic suction profile." Let the x- and z-axis lie along and perpendicular to the plate, and the corresponding components of velocity be u and w, respectively, and let the pressure p be independent of x. Then the equation of continuity $\frac{\partial u}{\partial x} + \frac{\partial w}{\partial z} = 0$ implies that $w = \text{const} = -W$, say, at the plate. Thus, the governing equation of the flow is

$$-W \frac{du}{dz} = \nu \frac{d^2 u}{dz^2}, \tag{8.45}$$

with $p = \text{const}$ throughout the flow, subject to the boundary conditions $u = 0$ at $z = 0$, and $\lim_{z \to \infty} u(z) = U < +\infty$, where U is the mean-stream velocity of the flow. Hence, the solution of this problem is

$$u = U \left(1 - e^{-Wz/\nu}\right). \tag{8.46}$$

Note that as $\nu \to 0$, the velocity approaches the mean-stream velocity which gets concentrated inside a boundary layer of the plate. ∎

EXAMPLE 8.14. Consider the circulating flow around a rotating circular cylinder with suction. Let an infinite cylinder that is immersed in a fluid at rest be suddenly rotated about its axis with constant angular velocity. Then the vorticity is given by $\omega_z = \frac{1}{r} \frac{\partial}{\partial r} (r u_\theta)$ (see Example 8.11). We know that initially this vorticity is concentrated at the surface of the cylinder, but it spreads out until $\omega_z = 0$ everywhere. Finally, the steady-state solution is given by

$$u_\theta = \frac{\Gamma_1}{2\pi r}, \tag{8.47}$$

where Γ_1 is the circulation around the cylinder. The solution (8.47) is known as the *asymptotic suction profile*.

Now, if there is suction throughout the surface of the cylinder, the velocity will eventually attain its steady-state value when the outward diffusion is balanced by the convection of vorticity toward the cylinder. Let a denote the radius of the cylinder, and let $-V$ be the component of the suction velocity perpendicular to the surface of the cylinder. Then the radial velocity u_r for the flow is given by

$$u_r = -\frac{aV}{r}. \tag{8.48}$$

This equation satisfies the equation of continuity div $\cdot$ $\mathbf{v} = 0$. Since the rate of diffusion of vorticity across a circle $r = a$ is given by $-2\pi a \nu r \dfrac{\partial \omega_z}{\partial r}$, and the rate of convection is $2\pi a r u_r \omega_z$, we find that ω_z satisfies the first-order equation

$$\frac{\partial \omega_z}{\partial r} + R\frac{\omega_z}{r} = 0, \tag{8.49}$$

where $R = aV/\nu$. The solution of Eq (8.49) is

$$\omega_z = \frac{1}{r}\frac{\partial}{\partial r}(r u_\theta) = A\left(\frac{a}{r}\right)^R, \tag{8.50}$$

where A is the value of ω_z at the cylinder, i.e., $A = \omega_z(a)$. Since the circulation $\Gamma = 2\pi r u_\theta$, we find from (8.50) that

$$\Gamma = \begin{cases} \Gamma_1 - \dfrac{2\pi a^2}{R-2} A\left(\dfrac{a}{r}\right)^{R-2} & \text{if } R \neq 2, \\[2ex] \Gamma_1 + 2\pi a^2 A \ln \dfrac{r}{a} & \text{if } r = 2. \end{cases}$$

Thus, for $R \leq 2$ the solution with finite circulation at $r = \infty$ is $\Gamma = \Gamma_1$, where $\omega_z = 0$ and $u_\theta = \Gamma_1/(2\pi r)$, as given in (8.47). However, for $R > 2$, the value of circulation at infinity is Γ_1, and A can be adjusted so as to give any prescribed value of circulation at the cylinder. Hence, to maintain different values of circulation at the cylinder and at infinity the suction velocity V must be such that $V > \dfrac{2\nu}{a}$. ∎

8.6. Additional Problems

We give more examples of initial and boundary value problems.

EXAMPLE 8.15. Consider the generalized Tricomi equation

$$y^c u_{xx} + u_{yy} = 0, \quad c \geq 0, \tag{8.51}$$

For $c = 0$ it represents the Laplace equation, and for $c = 1$ the Tricomi equation (see Exercise 1.17). If we set in (8.51)

$$y = \left[\frac{1}{2} (c+2)\, \eta \right]^{2/(c+1)},$$

i.e., $y^{(c+2)/2} = \dfrac{(c+2)\eta}{2}$, then Eq (8.51) reduces to

$$u_{xx} + u_{\eta\eta} + \frac{c}{(c+2)\eta}\, u_\eta = 0.$$

Eq (8.51) with the boundary conditions

$$\begin{aligned}
&u(0, y) = 0, \quad y > 0; \qquad u(x, 0) = f(x), \quad 0 < x < 1; \quad f(0) = 0; \\
&u_y(x, 0) = 0, \quad x > 1; \qquad u \to 0 \quad \text{as } x^2 + y^2 \to \infty,
\end{aligned} \tag{8.52}$$

represents a compressible irrotational fluid flow problem.

Now, we consider in detail the case when $c = 0$. Eq (8.51) reduces to

$$u_{xx} + u_{yy} = 0. \tag{8.53}$$

As in Example 6.17, the Fourier sine transform of Eq (8.53) with respect to x is

$$\tilde{u}_{yy} - \alpha^2\, \tilde{u} = 0,$$

which has the general solution

$$\tilde{u} = A(\alpha)\, e^{-\alpha y} + B(\alpha)\, e^{\alpha y},$$

where $B(\alpha) = 0$ since $e^{\alpha y}$ becomes unbounded as $y \to \infty$. Thus,

$$\tilde{u} = A(\alpha)\, e^{-\alpha y}.$$

Also,

$$\left. \frac{\partial \tilde{u}}{\partial y} \right|_{y=0} = -\alpha\, A(\alpha) = \sqrt{\frac{2}{\pi}} \int_0^\infty \tilde{u}_y(x, 0)\, \sin \alpha x\, dx,$$

which gives

$$A(\alpha) = -\frac{1}{\alpha} \sqrt{\frac{2}{\pi}} \int_0^\infty \tilde{u}_y(x, 0)\, \sin \alpha x\, dx,$$

or

$$\tilde{u} = -\sqrt{\frac{2}{\pi}} \frac{e^{-\alpha y}}{\alpha} \int_0^\infty \tilde{u}_y(x, 0)\, \sin \alpha x\, dx,$$

which on inversion yields

$$
\begin{aligned}
u(x, y) &= \sqrt{\frac{2}{\pi}} \int_0^\infty \tilde{u}(\alpha, y) \sin \alpha x \, dx \\
&= -\frac{2}{\pi} \int_0^\infty \frac{e^{-\alpha y}}{\alpha} \sin \alpha x \, d\alpha \int_0^\infty \tilde{u}_y(t, 0) \sin \alpha t \, dt \qquad (8.54) \\
&= -\frac{2}{\pi} \int_0^\infty \tilde{u}_y(t, 0) \, dt \cdot F(x, t, y),
\end{aligned}
$$

where

$$
F(x, t, y) = \int_0^\infty \frac{1}{\alpha} \sin \alpha x \, \sin \alpha t \, e^{-\alpha y} \, d\alpha.
$$

Since $F(0, t, y) = 0$, and

$$
\begin{aligned}
\frac{\partial F}{\partial x} &= \int_0^\infty \cos \alpha x \, \sin \alpha t \, e^{-\alpha y} \, d\alpha \\
&= \frac{1}{2} \int_0^\infty \left[\sin \alpha (x + t) - \sin \alpha (x - t) \right] e^{-\alpha y} \, d\alpha \\
&= \frac{1}{2} \left[\frac{x + t}{(x + t)^2 + y^2} - \frac{x - t}{(x - t)^2 + y^2} \right],
\end{aligned}
$$

we find that

$$
F(x, t, y) = \frac{1}{4} \left[\ln \left[(x + t)^2 + y^2 \right] - \ln \left[(x - t)^2 + y^2 \right] \right] + C, \qquad (8.55)
$$

where $C = 0$ because $F(0, t, y) = 0$. Thus, from (8.54) and (8.55) we get

$$
\begin{aligned}
u(x, y) &= \frac{2}{\pi} \frac{1}{4} \int_0^\infty \tilde{u}_y(t, 0) \left[\ln \left((x + t)^2 + y^2 \right) - \ln \left((x - t)^2 + y^2 \right) \right] dt \\
&= \frac{1}{2\pi} \int_0^1 u_y(t, 0) \ln \frac{(x + t)^2 + y^2}{(x - t)^2 + y^2} \, dt,
\end{aligned}
$$

$$(8.56)$$

because $u_y(x, 0) = 0$ for $x > 1$, and

$$
u(x, 0) \equiv f(x) = \frac{1}{\pi} \int_0^1 u_y(t, 0) \ln \left| \frac{x + t}{x - t} \right| \, dt, \qquad (8.57)
$$

which is a weakly singular integral. Also, the solution (8.56) satisfies the condition $u(0, y) = 0$ for $y > 0$. ■

EXAMPLE 8.16. Solve the two-dimensional steady heat conduction problem for the quadrant $x, y > 0$ with thermal conductivity a if the side $y = 0$ is maintained at zero temperature, while the other side $x = 0$ is thermally insulated except for the region $0 < y < b$ through which heat flows with constant density q (Fig. 8.9).

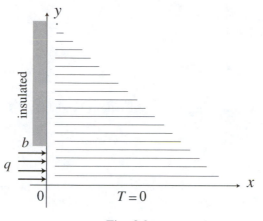

Fig. 8.9.

Thus, we solve the Laplace equation

$$\frac{\partial^2 T}{\partial x^2} + \frac{\partial^2 T}{\partial y^2} = 0, \quad 0 < x < \infty,\ 0 < y < \infty, \tag{8.58}$$

subject to the boundary conditions

$$T(x, 0) = 0, \quad \frac{\partial T}{\partial x}\bigg|_{x=0} = f(y) = \begin{cases} -\dfrac{q}{a}, & 0 < y < b, \\[2mm] 0, & b < y < \infty. \end{cases}$$

Applying the Fourier sine transform with respect to y, we get

$$\frac{d^2 \tilde{T}}{dx^2} - \alpha^2 \tilde{T} = 0,$$

which has the solution $\tilde{T}(x, \alpha) = \sqrt{\dfrac{\pi}{2}}\, B(\alpha)\, e^{-\alpha x}$. On inversion this gives

$$T(x, y) = \int_0^\infty B(\alpha)\, e^{-\alpha x} \sin \alpha y\, d\alpha, \tag{8.59}$$

where the coefficient $B(\alpha)$ is determined from the boundary condition

$$\frac{\partial T}{\partial x}\bigg|_{x=0} = f(y) = \int_0^\infty \alpha\, B(\alpha) \sin \alpha y\, d\alpha, \quad 0 < y < \infty,$$

which gives

$$B(\alpha) = -\frac{2}{\pi\alpha} \int_0^\infty f(y) \sin \alpha y \, dy = \frac{2q}{\pi a} \frac{1 - \cos \alpha b}{\alpha^2}.$$

Hence, the solution of the problem is

$$T(x, y) = \frac{2q}{\pi a} \int_0^\infty \frac{1 - \cos \alpha b}{\alpha^2} e^{-\alpha x} \sin \alpha y \, d\alpha. \tag{8.60}$$

Alternatively, we use the method of separation of variables, assuming that $T(x, y) = X(x)Y(y)$, and obtain

$$\frac{X''}{X} = -\frac{Y''}{Y} = \lambda^2.$$

The equation $Y'' + \lambda^2 Y = 0$ yields $y_\lambda = \sin \lambda y$, and the equation $X'' - \lambda^2 X = 0$ yields $X_\lambda = e^{-\lambda x}$. Hence,

$$T(x, y) = \sum_{\lambda > 0} A(\lambda) e^{-\lambda x} \sin \lambda y. \tag{8.61}$$

We partition the interval $[0, \infty)$ into subintervals $0 < \lambda_1 < \lambda_2 < \ldots.$ Let $A(\lambda)e^{-\lambda x} \sin \lambda y$ have average value $A(\lambda)e^{-\lambda x} \sin \lambda y$ on each subinterval $(\lambda_{i-1}, \lambda_i)$, $i = 1, 2, \ldots$, of uniform length $d\lambda$. Then summing over all these subintervals, we find from (8.61)

$$T(x, y) = \int_0^\infty A(\lambda) e^{-\lambda x} \sin \lambda y \, d\lambda,$$

which is the same as (8.59). Thus, the solution is given by (8.60) with α replaced by λ. ∎

EXAMPLE 8.17. Consider the eigenvalue problem $\nabla^2 u \equiv \dfrac{\partial^2 u}{\partial x^2} + \dfrac{\partial^2 u}{\partial y^2} + \lambda u = 0$ in a rectangle $R = \{0 < x < a, \, 0 < y < b\}$, such that $u(0, y) = 0 = u(a, y)$ for $0 < y < b$; and $u(x, 0) = 0 = u(x, b)$ for $0 < x < a$. For the definition of an eigenvalue problem, see §4.5. The solution of an eigenvalue problem involves finding the eigenpairs (i.e., the eigenvalues and the corresponding eigenfunctions). Using the method of separation of variables, let $u(x, y) = X(x)Y(y)$. Then we obtain

$$\frac{X''}{X} = -\frac{Y''}{Y} - \lambda.$$

Since both sides are equal for all values of $(x, y) \in R$, they must be equal to a constant $-k$, where $k > 0$. Thus,

$$X'' + kX = 0, \quad 0 < x < a; \quad X(0) = 0 = X(a),$$
$$Y'' + (\lambda - k)Y = 0, \quad 0 < y < b; \quad Y(0) = 0 = Y(b).$$

The first system has eigenvalues $n^2\pi^2/a^2$, $n = 1, 2, \ldots$, and the corresponding (normalized) eigenfunctions are (see Chapter 5)

$$X_n(x) = \sqrt{\frac{2}{a}}\, \sin \frac{n\pi x}{a}.$$

The nontrivial solutions of the second system are possible if $\lambda - k = m^2\pi^2/b^2$, $m = 1, 2, \ldots$. The corresponding eigenfunctions are

$$Y_m(y) = \sqrt{\frac{2}{b}}\, \sin \frac{m\pi y}{b}.$$

Hence, for this eigenvalue problem the eigenpairs are

$$\lambda_{n,m} = \frac{n^2\pi^2}{a^2} + \frac{m^2\pi^2}{b^2}, \quad u_{n,m}(x,y) = \frac{2}{\sqrt{ab}} \sin \frac{n\pi x}{a} \sin \frac{m\pi y}{b}. \; \blacksquare$$

EXAMPLE 8.18. We will solve the Cauchy problem in R^1, i.e., solve the initial value problem

$$u_t = k\, u_{xx}, \tag{8.62}$$

$$u(x,0) = f(x), \quad -\infty < x < \infty, \quad t > 0, \tag{8.63}$$

where $f(x)$ is a bounded and continuous function on the x-axis. Eq (8.62) is also known as the transient Fourier equation in R^1. We will prove that the function

$$u(x,t) = \frac{1}{\sqrt{4\pi kt}} \int_{-\infty}^{\infty} f(y)\, e^{-(y-x)^2/(4kt)}\, dy,$$

which belongs to the class C^∞ with respect to x and t for $t > 0$, satisfies Eq (8.62) such that $\lim_{t\to 0+} u(x,t) = f(x)$, $-\infty < x < \infty$.

METHOD 1: We use the method of separation of variables in the form $u(x,t) = X(x)T(t)$. Then we find that $T(t) = e^{-k\lambda^2 t}$ and $X(x) = A\cos\lambda x + B\sin\lambda x$, where λ, A, B are arbitrary. If we assume $A = A(\lambda)$ and $B = B(\lambda)$, then

$$u_\lambda(x,t) = [A(\lambda)\cos\lambda x + B(\lambda)\sin\lambda x]\, e^{-k\lambda^2 t}$$

is a solution of Eq (8.62). The integral form of this solution, which is

$$\int_{-\infty}^{\infty} u_\lambda(x,t)\, d\lambda,$$

with a proper choice of $A(\lambda)$ and $B(\lambda)$, is also a solution of Eq (8.62). Now, in view of the initial condition (8.63), we have

$$f(x) = \int_{-\infty}^{\infty} u_\lambda(x,0)\,d\lambda = \int_{-\infty}^{\infty} [A(\lambda)\cos\lambda x + B(\lambda)\sin\lambda x]\,d\lambda. \qquad (8.64)$$

The function $f(x)$, being continuous and bounded, has the Fourier integral representation*

$$f(x) = \frac{1}{2\pi} \int_{-\infty}^{\infty} d\lambda \int_{-\infty}^{\infty} f(y)\cos\lambda(y-x)\,dy. \qquad (8.65)$$

Comparing (8.64) and (8.65), we find that

$$A(\lambda) = \frac{1}{2\pi} \int_{-\infty}^{\infty} f(y)\cos\lambda y\,dy, \quad B(\lambda) = \frac{1}{2\pi} \int_{-\infty}^{\infty} f(y)\sin\lambda y\,dy.$$

Thus,

$$\begin{aligned}
u(x,t) &= \int_{-\infty}^{\infty} u_\lambda(x,t)\,d\lambda \\
&= \frac{1}{2\pi} \int_{-\infty}^{\infty} d\lambda \int_{-\infty}^{\infty} f(y)\cos\lambda(y-x)\,e^{-k\lambda^2 t}\,dy \\
&= \frac{1}{2\pi} \int_{-\infty}^{\infty} f(y)\,dy \int_{-\infty}^{\infty} \cos\lambda(y-x)\,e^{-k\lambda^2 t}\,d\lambda \\
&= \frac{1}{\sqrt{4\pi kt}} \int_{-\infty}^{\infty} f(y)\,e^{-(y-x)^2/(4kt)}\,dy.
\end{aligned}$$

METHOD 2: We apply the Fourier transform in space (with α as the transform variable) and the Laplace transform in time (with s as the transform variable). Then Eq (8.62) with the initial condition (8.63) is transformed into

$$(s + a\alpha^2)\tilde{u}(\alpha,s) - \tilde{f}(\alpha) = 0,$$

---

*Proof of (8.65): The Fourier transform of $f(x)$ is given by

$$\tilde{f}(\lambda) = \frac{1}{\sqrt{2\pi}} \int_{-\infty}^{\infty} f(x)e^{i\lambda x}\,dx,$$

and the inverse of $\tilde{f}(\lambda)$ is given by

$$\begin{aligned}
f(x) &= \frac{1}{\sqrt{2\pi}} \int_{-\infty}^{\infty} \tilde{f}(\lambda)e^{-i\lambda x}\,d\lambda = \frac{1}{2\pi} \int_{-\infty}^{\infty}\int_{-\infty}^{\infty} f(y)e^{i\lambda y}e^{-i\lambda x}\,dy\,d\lambda \\
&= \frac{1}{2\pi} \int_{-\infty}^{\infty}\int_{-\infty}^{\infty} f(y)e^{i\lambda(y-x)}\,dy\,d\lambda.
\end{aligned}$$

Since $f(x)$ is real, the imaginary part of the integral above is zero and the real part is equal to $f(x)$, which gives (8.65).

where $\tilde{u} = \mathcal{F}[u]$, and $\tilde{f} = \mathcal{F}[f]$. Since $\tilde{u}(\alpha, s) = \dfrac{\tilde{f}(\alpha)}{s + k\alpha^2}$, the function $\tilde{u}(\alpha, s)$ has a simple pole at $s = -k\alpha^2$. Hence, by applying the inverse Fourier and Laplace transforms, we get the solution of the Cauchy problem (7.61) as

$$u(x,t) = \frac{1}{\sqrt{2\pi}} \int_{-\infty}^{\infty} \left\{ \lim_{R\to\infty} \left[\frac{1}{2\pi i} \int_{u-iR}^{u+iR} \frac{\tilde{f}(\alpha)}{s + k\alpha^2} e^{st}\, ds \right] \right\} e^{-i\alpha x}\, d\alpha.$$

A direct inversion of the Laplace transform yields

$$\tilde{u}(\alpha, t) = e^{-k\alpha^2 t}\, \tilde{f}(\alpha),$$

which, after the inversion of the Fourier transform, gives

$$u(x,t) = \frac{1}{2\pi} \int_{-\infty}^{\infty} \tilde{u}(\alpha, t) e^{-i\alpha x}\, d\alpha$$

$$= \frac{1}{2\pi} \int_{-\infty}^{\infty} \tilde{f}(\alpha)\, e^{-(i\alpha x + k\alpha^2 t)}\, d\alpha.$$

If we identify $\tilde{g}(\alpha) = e^{-k\alpha^2 t}$, then for the convolution of the product $\tilde{f}(\alpha)\tilde{g}(\alpha)$, we have

$$\frac{1}{\sqrt{2\pi}} \int_{-\infty}^{\infty} \tilde{f}(\alpha)\tilde{g}(\alpha) e^{-i\alpha x}\, d\alpha = \frac{1}{\sqrt{2\pi}} \int_{-\infty}^{\infty} f(y)g(x - y)\, dy. \qquad (8.66)$$

Since $\tilde{f}(\alpha)$ is identified with the Fourier component of the Cauchy condition $f(x)$ in (8.63), we find that $f(y)$ in (8.66) corresponds to the initial condition and

$$g(x - y) = \frac{1}{\sqrt{4\pi kt}}\, e^{-(x-y)^2/(4kt)}.$$

Hence, by the convolution theorem we get the solution of the Cauchy problem from (8.66).

If $f(x) = \delta(x)$ in (8.63), then the solution of the Cauchy problem (7.61) is given by

$$u(x,t) = \frac{1}{\sqrt{4\pi kt}}\, e^{-x^2/(4kt)}, \quad t > 0.$$

METHOD 3: (By Laplace transform) Applying the Laplace Transform to (7.61), we get $s\bar{u} - f(x) = \bar{u}_{xx}$. Using the method of variation of parameters, we find the solution as

$$\bar{u} = \left[A + \frac{1}{2\sqrt{ks}} \int_0^x e^{y\sqrt{s/k}} f(y)\, dy \right] e^{-x\sqrt{s/k}}$$

$$+ \left[B - \frac{1}{2\sqrt{ks}} \int_0^x e^{-y\sqrt{s/k}} f(y)\, dy \right] e^{x\sqrt{s/k}}.$$

Note that $\lim\limits_{x \to \pm\infty} \bar{u} \to 0$. Then

$$A = \frac{1}{2\sqrt{ks}} \int_{-\infty}^{0} e^{y\sqrt{s/k}} f(y) dy, \quad B = \frac{1}{2\sqrt{ks}} \int_{0}^{\infty} e^{-y\sqrt{s/k}} f(y)\, dy.$$

Thus,

$$\begin{aligned}
\bar{u} &= \left[\frac{1}{2\sqrt{ks}} \int_{-\infty}^{0} e^{y\sqrt{s/k}} f(y)\, dy \right] e^{-x\sqrt{s/k}} \\
&\quad + \left[\frac{1}{2\sqrt{ks}} \int_{0}^{\infty} e^{-y\sqrt{s/k}} f(y)\, dy \right] e^{x\sqrt{s/k}} \\
&= \frac{1}{2\sqrt{ks}} \int_{-\infty}^{\infty} e^{|x-y|\sqrt{s/k}} f(y)\, dy.
\end{aligned}$$

Hence, by taking the Laplace inverse, we get

$$u(x,t) = \frac{1}{2\sqrt{\pi k t}} \int_{-\infty}^{\infty} f(y) e^{-(y-x)^2/(4kt)}\, dy. \quad \blacksquare$$

8.7. Ill-Posed Problems

The existence of the solution of a boundary value problem is generally difficult to establish. One of the methods to show that a solution exists is to construct it. We have already constructed solutions of different types of boundary value problems. For example, the existence of the solution of the Dirichlet problem $\nabla^2 u = f(\mathbf{x})$ in a domain Ω, such that $u = g(\mathbf{x}_s)$ on the boundary $\partial\Omega$, where $\mathbf{x}_s$ is an arbitrary point on the boundary, is established in §7.6.1 in terms of the Green's function. To understand what an ill-posed problem means, we will first note that a well-posed problem is the one for which a unique solution exists, and this solution depends continuously not only on the prescribed initial and boundary conditions but also on the coefficients in the equations, or boundary conditions, parameters, and geometry as well. A problem that is not well-posed is said to be *ill-posed*. For example, the Neumann problem $\nabla^2 u = F(\mathbf{x})$ in a domain Ω, such that $\dfrac{\partial u}{\partial n} = g(\mathbf{x}_s)$ on the boundary $\partial\Omega$, where $\mathbf{x}_s$ is an arbitrary point on the boundary, is solved in §7.6.1. It is shown that by the divergence theorem

$$\iiint_{\Omega} F(\mathbf{x})\, d\Omega = \iiint_{\Omega} \nabla^2 u\, d\Omega = \iint_{S} \frac{\partial u}{\partial n}\, dS = \iint_{S} g(\mathbf{x}_s)\, dS.$$

Thus, if the consistency condition $\iiint_{\Omega} F(\mathbf{x}) \, d\Omega = \iint_{S} g(\mathbf{x}_s) \, dS$ is not satisfied, the Neumann problem cannot have a solution. Examples of ill-posed problems arise, *inter alia*, in the areas of heat conduction, forced vibrations, bifurcations, Cauchy problems for hyperbolic systems, and singular perturbations. We will not discuss all these problems, but we will solve some specific problems only.

EXAMPLE 8.19.   The initial value problem for the Laplace equation $u_{xx} + u_{yy} = 0$, $x > 0$, $-\infty < y < \infty$, subject to the initial conditions $u(0, y) = f(y)$, $u_x(0, y) = g(y)$, $-\infty < y < \infty$, is ill-posed in the sense that the solution does not depend continuously on the data functions f and g. The details are as follows: For $f = f_1 = 0$ and $g = g_1 = 0$ we have the solution $u_1 = 0$. Also, for $f = f_2 = 0$ and $g = g_2 = \dfrac{\sin ny}{n}$ the solution is $u_2 = \dfrac{1}{n^2} \sin ny \sinh nx$. Since the data functions f_1 and f_2 are the same, and $\lim\limits_{n\to\infty} |g_1 - g_2| = \lim\limits_{n\to\infty} \left| -\dfrac{\sin ny}{n} \right| = 0$ uniformly for all $y \in R^1$, we conclude that the pairs of the data functions f_1, g_1 and f_2, g_2 can be made arbitrarily close by choosing n sufficiently large. We will compare the solutions u_1 and u_2 at $y = \pi/2$ for an arbitrary small, fixed value of $x > 0$ and odd positive values of n. Then

$$\lim_{n\to\infty} \left| u_1\left(x, \frac{\pi}{2}\right) - u_2\left(x, \frac{\pi}{2}\right) \right| = \lim_{n\to\infty} \frac{1}{n^2} \sinh nx = \lim_{n\to\infty} \frac{e^{nx} - e^{-nx}}{2n^2} = \infty.$$

Hence, by choosing n sufficiently large, the maximum difference between the data functions can be made arbitrarily small, but the maximum difference between the corresponding solutions becomes arbitrarily large. Thus, this initial value problem and all such problems for elliptic equations are, in general, ill-posed. ∎

EXAMPLE 8.20.   Consider the backward heat equation $v_t = v_{xx}$, $0 < x < 1$, $0 < t < T$, subject to the boundary and initial conditions $v(0, t) = 0 = v(1, t)$ for $0 < t < T$, and $v(x, T) = f(x)$ for $0 < x < 1$. Note that the initial condition of the forward condition has been replaced by a terminal condition, which specifies the state at the final time $t = T$. The initial value problem for the backward heat equation is completely identical to the terminal problem for the forward equation. The problem is to determine the previous states $v(x, t)$ for $t < T$, which have resulted into the state $f(x)$ at time T. This problem has no solution for arbitrary $f(x)$. Even if the solution exists, it does not depend continuously on the data. To see this, let $v_0(x)$ denote the initial state $v(x, 0)$. Then the function

$$v(x, t) = \sum_{n=1}^{\infty} C_n e^{-n^2 \pi^2 t} \sin n\pi x, \quad 0 < x < 1, \ 0 < t < T,$$

where

$$C_n = 2 \int_0^1 v_0(x) \sin n\pi x \, dx, \quad n = 1, 2, \ldots,$$

will be the solution of the problem provided the coefficients C_n are such that

$$f(x) = \sum_{n=1}^{\infty} C_n \, e^{-n^2 \pi^2 T} \sin n\pi x \, dx, \quad 0 < x < 1. \tag{8.67}$$

However, the series in (8.67) converges uniformly to a function in the class $C^{\infty}(0,1)$ irrespective of what values C_n have. Hence, no solution exists when $f(x) \notin C^{\infty}(0,1)$. Moreover, even if $f(x)$ is in the class $C^{\infty}(0,1)$, the solution does not depend continuously on the data. We will consider two special cases of $f(x)$.

CASE 1. $f(x) = \dfrac{\sin N\pi x}{N}$, where N is an integer. Then the unique solution of the problem is

$$v(x,t) = \frac{1}{N} \, e^{N^2 \pi^2 (T-t)} \sin N\pi x, \quad 0 < x < 1, \, 0 < t < T.$$

Note that $|f(x)| \to 0$ as $N \to \infty$, which means that this data function differs by an arbitrarily small quantity from the data function $f \equiv 0$, which yields the solution $v \equiv 0$. On the other hand, $|v(x,t)| \to \infty$ for $0 < t < T$ as $N \to \infty$, which means that the solution does not depend continuously on the data.

CASE 2. $f(x) = x$. Then

$$C_n \, e^{-n^2 \pi^2 T} = 2 \int_0^1 x \sin n\pi x \, dx = (-1)^{n+1} \frac{2}{n}.$$

Thus,

$$v_0(x) = \sum_{n=1}^{\infty} (-1)^{n+1} \frac{2}{n} \, e^{n^2 \pi^2 T} \sin n\pi x \, dx.$$

This series is divergent, and thus, $v_0(x) = v(x,0)$ cannot be determined. Moreover, the solution does not remain close to x for $x \neq 0$. For another case see Exercise 8.22. ∎

EXAMPLE 8.21. To show that the backward heat conduction problem in a finite rod is ill-posed, let the ends of the rod of length l be kept at zero temperature subject to the condition $v(x,1) = g(x)$, which is known, and assume that there are no sources/sinks. The problem reduces to finding the initial state $v(x,0) = f(x)$. From Example 5.2, it is known that

$$v(x,t) = \sum_{n=1}^{\infty} C_n \sin \frac{n\pi x}{l} \, e^{-n^2 \pi^2 t / l^2},$$

where

$$C_n = \frac{2}{l} \int_0^l f(\xi) \sin \frac{n\pi \xi}{l} \, d\xi.$$

Thus,

$$v(x,t) = \frac{2}{l} \sum_{n=1}^{\infty} \left\{ \int_0^l f(\xi) \sin \frac{n\pi\xi}{l}\, d\xi \right\} \sin \frac{n\pi x}{l}\, e^{-n^2\pi^2 t/l^2}.$$

At $t = 1$, we get

$$g(x) = v(x,1) = \frac{2}{l} \sum_{n=1}^{\infty} \left\{ \int_0^l f(\xi) \sin \frac{n\pi\xi}{l}\, d\xi \right\} \sin \frac{n\pi x}{l}\, e^{-n^2\pi^2/l^2}. \qquad (8.68)$$

Let the Fourier sine series for $g(x)$ be given by

$$g(x) = \sum_{n=1}^{\infty} G_n \sin \frac{n\pi x}{l}, \qquad (8.69)$$

where

$$G_n = \frac{2}{l} \int_0^l g(\xi) \sin \frac{n\pi\xi}{l}\, d\xi.$$

Comparing (8.68) and (8.69), we find that

$$G_n = \frac{2}{l} e^{-n^2\pi^2/l^2} \int_0^l f(\xi) \sin \frac{n\pi\xi}{l}\, d\xi,$$

which, being an integral equation of the first kind, is known to be ill-posed (Kythe and Puri, 2002, Ch. 11). ∎

8.8. Exercises

8.1. (a) Show that $e^{-|x|} = \dfrac{2}{\pi} \displaystyle\int_0^{\infty} \dfrac{\cos \alpha x}{1+\alpha^2}\, d\alpha,\ x \in R^1$.

(b) If

$$f(x) = \begin{cases} 0 & \text{if } x < 0, \\ \sin x & \text{if } 0 \le x \le \pi, \\ 0 & \text{if } x > \pi, \end{cases}$$

show that

$$f(x) = \frac{1}{\pi} \int_0^{\infty} \frac{\cos \alpha x + \cos[\alpha(\pi - x)]}{1 - \alpha^2}\, d\alpha, \qquad x \in R^1.$$

In particular, for $x = \pi/2$, show that

$$\int_0^{\infty} \frac{\cos (\alpha\pi/2)}{1 - \alpha^2}\, d\alpha = \frac{\pi}{2}.$$

HINT. Use the following result: The Fourier integral

$$\frac{1}{\pi} \int_0^\infty \int_{-\infty}^\infty f(s) \cos[\alpha(s - x)] \, ds \, d\alpha$$

converges to the value $\frac{1}{2} \left[f(x+) + f(x-) \right]$ at each point $x \in R^1$, where f satisfies the Dirichlet's conditions on R^1 (§6.1).

8.2. Consider a rod of length l with periodic boundary conditions at the end points. Use method of images and the result of Example 8.5 to determine Green's function $G(x, t)$ for the heat conduction problem in the following four cases of different types of boundary conditions. Note that this problem is regarded as that of an infinite sequence of reflections by placing parallel mirrors at the end points (as an optical example); then not only the primary heat pole (marked in Figs. 8.10(a)-(d)) but also all its images are reflected at both ends of the rod.

(a) $u = 0$ at both $x = 0$ and $x = l$.

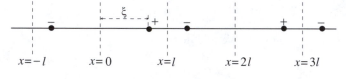

Fig. 8.10(a).

ANS. $f(x) = \sum B_n \sin \frac{n\pi x}{l}, \quad B_n = \frac{2}{l} \int_0^l f(x) \sin \frac{n\pi x}{l} \, dx,$

$$G(x, t) = \vartheta \left(\frac{x - \xi}{2l} \, \middle| \, \tau \right) - \vartheta \left(\frac{x + \xi}{2l} \, \middle| \, \tau \right).$$

(b) $\dfrac{\partial u}{\partial x} = 0$ at both $x = 0$ and $x = l$.

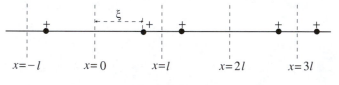

Fig. 8.10(b).

ANS. $f(x) = \sum A_n \cos \frac{n\pi x}{l},$

$$A_n = \frac{2}{l} \int_0^l f(x) \cos \frac{n\pi x}{l} \, dx, \quad A_0 = \frac{1}{l} \int_0^l f(x) \, dx,$$

$$G(x, t) = \vartheta \left(\frac{x - \xi}{2l} \, \middle| \, \tau \right) + \vartheta \left(\frac{x + \xi}{2l} \, \middle| \, \tau \right).$$

(c) $u = 0$ at $x = 0$, and $\dfrac{\partial u}{\partial x} = 0$ at $x = l$.

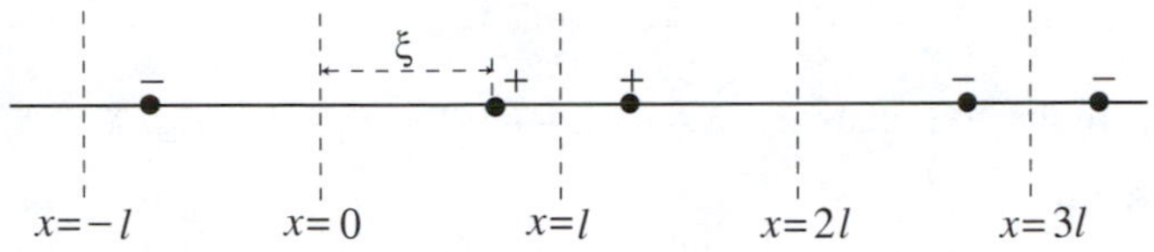

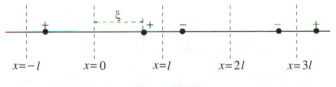

$$x=-l \qquad x=0 \qquad x=l \qquad x=2l \qquad x=3l$$

Fig. 8.10(c).

ANS. $f(x) = \sum B_n \sin \dfrac{(n + 1/2)\pi x}{l}$,

$$B_n = \frac{2}{l} \int_0^l f(x) \sin \frac{(n + 1/2)\pi x}{l}\, dx,$$

$$G(x, t) = \vartheta\left(\frac{x - \xi}{4l}\,\Big|\,\tau\right) - \vartheta\left(\frac{x + \xi}{4l}\,\Big|\,\tau\right)$$

$$+ \vartheta\left(\frac{x + \xi - 2l}{4l}\,\Big|\,\tau\right) - \vartheta\left(\frac{x + \xi - 2l}{4l}\,\Big|\,\tau\right).$$

(d) $\dfrac{\partial u}{\partial x} = 0$ at $x = 0$, and $u = 0$ at $x = l$

$$x=-l \qquad x=0 \qquad x=l \qquad x=2l \qquad x=3l$$

Fig. 8.10(d).

ANS. $f(x) = \sum A_n \cos \dfrac{(n + 1/2)\pi x}{l}$,

$$A_n = \frac{2}{l} \int_0^l f(x) \cos \frac{(n + 1/2)\pi x}{l}\, dx,$$

$$G(x, t) = \vartheta\left(\frac{x - \xi}{4l}\,\Big|\,\tau\right) + \vartheta\left(\frac{x + \xi}{4l}\,\Big|\,\tau\right)$$

$$- \vartheta\left(\frac{x + \xi - 2l}{4l}\,\Big|\,\tau\right) - \vartheta\left(\frac{x + \xi - 2l}{4l}\,\Big|\,\tau\right).$$

8.3. Solve the diffusion equation $u_t = k\, u_{xx}$, $0 < t$, $0 < x < \pi$, subject to the boundary conditions $u(0, t) = 0$, $u(\pi, t) = -\alpha\, u_x(\pi, t)$ and the initial condition $u(0, x) = f(x)$.

HINT. Use the results of Exercise 4.24.

ANS. $u(x,t) = \displaystyle\sum_{n=1}^{\infty} C_n e^{-\lambda^2 kt} \sin \lambda_n x$, where $C_n = \dfrac{\int_0^\pi f(x) \sin \lambda_n x\, dx}{\int_0^\pi \sin^2 \lambda_n x\, dx}$,

and λ_n are the consecutive positive roots of the equation $\tan \lambda = \lambda/\alpha$.

8.4. Consider an infinite rod , $-\infty < x < \infty$, with a unit positive source at $(x_0, 0)$ and a negative source at $(-x_0, 0)$, $x_0 > 0$. Then the temperature distribution for this problem is given by

$$u(x,t) = G\left(x, x_0; t, 0\right) - G\left(x, -x_0; t, 0\right).$$

Use (7.60) to show that

$$u(x,t) = \frac{H(t)}{2\sqrt{\pi t}} \left[e^{-(x-x_0)^2/(4t)} - e^{-(x+x_0)^2/(4t)} \right],$$

where u satisfies the conditions

$$u\left(x, x_0; t, 0\right) = 0, \quad t < 0, \quad 0 < x, x_0 < \infty,$$
$$u\left(0, x_0; t, 0\right) = 0, \quad -\infty < t < \infty, \quad 0 < x_0 < \infty.$$

8.5. Solve the problem: $u_t = u_{xx}$, $x > 0$, $t > 0$, such that $u(0,t) = 0$ and $u(x,0) = f(x)$. Also solve the particular case when $f(x) = 1$ and approximate the solution when $x/(2\sqrt{t})$ is small.

SOLUTION. From Exercise 8.4, we have

$$u(x,t) = \frac{1}{2\sqrt{\pi t}} \int_0^\infty \left[e^{-(x-\xi)^2/(4t)} - e^{-(x+\xi)^2/(4t)} \right] f(\xi)\, d\xi.$$

Set $z = \dfrac{\xi - x}{2\sqrt{t}}$ in the first integral, and $z = \dfrac{\xi + x}{2\sqrt{t}}$ in the second. Then

$$u(x,t) = \frac{1}{\sqrt{\pi}} \left[\int_{-x/(2\sqrt{t})}^\infty e^{-z^2} f\left(x + 2z\sqrt{t}\right) dz \right.$$
$$\left. - \int_{x/(2\sqrt{t})}^\infty e^{-z^2} f\left(-x + 2z\sqrt{t}\right) dz \right].$$

As $t \to 0^+$, the first integral approaches $\sqrt{\pi}\, f(x)$, while the second integral goes to zero. Hence, $u(x,0) = f(x)$, which matches with the prescribed initial condition. In the particular case when $f(x) = 1$, we get

$$u(x,t) = \frac{2}{\sqrt{\pi}} \int_0^{x/(2\sqrt{t})} e^{-z^2}\, dz = \operatorname{erf}\left(x/(2\sqrt{t})\right).$$

If $x/(2\sqrt{t})$ is small, let $y = x/(2\sqrt{t})$. Then, since $e^{-z^2} = \sum\limits_{n=0}^{\infty} \dfrac{(-1)^n z^{2n}}{n!}$, we

get $\displaystyle\int_0^y e^{-z^2}\, dz = \sum_{n=0}^{\infty} \dfrac{(-1)^n y^{2n+1}}{(2n+1)(n!)}$, and

$$u(x,t) = \frac{2}{\sqrt{\pi}} \sum_{n=0}^{\infty} \frac{(-1)^n}{(2n+1)(n!)} \left(\frac{x}{2\sqrt{t}}\right)$$

$$\approx \frac{2}{\sqrt{\pi}} \left[\frac{x}{2\sqrt{t}} - \frac{1}{3}\left(\frac{x}{2\sqrt{t}}\right)^3 + \cdots\right],$$

which gives a very good approximation when $x/(2\sqrt{t})$ is small; for example, for $x = 0.01$ and $t = 0.1$ the two terms in the above series give the result as 0.0178398, which matches with the value of erf $(0.01/0.2) = 0.0178398$.

SECOND METHOD. Applying the Laplace transform to the given problem, we get

$$s\bar{u} - \frac{d^2\bar{u}}{dx^2} = 1, \quad 0 < x < \infty, \quad \text{and } \bar{u}(0,s) = 0,$$

whose solution, which is finite at $x = \infty$ for $\Re\{s\} > 0$, is

$$\bar{u}(x,s) = \frac{1}{s}\left(1 - e^{-x\sqrt{s}}\right).$$

On inversion this gives

$$u(x,t) = 1 - \frac{2}{\sqrt{\pi}} \int_{x/(2\sqrt{t})}^{\infty} e^{-z^2}\, dz = \text{erf}\left(\frac{x}{2\sqrt{t}}\right).$$

8.6. Solve the one-dimensional Stefan problem. This problem deals with a change of phase in a heat conduction problem that is represented as follows: Consider the half-space $x > 0$, which is filled with ice maintained at zero temperature. At time t and thereafter, the wall $x = 0$ is kept at a constant temperature $T > 0$. Under this situation once the ice near the wall begins to melt, the melting front, described by $x = \xi(t)$ in the (x, t)-plane, behaves like an increasing function of time t, i.e., it propagates as t increases. It is obvious that for $x > \xi(t)$ the phase is ice at zero temperature, but for $x < \xi(t)$ the phase is water at temperature between 0 and T. The problem is to determine $x = \xi(t)$. For $x < \xi(t)$ the problem reduces to that of heat conduction in water such that we have $x = \xi(t)$ on the phase interface. At this interface let an element of ice melt in time Δt and the front move by $\Delta \xi$ in time Δt. Thus, the mass of ice melted per unit area in the yz-plane ($x = 0$) is $\rho\, \Delta\xi$, where ρ is the density of melted ice. If ν is the latent heat of melting, then the melting process is caused by conduction of an amount of heat $\nu\rho\, \Delta\xi$ per unit area. Since the temperature gradient is zero in the plane $x = 0$, we have

$$-a\left.\frac{\partial u}{\partial x}\right|_{x=\xi} \Delta t = \nu\rho\, \Delta\xi, \tag{8.70}$$

where $u(x,t)$ is the temperature of water and a the thermal conductivity (see Example 1.7). Thus, the Stefan problem is to solve the initial and boundary value problem:

$$u_t = k\, u_{xx}, \quad 0 < x < \xi(t),\ t > 0,$$
$$u(0,t) = T, \quad u(\xi, t) = 0, \tag{8.71}$$
$$-k\, u_x\left(\xi^-, t\right) = \nu\rho\xi'(t).$$

Since $\xi(t)$ is unknown, we must also find it as a part of the solution of the problem. Such problems are known as *free boundary value problems*. Since in an infinite medium the solution of the homogeneous heat equation is given by

$$u(x,t) = A + B\ \mathrm{erf}\left(\frac{x}{2\sqrt{kt}}\right). \tag{8.72}$$

Using the prescribed conditions in (8.71), we have

$$A = T, \quad A + B\ \mathrm{erf}\left(\frac{\xi}{2\sqrt{kt}}\right) = 0,$$
$$-\frac{Ba}{\sqrt{\pi kt}}\, e^{-\xi^2/(4kt)} = \nu\rho\xi'(t).$$

In view of the second of these conditions, we can take $\xi(t)$ proportional to $\sqrt{t}$, i.e., $\xi(t) = 2\alpha\sqrt{kt}$, where α is a constant to be determined. Using the first of the above conditions and this value of $\xi(t)$, the second and third conditions give

$$T + B\ \mathrm{erf}(\alpha) = 0,$$
$$-\frac{Ba}{\sqrt{\pi kt}}\, e^{-\alpha^2} = \nu\rho\sqrt{\frac{k}{t}},$$

respectively. Then we have $B = -T/\mathrm{erf}(\alpha)$, and

$$\frac{e^{-\alpha^2}}{\mathrm{erf}(\alpha)} = \frac{\nu\rho k\alpha\sqrt{\pi}}{aT}, \tag{8.73}$$

which gives the value of α. Once α is determined, we find from (8.72) that

$$u(x,t) = T - \frac{T}{\mathrm{erf}(\alpha)}\ \mathrm{erf}\left(\frac{x}{2\sqrt{kt}}\right), \quad 0 < x < \xi(t).$$

8.7. Solve the eigenvalue problem $\nabla^2 u + \lambda u = 0$ in the unit circle, where $\lambda > 0$, i.e., determine the eigenpairs for the problem (in polar coordinates)

$$\frac{1}{r}\frac{\partial}{\partial r}\left(r\frac{\partial u}{\partial r}\right) + \frac{1}{r^2}\frac{\partial^2 u}{\partial \theta^2} + \lambda u = 0, \quad r < 1,\ -\pi < \theta \le \pi;$$
$$u(1,\theta) = 0, \quad -\pi < \theta \le \pi.$$

SOLUTION. Using the method of separation of variables, let us assume that $u(r,\theta) = R(r)\Theta(\theta)$. Then we have

$$\frac{\Theta''}{\Theta} = -\frac{r\,(rR')' + \lambda r^2 R}{R}.$$

As in Example 8.17, taking each side equal to $-k$, $k > 0$, we get

$$\Theta'' + k\Theta = 0, \quad -\pi < \theta \le \pi;$$

$$(rR')' + \left(\lambda r - \frac{k}{r}\right) R = 0, \quad 0 < r < 1. \tag{8.74}$$

Since we need another boundary condition to solve this problem, note that the set of points $\theta = \pi$ and $\theta = -\pi$ represent the same radial line. Then, in order that both $R\Theta$ and $\nabla(R\Theta)$ are continuous in the unit circle, the following boundary conditions must be satisfied: $\Theta(-\pi) = \Theta(\pi)$ and $\Theta'(-\pi) = \Theta'(\pi)$. The nontrivial solutions of the first system in (8.74) are obtained for $k = n^2$, $n = 0, 1, 2, \ldots$. The eigenvalue $n = 0$ has the corresponding eigenfunction $\Theta_0 = \text{const.}$ The eigenvalues n^2, $n \ne 0$, correspond to the eigenfunctions $e^{in\theta}$ and $e^{-in\theta}$. As n takes all integral values, the eigenfunctions corresponding to $k = n^2$ are

$$\Theta_n(\theta) = e^{in\theta}, \quad n = \ldots, -2, -1, 0, 1, 2, \ldots.$$

Since the set $\{\Theta_n(\theta)\}$ is orthogonal on the interval $(-\pi, \pi)$ (to see this, check that $\int_{-\pi}^{\pi} \Theta_n(\theta)\,\Theta_m(\theta)\,d\theta = 0$, $m \ne n$), the normalized eigenfunctions are given by $\frac{1}{\sqrt{2\pi}}\,e^{in\theta}$. For the second system in (8.74), take $k = m^2$, and set $z = r\sqrt{\lambda}$. Then we get

$$z^2 \frac{d^2 R}{dr^2} + z\,\frac{1}{z}\frac{dR}{dr} + \left(z^2 - m^2\right) R = 0,$$

which is the Bessel equation of order m and has the general solution $R(z) = C\,J_m(z)$, where m is an integer and C a constant. Thus, $R(r) = C\,J_m\left(\sqrt{\lambda}\,r\right)$. Using the boundary condition at $r = 1$, we get $J_m\left(\sqrt{\lambda}\right) = 0$, which determines λ. For fixed n let $\alpha_1^{(n)}, \alpha_2^{(n)}, \ldots$ be consecutive positive roots of $J_n\left(\sqrt{\lambda}\right) = 0$. Hence, the eigenvalues for this eigenvalue problem are $\lambda_{n,m} = \left[\alpha_m^{(n)}\right]^2$, where n is an integer and $m = 1, 2, \ldots$. The corresponding eigenfunctions are $u_{n,m}(r,\theta) = e^{in\theta}\,J_n\left(\alpha_m^{(n)}\,r\right)$. Since for fixed n, the functions $J_n\left(\alpha_m^{(n)}\,r\right)$ form an orthogonal set with weight r in the interval $(0, 1)$, and since

$$\int_0^1 r\,J_n^2\left(\alpha_m^{(n)}\,r\right)\,dr = \frac{1}{2}\left[J_n'\left(\alpha_m^{(n)}\right)\right]^2,$$

the orthonormal eigenfunctions are

$$u_{n,m}(r,\theta) = \frac{2\,e^{in\theta}\,J_n\left(\alpha_m^{(n)}\,r\right)}{\left[J_n'\left(\alpha_m^{(n)}\right)\right]^2}.$$

8.8. Use (a) the method of separation of variables, and (b) the Laplace transform methods to solve the problem $u_t(x,t) = k\, u_{xx}(x,t)$, $0 < x < l$, $t > 0$, such that $u(0,t) = f_0(t)$, $u(l,t) = f_1(t)$ for $t > 0$, where $f_0(0) = 0 = f_1(0)$, and $u(x,0) = 0 = u_t(x,0)$.

SOLUTION. (a) Take $u(x,t) = v(x,t) + \left(1 - \dfrac{x}{l}\right) f_0(t) + \dfrac{x}{l} f_1(t)$. Then $v_t - k\, v_{xx} = F(x,t)$, $0 < x < l$, $t > 0$, such that $v(x,0) = 0$ for $0 < x < l$, and $v(0,t) = 0 = v(l,t)$ for $t > 0$, where $F(x,t) = -\left(1 - \dfrac{x}{l}\right) f_0'(t) - \dfrac{x}{l} f_1'(t)$. Then by Example 5.2,

$$v(x,t) = \sum_{n=1}^{\infty} \left[\int_0^t e^{-k(n\pi/l)^2 (t-\tau)} F_n(\tau)\, d\tau \right] \sin \frac{n\pi x}{l},$$

where

$$F_n(t) = -\frac{2}{l} \int_0^l \left[\left(1 - \frac{x}{l}\right) f_0'(t) + \frac{x}{l} f_1'(t) \right] \sin \frac{n\pi x}{l}\, dx$$

$$= \frac{2}{n\pi} \left[\cos n\pi\, f_1'(t) - f_0'(t) \right].$$

Hence,

$$u(x,t) = \int_0^t \left[g_x(x, t-\tau)\, f_0(\tau) + g_x(l-x, t-\tau)\, f_1(\tau) \right] d\tau,$$

where

$$g(x,t) = \sqrt{\frac{k}{\pi t}} \sum_{n=-\infty}^{\infty} e^{-(2nl-x)^2/(4kt)}.$$

(b) Taking the Laplace transform, we have $s\bar{u}(x,s) = k\, \bar{u}''(x,s)$, $0 < x < l$, such that $\bar{u}(0,s) = \bar{f}_0(s)$ and $\bar{u}(l,s) = \bar{f}_1(s)$. This system has the general solution $\bar{u}(x,s) = A \sinh \dfrac{xs}{k} + B \cosh \dfrac{xs}{k}$, where $A = \dfrac{\bar{f}_1(s)}{\sinh(ls/k)}$, and $B = \dfrac{\bar{f}_0(s)}{\sinh(ls/k)}$. Then use

$$\mathcal{L}^{-1}\left\{ \frac{\sinh(xs/k)}{\sinh(ls/k)} \right\} = -\sum_{n=-\infty}^{\infty} \frac{(2n+1)l + x}{\sqrt{4\pi k t^3}} e^{-\left[(2n+1)l+x\right]^2/(4kt)},$$

$$\mathcal{L}^{-1}\left\{ \frac{\sinh\left((l-x)s/k\right)}{\sinh(ls/k)} \right\} = -\sum_{n=-\infty}^{\infty} \frac{(2n+2)l - x}{\sqrt{4\pi k t^3}} e^{-\left[(2n+2)l-x\right]^2/(4kt)}.$$

8.9. Let R denote the exterior of the unit sphere. Using the spherical coordinates (ρ, θ, ϕ), solve the problem: $u_t = \nabla^2 u$, $\rho > 1$, $t > 0$, such that $u(\rho, 0) = 0$

and $u(1, t) = 1$. This problem represents heat conduction in an infinite three-dimensional medium with a spherical cavity of unit radius, such that the initial temperature is zero and the boundary of the cavity is kept at a constant temperature 1 for all $t > 0$. Thus, we solve

$$\frac{\partial u}{\partial t} - \frac{1}{\rho^2} \frac{\partial}{\partial \rho} \left(\rho^2 \frac{\partial u}{\partial \rho} \right) = 0, \quad \rho > 1, t > 0,$$

subject to the above initial and boundary conditions. Using the Laplace transform method, we have

$$s\bar{u} - \frac{1}{\rho^2} \frac{\partial}{\partial \rho} \left(\rho^2 \frac{\partial \bar{u}}{\partial \rho} \right) = 0, \quad \rho > 1; \quad \bar{u}(1, s) = \frac{1}{s}. \tag{8.75}$$

The solution $\bar{u}$ is analytic in the right half-plane $\Re\{s\} > 0$. Let $\sqrt{s}$ have a positive real part. Then $\sqrt{s}$ has a branch on the negative real axis and $\sqrt{s} > 0$ if s is real and positive. Then the general solution of Eq (8.75) is

$$\bar{u}(\rho, s) = A \frac{e^{-\rho \sqrt{s}}}{\rho} + B \frac{e^{\rho \sqrt{s}}}{\rho}.$$

Since $\bar{u}$ must be bounded as $\rho \to \infty$ in the half-plane $\Re\{s\} > 0$, we have $B = 0$. Then using the boundary condition in (8.75), we get $\bar{u} = \dfrac{e^{-\rho \sqrt{s}}}{\rho}$, which on inversion, gives

$$u(\rho, t) = \frac{1}{\rho} \left[1 - \text{erf} \left(\frac{\rho - 1}{2\sqrt{t}} \right) \right] = \frac{1}{\rho} \text{erfc} \left(\frac{\rho - 1}{2\sqrt{t}} \right).$$

Note that as $t \to \infty$, the steady-state temperature is $1/\rho$, which is the solution of the boundary value problem $\nabla^2 u = 0$, $\rho > 1$, and $u(1) = 1$.

8.10. Use the Laplace transform method to solve the following problem (see Example 7.12):
$u_{tt} - \nabla^2 u = e^{-k\rho - t}$, $u(\mathbf{x}, 0) = 0$, $u_t(\mathbf{x}, 0) = 0$, $\rho = |\mathbf{x}|$, $\mathbf{x} \in R^3$, $0 < t$.

ANS. $u(\rho, t) = \left[\dfrac{1}{2k(k+1)} e^{kt} + \dfrac{1}{2k(k-1)} e^{-kt} - \dfrac{1}{k^2 - 1} e^{-t} \right] e^{-k\rho}$

$+ \dfrac{2k}{\rho} \left[H(t - \rho) \left\{ \dfrac{e^{-t}}{(k^2 - 1)^2} + \dfrac{t - \rho}{4k^2(k+1)} e^{k(t-\rho)} - \dfrac{2k+1}{4k^3(k+1)^2} e^{k(t-\rho)} \right. \right.$

$\left. + \dfrac{t - \rho}{4k^2(k-1) e^{-k}(t-\rho)} - \dfrac{2k-1}{4k^3(k-1)^2} e^{-k(t-\rho)} \right\}$

$- \left\{ \dfrac{e^{-t}}{(k^2 - 1)^2} + \dfrac{t - \rho}{4k^2(k+1)} e^{k(t-\rho)} - \dfrac{2k+1}{4k^3(k+1)^2} e^{k(t-\rho)} \right.$

$\left. \left. + \dfrac{t - \rho}{4k^2(k-1) e^{-k}(t-\rho)} - \dfrac{2k-1}{4k^3(k-1)^2} e^{-k(t-\rho)} \right\} e^{-k\rho} \right].$

8.11. A simple form of one-dimensional Fokker-Planck equation is

$$\frac{\partial u}{\partial t} + 2\gamma \frac{\partial u}{\partial x} - \frac{\partial^2 u}{\partial x^2} = 0, \qquad (8.76)$$

where 2γ is the drift velocity in the x direction. Solve Eq (8.76) subject to the initial condition $u(x,0) = \delta(x - x_0)$.

SOLUTION. Let $u = v\, e^{\gamma x - \gamma^2 t}$. Then Eq (8.76) reduces to $v_t = v_{xx}$, subject to the initial condition $v(x,0^+) = e^{-\gamma x_0}\,\delta(x - x_0)$, $-\infty < x < \infty$, $t > 0$. Using (7.60) (from heat equation), we get

$$v(x,t) = \frac{e^{-\gamma x_0}}{2\sqrt{\pi t}}\, e^{-(x-x_0)^2/(4t)},$$

and hence,

$$u(x,t) = \frac{e^{-\gamma(x-x_0)-\gamma^2 t}}{2\sqrt{\pi t}}\, e^{-(x-x_0)^2/(4t)}.$$

Physically, this solution represents the concentration of a solvent in an infinite medium when the solute is localized at a point $x = x_0$ at time $t = 0$ (initial condition).

8.12. Derive the results for $f(y, z)$ and C in the parts (a)–(e) of Example 8.9.

8.13. Consider the backward heat equation $v_t = -(v_{xx} + v_{yy})$, subject to the initial condition $v(x,y,0) = u(x,y)$. Show that $v(x,y,t) = u(x,y)\, e^{(m^2+n^2)t}$ is a solution of this initial value problem, where

$$u(x,y) = (a\cos mx + b\sin nx)(c\cos my + d\sin ny).$$

Note that the ill-posedness of this problem is an example of discontinuous dependence on the initial function $u(x,y)$, because although this initial function can be made arbitrarily small by taking the coefficients a, b, c, d arbitrarily small, the solution $v(x,y,t)$ becomes exponentially large for $t > 0$ if m and n are large.

8.14. Solve the problem of temperature distribution in an infinite rod subject to the initial condition, i.e., $T_t = k\, T_{xx}$, subject to the initial condition

$$T(x,0) = \begin{cases} T_0 & \text{if } |x| < x_0, \\ 0 & \text{if } |x| > x_0. \end{cases}$$

HINT. Use the Fourier transform. Note that $\mathcal{F}^{-1}\left\{ T_0 \sqrt{\dfrac{2}{\pi}}\, \dfrac{\sin \alpha x_0}{\alpha}\, e^{-k\alpha^2 t} \right\} =$

$$\frac{T_0}{\sqrt{2\pi}} \int_{-\infty}^{\infty} f(x-\eta)g(\eta)\, d\eta = \frac{T_0}{\sqrt{2\pi}} \int_{x-x_0}^{x+x_0} \frac{e^{-\eta^2/(4kt)}}{2\sqrt{kt}}\, d\eta$$

$$= \frac{T_0}{2}\sqrt{\frac{2}{\pi}} \left[\int_0^{x+x_0} e^{-u^2}\, du - \int_0^{x-x_0} e^{-u^2}\, du \right], \text{ where } u = \eta/(2\sqrt{kt}).$$

ANS. $T(x,t) = \dfrac{T_0}{\sqrt{2}} \left[\operatorname{erf} \left(\dfrac{x+x_0}{2\sqrt{kt}} \right) - \operatorname{erf} \left(\dfrac{x-x_0}{2\sqrt{kt}} \right) \right].$

8.15. Solve the wave equation $u_{tt} = c^2 u_{xx}$, subject to the initial conditions

$$u(x,0) = f(x) = \begin{cases} \dfrac{hx}{x_0} & \text{if } 0 \le x \le x_0, \\[2mm] \dfrac{h(l-x)}{l-x_0} & \text{if } x_0 \le x \le l, \end{cases} \quad \text{and} \quad u_t(x,0) = 0,$$

and the boundary conditions $u(0,t) = 0 = u(l,t)$. This problem describes the displacement $u(x,t)$ of a string of length l, which has its ends fastened at the points $x = 0$ and $x = l$, is plucked at time $t = 0$ at a point x_0, $0 < x_0 < l$, and then released without initial velocity.

SOLUTION. Using the method of separation of variables, we take $u(x,t) = X(x)\,T(t)$. Then we get $X'' + \lambda X = 0$, and $T'' + \lambda c^2 T = 0$, with the corresponding boundary conditions $X(0) = 0 = X(l)$, and $T'(0) = 0$, respectively. The solutions of these problems yield the eigenpairs

$$\lambda_n = \frac{n^2 \pi^2}{l^2}, \quad X_n(x) = \sin \frac{n\pi x}{l}, \quad T_n(t) = \cos \frac{n\pi ct}{l}, \quad n = 1,2,\dots.$$

Hence, the displacement $u(x,t)$ is given by

$$u(x,t) = \sum_{n=1}^{\infty} A_n \sin \frac{n\pi x}{l} \cos \frac{n\pi ct}{l},$$

where the coefficients A_n are determined from the first initial condition. Thus, let $f(x)$ have the Fourier series expansion

$$f(x) = \sum_{n=1}^{\infty} A_n \sin \frac{n\pi x}{l}, \quad 0 < x < l.$$

Then

$$\begin{aligned} A_n &= \frac{2}{l} \int_0^l f(x) \sin \frac{n\pi x}{l}\,dx \\ &= \frac{2}{l} \left[\int_0^{x_0} \frac{hx}{x_0} \sin \frac{n\pi x}{l}\,dx + \int_{x_0}^l \frac{h(l-x)}{l-x_0} \sin \frac{n\pi x}{l}\,dx \right] \\ &= \frac{2h\,l^2}{n^2\,\pi^2 x_0\,(l-x_0)} \sin \frac{n\pi x_0}{l}. \end{aligned}$$

Hence,

$$u(x,t) = \frac{2h\,l^2}{\pi^2 x_0\,(l-x_0)} \sum_{n=1}^{\infty} \frac{1}{n^2} \sin \frac{n\pi x_0}{l} \sin \frac{n\pi x}{l} \cos \frac{n\pi ct}{l}.$$

8.16. Solve

$$\frac{\partial^2 u}{\partial t^2} = c^2 \frac{1}{y(x)} \frac{\partial}{\partial x}\Big[y(x)\frac{\partial u}{\partial x}\Big],$$

where $y(x) = a - x\tan\alpha$ is the variable height of the cross section of a pyramid-shaped cantilever of rectangular cross section and constant thickness (Fig. 8.11), subject to the initial deformation $u(x,0) = f(x)$, and the boundary conditions $u(0,t) = 0$ and $u_x(l,t) = 0$. The function $u(x,t)$ describes the longitudinal oscillations of the cantilever.

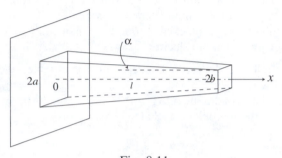

Fig. 8.11.

SOLUTION. Using the method of separation of variables and assuming that $u(x,t) = X(x)\,T(t)$, we obtain

$$X'' - \frac{\tan\alpha}{y}X' + \lambda X = 0, \quad T'' + \lambda c^2 T = 0.$$

We introduce a new independent variable y defined by $y = a - x\tan\alpha$, so that $X' = -\tan\alpha\,\dfrac{dX}{dy}$, and $X'' = \tan^2\alpha\,\dfrac{d^2 X}{dy^2}$, the first equation reduces to

$$\frac{d^2 X}{dy^2} + \frac{1}{y}\frac{dX}{dy} + \Big(\frac{\sqrt{\lambda}}{\tan\alpha}\Big)^2 X = 0,$$

which is the Bessel equation in the variable y. Its solution is

$$X = A\,J_0\Big(\frac{y\sqrt{\lambda}}{\tan\alpha}\Big) + B\,Y_0\Big(\frac{y\sqrt{\lambda}}{\tan\alpha}\Big).$$

The boundary conditions become $X\Big|_{y=0} = 0$ and $\dfrac{dX}{dy}\Big|_{y=b} = 0$, respectively. The eigenpairs are given by

$$\lambda_n = \Big(\frac{\gamma_n\tan\alpha}{a}\Big)^2, \quad X_\gamma(y) = Y_0(\gamma)\,J_0\Big(\frac{\gamma y}{a}\Big) - J_0(\gamma)\,Y_0\Big(\frac{\gamma y}{a}\Big),$$

where γ_n are the consecutive positive roots of the equation $X'_\gamma(b) = 0$. The equation for T after integration gives

$$T = C_n \cos \frac{\gamma_n \, c t \tan \alpha}{a}.$$

Hence, the solution is given by

$$u(x,t) = \sum_{n=1}^{\infty} C_n \, X_{\gamma_n}(y) \, \cos \frac{\gamma_n \, c t \tan \alpha}{a},$$

where the coefficients C_n are determined from the initial condition

$$u(x,0) = f(x) = \sum_{n=1}^{\infty} C_n \, X_{\gamma_n}(y), \quad b < y < a.$$

Using the formula

$$\int_b^a X_{\gamma_m}(y) \, X_{\gamma_n}(y) \, y \, dy = \begin{cases} 0 & \text{if } m \neq n, \\ \dfrac{b^2}{2} \left[\dfrac{a^4}{b^2 \gamma_n} {X'_{\gamma_n}}^2(a) - X_{\gamma_n}^2(b) \right], & \text{if } m = n, \end{cases}$$

and noting that $X'_{\gamma_n}(a) = -2/(\pi a)$, we find that

$$C_n = 2\pi^2 \gamma_n^2 \frac{\int_a^b f(x) \, X_{\gamma_n}(y) \, y \, dy}{4a^2 - \pi^2 \gamma_n^2 \, b^2 \, X_{\gamma_n}^2(b)}.$$

8.17. Solve the problem

$$\frac{\partial^2 u}{\partial t^2} = c^2 \frac{1}{r} \frac{\partial}{\partial r} \left(r \frac{\partial u}{\partial r} \right),$$

subject to the initial conditions

$$u(r,0) = 0, \quad \text{and} \quad u_t(r,0) = f(r) = \begin{cases} \dfrac{P}{\pi \varepsilon^2 \rho} & \text{if } 0 \leq r \leq \varepsilon, \\ 0 & \text{if } \varepsilon < r \leq a, \end{cases}$$

and the boundary condition $u(a,t) = 0$. This problem represents the axially symmetric vibrations of a circular membrane of radius a, where the vibrations are caused by an impulse P applied at time $t = 0$ and distributed over a disk of radius $\varepsilon < a$.

SOLUTION. Using the separation of variables method and assuming $u(r,t) = R(r) \, T(t)$, we get

$$\frac{1}{r} \frac{d}{dr} \left(r \frac{dR}{dt} \right) + \lambda R = 0, \quad T'' + \lambda c^2 T = 0.$$

Since the solution of the first of these equations must be bounded over the region $0 \le r < a$ and satisfy the boundary condition $R(a) = 0$, we find the eigenpairs as

$$\lambda_n = \frac{\gamma_n^2}{a^2}, \quad R_n(r) = J_0\left(\frac{\gamma_n r}{a}\right), \quad n = 1, 2, \ldots,$$

where γ_n are the consecutive positive roots of the equation $J_0(x) = 0$. The solution of the second equation is $T = B_n \sin \frac{\gamma_n ct}{a}$. Thus, the general solution of the problem is

$$u(r,t) = \sum_{n=1}^{\infty} A_n \sin \frac{\gamma_n ct}{a} J_0\left(\frac{\gamma_n r}{a}\right),$$

where the coefficients A_n are determined from the initial condition, i.e., from

$$f(r) = \frac{c}{a} \sum_{n=1}^{\infty} A_n \gamma_n J_0\left(\frac{\gamma_n r}{a}\right), \quad 0 < r < a,$$

which gives

$$A_n = \frac{2a}{c\gamma_n a^2 J_1^2(\gamma_n)} \int_0^a f(r) J_0\left(\frac{\gamma_n r}{a}\right) r \, dr$$

$$= \frac{2a}{c\gamma_n a^2 J_1^2(\gamma_n)} \int_0^\varepsilon \frac{P}{\pi\varepsilon^2} J_0\left(\frac{\gamma_n r}{a}\right) r \, dr = \frac{2P J_1(\gamma_n \varepsilon/a)}{\pi\varepsilon a\gamma_n J_1^2(\gamma_n)}.$$

8.18. Solve the heat conduction problem

$$\frac{\partial T}{\partial t} = \frac{1}{\rho^2} \frac{\partial}{\partial \rho}\left(\rho^2 \frac{\partial T}{\partial \rho}\right),$$

subject to the initial condition $T(\rho, 0) = f(\rho)$ and the boundary condition $T(a, t) = 0$. This problem describes the cooling of a sphere of radius a with the above initial temperature distribution, where the temperature of its surface is kept at zero.

HINT. Use the separation of variables method, assuming $u(\rho, t) = R(\rho)\Theta(t)$. This leads to $\frac{1}{\rho^2}\left(\rho^2 R'\right)' + \lambda R = 0$, and $\Theta' + \lambda\Theta = 0$. The general solutions are $R = A \frac{\sin\sqrt{\lambda}\rho}{\rho} + B \frac{\cos\sqrt{\lambda}\rho}{\rho}$, and $\Theta = C e^{-\lambda t}$. The eigenpairs are $\lambda_n = \frac{n^2\pi^2}{a^2}$, and $R_n = \frac{1}{\rho} \sin \frac{n\pi\rho}{a}$. Hence,

$$T(\rho, t) = \frac{1}{\rho} \sum_{n=1}^{\infty} A_n e^{-n^2\pi^2 t/a^2} \sin \frac{n\pi\rho}{a},$$

where, using the initial condition, we obtain $A_n = \dfrac{2}{a} \displaystyle\int_0^a f(\rho) \sin \dfrac{n\pi\rho}{a}\, \rho\, d\rho.$

ANS. $T(\rho, t) = \dfrac{2}{a\rho} \displaystyle\sum_{n=1}^{\infty} A_n\, e^{-n^2\pi^2 t/a^2}\, sin\dfrac{n\pi\rho}{a} \displaystyle\int_0^a f(\rho) \sin \dfrac{n\pi\rho}{a}\, \rho\, d\rho.$

8.19. Solve the nonhomogeneous wave equation

$$\frac{\partial^2 u}{\partial x^2} - \frac{1}{c^2}\frac{\partial^2 u}{\partial t^2} = -\frac{q(x,t)}{T_0}, \tag{8.77}$$

subject to the initial conditions $u(x,0) = 0 = u_t(x,0)$ and the boundary conditions $u(0,t) = 0 = u(l,t)$. This problem describes the vibrations of a string of length l, which is initially at rest and is under the action of an external load $q(x,t)$, where T_0 denotes the tension in the string.

SOLUTION. First, we solve the corresponding homogeneous equation by using the separation of variables method and assuming $u(x,t) = X(x)T(t)$ (see Example 5.1). The functions $X_n(x)$ are the eigenfunctions for the equation $X'' + \lambda X = 0$, under the boundary conditions $X(0) = 0 = X(l)$; the eigenpairs are $\lambda_n = \dfrac{n^2\pi^2}{l^2}$, and $X_n = \sin\dfrac{n\pi x}{l}$, $n = 1, 2, \ldots$. Now, for the nonhomogeneous problem we seek the solution in the form

$$u(x,t) = \sum_{n=1}^{\infty} \frac{g_n(x,t)\, X_n(x)}{\int_0^l X_n^2(x)\, dx},$$

where the coefficients g_n are given by $g_n(x,t) = \displaystyle\int_0^l u(x,t)\, X_n(x)\, dx.$ To determine the functions g_n, we multiply (8.77) by $X_n(x)$ and integrate with respect to x from 0 to l. Integration by parts twice and using the boundary conditions, we get

$$g_n'' + \left(\frac{n\pi c}{l}\right)^2 g_n - \frac{c^2}{T_0}\int_0^l q(x,t)\sin\frac{n\pi x}{l}\, dx = 0.$$

The solution of this equation is found by the method of variation of parameters as

$$g_n(x,t) = A_n\, \cos\frac{n\pi ct}{l} + B_n\, \sin\frac{n\pi ct}{l}$$

$$+ \frac{cl}{n\pi T_0}\int_0^l \sin\frac{n\pi c(t-\tau)}{l}\, d\tau \int_0^l q(\xi, \tau)\sin\frac{n\pi\xi}{l}\, d\xi.$$

The constants A_n and B_n are determined from the initial conditions, which are $g_n(0) = 0 = g_n'(0)$. Thus, we get $A_n = 0 = B_n$. Hence, the solution of the problem is

$$u(x,t) = \frac{2c}{\pi T_0}\sum_{n=1}^{\infty}\frac{1}{n}\sin\frac{n\pi x}{l}\int_0^l \sin\frac{n\pi c(t-\tau)}{l}\, d\tau \int_0^l q(\xi, \tau)\sin\frac{n\pi\xi}{l}\, d\xi,$$

$$0 < x < l.$$

8.20. Use the Laplace transform method to solve the heat conduction problem

$$\frac{\partial T}{\partial t} = \frac{1}{r}\frac{\partial}{\partial r}\left(r\,\frac{\partial T}{\partial r}\right), \quad 0 \le r < a,\ t > 0,$$

subject to the initial conditions

$$T(r,0) = \begin{cases} 0 & \text{if } r < a, \\ T_0 & \text{if } r = a, \end{cases}$$

and the mixed boundary condition $\left(\alpha\,\dfrac{\partial T}{\partial t} + \dfrac{1}{a}\dfrac{\partial T}{\partial r}\right)\Big|_{r=a,\,t>0} = 0$, such that $\lim\limits_{r\to 0} T < \infty$. This problem describes the temperature distribution in a cold cylinder of radius a which at time $t = 0$ is encased in a thin heated cylindrical sleeve which is thermally insulated outside, and α denotes its thermal capacitance (Fig. 8.12). It is assumed that initially the cylinder and the sleeve have temperature zero and T_0, respectively, and any temperature drop inside the sleeve is neglected.

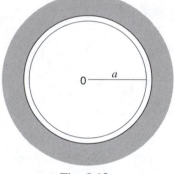

Fig. 8.12.

HINT. Taking the Laplace transform and using the initial condition, we get

$$\frac{1}{r}\frac{d}{dr}\left(r\,\frac{d\overline{T}}{dr}\right) - s\overline{T} = 0, \quad 0 \le r < a,$$

which is subjected to the boundary condition $\left[\alpha\left(s\overline{T} - T_0\right) + \dfrac{1}{a}\overline{T}'\right]_{r=a} = 0$. Solving this system we obtain

$$\overline{T}(r,s) = \frac{\alpha a T_0\, I_0\left(r\sqrt{s}\right)}{\sqrt{s}\left[\alpha a\sqrt{s}\, I_0\left(a\sqrt{s}\right) + I_1\left(a\sqrt{s}\right)\right]}.$$

ANS. $T(r,t) = T_0\left[\dfrac{2\alpha}{1+2\alpha} + \sum\limits_{n=1}^{\infty} \dfrac{J_0\left(\gamma_n r/a\right)\, e^{-\gamma_n^2 t/a^2}}{\left(1 + \dfrac{1}{2\alpha} + \dfrac{\alpha\gamma_n^2}{2}\right) J_0(\gamma_n)}\right]$, where γ_n are the consecutive positive roots of the equation $J_1(x) + \alpha\gamma\, J_0(x) = 0$. Use tables of integral transforms, e.g., Erdélyi et al. (1954), for the Laplace inversion.

8.21. Consider the Dirichlet problem for the wave equation $u_{tt} = u_{xx}$, $0 < x < 1$, $0 < t < T$, such that $u(x,0) = 0 = u(x,T)$ for $0 < x < 1$, and $u(0,t) = 0 = u(1,t)$ for $0 < t < T$. Show that if T is irrational, the only solution is $u(x,t) = 0$, but if T is rational, then the problem has infinitely many nonzero solutions. Hence, deduce that this problem is ill-posed because the solutions do not depend continuously on the data.

8.22. Show that the problem in Example 8.20 for $f(x) = x(1-x)$, $0 < x < 1$, is ill-posed.

9

Weighted Residual Methods

Variational formulation of boundary value problems originates from the fact that weighted variational methods provide approximate solutions of such problems. Variational methods for solving boundary value problems are based on the techniques developed in the calculus of variations. They deal with the problem of minimizing a functional and thus reducing the given problem to the solution of Euler-Lagrange differential equations. If the functional to be minimized has more than one independent variable, the Euler-Lagrange equations are partial differential equations. Conversely, a boundary value problem can be formulated as a minimizing problem. The functional which corresponds to the partial differential equation is generally known as the energy function. In the case when the solution is not available in a simple form, an approximate solution is found such that it minimizes the energy equation. The approximating function is a linear combination of the form $\sum_{i=0}^{\infty} c_i \phi_i$, $c_0 = 1$, of specially chosen functions ϕ_i which are known as the test functions or interpolation functions. The function ϕ_0 satisfies the same boundary conditions as the original unknown function, while the remaining functions ϕ_i, $i \neq 0$, satisfy the homogeneous boundary conditions. The constants c_i, $i \neq 0$, are then determined by minimizing the energy function.

9.1. Line Integrals

In many types of boundary value problems, the variational methods are used to provide precise formulations, which can be applied in any prescribed system of coordinates.

We will derive the Euler equation, which is the necessary condition for the solution of the following problem: Find a function $u(x)$ for which the integral

$$I(u) = \int_a^b F(x, u, u') \, dx \tag{9.1}$$

is a minimum, where F is twice-differentiable with respect to x, u, u'.

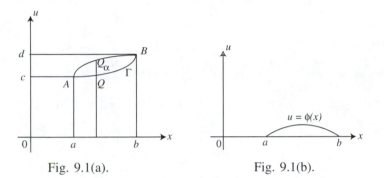

Fig. 9.1(a).   Fig. 9.1(b).

Let two fixed points $A(a, c)$ and $B(b, d)$ in the xu-plane be joined by a curve $\Gamma : x \longmapsto u(x)$ (see Fig. 9.1(a)). Then $c = u(a), d = u(b)$, and $u' = du/dx = u'(x)$ at each point Q of Γ. Thus, the curve Γ determines a value of the integral in (9.1). However, the value of the integral I changes if we replace Γ by a new curve joining A and B. To investigate the variation of I with Γ, i.e., to determine the curve Γ for which the integral I has a minimum (or a maximum) value, we confine to a set of curves Γ_α, which are defined as follows: First, we select an admissible function $u = \phi(x)$ such that

$$\phi(a) = 0 = \phi(b) \tag{9.2}$$

(see Fig. 9.1(b)). Then for any value of the parameter α, the curve Γ_α is, in view of (9.2), defined by

$$U = U(x) = u(x) + \alpha\phi(x), \quad U(a) = c, \quad U(b) = d, \tag{9.3}$$

where $u(x)$ is assumed to minimize $I(u)$. Hence, it passes through A and B. From (9.3) we find that on the curve Γ_α

$$U' = \frac{dU}{dx} = u'(x) + \alpha\phi'(x). \tag{9.4}$$

Thus, for any value of α, we have a curve Γ_α, and, by substituting the values from (9.3) and (9.4), we form the value of I along Γ_α as

$$I(U, \alpha) = \int_a^b F(x, U, U') \, dx. \tag{9.5}$$

Using the differentiation formula

$$\frac{\partial F}{\partial \alpha} = \frac{\partial F}{\partial U}\frac{\partial U}{\partial \alpha} + \frac{\partial F}{\partial U'}\frac{\partial U'}{\partial \alpha},$$

and noting from (9.3) and (9.4) that $\dfrac{\partial U}{\partial \alpha} = \phi(x)$, $\dfrac{\partial U'}{\partial \alpha} = \phi'(x)$, we find that

$$\frac{\partial F}{\partial \alpha} = \phi(x)\frac{\partial F}{\partial U} + \phi'(x)\frac{\partial F}{\partial U'}.$$

Hence,

$$\frac{\partial I}{\partial \alpha} = \int_a^b \left[\phi(x)\frac{\partial F}{\partial U} + \phi'(x)\frac{\partial F}{\partial U'} \right] dx. \tag{9.6}$$

If we integrate by parts the second term in the integrand in (9.6), and use (9.2), we get

$$\frac{\partial I}{\partial \alpha} = \int_a^b \phi(x)\left[\frac{\partial F}{\partial U} - \frac{d}{dx}\left(\frac{\partial F}{\partial U'}\right) \right] dx. \tag{9.7}$$

Let us assume that there is a twice-differentiable curve, say Γ, for which the value of I is a minimum. Then the value of I on Γ is less than the value of I on any other curve Γ_α. Thus, $I(\alpha)$ assumes a minimum value for $\alpha = 0$, since $\partial I/\partial \alpha$ is continuous. But, then $U = u$ and $U' = u'$ when $\alpha = 0$. Hence, taking $\alpha = 0$ in (9.7) and $I'(0) = 0$, we find that

$$\int_a^b \phi(x)\left[\frac{\partial F}{\partial u} - \frac{d}{dx}\left(\frac{\partial F}{\partial u'}\right) \right] dx = 0. \tag{9.8}$$

Since the function $\phi(x)$ is arbitrary, except for the conditions (9.2), the factor in the square brackets in (9.8) is continuous, which implies that

$$\frac{\partial F}{\partial u} - \frac{d}{dx}\left(\frac{\partial F}{\partial u'}\right) = 0 \tag{9.9}$$

for all $x \in [a, b]$. In fact, if the expression within the square brackets in (9.8) were nonzero at any point, say x_0, there would be some small interval including this point, say $x_1 < x_0 < x_2$, in which it remains nonzero; then, by taking the function $\phi(x)$ in Fig. 9.2, the integrand in (9.9) would be zero except for the interval (x_1, x_2) where it is nonzero. Thus, we have proved: If the integral I, defined by (9.1), has a minimum (or a maximum) along any sufficiently smooth curve Γ joining A and B, then $u = u(x)$ will be a solution of the differential Eq (9.9). This is known as the *Euler-Lagrange equation* and its solution $u(x)$ as the *extremal*. Note that the derivative d/dx in (9.9) is computed by recalling that $u = u(x)$ and $u' = u'(x) = du/dx$ are functions of x. After this differentiation is carried out, Eq (9.9) becomes

$$\frac{\partial F}{\partial u} - \frac{\partial^2 F}{\partial x \partial u'} - \frac{\partial^2 F}{\partial u \partial u'}\frac{du}{dx} - \frac{\partial^2 F}{\partial u'^2}\frac{d^2 u}{dx^2} = 0. \tag{9.10}$$

This is a second-order differential equation, and its solution contains two arbitrary constants which must be determined from the conditions that the curve passes through A and B.

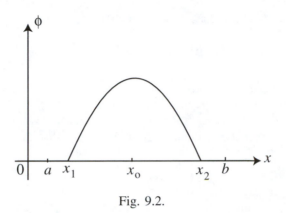

Fig. 9.2.

Similarly, by following the above procedure and using integration by parts twice in the third term appearing in the integrand of

$$\frac{dI}{d\alpha} = \int_a^b \left[\phi(x)\frac{\partial F}{\partial U} + \phi'(x)\frac{\partial F}{\partial U'} + \phi''(x)\frac{\partial^2 F}{\partial U''} \right] dx, \tag{9.11}$$

and taking the additional requirement that $\phi'(a) = 0 = \phi'(b)$, we find that the Euler equation which provides a necessary condition for the functional

$$I(u) = \int_a^b F(x, u, u', u'') \, dx \tag{9.12}$$

to be a minimum is

$$\frac{\partial F}{\partial u} - \frac{d}{dx}\left(\frac{\partial F}{\partial u'}\right) + \frac{d^2}{dx^2}\left(\frac{\partial F}{\partial u''}\right) = 0 \tag{9.13}$$

(see Exercise 9.1.).

If the integral I depends on two functions u and v, i.e., if

$$I(u, v) = \int_a^b F(x, u, v, u', v') \, dx, \tag{9.14}$$

then there are two Euler equations:

$$\frac{\partial F}{\partial u} - \frac{d}{dx}\left(\frac{\partial F}{\partial u'}\right) = 0, \quad \frac{\partial F}{\partial v} - \frac{d}{dx}\left(\frac{\partial F}{\partial v'}\right) = 0. \tag{9.15}$$

To derive (9.15), introduce two functions $\phi(x)$ and $\psi(x)$ and two parameters α and β such that $U = u + \alpha\phi(x)$, $V = v + \beta\psi(x)$. Then $\partial I/\partial\alpha = 0$ and $\partial I/\partial\beta = 0$.

9.2. Variational Notation

Eq (9.3) implies that the difference $U - u = \alpha\phi(x)$, or QQ_α in Fig. 9.1, is the change in U, starting from u and the value of $\alpha = 0$. Hence, $\alpha\phi(x) = U - u$ is called the *(first) variation* in u and is denoted by δu. It is also regarded as the differential dU, since $d\alpha = \alpha - 0 = \alpha$ at $\alpha = 0$; then $dU/d\alpha = \phi(x)$ so that

$$dU = \left(\frac{dU}{d\alpha}\right)d\alpha = \phi(x)\alpha = \delta u. \tag{9.16}$$

Also, Eq (9.4) is rewritten as $U' = u' + \alpha\phi'(x)$ or

$$U' - u' = \alpha\phi'(x). \tag{9.17}$$

The term $\alpha\phi'(x) = U' - u'$ is the variation of u' and is denoted by $\delta u'$. It is also the differential

$$dU' = \left(\frac{dU'}{d\alpha}\right)d\alpha = \phi'(x)\alpha = \delta u'. \tag{9.18}$$

From (9.16) and (9.18), it follows that $\dfrac{d(\delta u)}{dx} = \delta\left(\dfrac{du}{dx}\right)$ or

$$(\delta u)' = \delta(u'), \tag{9.19}$$

i.e., the operator δ and d/dx are commutative. The variation of u'' and higher derivatives of u are defined analogously.

If u is changed to $u + \delta u = u + \alpha w$, then the corresponding change in a function $F = F(x, u, u')$ at u is defined by

$$\delta F = \frac{\partial F}{\partial u}\delta u + \frac{\partial F}{\partial u'}\delta u'. \tag{9.20}$$

Since δ behaves like a differential operator, it has the following properties:

(i) $\delta(f_1 \pm f_2) = \delta f_1 \pm \delta f_2$,
(ii) $\delta(f_1 f_2) = f_1\,\delta f_2 + f_2\,\delta f_1$,
(iii) $\delta(f_1/f_2) = (f_2\,\delta f_1 - f_1\,\delta f_2)/f_2^2$,
(iv) $\delta(f)^n = n(f)^{n-1}\,\delta f$,
(v) $D(\delta u) = \delta(Du)$,
(vi) $\delta\int_a^b u(x)\,dx = \int_a^b \delta u(x)\,dx$,

where $' \equiv D \equiv d/dx$, and f_1, f_2 are functions of x, u, u'. Note that (v) and (vi) are commutative properties under differential and integral operators. The above six properties (i)–(vi) can be easily proved by using (9.20).

Now, we find from (9.7) that

$$\delta I = \left(\frac{dI}{d\alpha}\right)_{\alpha=0} = \int_a^b \left[\frac{\partial F}{\partial u} - \frac{d}{dx}\left(\frac{\partial F}{\partial u'}\right)\right] \delta u \, dx, \qquad (9.21)$$

where the variations at the end points are zero. Similarly, for I defined by (9.12), we have

$$\delta I = \int_a^b \left[\frac{\partial F}{\partial u} - \frac{d}{dx}\left(\frac{\partial F}{\partial u'}\right) + \frac{d^2}{dx^2}\left(\frac{\partial F}{\partial u''}\right)\right] \delta u \, dx, \qquad (9.22)$$

where the variations are restricted by the conditions: $\phi'(a) = 0 = \phi'(b)$.

For the integral I in (9.14) whose integrand $F(x, u, u', v, v')$ contains two independent variables u and v, we define the variations as total differentials so that

$$\delta u = \alpha\phi(x), \quad \delta u' = \alpha\phi'(x), \quad \delta v = \alpha\psi(x), \quad \delta v' = \alpha\psi'(x).$$

Then the definition in this case becomes

$$\delta F = \frac{\partial F}{\partial u} \delta u + \frac{\partial F}{\partial u'} \delta u' + \frac{\partial F}{\partial v} \delta v + \frac{\partial F}{\partial v'} \delta v'. \qquad (9.23)$$

For the integral I in (9.14) we have

$$\delta I = \int_a^b \left[\frac{\partial F}{\partial u} - \frac{d}{dx}\left(\frac{\partial F}{\partial u'}\right)\right] \delta u \, dx + \int_a^b \left[\frac{\partial F}{\partial v} - \frac{d}{dx}\left(\frac{\partial F}{\partial v'}\right)\right] \delta v \, dx. \qquad (9.24)$$

In the parametric case, we can introduce a new parameter t as the independent variable and regard x and u (or x, u and v) as the dependent variables. For example, let us consider the integrand $F(x, u, u')$. Using the circumflex to denote differentiation with respect to the parameter t, we have: $u' = \hat{u}/\hat{x}$ and $F(x, u, u') \, dx = F(x, u, \hat{u}/\hat{x})\hat{x} \, dt$. Let $x = a$ when $t = t_1$, and $x = b$ when $t = t_2$, and denote $F(x, u, \hat{u}/\hat{x})\hat{x}$ by $G(x, u, \hat{x}, \hat{u})$. Then, the integral (9.1) is written as

$$I = \int_a^b F(x, u, u') \, dx = \int_{t_1}^{t_2} G(x, u, \hat{x}, \hat{u}) \, dt,$$

and (9.21) becomes (using (9.24) with x, u, v replaced by t, x, u, respectively)

$$\delta I = \int_{t_1}^{t_2} \left[\frac{\partial G}{\partial x} - \frac{d}{dt}\left(\frac{\partial G}{\partial \hat{x}}\right)\right] \delta x \, dt + \int_{t_1}^{t_2} \left[\frac{\partial G}{\partial u} - \frac{d}{dt}\left(\frac{\partial G}{\partial \hat{u}}\right)\right] \delta u \, dt, \qquad (9.25)$$

where δx and δu are zero at t_1 and t_2.

EXAMPLE 9.1. Show that in R^2

$$\nabla(\delta u) \cdot \nabla u = \frac{1}{2}\delta|\nabla u|^2. \tag{9.26}$$

A direct computation shows that

$$
\begin{aligned}
\nabla(\delta u) \cdot \nabla u &= \left(\mathbf{i}\frac{\partial}{\partial x}(\delta u) + \mathbf{j}\frac{\partial}{\partial y}(\delta u)\right) \cdot (\mathbf{i}\,u_x + \mathbf{j}\,u_y) \\
&= (\mathbf{i}\,u_x + \mathbf{j}\,u_y) \cdot \delta\,(\mathbf{i}\,u_x + \mathbf{j}\,u_y) \\
&= \frac{1}{2}\delta\left[(\mathbf{i}\,u_x + \mathbf{j}\,u_y) \cdot (\mathbf{i}\,u_x + \mathbf{j}\,u_y)\right] = \frac{1}{2}\delta|\nabla u|^2. \blacksquare
\end{aligned}
$$

9.3. Multiple Integrals

Let $F(a, y, u, p, q)$ be a twice-differentiable function of five variables, where the dependent variable u is a function of x and y and $p = u_x = \partial u/\partial x$, $q = u_y = \partial u/\partial y$. We study the variation of the integral

$$I(u) = \iint_S F(x, y, u, p, q)\,dS \tag{9.27}$$

over a surface S, which passes through a fixed boundary curve B, and determine for which surface S the integral $I(u)$ is a minimum. Following the method of §9.1, we first choose any surface S defined by $u = \phi(x, y)$ such that $\phi(x, y) = 0$ on the curve Γ, where Γ is the projection of the boundary curve B in the xy-plane. Since $u(x, y)$ passes through Γ, we define a surface S_α for any α by

$$U(x, y) = u(x, y) + \alpha\phi(x, y).$$

The surface S_α also passes through Γ. On this surface $P \equiv U_x = p + \alpha\phi_x$, $Q \equiv U_y = q + \alpha\phi_y$, so that the integral I is written as

$$I = \iint_{S_\alpha} F(x, y, U, P, Q)\,dx\,dy. \tag{9.28}$$

Hence,

$$\frac{dI}{d\alpha} = \iint \left[\frac{\partial F}{\partial U}\phi + \frac{\partial F}{\partial P}\phi_x + \frac{\partial F}{\partial Q}\phi_y\right] dx\,dy. \tag{9.29}$$

Using formula (A.3), we find from (9.29) that

$$\frac{dI}{d\alpha}\bigg|_{\alpha=0} = \iint \left[\frac{\partial F}{\partial U} - \frac{\partial}{\partial x}\left(\frac{\partial F}{\partial P}\right) - \frac{\partial}{\partial y}\frac{\partial F}{\partial Q}\right]\phi\,dx\,dy,$$

since $U = u$, $P = p$, $Q = q$ when $\alpha = 0$. If we multiply the value of $dI/d\alpha\big|_{\alpha=0}$ by $d\alpha = \alpha - 0 = \alpha$ and write δu for $\alpha\phi$, we find, as in (9.21), that

$$\delta I = \iint \left[\frac{\partial F}{\partial u} - \frac{\partial}{\partial x}\left(\frac{\partial F}{\partial p}\right) - \frac{\partial}{\partial y}\left(\frac{\partial F}{\partial q}\right) \right] \delta u \, dx \, dy. \qquad (9.30)$$

If $u(x, y)$ is the surface for which I is a minimum, then $\delta I = 0$ for arbitrary δu. Since the expression in the square brackets in (9.30) is continuous, it must be zero. This gives a necessary condition for I to be a minimum as

$$\frac{\partial F}{\partial u} - \frac{\partial}{\partial x}\left(\frac{\partial F}{\partial p}\right) - \frac{\partial}{\partial y}\left(\frac{\partial F}{\partial q}\right) = 0. \qquad (9.31)$$

This is the Euler equation for (9.27), and any surface corresponding to a solution of this equation is an extremal.

If $F = F(x, y, u, p, q, r, s, t)$, where p and q are defined above and $r = u_{xx}$, $s = u_{xy}$, and $t = u_{yy}$, then the Euler equation to minimize the integral $I = \iint F(x, y, u, p, q, r, s, t) \, dx \, dy$ is given by

$$\frac{\partial F}{\partial u} - \frac{\partial}{\partial x}\left(\frac{\partial F}{\partial p}\right) - \frac{\partial}{\partial y}\left(\frac{\partial F}{\partial q}\right) + \frac{\partial^2}{\partial x^2}\left(\frac{\partial F}{\partial r}\right) + \frac{\partial^2}{\partial x \partial y}\left(\frac{\partial F}{\partial s}\right) + \frac{\partial^2}{\partial y^2}\left(\frac{\partial F}{\partial t}\right) = 0. \qquad (9.32)$$

EXAMPLE 9.2. The condition that the Dirichlet integral

$$I(u) = \iint \left[\left(\frac{\partial u}{\partial x}\right)^2 + \left(\frac{\partial u}{\partial y}\right)^2 \right] dx \, dy$$

be a minimum is $\nabla^2 u = \dfrac{\partial^2 u}{\partial x^2} + \dfrac{\partial^2 u}{\partial y^2} = 0$, which is the Laplace equation. ∎

9.4. Weak Variational Formulation

The weak variation formulation of boundary value problems is derived from the fact that variational methods for finding approximate solutions of boundary value problems, viz., Galerkin, Rayleigh-Ritz, collocation, or other weighted residual methods, are based on the weak variational statements of the boundary value problems. In fact, the weak variational formulation is more general than the corresponding strong formulation, since even the irregular boundary conditions are easily managed in the weak formulation. We will not discuss the evolution of the strong formulation, but rather explain the method of the weak variational formulation, which in turn defines the underlying concept.

We consider a general form of a second-order mixed boundary value problem, defined by Eq (9.31), in a two-dimensional region Ω with the prescribed boundary conditions

$$u = u_0 \quad \text{on } \Gamma_1, \tag{9.33a}$$

and

$$\frac{\partial F}{\partial p} n_x + \frac{\partial F}{\partial q} n_y = q_0 \quad \text{on } \Gamma_2, \tag{9.33b}$$

where $F = F(x, y, u, p, q)$, and n_x, n_y are the direction cosines of the unit vector $\hat{n}$ normal to the boundary $\Gamma = \Gamma_1 \cup \Gamma_2$ of the region Ω such that $\Gamma_1 \cap \Gamma_2 = \emptyset$.

For example, a special case of (9.31) is when F is defined as

$$F = \frac{1}{2} \left[k_1 \left(\frac{\partial u}{\partial x} \right)^2 + k_2 \left(\frac{\partial u}{\partial y} \right)^2 \right] - f u.$$

This equation arises in heat conduction problems in a two-dimensional region with k_1, k_2 as thermal conductivities in the x, y directions, and f is the heat source (sink). Here

$$\frac{\partial F}{\partial p} = k_1 \frac{\partial u}{\partial x}, \quad \frac{\partial F}{\partial q} = k_2 \frac{\partial u}{\partial y}, \quad \frac{\partial F}{\partial u} = -f,$$

and Eq (9.31) becomes

$$-\frac{\partial}{\partial x} \left(k_1 \frac{\partial u}{\partial x} \right) - \frac{\partial}{\partial y} \left(k_2 \frac{\partial u}{\partial y} \right) = f \quad \text{in} \quad \Omega.$$

If $k_1 = k_2 = 1$, then we get the Poisson's equation $-\nabla^2 u = f$ with appropriate boundary conditions.

The weak variational formulation for Eq (9.31) is obtained by the following three steps:

STEP 1. Multiply Eq (9.31) by a test function $w \, (\equiv \delta u)$ and integrate the product over the region Ω:

$$\iint_\Omega \left[\frac{\partial F}{\partial u} - \frac{\partial}{\partial x} \left(\frac{\partial F}{\partial p} \right) - \frac{\partial}{\partial y} \left(\frac{\partial F}{\partial q} \right) \right] w \, dx \, dy = 0. \tag{9.34}$$

The test function w is arbitrary, but it must satisfy the homogeneous essential boundary conditions (9.33a) on u.

STEP 2. Use formula (A.3) componentwise to the second and third terms in (9.34) for transferring the differentiation from the dependent variable u to the test function w, and identify the type of the boundary conditions admissible by the variational form:

$$\iint_\Omega \left[w \frac{\partial F}{\partial u} + \frac{\partial w}{\partial x} \frac{\partial F}{\partial p} + \frac{\partial w}{\partial y} \frac{\partial F}{\partial q} \right] dx \, dy - \int_{\Gamma = \Gamma_1 \cup \Gamma_2} \left(\frac{\partial F}{\partial p} n_x + \frac{\partial F}{\partial q} n_y \right) w \, ds = 0. \tag{9.35}$$

Note that the formula (A.3) does not apply to the first term in the integrand in (9.34). This step also yields boundary terms which determine the nature of the essential and natural boundary conditions for the problem. The general rule to identify the essential and natural boundary conditions for (9.31) is as follows: The essential boundary condition is prescribed on the dependent variable (u in this case), i.e.,

$$u = u_0 \quad \text{on} \quad \Gamma_1$$

is the essential boundary condition for (9.31). The test function w in the boundary integral (9.35) satisfies the homogeneous form of the same boundary condition as that prescribed on u. The natural boundary condition arises by specifying the coefficients of w and its derivatives in the boundary integral in (9.35). Thus,

$$\frac{\partial F}{\partial p} n_x + \frac{\partial F}{\partial q} n_y = q_0 \quad \text{on} \quad \Gamma_2$$

is the natural boundary condition in a Neumann boundary value problem. In one-dimensional problems, use integration by parts instead of the divergence formula (A.3).

To equalize the continuity requirements on u and w, the differentiation in the divergence formula (A.3) is transferred from F to w. It imparts weaker continuity requirements on the solution u in the variational problem than in the original equation.

STEP 3. Simplify the boundary terms by using the prescribed boundary conditions. This affects the boundary integral in (9.35), which is split into two terms, one on Γ_1 and the other on Γ_2:

$$\iint_\Omega \left[w \frac{\partial F}{\partial u} + \frac{\partial w}{\partial x} \frac{\partial F}{\partial p} + \frac{\partial w}{\partial y} \frac{\partial F}{\partial q} \right] dx\, dy - \int_{\Gamma_1 \cup \Gamma_2} \left(\frac{\partial F}{\partial p} n_x + \frac{\partial F}{\partial u_y} n_y \right) w\, ds = 0.$$

$$(9.36)$$

The integral on Γ_1 vanishes since $w = \delta u = 0$ on Γ_1. The natural boundary condition is substituted in the integral on Γ_2. Then (9.36) reduces to

$$\iint_\Omega \left[w \frac{\partial F}{\partial u} + \frac{\partial w}{\partial x} \frac{\partial F}{\partial p} + \frac{\partial w}{\partial y} \frac{\partial F}{\partial q} \right] dx\, dy - \int_{\Gamma_2} w\, q_0\, ds = 0. \qquad (9.37)$$

This is the weak variational form for the problem (9.31). We write (9.37) in terms of the bilinear and linear differential forms as

$$b(w, u) = l(w), \qquad (9.38)$$

where

$$b(w, u) = \iint_\Omega \left[\frac{\partial w}{\partial x} \frac{\partial F}{\partial p} + \frac{\partial w}{\partial y} \frac{\partial F}{\partial q} \right] dx\, dy,$$

$$l(w) = -\iint_\Omega w \frac{\partial F}{\partial u} dx\, dy + \int_{\Gamma_2} w q_0\, ds.$$

$$(9.39)$$

Formula (9.38) defines the weak variational form for Eq (9.31) subject to the boundary conditions (9.33). The quadratic functional associated with this variational form is given by

$$I(u) = \frac{1}{2}b(u, u) - l(u). \tag{9.40}$$

EXAMPLE 9.3. Consider the system of Navier-Stokes equations for a two-dimensional flow of a viscous, incompressible fluid (pressure-velocity fields):

$$u\frac{\partial u}{\partial x} + v\frac{\partial u}{\partial y} = -\frac{1}{\rho}\frac{\partial p}{\partial x} + \nu\left(\frac{\partial^2 u}{\partial x^2} + \frac{\partial^2 u}{\partial y^2}\right),$$

$$u\frac{\partial v}{\partial x} + v\frac{\partial v}{\partial y} = -\frac{1}{\rho}\frac{\partial p}{\partial y} + \nu\left(\frac{\partial^2 v}{\partial x^2} + \frac{\partial^2 v}{\partial y^2}\right),$$

$$\frac{\partial u}{\partial x} + \frac{\partial v}{\partial y} = 0,$$

in a region Ω, with boundary conditions $u = u_0$, $v = v_0$ on Γ_1, and

$$\nu\left(\frac{\partial u}{\partial x}n_x + \frac{\partial u}{\partial y}n_y\right) - \frac{1}{\rho}pn_x = \hat{t}_x,$$

$$\nu\left(\frac{\partial v}{\partial x}n_x + \frac{\partial v}{\partial y}n_y\right) - \frac{1}{\rho}pn_y = \hat{t}_y,$$

on Γ_2, where (u, v) denotes the velocity field, p the pressure, and $\hat{t}_x, \hat{t}_y$ the prescribed values of the secondary variables. Let w_1, w_2, w_3 be the test functions, one for each equation, such that w_1 and w_2 satisfy the essential boundary conditions on u and v, respectively, and w_3 does not satisfy any essential condition. Then

$$0 = \iint_\Omega \left[w_1\left(u\frac{\partial u}{\partial x} + v\frac{\partial u}{\partial y}\right) - \frac{p}{\rho}\frac{\partial w_1}{\partial x} + \nu\left(\frac{\partial w_1}{\partial x}\frac{\partial u}{\partial x} + \frac{\partial w_1}{\partial y}\frac{\partial u}{\partial y}\right)\right] dx\,dy$$

$$- \int_{\Gamma_2} w_1\hat{t}_x\, ds,$$

$$0 = \iint_\Omega \left[w_2\left(u\frac{\partial v}{\partial x} + v\frac{\partial v}{\partial y}\right) - \frac{p}{\rho}\frac{\partial w_2}{\partial y} + \nu\left(\frac{\partial w_2}{\partial x}\frac{\partial v}{\partial x} + \frac{\partial w_2}{\partial y}\frac{\partial v}{\partial y}\right)\right] dx\,dy$$

$$- \int_{\Gamma_1} w_2\hat{t}_y\, ds,$$

$$0 = \iint_\Omega w_3\left(\frac{\partial u}{\partial x} + \frac{\partial v}{\partial y}\right) dx\,dy.$$

Then

$$b((w_1, w_2, w_3), (u, v))$$

$$= \iint_\Omega \left[w_1 \left(u\frac{\partial u}{\partial x} + v\frac{\partial u}{\partial y} \right) + w_2 \left(u\frac{\partial v}{\partial x} + v\frac{\partial v}{\partial y} \right) + w_3 \left(u\frac{\partial u}{\partial x} + v\frac{\partial v}{\partial y} \right) \right] dx\, dy$$

$$+ \nu \iint_\Omega \left(\frac{\partial w_1}{\partial x}\frac{\partial u}{\partial x} + \frac{\partial w_1}{\partial y}\frac{\partial u}{\partial y} + \frac{\partial w_2}{\partial x}\frac{\partial v}{\partial x} + \frac{\partial w_2}{\partial y}\frac{\partial v}{\partial y} \right) dx\, dy$$

$$+ \frac{p}{\rho} \iint_\Omega \left(\frac{\partial w_1}{\partial x} + \frac{\partial w_2}{\partial y} \right) dx\, dy,$$

$$l(w_1, w_2, w_3) = \int_{\Gamma_2} \left(w_1 \hat{t}_x + w_2 \hat{t}_y \right) ds.$$

Note that the boundary integral in the linear form $l(w_1, w_2, w_3)$ has no term containing w_3. ∎

9.5. Galerkin Method

We discuss two frequently used methods, which are the Galerkin and the Rayleigh-Ritz methods, for obtaining approximate numerical solutions of boundary value problems. These methods give the same results for homogeneous boundary value problems. Consider the boundary value problem

$$L\, u = f \qquad \text{in } \Omega, \tag{9.41}$$

subject to the boundary conditions

$$u = g(s) \qquad \text{on } \Gamma_1, \tag{9.42}$$

$$\frac{\partial u}{\partial n} + k(s)\, u = h(s) \qquad \text{on } \Gamma_2, \tag{9.43}$$

where $\Gamma = \Gamma_1 \cup \Gamma_2$ is the boundary of the region Ω. Let us choose an approximate solution $\tilde{u}$ of the form

$$\tilde{u} = \sum_{i=1}^N c_i \phi_i. \tag{9.44}$$

An approximate solution does not, in general, satisfy the system (9.41)–(9.43). The residual (error) associated with an approximate solution is defined by

$$r(\tilde{u}) \equiv L\, \tilde{u} - f = L\left(\sum_{i=1}^N c_i \phi_i \right) - f. \tag{9.45}$$

Note that if u_0 is an exact solution of (9.41)–(9.43), then $r(u_0) = 0$. The Galerkin method requires that the residual be orthogonal with respect to the basis functions ϕ_i (also called the trial functions) used in (9.44), i.e.,

$$\langle r, \phi_i \rangle = 0. \tag{9.46}$$

Hence,

$$\iint_\Omega \{ L(\tilde{u}) - f \} \phi_i \, dx \, dy = 0, \qquad i = 1, \cdots, N, \tag{9.46a}$$

or

$$\sum_{j=1}^n c_j \iint_\Omega \phi_i \, L\phi_j \, dx \, dy = \iint_\Omega f \phi_i \, dx \, dy,$$

which in the matrix form is written as

$$[A] \{c\} = \{b\}, \tag{9.46b}$$

where

$$A_{ij} = \iint_\Omega \phi_i \, L\phi_j \, dx \, dy, \qquad b_i = \iint_\Omega f \phi_i \, dx \, dy. \tag{9.46c}$$

In the examples given below, we choose different values of N in (9.34) for the trial function $\tilde{u}$. There is some guidance from geometry for such choices; they also satisfy the essential conditions and exhibit the nature of the approximation solutions vis-á-vis the exact solutions (see §9.7 for some choices). However, the larger the N, the better the approximation becomes.

EXAMPLE 9.4. Consider the Poisson's equation

$$-\nabla^2 u \equiv -\left(\frac{\partial^2 u}{\partial x^2} + \frac{\partial^2 u}{\partial y^2} \right) = c, \quad 0 < x < a, \quad 0 < y < b,$$

such that $u = 0$ at $x = 0, a$ and $y = 0, b$. First, we choose the first-order approximate solution as

$$\tilde{u}_1^{(1)} = \alpha xy(x - a)(y - b).$$

Note that this choice satisfies all four Dirichlet boundary conditions. The Galerkin equation (9.46a) gives

$$\int_0^b \int_0^a \left[-2\alpha(y^2 - by + x^2 - ax) - c \right] xy(x - a)(y - b) \, dx \, dy = 0$$

which simplifies to

$$\frac{\alpha}{90} \left[a^3 b^3 (a^2 + b^2) \right] - \frac{a^3 b^3 c}{36} = 0.$$

Thus,

$$\alpha = \frac{5c}{2(a^2 + b^2)}.$$

Hence,

$$\tilde{u}_1^{(1)} = \frac{5c}{2(a^2 + b^2)} xy(x - a)(y - b).$$

Alternatively, we solve this problem by choosing the first-order approximate solution as

$$\tilde{u}_1^{(2)} = \sum_{j,k=1}^{N} \alpha_{jk} \sin \frac{j\pi x}{a} \sin \frac{k\pi y}{b},$$

which is an orthogonal trigonometric series with a finite number of terms. Note that $u_1^{(2)}$ satisfies the boundary conditions. Also note the orthogonality condition

$$\int_0^a \sin \frac{m\pi x}{a} \sin \frac{n\pi x}{a} \, dx = \begin{cases} 0, & m \neq n \\ a/2, & m = n. \end{cases}$$

The Galerkin equation (9.46a) in this case gives

$$\int_0^b \int_0^a \left[\alpha_{jk} \left(\frac{j^2 \pi^2}{a^2} + \frac{k^2 \pi^2}{b^2} \right) \sin \frac{j\pi x}{a} \sin \frac{k\pi y}{b} + c \right] \sin \frac{j\pi x}{a} \sin \frac{k\pi y}{b} \, dx \, dy = 0.$$

Hence,

$$\alpha_{jk} \frac{\pi^2}{4} \left(\frac{j^2}{a^2} + \frac{k^2}{b^2} \right) = \frac{c}{jk\pi^2}(1 - \cos j\pi)(1 - \cos k\pi),$$

or

$$\alpha_{jk} = \frac{4c(1 - \cos j\pi)(1 - \cos k\pi)a^2 b^2}{\pi^4 jk(b^2 j^2 + a^2 k^2)}.$$

Thus, this approximate solution is

$$\tilde{u}_1^{(2)} = \sum_{j,k=1}^{N} \frac{4a^2 b^2 c(1 - \cos j\pi)(1 - \cos k\pi)}{jk\pi^4 (a^2 k^2 + b^2 j^2)} \sin \frac{j\pi x}{a} \sin \frac{k\pi y}{b}.$$

If the number of terms in each sum is infinite, then $\tilde{u}_1^{(2)}$ becomes the exact solution u_0.

At the center point (a/2,b/2), we have

$$\tilde{u}_{1,c}^{(2)} = \sum_{j,k=1}^{N} \frac{4a^2 b^2 c(1 - \cos j\pi)(1 - \cos k\pi)}{jk\pi^4 (a^2 k^2 + b^2 j^2)} \sin \frac{j\pi}{2} \sin \frac{k\pi}{2}.$$

If $a = b$, then at the center point $(a/2, a/2)$

$$\tilde{u}_{1,c}^{(2)} = \sum_j \sum_k \frac{4a^2 c(1 - \cos j\pi)(1 - \cos k\pi)}{jk\pi^4(j^2 + k^2)} \sin \frac{j\pi}{2} \sin \frac{k\pi}{2}$$

$$= \frac{a^2 c}{\pi^4} \left[8 + \frac{8}{15} + \frac{8}{15} + \frac{8}{81} + \cdots \right] \equiv u_0$$

$$\approx \frac{36.64}{\pi^4} c \left(\frac{a}{2} \right)^2.$$

For the N-th approximation, the trial functions are chosen as $\phi_{jk}(x, y) = f_j(x) g_k(y)$, where

$$f_j(x) = x^j(x - a), \quad g_k(y) = y^k(y - b).$$

Then the N-th approximate solution is

$$\tilde{u}_N(x, y) = \sum_{j,k=1}^{N} \alpha_{jk} \phi_{jk}(x, y),$$

and the residual is

$$r = -c - \sum_{j,k=1}^{N} \left[f_j''(x) g_k(y) + f_j(x) g_k''(y) \right].$$

Hence, since the Galerkin method requires that $\langle \phi_{mn}, r \rangle = 0$ for $m, n = 1, 2, \cdots, N$, we get

$$0 = \int_0^a \int_0^b \left\{ -c - \sum_{j,k=1}^{N} \left[f_j''(x) g_k(y) + f_j(x) g_k''(y) \right] \right\} f_m(x) g_n(y) \, dx \, dy,$$

which after integration yields

$$\sum_{j,k=1}^{N} \alpha_{jk} \left[j \, p(j, m, a) \, q(k, n, b) + k \, p(k, n, b) \, q(j, m, a) \right]$$

$$+ c a^{m+2} b^{n+2} h(m, n) = 0, \quad (m, n = 1, 2, \cdots, N),$$

where

$$p(j, m, a) = a^{j+m+1} \left[\frac{j-1}{j+m-1} - \frac{2}{j+m} + \frac{j+1}{j+m+1} \right],$$

$$q(k, n, b) = b^{k+n+3} \left[\frac{1}{k+n+1} - \frac{2}{k+n+2} + \frac{1}{k+n+3} \right],$$

$$h(m, n) = \frac{1}{(m+1)(m+2)(n+1)(n+2)}.$$

The coefficients α_{jk} for $j, k = 1, 2, \cdots, N$ are determined from the above system of equations. The result for $\tilde{u}_1$ found earlier follows from this general case.

The trial functions $\phi_{jk}(x, y) = \sin \dfrac{j\pi x}{a} \sin \dfrac{k\pi y}{b}$, used in the approximation $\tilde{u}_1^{(2)}$, belong to the set of orthogonal functions obtained by solving the given boundary value problem by the separation of variables method (see Table 4.1, Dirichlet-Dirichlet case, and Exercise 5.23). ■

9.6. Rayleigh-Ritz Method

Consider the Poisson's equation $-\nabla^2 u = f$, with the homogeneous boundary conditions $u = 0$ on Γ_1 and $\partial u / \partial n = 0$ on Γ_2. Then the weak variational formulation leads to

$$I(u) = \iint_\Omega \left\{ \frac{1}{2} |\nabla u|^2 - fu \right\} dx\, dy = 0. \tag{9.47}$$

A generalization of the result in (9.47) for the case of the system $u\, Lu = f$ with the above homogeneous boundary conditions, where L is a linear self-adjoint and positive definite operator, leads to the functional

$$I(u) = \frac{1}{2} \iint_\Omega \{ uLu - 2fu \} \, dx\, dy. \tag{9.48}$$

Theorem 9.1. *If the operator L is self-adjoint and positive definite, then the unique solution of $Lu = f$ with homogeneous boundary conditions occurs at a minimum value of $I(u)$.*

An application of Theorem 9.1 is the Rayleigh-Ritz method, where we find the direct solution of the variational problem for the system $Lu = f$ by constructing minimizing sequences and securing the approximate solutions by a limiting process based on such sequences. Thus, we choose a complete set of linearly independent basis (test) functions ϕ_i, $i = 1, \cdots$, and then approximate the exact solution u_0 by taking the approximate solution $\tilde{u}$ in the form

$$\tilde{u} = \sum_{i=1}^{n} c_i \phi_i, \tag{9.49}$$

where the constants c_i are chosen such that the functional $I(\tilde{u})$ is minimized at each stage. If $\tilde{u} \to u_0$ as $n \to \infty$, then the method yields a convergent solution. At each stage the method reduces the problem to that of solving a set of linear algebraic equations. The details for the boundary value problem $-\nabla^2 u = f$ with homogeneous boundary conditions are as follows: Using (9.49) in the functional (9.48), we get

$$I(\tilde{u}) = I(c_1, \cdots, c_n)$$

$$= \iint_{\Omega} \left\{ \left(\frac{\partial \tilde{u}}{\partial x}\right)^2 + \left(\frac{\partial \tilde{u}}{\partial y}\right)^2 - 2\tilde{u}f \right\} dx\, dy$$

$$= \iint_{\Omega} \left\{ \left(\sum c_i \frac{\partial \phi_i}{\partial x}\right)^2 + \left(\sum c_i \frac{\partial \phi_i}{\partial y}\right)^2 - 2f \sum c_i \phi_i \right\} dx\, dy.$$

Thus,

$$I(c_i) = c_i^2 \iint_{\Omega} \left\{ \left(\frac{\partial \phi_i}{\partial x}\right)^2 + \frac{\partial \phi_i}{\partial y}\right)^2 \right\} dx\, dy$$

$$+ 2 \sum_{i \neq j} c_i c_j \iint_{\Omega} \left(\frac{\partial \phi_i}{\partial x} \frac{\partial \phi_j}{\partial x} + \frac{\partial \phi_i}{\partial y} \frac{\partial \phi_j}{\partial y} \right) dx\, dy - 2c_i \iint_{\Omega} \phi_i f\, dx\, dy.$$

Hence,

$$\frac{\partial I}{\partial c_i} = 2A_{ii}c_i + 2 \sum_{i \neq j} A_{ij}c_j - 2h_i, \tag{9.50}$$

and

$$A_{ij} = \iint_{\Omega} \left(\frac{\partial \phi_i}{\partial x} \frac{\partial \phi_j}{\partial x} + \frac{\partial \phi_i}{\partial y} \frac{\partial \phi_j}{\partial y} \right) dx\, dy, \tag{9.51}$$

$$h_i = \iint_{\Omega} \phi_i f\, dx\, dy. \tag{9.52}$$

Now, if we choose c_i such that $I(c_i)$ is a minimum (i.e., $\partial I / \partial c_i = 0$), then from (9.50) we get

$$\sum_{j=1}^{n} A_{ij}c_i = h_i, \quad i = 1, \cdots, n, \tag{9.53}$$

which in the matrix notation is

$$[A]\{c\} = \{h\}, \tag{9.54}$$

where the matrix $[A]$ has elements A_{ij} given by (9.51), $\{h\}$ has elements h_i given by (9.52), and $\{c\} = [c_1, \cdots, c_n]^T$. Note that (9.54) is a system of linear algebraic equations to be solved for the unknown parameter c_i, and $[A]$ is nonsingular if L is positive definite.

The Rayleigh-Ritz method is alternatively developed by solving for u the equation (9.38), where we require that w satisfy the homogeneous essential conditions only.

Then this problem is equivalent to minimizing the functional (9.40). In other words, we will find an approximate solution of (9.38) in the form

$$u_n = \sum_{j=1}^{n} c_j \phi_j + \phi_0, \tag{9.55}$$

where the functions ϕ_j, $j = 1, \ldots, n$, satisfy the homogeneous boundary conditions while the function ϕ_0 satisfies the nonhomogeneous boundary condition, and the coefficients c_j are chosen such that Eq (9.38) is true for $w = \phi_i$, $i = 1, \cdots, n$, i.e., $b(\phi_i, u_n) = l(\phi_i)$, or for $i = 1, \cdots, n$,

$$b\left(\phi_i, \sum_{j=1}^{n} c_j \phi_j + \phi_0\right) = l(\phi_i).$$

Thus,

$$\sum_{j=1}^{n} c_j b(\phi_i, \phi_j) = l(\phi_i) - b(\phi_i, \phi_0), \quad i = 1, \cdots, n. \tag{9.56}$$

This is a system of n linear algebraic equations in n unknowns c_j and has a unique solution if the coefficient matrix in (9.56) is nonsingular and thus has an inverse.

The functions ϕ_i must satisfy the following requirements: (i) ϕ_i be well defined such that $b(\phi_i, \phi_j) \neq 0$, (ii) ϕ_i satisfy at least the essential homogeneous boundary condition, (iii) the set $\{\phi_i\}_{i=1}^{n}$ be linearly independent, and (iv) the set $\{\phi_i\}_{i=1}^{n}$ be complete. The term ϕ_0 in the representation (9.55) is dropped if all boundary conditions are homogeneous.

EXAMPLE 9.5. Consider the Bessel equation

$$x^2 u'' + x u' + (x^2 - 1)u = 0, \quad u(1) = 1, u(2) = 2.$$

Set $u = v + x$. Then the given equation and the boundary conditions become

$$x^2 v'' + x v' + (x^2 - 1)v + x^3 = 0, \quad v(1) = 0 = v(2).$$

In the self-adjoint form this equation is written as

$$x v'' + v' + \frac{x^2 - 1}{x} v + x^2 = 0.$$

For the first approximation, we take

$$v_1 = a_1 \phi_1 = a_1 (x - 1)(x - 2).$$

Then using (9.46) we get $\int_1^2 (Lv_1 - f)\phi_1 \, dx = 0$, which gives

$$\int_1^2 \left[2a_1 x - (3 - 2x)a_1 + \frac{x^2 - 1}{x}(x - 1)(x - 2)a_1 + x^2 \right] (x - 1)(x - 2) \, dx = 0,$$

which, on integration, yields $a_1 = -0.811$, and thus,

$$u_1 = v_1 + x = -0.811(x - 1)(x - 2) + x.$$

The exact solution is $u = c_1 J_1(x) + c_2 Y_1(x)$, where $c_1 = 3.60756$, $c_2 = 0.75229$. A comparison with the exact solution in the following table shows that u_1 is a good approximation:

x	u_1	u_{exact}
1.3	1.4703	1.4706
1.5	1.7027	1.7026
1.8	1.9297	1.9294 ∎

EXAMPLE 9.6. Consider the fourth-order equation

$$[(x + 2l)u'']'' + bu - kx = 0, \quad 0 < x < l,$$

with the boundary conditions: $u(l) = 0 = u'(l)$, $(x + 2l)u''(0) = 0$, $[(x + 2l)u'']'(0) = 0$. We choose the test functions

$$\phi_1(x) = (x - l)^2(x^2 + 2lx + 3l^2),$$
$$\phi_2(x) = (x - l)^3(3x^2 + 4lx + 3l^2).$$

For the first approximation, we have $u_1 = a_1\phi_1(x)$. Then $\int_0^l (Lu_1 - f)\phi_1(x)\,dx = 0$, which gives

$$\left[a_1 \left(\frac{408}{5} + \frac{104}{45} bl^3 \right) - \frac{k}{3} \right] l^6 = 0.$$

If, for example, we take $l = 1 = b$ and $k = 3$, then $a_1 = 0.0119174$, and thus,

$$u_1 = 0.0119174\,(x - 1)^2(x^2 + 2x + 3).$$

For the second approximation, we take $u_2 = a_1\phi_1(x) + a_2\phi_2(x)$. Then

$$\int_0^l (Lu_2 - f)\phi_1\,dx = 0, \quad \text{and} \quad \int_0^l (Lu_2 - f)\phi_2(x)\,dx = 0,$$

which, with $l = b = 1$, $k = 3$, yield

$$\begin{bmatrix} 3776/45 & -21172/315 \\ -21172/315 & 63674/693 \end{bmatrix} \begin{Bmatrix} a_1 \\ a_2 \end{Bmatrix} = \begin{Bmatrix} 1 \\ -5/7 \end{Bmatrix}.$$

Thus, $a_1 = 0.013743$, $a_2 = 0.002279$, and

$$u_2 = 0.013743(x - 1)^2(x^2 + 2x + 3) + 0.002279(x - 1)^3(3x^2 + 4x + 3).$$

Instead of determining the exact solution, we can compare u_1 and u_2. Thus, e.g., $u_1(0.5) = 0.012662$, and $u_2(0.5) = 0.012964$, which give good results. ∎

9.7. Choice of Test Functions

The test functions $\phi_i(x, y)$ are suitably chosen by taking linear combinations of polynomials, or trigonometric functions, such that they satisfy the homogeneous boundary conditions. For example, we can choose a system of functions

$$\phi_0 = g, \quad \phi_1 = gx, \quad \phi_2 = gy, \quad \phi_3 = gx^2, \quad \phi_4 = gxy, \ldots, \tag{9.57}$$

where $g = g(x, y)$. The system (9.57) is complete.

Some practical rules for constructing the functions $g(x, y)$ in (9.57) are as follows:

(i) For the rectangle $[-a, a; -b, b]$:

$$g(x, y) = (x^2 - a^2)(y^2 - b^2).$$

(ii) For a circle of radius r and center at origin:

$$g(x, y) = r^2 - x^2 - y^2.$$

(iii) If the boundary Γ of a region Ω is defined by $F(x, y) = 0$, where $F \in C^n$, then

$$g(x, y) = \pm F(x, y).$$

See (ii) above if Γ is a circle.

(iv) For the case of a convex polygon whose sides are defined by $a_1 x + b_1 y + c_1 = 0, \cdots, a_m x + b_m y + c_m = 0$, we have

$$g(x, y) = \pm(a_1 x + b_1 y + c_1) \cdots (a_m x + b_m y + c_m).$$

See (i) above for a rectangle. (v) The choice in (iv) is also suitable in different types of regions bounded by curved lines; e.g., for a sector formed by the circles of radii r and $r/2$, as in Fig. 9.3, we have

$$g(x, y) = (r^2 - x^2 - y^2)(x^2 - rx + y^2).$$

(vi) For nonconvex polygons, the function $g(x, y)$ must be assigned piecewise in different parts of the region, and we must introduce moduli for any re-entrant angles. Thus, for the region in Fig. 9.4,

$$g(x, y) = \big(|x| + |y| - x - y\big)(x + p)(y + q)(l - x)(h - y)$$

$$= \begin{cases} -2y(x + p)(y + q)(l - x)(h - y), & \text{in } [0, l; -q, 0] \\ -2(x + y)(x + p)(y + q)(l - x)(h - y), & \text{in } [-p, 0; -q, 0] \\ -2x(x + p)(y + q)(l - x)(h - y) & \text{in } [-p, 0; 0, h]. \end{cases}$$

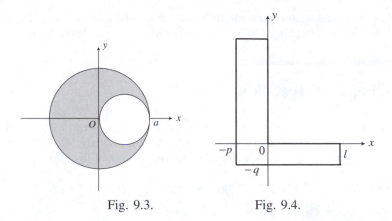

Fig. 9.3.   Fig. 9.4.

As a second choice we take

$$g(x, y) = \left(x^2 + y^2 - x|x| - y|y|\right)(x + p)(y + q)(l - x)(h - y).$$

In this case $g \in C^1$. A third choice is to assign the functions $u_n(x, y)$ separately in the three parts of the corner region of Fig. 9.4:

$$u_n(x, y) = \begin{cases} (x + p)x(h - y)(a_1 + a_2x + a_3y + \cdots + a_ny^m) & \text{in } [-p, 0; 0, h] \\ (x + p)(y + q)(b_1 + b_2x + b_3y + \cdots + b_ny^m) & \text{in } [-p, 0; -q, 0] \\ (y + q)(l - x)(c_1 + c_2x + c_3y + \cdots + c_ny^m) & \text{in } [0, l; -q, 0], \end{cases}$$

where a_k, b_k, c_k ($k = 1, \cdots, n$) are parameters which must be connected by the following condition on the axes $x = 0$ and $y = 0$:

$$(x + p)xh(a_1 + a_2x + a_4x^2 + \cdots) = (x + p)q(b_1 + b_2x + b_4x^2 + \cdots)$$
$$p(y + q)(b_1 + b_2y + \cdots + b_ny^m) = (y + q)yl(c_1 + c_2y + \cdots + c_ny^m).$$

In view of the above considerations, the test functions $\phi_i(x, y)$ are also called the shape functions for the region Ω.

EXAMPLE 9.7. The torsion of a prismatic rod of rectangular cross section of length $2a$ and width $2b$ is defined by

$$\nabla^2 u = 2, \quad u = 0 \quad \text{on } \Gamma.$$

For the rectangle shown in Fig. 9.5, we choose

$$\phi(x, y) = (a^2 - x^2)(b^2 - y^2),$$

and seek an approximate solution of the form

$$u_n(x, y) = (a^2 - x^2)(b^2 - y^2)\left(A_1 + A_2x^2 + A_3y^2 + \cdots + A_nx^{2i}y^{2j}\right).$$

First, for $n = 1$, we use (9.46a) with $f = 2$ and $\phi(x) = (a^2 - x^2)(b^2 - x^2)$, and find that

$$
\begin{aligned}
0 &= \iint_\Omega \left(-\frac{\partial^2 u_1}{\partial x^2} - \frac{\partial^2 u_1}{\partial y^2} - 2 \right) \phi \, dx \, dy \\
&= 2 \int_{-a}^{a} \int_{-b}^{b} \left[1 - A_1(a^2 - x^2) - A_1(b^2 - y^2) \right] (a^2 - x^2)(b^2 - y^2) \, dy \, dx \\
&= -\frac{128}{45} a^3 b^3 (a^2 + b^2) A_1 + \frac{32}{9} a^2 b^3.
\end{aligned}
$$

Thus, $A_1 = 5/(4(a^2 + b^2))$, and

$$
u_1 = \frac{5}{4} \frac{(a^2 - x^2)(b^2 - y^2)}{a^2 + b^2}.
$$

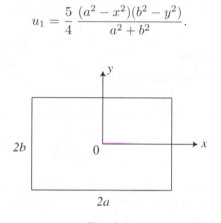

Fig. 9.5.

The torsional moment is given by

$$
M = 2G\theta \int_{-a}^{a} \int_{-b}^{b} u_1 \, dy \, dx = \frac{40}{9} \frac{G\theta a^3 b^3}{a^2 + b^2},
$$

where G is the shear modulus, and θ is the angle of twist per unit length. The tangential stresses τ_{zx} and τ_{zy} are given by

$$
\tau_{zx} = G\theta \frac{\partial u_1}{\partial y}, \quad \tau_{zy} = G\theta \frac{\partial u_1}{\partial x}.
$$

For $a = b$, we find that $M = 20G\theta a^4/9 \approx 0.1388(2a)^4 G\theta$. The exact classical solution is given by

$$
u = ax - x^2 - \frac{8a^2}{\pi^3} \sum_{n=1}^{\infty} \frac{\cosh \frac{(2n-1)\pi y}{2a}}{(2n-1)^3 \cosh \frac{(2n-1)\pi b}{a}} \sin \frac{(2n-1)\pi x}{a},
$$

which yields

$$M = 2G\theta \left\{ \frac{a^3 b}{6} - \frac{32a^4}{\pi^5} \sum_{n=1}^{\infty} \frac{1}{(2n-1)^5} \tanh \frac{(2n-1)\pi b}{2a} \right\}.$$

For $a = b$, the exact value of M is $0.1406(2a)^4 G\theta$, which compares very well with the approximate value obtained above by the Galerkin method. ∎

EXAMPLE 9.8. Solve $\nabla^4 u = 0$ on the rectangle $[-a, a; -b, b]$ (Fig. 9.5) under the boundary conditions

$$\frac{\partial^2 u}{\partial x \partial y} = 0, \quad \frac{\partial^2 u}{\partial y^2} = c\left(1 - \frac{y^2}{b^2}\right) \quad \text{at } x = \pm a,$$

$$\frac{\partial^2 u}{\partial x \partial y} = 0, \quad \frac{\partial^2 u}{\partial x^2} = 0 \quad \text{at } y = \pm b,$$

where c is a constant. This problem pertains to the expansion of a rectangular plate under tensile forces.

First, we reduce the above boundary conditions to homogeneous boundary conditions: The function

$$u_0 = \frac{c}{2} y^2 \left(1 - \frac{y^2}{6b^2}\right)$$

obviously satisfies the given boundary conditions (it follows by integrating each one of the above boundary conditions). Set $u = u_0 + \hat{u}$. Then $\nabla^4 \hat{u} = 2c/b^2$, and the boundary conditions become

$$\frac{\partial \hat{u}}{\partial x \partial y} = 0, \quad \frac{\partial \hat{u}}{\partial y^2} = 0 \quad \text{at } x = \pm a,$$

$$\frac{\partial^2 \hat{u}}{\partial x \partial y} = 0, \quad \frac{\partial^2 \hat{u}}{\partial x^2} = 0 \quad \text{at } y = \pm b.$$

These boundary conditions are satisfied if the following conditions are met:

$$\hat{u} = 0, \quad \frac{\partial \hat{u}}{\partial x} = 0 \quad \text{at } x = \pm a,$$

$$\hat{u} = 0, \quad \frac{\partial \hat{u}}{\partial y} = 0 \quad \text{at } y = \pm b.$$

Thus, the given problem reduces to that of minimizing the integral

$$I(u) = \iint_{\Omega} \left[(\nabla^2 \hat{u})^2 - \frac{4c}{b^2} \hat{u} \right] dx\, dy.$$

Then, by Rayleigh-Ritz (or Galerkin) method, we have

$$\iint_\Omega \left(\nabla^4 u_n - f \right) \phi_j \, dx \, dy = 0, \quad j = 1, \ldots, n, \tag{9.58}$$

where u_n is the n-th approximate solution, which, in view of the geometric symmetry of the rectangle, is taken as

$$u_n = (x^2 - a^2)^2 (y^2 - b^2)^2 (a_1 + a_2 x^2 + a_3 y^2 + \cdots).$$

For $n = 1$, we find from (9.58) that

$$\int_{-a}^{a} \int_{-b}^{b} [24a_1(y^2 - b^2)^2 + 16a_1(3x^2 - a^2)(3y^2 - b^2)$$

$$+24a_1(x^2 - a^2)^2 - \frac{2c}{b}](x^2 - a^2)^2 (y^2 - b^2)^2 \, dy \, dx = 0,$$

or

$$\left(\frac{54}{7} + \frac{256}{49} \frac{b^2}{a^2} + \frac{64}{7} \frac{b^4}{a^4} \right) a_1 = \frac{c}{a^4 v^2},$$

which gives $a_1 = 0.043253 c / a^6$, and

$$u_1 = u_0 + \hat{u}_1 = \frac{c}{2} y^2 \left(1 - \frac{y^2}{6b^2} \right) + \frac{0.04253c}{a^6} (x^2 - a^2)^2 (y^2 - b^2)^2. \ \blacksquare$$

9.8. Transient Problems

For time-dependent problems the semi-discrete formulation is used to choose the basis functions. Thus, for one-dimensional problems the N-th approximate solution is taken as

$$\tilde{u}_N(x, t) = \phi_0 + \sum_{j=1}^{N} c_j(t) \, \phi_j(x), \tag{9.59}$$

where, as before, the functions ϕ_j satisfy the homogeneous boundary conditions and ϕ_0 is chosen as in (9.55). Then, using the Galerkin or Rayleigh-Ritz method such that the residual is orthogonal to the first N basis functions ϕ_i, $i = 1, 2, \cdots, N$, we obtain the N first-order ordinary differential equations in t. For example, for the diffusion equation $u_t = \nabla^2 u$, this system is

$$\sum_{j=1}^{N} \dot{c}_j(t) \langle \phi_j, \phi_i \rangle = \sum_{j=1}^{N} c_j(t) \langle \phi_j, \phi_i \rangle + \langle \phi_j, \phi_0 \rangle,$$

where the dot denotes the time derivative. The initial conditions for this system are subject to another Galerkin approximation such that its residual $R = u(x, 0) - \tilde{u}_N(x, 0)$ is orthogonal to the first N basis functions ϕ_j. This yields the system of N algebraic equations

$$\sum_{j=1}^{N} c_j(0) \langle \phi_j, \phi_i \rangle = \langle \phi_j, u(x, 0) - \phi_0(r) \rangle,$$

which is generally solved for the unknowns $c_i(0)$ by numerical methods.

EXAMPLE 9.9. Consider the heat conduction equation

$$\frac{\partial u}{\partial t} = \frac{\partial^2 u}{\partial r^2} + \frac{1}{r}\frac{\partial u}{\partial r}, \quad 0 < r < 1,$$

subject to the boundary conditions $u(1, t) = 0 = u_r(0, t)$ and the initial condition $u(r, 0) = \ln r$. For a general formulation , we will first consider the annular region $a < r < 1$ $(a > 0)$, and then let $a \to 0$. Now the boundary conditions become $u(1, t) = 0 = u_r(a, t)$. For the first-order approximation, we take the basis function as $\phi_1(r) = c_0 + c_1 r + c_2 r^2$. To determine the coefficients c_0, c_1, and c_2, we require that ϕ_1 satisfies the above boundary conditions. Thus, $\phi_1(1) = c_0 + c_1 + c_2 = 0$, and $\dfrac{\partial \phi_1(a)}{\partial r} = c_1 + 2c_2 a = 0$. By solving these two equations in terms of c_0, we find that $c_1 = 2ac_0/(1 - 2a)$, and $c_2 = -c_0/(1 - 2a)$. If we take $c_0 = 1 - 2a$, then $c_1 = 2a$, and $c_2 = -1$, and the basis function becomes $\phi_1(r) = 1 - 2a + 2ar - r^2$, or $\phi_1(r) = 1 - b + br - r^2$, with $b = 2a$. This suggests that for the N-th approximation we should choose the basis functions as

$$\phi_j(r) = 1 - b_j + b_j r - r^{j+1}, \quad b_j = (j+1)a^j, \quad j = 1, 2, \cdots, N,$$

with $\phi_0 = 0$. The N-th order approximate solution is then taken in the semi-discrete form as

$$\tilde{u}_N(r, t) = \sum_{j=1}^{N} c_j(t)\, \phi_j(r).$$

The residual is given by

$$\sum_{j=1}^{N} \left\{ \dot{c}_j(t)\, \phi_j + c_j(t) \left[(j+1)^2 r^{j-1} - \frac{b_j}{r} \right] \right\}.$$

Then for the Galerkin method, as $a \to 0$, we have

$$0 = \sum_{j=1}^{N} \int_0^1 \left\{ \dot{c}_j(t)\, \phi_j + c_j(t) \left[(j+1)^2 r^{j-1} - \frac{b_j}{r} \right] \right\} \left(1 - b_i + b_i r - r^{i+1} \right\} r\, dr$$

$$= \sum_{j=1}^{N} \int_0^1 \left\{ \dot{c}_j(t)\, f(i, j) + c_j(t)\, g(i, j) \right\},$$

for $i = 1, 2, \cdots, N$, where

$$f(i, j) = \frac{(1 - b_i)(1 - b_j)}{2} + \frac{b_i + b_j - b_i b_j}{3} + \frac{b_i b_j}{4} - \frac{1 - b_j}{i + 3}$$
$$- \frac{1 - b_i}{j + 3} - \frac{b_j}{i + 4} - \frac{b_i}{j + 4} + \frac{1}{i + j + 4},$$
$$g(i, j) = (j + 1)^2 \left[\frac{1 - b_i}{j + 1} + \frac{b_i}{j + 2} - \frac{1}{i + j + 2} \right] - b_j + \frac{b_i b_j}{2} + \frac{b_j}{i + 2}.$$

The initial condition is, in general, satisfied approximately. This is accomplished by requiring that the residual

$$R = \sum_{j=1}^{N} c_j(0)\, \phi_j(r) - \ln r$$

be orthogonal to the basis functions $\phi_i(r)$, i.e., $\langle \phi_i, R \rangle = 0$ for $i = 1, 2, \cdots, N$. This means that

$$\lim_{\varepsilon \to 0} \int_{\varepsilon}^{1} R\, \phi_i(r)\, r\, dr = 0 \quad \text{for } i = 1, 2, \cdots, N,$$

since R has a logarithmic singularity at $r = 0$. After evaluating this improper integral, we obtain a system of N algebraic equations:

$$\sum_{j=1}^{N} c_j(0) f(i, j) = \frac{1}{4} + \frac{5}{36} b_j + \frac{1}{(i + 3)^2}, \quad i = 1, 2, \cdots, N,$$

which is solved for the unknowns $c_j(0)$. ∎

Note that there are other weighted residual methods, like the collocation method, least-squares method, and the method of moments, which are sometimes used, but we will not discuss them. Interested readers can find detailed information on these methods in Connor and Brebbia (1973), Davies (1980), Kantorovitch and Krylov (1958), and Reddy (1984).

9.9. Mathematica Projects

PROJECT 9.1. For this chapter, refer to the Mathematica Notebook galerkin.nb.

9.10. Exercises

9.1. Fill in the details in the derivation of the Euler equation (9.13).

ANS. From (9.11), integrating by parts twice and using $\phi(x)\dfrac{\partial F}{\partial U'}\Big|_a^b = 0$,

$$\phi'(x)\frac{\partial F}{\partial U'} - \phi(x)\frac{d}{dx}\left(\frac{\partial F}{\partial U''}\right)\Big|_a^b = 0, \text{ we get}$$

$$0 = \frac{\partial I}{\partial \alpha} = \int_a^b \phi(x)\left[\frac{\partial F}{\partial u} - \frac{d}{dx}\left(\frac{\partial F}{\partial u'}\right) + \frac{d^2}{dx^2}\left(\frac{\partial F}{\partial u''}\right)\right] dx.$$

9.2. Fill in the details in the derivation of the Euler equation (9.15).

ANS. Introduce two functions $\phi(x)$ and $\psi(x)$ and two parameters α and β, respectively, such that $U = u + \alpha\phi(x)$, $V = v + \beta\psi(x)$. Then $\partial I/\partial \alpha = 0$ and $\partial I/\partial \beta = 0$ lead to (9.15).

9.3. Find the geodesics for the following problems:

(a) On the xy-plane, take $I = \displaystyle\int ds = \int \sqrt{1+y'^2}\,dx$.

(b) On the xy-plane, take $I = \displaystyle\int ds = \int \sqrt{1+r^2(d\theta/dr)^2}\,dr$.

(c) On the cylinder $x^2 + y^2 = a^2$, $-\infty < z < \infty$, take $x = a\cos t$, $y = a\sin t$, and $I = \displaystyle\int ds = \int \sqrt{a^2 + (dz/dt)^2}\,dt$.

ANS. (a) Straight lines $y = c_1 x + c_2$; (b) Straight lines $r\cos(\theta - c_1) = c_2$; (c) $z = c_1 t + c_2$.

9.4. A ray of light moves between two fixed points in the xy-plane with variable velocity $v(x, y)$. By Fermat's law, its travel time is $\displaystyle\int \frac{ds}{v} = \int \frac{\sqrt{1+y'^2}}{v}\,dx$. Show that the paths for a minimum travel time are given by

$$\frac{vy''}{\sqrt{1+y'^2}} - \frac{\partial v}{\partial x}y' + \frac{\partial v}{\partial y} = 0.$$

9.5. Find the extremal when the following integral is minimized:

(a) $\displaystyle\int (y'^2 + y^2)\,dx$, and (b) $\displaystyle\int (y''^2 + y^2)\,dx$.

ANS. (a) $y = c_1 e^x + c_2 e^{-x}$;

(b) $y = \left(c_1 e^{x/\sqrt{2}} + c_2 e^{-x/\sqrt{2}}\right)\cos(y/\sqrt{2}) + \left(c_3 e^{x/\sqrt{2}} + c_4 e^{-x/\sqrt{2}}\right)\sin(y/\sqrt{2}).$

9.6. If the end points are not fixed, derive

$$\int_a^b F(x, u, u')\,dx = \delta u\frac{\partial F}{\partial u'}\Big|_a^b + \int_a^b \left[\frac{\partial F}{\partial u} - \frac{d}{dx}\left(\frac{\partial F}{\partial u'}\right)\right]\delta u\,dx.$$

9.7. Take $\delta u = \alpha\phi(x) + \beta\psi(x)$ in the integral (9.1), and let δI denote the total differential of I at $\alpha = 0 = \beta$, where $\delta\alpha = \alpha$, $\delta\beta = \beta$. Set $H \equiv \dfrac{\partial F}{\partial u} - \dfrac{d}{dx}\left(\dfrac{\partial F}{\partial u'}\right)$, and show that the variation δI in this case is

$$\delta I = d\alpha \int_a^b H\phi \, dx + d\beta \int_a^b H\psi \, dx.$$

9.8. Find the extremal for the problem of determining a curve Γ of prescribed length l joining AB and maximizing the area $A = \int y \, dx$, bounded by Γ, x-axis and two fixed ordinates.

ANS. $(x - c_1)^2 + (y - c_2)^2 = k^2$, where c_1, c_2, and k make the arc Γ pass through A and B and have length l.

Derive the variational formulation for the boundary value problems 9.9– 9.14 (here a, b, f, g are functions of x; u_0, h_0, m_0, q_0, T_∞, u_∞ are constants):

9.9. $-\dfrac{d}{dx}\left(a\dfrac{du}{dx}\right) - f = 0,\ u(0) = u_0,\ a\dfrac{du}{dx}(l) = q_0,\ 0 < x < l$

(one-dimensional heat conduction).

ANS. $b(w, u) = \displaystyle\int_0^1 a\dfrac{dw}{dx}\dfrac{du}{dx}\,dx,$

$l(w) = \displaystyle\int_0^1 wf\,dx - w(0)\left[a\dfrac{du}{dx}\right]_{x=0} + q_0 w(l).$

9.10. $-\dfrac{d}{dx}\left(a\dfrac{du}{dx}\right) - cu + x^2 = 0;\ u(0) = 0,\ a\dfrac{du}{dx}(1) = 1,\ 0 < x < 1$

(one-dimensional deformation of a bar).

ANS. $b(w, u) = \displaystyle\int_0^1 \left(a\dfrac{dw}{dx}\dfrac{du}{dx} - cwu\right) dx,\ l(w) = -\int_0^1 wx^2\,dx + w(1).$

9.11. $-\dfrac{\partial}{\partial x}\left(c_{11}\dfrac{\partial u}{\partial y} + c_{12}\dfrac{\partial u}{\partial x}\right) - \dfrac{\partial}{\partial y}\left(c_{21}\dfrac{\partial u}{\partial x} + c_{22}\dfrac{\partial u}{\partial y}\right) + f = 0$ in Ω with boundary conditions $u = u_0$ on Γ_1, and $q_n \equiv \left(c_{11}\dfrac{\partial u}{\partial x} + c_{12}\dfrac{\partial u}{\partial y}\right)n_x + \left(c_{21}\dfrac{\partial u}{\partial x} + c_{22}\dfrac{\partial u}{\partial y}\right)n_y = q_0$, on Γ_2, where c_{ij}, u_0, and q_0 are prescribed.

ANS. $B(w, u) = \displaystyle\iint_\Omega \left[\dfrac{dw}{dx}\left(c_{11}\dfrac{du}{dx} + c_{12}\dfrac{du}{dy}\right) + \dfrac{dw}{dx}\left(c_{21}\dfrac{du}{dx} + c_{22}\dfrac{du}{dy}\right)\right.$

$\left. + wf\right] dx\,dy,\quad l(w) = \displaystyle\int_\Gamma wq_n\,ds,$

where $q_n = \left(c_{11}\dfrac{du}{dx} + c_{12}\dfrac{du}{dy}\right)n_x + \left(c_{21}\dfrac{du}{dx} + c_{22}\dfrac{du}{dy}\right)n_y.$

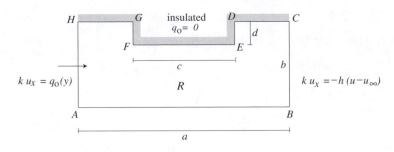

Fig. 9.6.

9.12. $-k\left(\dfrac{\partial^2 T}{\partial x^2} + \dfrac{\partial^2 T}{\partial y^2}\right) = f$ in the region Ω with boundary conditions as shown in Fig. 9.6. The following boundary conditions are prescribed: $ku_x = q_0(y)$ on HA; $ku_x = -h(u - u_\infty)$ on BC; $u = u_0(x)$ on AB, and $\partial u/\partial n = q_0 = 0$ on $CDEFGH$ (insulated), where k is the thermal conductivity of the material of the region Ω, h and u_∞ are ambient quantities, and $\partial u/\partial n = -\partial u/\partial x = -u_x$ on HA (two-dimensional heat conduction).

ANS. $0 = -\displaystyle\iint_\Omega k\left(\dfrac{d^2 T}{dx^2} + \dfrac{d^2 T}{dy^2}\right) w\, dx\, dy,$

$= \displaystyle\iint_\Omega k\left(\dfrac{dw}{dx}\dfrac{dT}{dx} + \dfrac{dw}{dy}\dfrac{dT}{dy}\right) dx\, dy - \int_C kw\left(\dfrac{dT}{dx}n_x + \dfrac{dT}{dy}n_y\right) ds.$

The boundary conditions on $C_1 = AB$ (prescribed temperature T_0): $n_x = 0$, $n_y = -1$; on $C_2 = BC$ (convective boundary, T_∞): $n_x = 1$, $n_y = 0$; on $C_3 = CDEFGH$ (insulated boundary): $q = \partial T/\partial n = 0$; and on $C_4 = HA$ (prescribed conduction $q_0(y)$): $n_x = -1$, $n_y = 0$. Thus,

$$b(w, u) = \iint_\Omega k\left(\frac{dw}{dx}\frac{dT}{dx} + \frac{dw}{dy}\frac{dT}{dy}\right) dx\, dy + h\int_0^b w(a, y)T(a, y)\, dy,$$

$$l(w) = -\int_0^b w(0, y)q_0(y)\, dy + hT_\infty\int_0^b w(a, y)\, dy.$$

9.13. $-\dfrac{d}{dx}\left\{a\left[\dfrac{du}{dx} + \dfrac{1}{2}\left(\dfrac{dv}{dx}\right)^2\right]\right\} + g = 0,$

$\dfrac{d^2}{dx^2}\left(b\dfrac{d^2 v}{dx^2}\right) - \dfrac{d}{dx}\left\{a\dfrac{dv}{dx}\left[\dfrac{du}{dx} + \dfrac{1}{2}\left(\dfrac{dv}{dx}\right)^2\right]\right\} + f = 0;$

$u = v = 0$ at $x = 0, l$; $\left.\dfrac{dv}{dx}\right|_{x=0} = 0$, $\left[b\dfrac{d^2 v}{dx^2}\right]_{x=l} = m_0$ (large-deflection bending of a beam).

ANS. Let w_1 and w_2 be the two test functions, one for each equation, such that they satisfy the essential boundary conditions on u and v. Then

$$0 = \int_0^l \left[a\frac{dw_1}{dx}\left\{ \frac{du}{dx} + \frac{1}{2}\left(\frac{dv}{dx}\right)^2 \right\} + w_1 g \right] dx,$$

$$0 = \int_0^l \left[b\frac{d^2 w_2}{dx^2}\frac{d^2 v}{dx^2} + a\frac{dw_2}{dx}\frac{dv}{dx}\left\{ \frac{du}{dx} + \frac{1}{2}\left(\frac{dv}{dx}\right)^2 \right\} + w_2 f \right] dx$$

$$- m_0 \frac{dw_2}{dx}(l).$$

Thus,

$$b((w_1, w_2), (u, v)) = \int_0^l \left[a\frac{dw_1}{dx}\left\{ a\frac{du}{dx} + \frac{1}{2}\left(\frac{dv}{dx}\right)^2 \right\} + b\frac{d^2 w_2}{dx^2}\frac{d^2 v}{dx^2} \right.$$

$$\left. + a\frac{dw_2}{dx}\frac{dv}{dx}\left\{ \frac{du}{dx} + \frac{1}{2}\left(\frac{dv}{dx}\right)^2 \right\} \right] dx,$$

$$l((w_1, w_2)) = -\int_0^l \left(w_1 g + w_2 f \right) dx + m_0 \frac{dw_2}{dx}(l),$$

$$I[(u, v)] = \int_0^l \left[\frac{1}{2}\left\{ a\left(\frac{du}{dx}\right)^2 + b\left(\frac{d^2 v}{dx^2}\right)^2 + a\frac{du}{dx}\left(\frac{dv}{dx}\right)^2 + \frac{a}{4}\left(\frac{dv}{dx}\right)^4 \right\} \right.$$

$$\left. + w_1 g + w_2 f \right] dx - m_0 \frac{dw_2}{dx}(l).$$

9.14. Find the functional $I(u)$ for the transverse deflection u of a membrane stretched across a frame, in the shape of a curve C, subjected to a pressure loading $f(x, y)$ per unit area. Assume that the tension T in the membrane is constant. Note that u satisfies the equation

$$-\nabla^2 u = \frac{f}{T}.$$

ANS. The variation of the total work done by the force f/T is

$$\delta \iint_\Omega \frac{fu}{T}\, dx\, dy = \iint_\Omega \frac{f\delta u}{T}\, dx\, dy = -\iint_\Omega \nabla^2 u\, \delta u\, dx\, dy$$

$$= -\iint_\Omega [\nabla \cdot (\nabla u \delta u) - \nabla(\delta u)\cdot \nabla u]\, dx\, dy$$

$$= \frac{1}{2}\iint_\Omega \delta|\nabla u|^2\, dx\, dy - \int_C \frac{\partial u}{\partial n}\delta u\, ds,$$

which leads to

$$I[u] = \frac{1}{2}\iint_\Omega \{|\nabla u|^2 - 2fu\}\, dx\, dy.$$

Note that Example 9.1 is used. Thus,

$$\nabla \cdot (\nabla u \delta u) = \left(\hat{\mathbf{i}} \frac{\partial}{\partial x} + \hat{\mathbf{j}} \frac{\partial}{\partial y} \right) \cdot \left(\hat{\mathbf{i}} \frac{\partial u}{\partial x} \delta u + \hat{\mathbf{j}} \frac{\partial u}{\partial y} \delta u \right)$$

$$= \nabla^2 u \delta u = -\frac{f}{T} \delta u.$$

9.15. Consider the Poisson's boundary value problem: $-\nabla^2 u = f$ in Ω, with the boundary conditions $u = 0$ on C_1 and $\partial u / \partial n = 0$ on C_2. Show that

$$I(u) = \frac{1}{2} \iint_\Omega \left\{ \left(\frac{\partial u}{\partial x} \right)^2 + \left(\frac{\partial u}{\partial y} \right)^2 - 2 f u \right\} dx\, dy.$$

ANS. $b(w, u) = \displaystyle\iint_\Omega \left(\frac{\partial w}{\partial x} \frac{\partial u}{\partial x} + \frac{\partial w}{\partial y} \frac{\partial u}{\partial y} \right) dx\, dy,$

$l(w) = \displaystyle\iint_\Omega fw\, dx\, dy - \int_C w \frac{\partial u}{\partial n} ds.$

9.16. Use both Galerkin and Rayleigh-Ritz methods to solve:

$$\frac{d^2}{dx^2} \left(EI \frac{d^2 u}{dx^2} \right) + f = 0, \quad 0 < x < l, \ EI > 0, \ f = \text{const},$$

where EI is called the flexural rigidity of the beam, with

$$u(0) = 0 = \frac{du}{dx}(0), \quad EI \frac{d^2 u}{dx^2} \Big|_{x=l} = M_0, \quad \frac{d}{dx} \left(EI \frac{d^2 u}{dx^2} \right) \Big|_{x=l} = 0.$$

HINT. Take $w = \phi_i = x^{i+1}$. The exact solution is

$$EI\, u = \frac{fl^4}{24} - \frac{fl^3}{6} x + \frac{M_0}{2} x^2 - \frac{f}{24} (l - x)^4.$$

ANS. $b(w, u) = \displaystyle\int_0^l EI \frac{d^2 w}{dx^2} \frac{d^2 u}{dx^2} dx,$

$l(w) = -\displaystyle\int_0^l wf\, dx + w(0) \left[\frac{d}{dx} \left(EI \frac{dw}{dx} \frac{du}{dx} \right) \right]_{x=0}$

$\qquad - \left[\frac{dw}{dx} \right]_{x=0} \left[EI \frac{du}{dx} \right]_{x=0} - f_0 w(l) + m_0 \left[\frac{dw}{dx} \right]_{x=l}.$

Exact solution is obtained by direct integration.

9.17. $-\nabla^2 u = 1$ in $\Omega = \{(x, y) : 0 < x, y < 1\}$ such that

$$u(1, y) = 0 = u(x, 1), \quad \frac{\partial u}{\partial n}(0, y) = 0 = \frac{\partial u}{\partial n}(x, 0).$$

ANS. If we take $w = \phi_i = (1 - x^i)(1 - y^i), i = 1, \cdots, n$, then this choice satisfies the essential boundary conditions but not the natural boundary conditions. Hence, we assume the first approximate solution as $u_1 = a\phi_1$, $\phi_1 = (1-x^2)(1-y^2)$. Alternatively, we can take $w = \phi_i = \cos\dfrac{(2i-1)\pi x}{2}\cos\dfrac{(2i-1)\pi y}{2}$, $i = 1, \cdots, n$. The exact solution is

$$u(x, y) = \frac{1}{2}\Big\{(1 - y^2)$$
$$+ \frac{32}{\pi^3}\sum_{k=1}^{\infty}\frac{(-1)^k\cos[(2k-1)\pi y/2]\cosh[(2k-1)\pi x/2]}{(2k-1)^3\cosh[(2k-1)\pi/2]}\Big\}.$$

9.18. Find the N-th approximate solution of Example 9.4 by taking the basis functions as $\phi_{jk}(x, y) = \sin\dfrac{j\pi x}{a}\sin\dfrac{k\pi y}{b}$.
ANS.

$$\sum_{j,k=1}^{N}\alpha_{jk}\frac{ab\pi^2}{4}\left(\frac{j^2}{a^2} + \frac{k^2}{b^2}\right) = \frac{4cab}{jk\pi^2}$$

for both j, k odd.

9.19. Find an approximate solution by the Galerkin method for the nonlinear problem $u_t = u_{xx} + \varepsilon u^2$ on $0 < x < 1$, subject to the boundary conditions $u(0, t) = 0 = u(1, t)4$ and the initial condition $u(x, 0) = 1$.
HINT. Choose $\phi_j(x) = \sin j\pi x$.

ANS. $\displaystyle\sum_{j,k=1}^{N}\Big\{\frac{1}{2}\dot{c}_j(t) + \frac{1}{2}j^2\pi^2 c_j(t) - \varepsilon\Big[\frac{2(1 - \cos j\pi)}{3j\pi}c_j^2(t)$

$$+ \frac{1}{2\pi}\sum_{m\neq n}c_m(t)f(m, n, j)\Big]\Big\}, \quad m, n = 1, 2, \cdots, N,$$

where

$$f(m, n, j) = \frac{1 - \cos(m - n + j)\pi}{m - n + j} - \frac{1 - \cos(m - n - j)\pi}{m - n - j}$$
$$- \frac{1 - \cos(m + n + j)\pi}{m + n + j} + \frac{1 - \cos(m + n - j)\pi}{m + n - j}.$$

To find $c_j(0)$, solve $\langle\phi_j, R\rangle = 0$, where $R = \displaystyle\sum_{j,k=1}^{N}c_j(0)\phi_j(x) - 1$.

9.20. Use the Galerkin method to solve the Poisson's equation $\nabla^2 u = 2$ subject to the Dirichlet boundary condition $u = 0$ along the boundary of the square $\{-a \leq x, y \leq a\}$ (Fig. 9.5).
HINT: Use the basis functions $\phi(x, y) = (a^2 - x^2)(a^2 - y^2)$, and consider the approximate solution

$$\tilde{u}_N(x, y) = (a^2 - x^2)(a^2 - y^2)(A_1 + A_2 x^2 + A_3 y^3 + \cdots + A_n x^{2i}y^{2j}).$$

ANS. For $N = 1$, we have

$$\int_{-a}^{a} \int_{-a}^{a} [-2(a^2 - y^2)A_1 - 2(a^2 - x^2)A_1 + 2](a^2 - x^2)(a^2 - y^2)\, dx\, dy = 0.$$

This yields

$$A_1 = \frac{5}{8a^2}, \quad \tilde{u}_1 = \frac{5(a^2 - x^2)(a^2 - y^2)}{8a^2}.$$

We must have $A_2 = A_3$. Then for $N = 3$, take

$$\tilde{u}_2 = (a^2 - x^2)(a^2 - y^2)[A_1 + A_2(x^2 + y^2)],$$

where

$$A_1 = \frac{1295}{1416a^2}, \quad A_2 = \frac{525}{4432a^4},$$

and

$$\tilde{u}_2(x, y) = \frac{35}{4432a^2}(a^2 - x^2)(a^2 - y^2)\left[74 + \frac{15}{a^2}(x^2 + y^2)\right].$$

Alternately, if we choose the basis functions as $\phi_{jk} = \cos\dfrac{j\pi x}{2a} \cos\dfrac{k\pi y}{a}$, j, k odd, then

$$\tilde{u}_N = \sum_{\substack{j,k=1 \\ j,k\ \text{odd}}} \alpha_{jk} \cos\frac{j\pi x}{a} \cos\frac{k\pi y}{2a},$$

which leads to

$$\int_{-a}^{a} \int_{-a}^{a} \left[\sum_{j,k} \alpha_{jk}\left(\frac{j^2\pi^2}{4a^2} + \frac{k^2\pi^2}{4a^2}\right) \cos\frac{j\pi x}{a} \cos\frac{k\pi xy}{2a}\right]$$
$$\times \cos\frac{m\pi x}{2a} \cos\frac{k\pi y}{2a}\, dx\, dy = 0.$$

Hence, for $j = m$ and $k = n$,

$$\alpha_{jk} = \frac{128a^2(-1)^{j+k-2)/2}}{jk(j^2 + k^2)\pi^4}.$$

9.21. Use the Galerkin method to solve the eigenvalue problem $\nabla^2 u + \lambda u = 0$ in the polar coordinates for $0 < rA$.

HINT: Solve $\dfrac{1}{r}\dfrac{d}{dr}\left(r\dfrac{du}{dr}\right) + \lambda u = 0$, $\quad 0 < r < a$.

ANS. Take $\phi_j(r) = \cos\dfrac{j\pi r}{2a}$. For the first-order approximation, $\phi_1 = \cos\dfrac{\pi r}{2a}$ and $\tilde{u}_1 = \alpha_1 \cos\dfrac{\pi r}{2a}$, which lead to

$$2\pi \int_0^a \left\{\frac{1}{r}\frac{d}{dr}\left[\frac{r\pi}{2a}\left(-\sin\frac{\pi r}{2a}\right)\right]\alpha_1 + \lambda\alpha_1 \cos\frac{\pi r}{2a}\right\} r\, dr = 0.$$

This gives the equation for the eigenvalue λ as

$$\frac{\pi^2}{4}\left(\frac{1}{2} + \frac{2}{\pi^2}\right) - \lambda a^2 \left(\frac{1}{2} - \frac{2}{\pi^2}\right) = 0.$$

Hence,

$$\lambda_1 = \frac{\pi^2(\pi^2 + 4)}{4a^2(\pi^2 - 4)} \approx \frac{5.8304}{a^2}.$$

The exact value is $\lambda_1 = \dfrac{5.779}{a^2}$. For the second-order approximation, $\tilde{u}_2 =$ $\alpha_1 \cos \dfrac{\pi r}{2a} + \alpha_2 \cos \dfrac{3\pi r}{2a}$, which gives $\lambda_2 = \dfrac{5.792}{a^2}$.

9.22. Use the Galerkin method to determine the lowest frequency (fundamental tone) of the vibration of a homogeneous circular plate Ω of radius a and center at the origin of cylindrical polar coordinates, clamped at the entire edge, i.e., solve $\nabla^4 u = \lambda u$ subject to the conditions $u(a) = 0 = u_r(a)$.

HINT: Minimize the variational problem $I(u) = \displaystyle\iint_\Omega \nabla^4 u \, dx \, dy$, such that $\displaystyle\iint_\Omega u^2 \, dx \, dy = 1$, subject to the given conditions.

ANS. Solve

$$\left(\frac{d^2}{dr^2} + \frac{1}{r}\frac{d}{dr}\right)\left(\frac{d^2 u}{dr^2} + \frac{1}{r}\frac{du}{dr}\right) = \lambda u.$$

Take $\tilde{u}_N = \displaystyle\sum_{j=1}^{N} \alpha_j \left(1 - \frac{r^2}{a^2}\right)^{j+1}$. Then, e.g., for $\tilde{u}_2$ we have

$$\alpha_1\left(\frac{192}{9} - \frac{\lambda a^4}{5}\right) + \alpha_2\left(\frac{144}{9} - \frac{\lambda a^4}{6}\right) = 0,$$

$$\alpha_1\left(\frac{144}{9} - \frac{\lambda a^4}{6}\right) + \alpha_2\left(\frac{96}{5} - \frac{\lambda a^4}{7}\right) = 0,$$

and the equation for λ is

$$(\lambda a^4)^2 - \frac{9792}{5}\lambda a^4 + 435456 = 0,$$

which has the smaller root as $\lambda = \dfrac{104.387654}{a^4}$. Using this value of λ in the above system of two equations, we find $\alpha_2 = 0.325\,\alpha_1$, and

$$\tilde{u}_2 = \alpha_1\left[\left(1 - \frac{r^2}{a^2}\right)^2 + 0.325\left(1 - \frac{r^2}{a^2}\right)^2\right],$$

where α_1 can be found from the above system of two equations.

10

Perturbation Methods

In this chapter we discuss some elementary perturbation methods for solving initial and boundary value problems. Only regular perturbation methods are presented. We leave singular perturbation techniques for more advanced textbooks. The solution of a perturbed problem in a regular perturbation approaches the solution of the unperturbed problem as the perturbation parameter approaches a specific limit. A perturbed problem may arise either due to a small variation in the governing differential equation or due to a slight perturbation of the boundary of the domain under consideration.

10.1. Perturbation Problem

The perturbation methods provide approximate solutions for boundary value and initial value problems. These methods are used when such problems contain a small parameter, say ε, and the solution for $\varepsilon = 0$ is known. This parameter occurs, in general, in a partial differential equation of the form

$$L\,u + \varepsilon N\,u = 0, \tag{10.1}$$

where L is a linear partial differential operator, and N is either a nonlinear or a linear differential operator, which makes the solution of Eq (10.1) difficult. If, by taking $\varepsilon = 0$, Eq (10.1) reduces to an ordinary differential equation, then the perturbation method fails.

Another kind of perturbation problems arise by perturbing the boundary. In this case the parameter ε will appear in the boundary conditions. The two common perturbation methods discussed here are: (1) series expansion in powers of ε, and

(2) successive approximations. These methods apply when the partial differential equation or the boundary is perturbed.

Quite often these two methods yield the same approximate solution. Frequently the perturbation solution is equivalent to the iterative solution of the corresponding integral equation. For the perturbation method to be successful, it is assumed that u is a continuous function of the perturbation parameter and that the differences in the two problems (perturbed and unperturbed) are not singular in character. Thus, these methods are applied to problems in which the solution of an ideal simple problem is known and, for a more realistic situation, the pertubation method occurs in the differential equation, or the boundary conditions, or the boundary of the region.

There are other more complicated methods which deal with singular perturbation problems. We will, however, not discuss such methods in this book. For more information about them, the reader is referred to Kevorkian (1990).

10.2. Taylor Series Expansions

The general scheme for this method is as follows: Consider Eq (10.1) subject to some prescribed boundary conditions and/or initial conditions. Assume that the solution of the homogeneous equation $Lu = 0$ subject to the same prescribed conditions is known. Then to solve the given problem, we further assume that u possesses a series expansion in powers of ε of the form

$$u = u_0 + \varepsilon u_1 + \varepsilon^2 u_2 + \cdots = \sum_{n=0}^{\infty} \varepsilon^n u_n. \tag{10.2}$$

If we substitute this power series into Eq (10.1), we get

$$L\left(\sum_{n=0}^{\infty} \varepsilon^n u_n \right) + \varepsilon N\left(\sum_{n=0}^{\infty} \varepsilon^n u_n \right) = 0. \tag{10.3}$$

Assuming that u_0 satisfies the prescribed conditions and that u_n, $n \neq 0$, satisfy homogeneous conditions, we can obtain a system of partial differential equations by comparing coefficients of various powers of ε on both sides of Eq (10.3). These equations are such that a partial differential equation in u_n depends only on u_0, $u_1, \dots, u_{n-1}$. The solutions for $u_0, u_1, \dots, u_{n-1}$ are known successively, i.e., first we solve for u_0, which enables us to solve for u_1, and so on, until we finally solve for u_n. We will demonstrate this method by some examples.

EXAMPLE 10.1. Consider

$$\nabla^2 u + \varepsilon f(u) = 0, \quad u(1, \theta) = g(\theta), \quad \lim_{r \to 0} u(r, \theta) < \infty. \tag{10.4}$$

We consider two cases:

(a) For $f(u) = u$, let us assume the power series (10.2). The partial differential equation and the boundary conditions for each u_n, $n = 0, 1, 2, \ldots$, are given by

$$
\begin{aligned}
\nabla^2 u_0 &= 0, & u_0(1, \theta) &= g(\theta), \\
\nabla^2 u_1 + u_0 &= 0, & u_1(1, \theta) &= 0, \\
\nabla^2 u_{n+1} + u_n &= 0, & u_{n+1}(1, \theta) &= 0.
\end{aligned}
\tag{10.5}
$$

The general solution for $\nabla^2 u_0 = 0$, by the method of separation of variables, is

$$
u_0 = \sum_0^\infty r^n (A_n \cos n\theta + B_n \sin n\theta),
\tag{10.6}
$$

where

$$
\sum_{n=0}^\infty (A_n \cos n\theta + B_n \sin n\theta) = g(\theta).
\tag{10.7}
$$

If $g(\theta)$ is a periodic function that satisfies the Dirichlet's conditions (see §6.1), then A_n and B_n are determined, and the subsequent equations for u_n are solved. For example, let $g(\theta) = \cos\theta$. In this case $u_0 = r\cos\theta$, and the partial differential equation for u_1 becomes

$$
\nabla^2 u_1 + r\cos\theta = 0.
\tag{10.8}
$$

The particular integral for a partial differential equation of the type

$$
\nabla^2 v + c_1 r^n \cos p\theta + c_2 r^m \sin q\theta = 0
$$

is of the form

$$
v_p = c_{11} r^{n+2} \cos p\theta + c_{22} r^{m+2} \sin q\theta.
$$

We find the solution for u_1 as follows: Writing Eq (10.8) as

$$
\frac{\partial^2 u_1}{\partial r^2} + \frac{1}{r}\frac{\partial u_1}{\partial r} + \frac{1}{r^2}\frac{\partial^2 u_1}{\partial \theta^2} = -r\cos\theta,
$$

we find that its particular integral is $u_{1p} = A r^3 \cos\theta$, which, when substituted into the above equation, gives $A = -1/8$. Let

$$
u_1(r, \theta) = \sum_{n=0}^\infty r^n (A_n \cos n\theta + B_n \sin n\theta) - \frac{1}{8} r^3 \cos\theta.
$$

Then the condition $u_1(1, \theta) = 0$ yields

$$
\sum_{n=0}^\infty (A_n \cos n\theta + B_n \sin n\theta) - \frac{1}{8}\cos\theta = 0.
$$

Thus, $A_n = 0$ for $n \neq 1$, $A_1 = 1/8$, and $B_n = 0$. Hence,

$$u_1 = \frac{1}{8}r(1 - r^2)\cos\theta. \tag{10.9}$$

Similarly, the solution for u_2 is given by

$$u_2 = \frac{1}{192}r(2 - 3r^2 + r^4)\cos\theta. \tag{10.10}$$

Other terms are similarly obtained.

(b) For $f(u) = u^2$, we use the series (10.2) for u in powers of ε, where u_0 satisfies the boundary condition $u_0(1, \theta) = \cos\theta$, and u satisfies the homogeneous boundary condition. Thus, the partial differential equations for $u_0, u_1, u_2, \ldots$ are given by

$$u_{0rr} + \frac{1}{r}u_{0r} + \frac{1}{r^2}u_{0\theta\theta} = 0,$$

$$u_{1rr} + \frac{1}{r}u_{1r} + \frac{1}{r^2}u_{1\theta\theta} + u_0^2 = 0, \tag{10.11}$$

$$u_{2rr} + \frac{1}{r}u_{2r} + \frac{1}{r^2}u_{2\theta\theta} + 2u_0u_1 = 0,$$

and so on. The solution for u_0 is clearly $u_0 = r\cos\theta$, and the equation for u_1 becomes

$$u_{1rr} + \frac{1}{r}u_{1r} + \frac{1}{r^2}u_{1\theta\theta} + r^2\cos^2\theta = 0,$$

or

$$\frac{\partial^2 u_1}{\partial r^2} + \frac{1}{r}\frac{\partial u_1}{\partial r} + \frac{1}{r^2}\frac{\partial^2 u_1}{\partial\theta^2} = -\frac{1}{2}r^2(1 + \cos\theta).$$

The particular solution for this equation is of the form

$$u_{1p} = A r^4 + B r^4 \cos 2\theta,$$

which, after substitution into the above equation, gives $A = -1/32$, and $B = -1/24$. Thus,

$$u_1(r, \theta) = \sum_{n=0}^{\infty} r^n\left(A_n \cos n\theta + B_n \sin n\theta\right) - \frac{1}{32}r^4 - \frac{1}{24}r^4 \cos 2\theta.$$

Then the condition $u_1(1, \theta) = 0$ yields

$$\sum_{n=0}^{\infty}\left(A_n \cos n\theta + B_n \sin n\theta\right) - \frac{1}{32} - \frac{1}{24}\cos 2\theta = 0.$$

Thus, $A_0 = \dfrac{1}{32}$, $A_2 = \dfrac{1}{24}$, $A_n = 0$ for $n \neq 0$ and 2, and $B_n = 0$. Hence,

$$u_1 = \frac{1}{32}(1 - r^4) + \frac{1}{24}(r^2 - r^4)\cos 2\theta. \tag{10.12}$$

We continue this process to find $u_2, u_3, \ldots$. ∎

EXAMPLE 10.2. To solve

$$u_{rr} + \frac{1}{r}u_r + \frac{1}{r^2}u_{\theta\theta} + \varepsilon u u_\theta = 0, \quad u(1, \theta) = \cos\theta, \tag{10.13}$$

up to the first three terms of the power series solution, we use the series (10.2) for u. Substituting this series into the partial differential equation and comparing coefficients of different powers of ε on both sides, we get

$$u_{0rr} + \frac{1}{r}u_{0r} + \frac{1}{r^2}u_{0\theta\theta} = 0,$$

$$u_{1rr} + \frac{1}{r}u_{1r} + \frac{1}{r^2}u_{1\theta\theta} + u_0 u_{0\theta} = 0, \tag{10.14}$$

$$u_{2rr} + \frac{1}{r}u_{2r} + \frac{1}{r^2}u_{2\theta\theta} + u_1 u_{0\theta} + u_0 u_{1\theta} = 0.$$

The new boundary conditions are $u_0(1, \theta) = \cos\theta$, and $u_n(1, \theta) = 0$, $n \geq 1$. It is easy to see that $u_0(1, \theta) = r\cos\theta$, and the partial differential equation for u_1 is

$$u_{1rr} + \frac{1}{r}u_{1r} + \frac{1}{r^2}u_{1\theta\theta} = r^2 \cos\theta \sin\theta = \frac{1}{2}r^2 \sin 2\theta. \tag{10.15}$$

Its solution is $u_1 = \dfrac{1}{24}r^2(r^2 - 1)\sin 2\theta$. The partial differential equation for u_2 becomes

$$u_{2rr} + \frac{1}{r}u_{2r} + \frac{1}{r^2}u_{2\theta\theta} = \frac{1}{24}r^3(r^2 - 1)(\sin 2\theta \sin\theta - 2\cos 2\theta \cos\theta),$$

and its solution is given by

$$u_2 = \frac{1}{256}r^3(r^2 - 1)\cos 3\theta + \frac{1}{1152}r(r^4 - 1)\cos\theta$$

$$- \frac{1}{640}r^3(r^4 - 1)\cos 3\theta - \frac{1}{2304}r(r^6 - 1)\cos\theta. \blacksquare \tag{10.16}$$

We continue this process as long as needed to obtain the required degree of accuracy. There will, of course, be the question of convergence, but we observe that the coefficients are getting smaller, so convergence for values of $\varepsilon < 1$ appears likely.

10.3. Successive Approximations

The general scheme for this method is as follows: We assume the first approximation to be u_0, which satisfies the given boundary and initial conditions and the homogeneous equation $Lu = 0$. Then the second approximation is u_1, which satisfies the equation $Lu_1 + \varepsilon N u_0 = 0$, together with the given boundary and initial conditions. The process is continued until the required degree of accuracy is achieved. It is obvious that the order of difficulty is directly proportional to the order of the approximation. We demonstrate this method by solving the previous examples again.

EXAMPLE 10.3. We now solve Example 10.1(b) for $f(u) = u^2$ by the method of successive approximations. The partial differential equation for u_0 is the same as in (10.11), i.e.,

$$u_{0rr} + \frac{1}{r} u_{0r} + \frac{1}{r^2} u_{0\theta\theta} = 0,$$

and its solution is $u_0 = r \cos\theta$. Since $N u_0 = u_0^2$, the partial differential equation for u_1 is

$$u_{1rr} + \frac{1}{r} u_{1r} + \frac{1}{r^2} u_{1\theta\theta} + \varepsilon r^2 \cos^2\theta = 0, \quad u_1(1, \theta) = \cos\theta. \tag{10.17}$$

Its solution is

$$u_1 = r \cos\theta + \varepsilon \left[\frac{1}{32}(1 - r^4) + \frac{1}{24}(r^2 - r^4)\cos 2\theta \right]. \tag{10.18}$$

Notice that this u_1 is actually the sum $u_0 + \varepsilon u_1$ of Example 10.1(b). ∎

EXAMPLE 10.4. We now solve Example 10.2 by the method of successive approximations. Obviously, the solution for u_0 is $u_0 = r \cos\theta$. The next approximation u_1 satisfies the given boundary condition and the partial differential equation

$$u_{1rr} + \frac{1}{r} u_{1r} + \frac{1}{r^2} u_{1\theta\theta} + \varepsilon u_0 u_{0\theta} = 0, \tag{10.19}$$

or

$$u_{1rr} + \frac{1}{r} u_{1r} + \frac{1}{r^2} u_{1\theta\theta} - \frac{\varepsilon}{2} r^2 \sin 2\theta = 0. \tag{10.20}$$

The solution for u_1 is easily seen to be

$$u_1 = r \cos\theta + \frac{\varepsilon}{24}(r^4 - r^2) \sin 2\theta. \tag{10.21}$$

The next approximation u_2 once again satisfies the same boundary conditions. Since $N u_0 = u\, u_0$, the new partial differential equation is

$$u_{2rr} + \frac{1}{r} u_{2r} + \frac{1}{r^2} u_{2\theta\theta} + \varepsilon u_1 u_{1\theta} = 0. \tag{10.22}$$

Since

$$u_1 u_{1\theta} = -\frac{r^2}{2}\sin 2\theta + \frac{\varepsilon}{48}r^3(r^2-1)(\cos\theta+3\cos\theta) + \frac{\varepsilon^2}{576}r^4(r^2-1)^2\sin 4\theta,$$

$$(10.23)$$

the equation for u_2 becomes

$$u_{2rr} + \frac{1}{r}u_{2r} + \frac{1}{r^2}u_{2\theta\theta}$$

$$= \frac{r^2}{2}\sin 2\theta - \frac{\varepsilon}{48}r^3(r^2-1)(\cos\theta+3\cos\theta) - \frac{\varepsilon^2}{576}r^4(r^2-1)^2\sin 4\theta. \quad (10.24)$$

Its solution is

$$u_2 = r\cos\theta + \frac{\varepsilon}{24}r^2(r^2-1)\sin 2\theta + \frac{\varepsilon^2}{1152}r(r^4-1)\cos\theta$$

$$- \frac{\varepsilon^2}{2304}r(r^6-1)\cos\theta + \frac{\varepsilon^2}{256}r^3(r^2-1)\cos 3\theta - \frac{\varepsilon^2}{640}r^3(r^4-1)\cos 3\theta$$

$$- \varepsilon^3\left[\frac{1}{48384}r^4(r^6-1) - \frac{1}{13824}r^4(r^4-1) + \frac{1}{11520}r^4(r^2-1)\right]\sin 4\theta.$$

$$(10.25)$$

It is clear that the solution (10.25) is almost similar to the one obtained in Example 10.2. These two solutions would be exactly the same if enough terms in the series solution and a sufficiently large number of approximations are taken. ∎

10.4. Boundary Perturbations

The series (10.2) for u is used for problems involving boundary perturbation. The following examples illustrate this method.

EXAMPLE 10.5. Solve the Laplace equation $u_{xx} + u_{yy} = 0$, $0 \le x \le \pi$, $y \ge 0$, such that

$$u(\varepsilon\sin\omega y, y) = 0, \quad u(\pi, y) = 0, \quad u(x, 0) = \sin x,$$

and $\lim_{y\to\infty} u(x, y)$ is bounded. We assume the series (10.2) for u and find that $\nabla u_n = 0$ for all n. Also,

$$u(\varepsilon\sin\omega y, y) = u(0, y) + \varepsilon\sin\omega y u_x(0, y) + \frac{1}{2}(\varepsilon\sin\omega y)^2 u_{xx}(0, y) + \cdots,$$

$$(10.26)$$

$$u(\varepsilon\sin\omega y, y) = u_0(0, y) + \varepsilon u_1(0, y) + \varepsilon^2 u_2(0, y) + \cdots$$

$$+ \varepsilon\sin\omega y\left[u_{0,x}(0, y) + \varepsilon u_{1,x}(0, y) + \varepsilon^2 u_{2,x}(0, y) + \cdots\right]$$

$$+ \frac{1}{2}(\varepsilon\sin\omega y)^2\left[u_{0,xx}(0, y) + \varepsilon u_{1,xx}(0, y) + \varepsilon^2 u_{2,xx}(0, y) + \cdots\right] + \cdots.$$

$$(10.27)$$

By comparing (10.26) and (10.27), we find that $u_0(x, y)$ satisfies the conditions

$$u_0(0, y) = 0, \quad u_0(\pi, y) = 0, \quad \text{and} \quad u_0(x, 0) = \sin x, \tag{10.28}$$

where the function $u_0(x, y)$ that satisfies the given equation and the conditions (10.28) is given by $u_0 = e^{-y} \sin x$. Now, $u_1(x, y)$ must satisfy the conditions

$$u_1(0, y) = \sin \omega y \, u_{0x}(0, y) = -e^{-y} \sin \omega y, \quad \text{and} \quad u_1(\pi, y) = 0 = u_1(x, 0). \tag{10.29}$$

Since the solution for u_1 is complicated, we will separate it into two parts: The first part satisfies the Laplace equation and the boundary conditions at $x = 0, \pi$, while the second part, although satisfying the Laplace equation, satisfies the homogeneous boundary conditions at $x = 0, \pi$. Then the sum of both parts will satisfy the conditions at $y = 0$. Thus, let $u_1 = v_1 + v_2$, where $\nabla v_1 = 0$, subject to the conditions $v_1(0, y) = -e^{-y} \sin \omega y$, $v_1(\pi, y) = 0$. Then

$$\begin{aligned} v_1 &= [f(x) \cos \omega y + g(x) \sin \omega y] \, e^{-y}, \\ f(0) &= f(\pi) = g(\pi) = 0, \quad g(0) = -1. \end{aligned} \tag{10.30}$$

After substituting v_1 into the Laplace equation and comparing the coefficients of $\sin \omega x$ and $\cos \omega x$ on both sides, we get

$$\begin{aligned} f'' + (1 - \omega^2)f - 2\omega g &= 0, \\ g'' + (1 - \omega^2)g + 2\omega f &= 0. \end{aligned} \tag{10.31}$$

Let $z = f + ig$. Then, multiplying the second equation in (10.31) by i and adding it to the first, we get

$$z'' + (1 + 2i\omega - \omega^2)z = 0,$$

or

$$z'' + (1 + i\omega)^2 z = 0, \tag{10.32}$$

with the boundary conditions $z(0) = -i$, $z(\pi) = 0$. Its solution is

$$z = A \, e^{i(1+i\omega)x} + B \, e^{-i(1+i\omega)x}, \tag{10.33}$$

which, after applying the boundary conditions, yields

$$A = \frac{-ie^{-i(1+i\omega)\pi}}{e^{-i(1+i\omega)\pi} - e^{i(1+i\omega)\pi}}, \quad B = \frac{ie^{i(1+i\omega)\pi}}{e^{-i(1+i\omega)\pi} - e^{i(1+i\omega)\pi}}. \tag{10.34}$$

Now, solving for the real and imaginary parts of z, we get

$$f(x) = \frac{\sin x \cosh \omega(\pi - x)}{\sinh \omega \pi}, \quad g(x) = -\frac{\cos x \sinh \omega(\pi - x)}{\sinh \omega \pi}. \tag{10.35}$$

Next, we determine v_2 by the separation of variables method with homogeneous boundary conditions with respect to x. The solution is of the form

$$v_2 = \sum_0^\infty A_n e^{-ny} \sin nx. \tag{10.36}$$

Then adding the solutions (10.30) and (10.36), and using (10.35), the complete solution for u_1 is given by

$$u_1 = \sum_0^\infty A_n e^{-ny} \sin nx + \big[\cosh \omega(\pi - x) \sin x \cos \omega y$$

$$- \sinh \omega(\pi - x) \cos x \sin \omega y \big] \frac{e^{-y}}{\sinh \omega \pi}. \tag{10.37}$$

On applying the condition $u_1(x, 0) = 0$ and using the Fourier series expansion (4.22) in x at $y = 0$, we find from (10.37) that

$$A_n = -\frac{2}{\pi} \int_0^\pi \cosh \omega(\pi - x) \, \sin x \sin nx \, dx$$

$$= -\Big[\frac{1}{(n-1)^2 + \omega^2} - \frac{1}{(n+1)^2 + \omega^2} \Big] \frac{\omega}{\sinh \omega \pi}.$$

Hence, the solution for the problem is

$$u_1(x, y) = \Big[\cosh \omega(\pi - x) \sin x \cos \omega y - \sinh \omega(\pi - x) \cos x \sin \omega y \Big] \frac{e^{-y}}{\sinh \omega \pi}$$

$$- \sum_{n=0}^\infty \Big[\frac{1}{(n-1)^2 + \omega^2} - \frac{1}{(n+1)^2 + \omega^2} \Big] \frac{\omega \sin nx}{\sinh \omega \pi} e^{-ny}. \ \blacksquare$$

$$\tag{10.38}$$

EXAMPLE 10.6. Consider

$$\nabla^2 u = 0, \, u(1 + \varepsilon \sin \theta, \theta) = f(\theta). \tag{10.39}$$

The Taylor's series for $u(r, \theta)$, $r = 1 + \varepsilon \sin \theta$, about $r = 1$ is

$$u(1 + \varepsilon \sin \theta, \theta) = f(\theta) = u(1, \theta) + \varepsilon \sin \theta u_r(1, \theta) + \frac{1}{2} \varepsilon^2 \sin^2 \theta u_{rr}(1, \theta) + \cdots. \tag{10.40}$$

We also assume that u has a series expansion in powers of ε of the form (10.2), that is,

$$u(r, \theta) = u_0 + \varepsilon u_1 + \varepsilon^2 u_2 + \cdots. \tag{10.41}$$

Combining (10.40) and (10.41), we get

$$u_0(1, \theta) = f(\theta),$$
$$u_1(1, \theta) + \sin \theta u_{0r}(1, \theta) = 0,$$
$$u_2(1, \theta) + \sin \theta u_{1r}(1, \theta) + \frac{1}{2} \sin^2 \theta u_{0rr}(1, \theta) = 0,$$

(10.42)

and so on. The partial differential equation to be satisfied by u_n for all n is $\nabla^2 u_n = 0$. Using the general solution (10.6) and applying the boundary conditions for u_0, we get

$$u_0(1, \theta) = f(\theta) = \sum_0^\infty (A_n \cos n\theta + B_n \sin n\theta).$$

(10.43)

If $f(\theta)$ is a periodic function that satisfies the Dirichlet's conditions (§6.1), then

$$A_0 = \frac{1}{2\pi} \int_{-\pi}^{\pi} f(\theta) \, d\theta, \quad A_n = \frac{1}{\pi} \int_{-\pi}^{\pi} f(\theta) \cos n\theta \, d\theta,$$
$$B_n = \frac{1}{\pi} \int_{-\pi}^{\pi} f(\theta) \sin n\theta \, d\theta.$$

(10.44)

Case 1. If $f(\theta) = \cos \theta$, then

$$u_0 = r \cos \theta, \quad u_1 = -\frac{1}{2} r^2 \sin 2\theta, \quad u_2 = \frac{1}{2}(r \cos \theta - r^3 \cos 3\theta). \quad (10.45)$$

Case 2. If $f(\theta) = \sin \theta$, then

$$u_0 = r \sin \theta, \quad u_1 = \frac{1}{2}(r^2 \cos \theta - 1), \quad u_2 = \frac{1}{2}\left(r \sin \theta - r^3 \sin 3\theta\right). \quad (10.46)$$

For the solution of this example by the method of successive approximations, see Exercise 10.4. ∎

10.5. Fluctuating Flows

In this section we discuss an important application of the perturbation theory. The behavior of fluctuating flows of various kinds of fluids is of considerable significance to industry. The flight of a plane during turbulence can be idealized to the fluctuating flow of air past a flat plate or a slender body, or the flow of oil through pipes with a fluctuating velocity are examples of such flows. We will discuss the simplest case of the flow of a viscous fluid past a porous flat plate when the fluid is sucked out of the

plate at a constant rate. The mathematical analysis of this problem and its solution
are presented in the next example.

EXAMPLE 10.7. Find the steady-state solution of

$$\frac{\partial u}{\partial t} - s_0 \frac{\partial u}{\partial z} = \frac{\partial U}{\partial t} + \nu \frac{\partial^2 u}{\partial z^2}, \tag{10.47}$$

subject to the conditions $u(0, t) = 0$, and $\lim\limits_{z \to \infty} u(z, t) = U(t) = U_0 \left(1 + \varepsilon e^{i\omega t}\right)$,
where ε is the perturbation parameter, s_0 is the constant rate of suction, ν is the
kinematic viscosity, and U_0 denotes the constant mean free-stream velocity (Stuart,
1955, and Puri, 1975). We assume a solution of the form

$$u(z, t) = U_0 \left[f_1(z) + \varepsilon f_2(z) e^{i\omega t}\right].$$

Substituting it into Eq(10.47) and dividing by U_0, we get

$$i\omega \varepsilon f_2(z) e^{i\omega t} - s_0 \left[f_1'(z) + \varepsilon f_2'(z) e^{i\omega t}\right]$$
$$= i\omega \varepsilon e^{i\omega t} + \nu \left[f_1''(z) + \varepsilon f_2''(z) e^{i\omega t}\right]. \tag{10.48}$$

If we introduce the nondimensional quantities

$$\zeta = \frac{z\, s_0}{\nu}, \quad \lambda = \frac{\omega \nu}{s_0^2}, \quad \text{and} \quad \tau = \frac{t\, s_0^2}{\nu},$$

into Eq(10.48), then, on comparing the nonperiodic and harmonic terms, we get

$$f_1'' + f_1' = 0, \quad \text{and} \quad f_2'' + f_2' - i\lambda f_2 = -i\lambda,$$

where the primes denote differentiation with respect to ζ, such that

$$f_1 = f_2 = 0 \quad \text{at } \zeta = 0, \quad \text{and} \quad f_1, f_2 \to 1 \quad \text{as } \zeta \to \infty.$$

The solutions for f_1 and f_2 are given by

$$f_1(\zeta) = 1 - e^{-\zeta}, \quad f_2(\zeta) = 1 - e^{-m\zeta},$$

where $m = \dfrac{1}{2}\{1 + \sqrt{(1 + 4i\lambda)}\}$. Hence,

$$u(z, t) = u(\zeta, \tau) = U_0 \left[\left(1 - e^{-\zeta}\right) + \varepsilon \left(1 - e^{-m\zeta}\right) e^{i\lambda\tau}\right]. \; \blacksquare \tag{10.49}$$

EXAMPLE 10.8. In this example we discuss the previous problem in the case
when s_0 is the mean suction velocity, and the rate of suction is given by $s_0 \left(1 + \varepsilon A\, e^{i\omega t}\right)$,

where A is a known constant such that $\varepsilon A < 1$ (Messiha, 1966). The governing equation in this case is

$$\frac{\partial u}{\partial t} - s_0 \left(1 + \varepsilon A\, e^{i\omega t}\right) \frac{\partial u}{\partial z} = \frac{\partial U}{\partial t} + \nu \frac{\partial^2 u}{\partial z^2}. \tag{10.50}$$

The boundary conditions remain the same as in Example 10.7. Proceeding as before, we assume a solution of the form

$$u(z,t) = U_0 \left[f_1(z) + \varepsilon f_2(z)\, e^{i\omega t}\right].$$

Substituting this solution into Eq (10.4), and dividing by U_0, we get

$$\begin{aligned}
i\omega\varepsilon f_2(z)\, e^{i\omega t} &- s_0 \left(1 + \varepsilon A e^{i\omega t}\right) \left[f_1'(z) + \varepsilon f_2'(z)\, e^{i\omega t}\right] \\
&= i\omega\varepsilon\, e^{i\omega t} + \nu \left[f_1''(z) + \varepsilon f_2''(z)\, e^{i\omega t}\right].
\end{aligned} \tag{10.51}$$

Introducing the nondimensional quantities

$$\zeta = \frac{z\, s_0}{\nu}, \qquad \lambda = \frac{\omega\nu}{s_0^2}, \quad \text{and} \quad \tau = \frac{t\, s_0^2}{\nu},$$

into Eq (10.51), and comparing the nonperiodic and harmonic terms and neglecting the terms containing ε^2, we have

$$f_1'' + f_1' = 0,$$
$$f_2'' + f_2' - i\lambda f_2 = -i\lambda - A f_1',$$

where the primes denote differentiation with respect to ζ. Since the conditions are the same as in the previous example, the solutions for f_1 and f_2 are given by

$$f_1(\zeta) = 1 - e^{-\zeta},$$
$$f_2(\zeta) = \left(1 - \frac{iA}{\lambda} e^{-\zeta}\right) - \left(1 - \frac{iA}{\lambda}\right) e^{-m\zeta},$$

where $m = \dfrac{1}{2}\left\{1 + \sqrt{(1 + 4i\lambda)}\right\}$. Hence,

$$\begin{aligned}
u(z,t) &= u(\zeta,\tau) \\
&= U_0 \left[\left(1 - e^{-\zeta}\right) + \varepsilon\left\{\left(1 - \frac{iA}{\lambda} e^{-\zeta}\right) - \left(1 - \frac{iA}{\lambda}\right) e^{-m\zeta}\right\} e^{i\lambda\tau}\right]. \blacksquare
\end{aligned} \tag{10.52}$$

EXAMPLE 10.9. We will consider the problem discussed in Example 10.7 for a viscoelastic fluid. Thus, we find the steady-state solution of the equation

$$\frac{\partial u}{\partial t} - s_0 \frac{\partial u}{\partial z} = \frac{\partial U}{\partial t} + \nu \frac{\partial^2 u}{\partial z^2} - a\left[\frac{\partial^3 u}{\partial z^2 \partial t} - s_0 \frac{\partial^3 u}{\partial z^3}\right], \tag{10.53}$$

subject to the conditions

$$u(0,t) = 0, \quad \text{and} \quad \lim_{z \to \infty} u(z,t) = U(t) = U_0 \left(1 + \varepsilon\, e^{i\omega t} \right),$$

where a is the viscoelastic parameter, ε is, as before, the perturbation parameter, and U_0 denotes the constant mean free stream velocity (Kaloni, 1967, and Soundalgekar and Puri, 1969).

We assume a solution of the form

$$u(z,t) = U_0 \left[f_1(z) + \varepsilon f_2(z)\, e^{i\omega t} \right].$$

Substituting this solution into the Eq (10.53) and after dividing by U_0, we get

$$\begin{aligned}
i\omega\varepsilon f_2(z)\, e^{i\omega t} &- s_0 \left[f_1'(z) + \varepsilon f_2'(z)\, e^{i\omega t} \right] \\
&= i\omega\varepsilon\, e^{i\omega t} + \nu \left[f_1''(z) + \varepsilon f_2''(z)\, e^{i\omega t} \right] \\
&\quad - a\left\{ \varepsilon i\omega f_2''(z)\, e^{i\omega t} - s_0 \left[f_1''' + \varepsilon f_2'''(z)\, e^{i\omega t} \right] \right\}.
\end{aligned} \qquad (10.54)$$

Introducing the nondimensional quantities

$$\zeta = \frac{z\, s_0}{\nu}, \quad \lambda = \frac{\omega\nu}{s_0^2}, \quad \tau = \frac{t\, s_0^2}{\nu}, \quad \alpha = \frac{a\, s_0^2}{\nu^2},$$

into Eq (10.54) and comparing the nonperiodic and harmonic terms, we obtain the equations

$$\alpha f_1''' + f_1'' + f_1' = 0, \qquad (10.55)$$
$$\alpha f_2''' + f_2''(1 - i\alpha\lambda) + f_2' - \lambda f_2 = -i\lambda, \qquad (10.56)$$

where the primes denote differentiation with respect to ζ, under the conditions

$$f_1 = f_2 = 0 \quad \text{at } \zeta = 0, \quad \text{and } f_1, f_2 \to 1 \text{ as } \zeta \to \infty.$$

Now, we regard α as another perturbation parameter and assume perturbation solutions of the form

$$f_1(\zeta) = p_1(\zeta) + \alpha p_2(\zeta) + O(\alpha^2),$$
$$f_2(\zeta) = q_1(\zeta) + \alpha q_2(\zeta) + O(\alpha^2).$$

Then, substituting these expressions for f_1 and f_2 into Eqs (10.55) and (10.56), we get

$$\begin{array}{ll}
p_1'' + p_1' = 0, & p_2'' + p_2' = -p_1''', \qquad (10.57) \\
q_1'' + q_1' - i\lambda q_1 = -i\lambda, & q_2'' + q_2' - i\lambda q_2 = i\lambda q_1'' - q_1''', \qquad (10.58)
\end{array}$$

and the corresponding boundary conditions become

$$p_1 = p_2 = q_1 = q_2 = 0 \quad \text{at } \zeta = 0$$
$$p_1, q_1 \to 1 \quad \text{and} \quad p_2, q_2 \to 0 \quad \text{as } \zeta \to \infty.$$

Solving Eqs (10.57) and (10.58) under these boundary conditions, we have

$$p_1 = 1 - e^{-\zeta}, \qquad p_2 = \zeta e^{-\zeta},$$
$$q_1 = 1 - e^{-m\zeta}, \qquad q_2 = \beta \zeta e^{-m\zeta}.$$

Thus,

$$f_1 = 1 - e^{-\zeta} + \alpha \zeta e^{-\zeta}, \quad f_2 = 1 - e^{-m\zeta} + \alpha \beta \zeta e^{-m\zeta},$$

where

$$m = \frac{1}{2}\{1 + \sqrt{(1 + 4i\lambda)}\}, \quad \text{and} \quad \beta = \frac{m^2(m + i\lambda)}{\sqrt{1 + 4i\lambda}}.$$

Hence,

$$u(z,t) = u(\zeta,\tau) = U_0 \left[\left(1 - e^{-\zeta} + \alpha \zeta e^{-\zeta}\right) + \varepsilon e^{i\lambda\tau} \left(1 - e^{-m\zeta} + \alpha \beta \zeta e^{-m\zeta}\right) \right].$$

This solution is valid for $\alpha \zeta < 1$. ∎

10.6. Mathematica Projects

PROJECT 10.1. For this chapter the reader should refer to the Mathematica Notebook `perturbationmethods.nb`.

10.7. Exercises

10.1. Obtain a perturbation solution for the problem: $u_t = u_{yy} + ku_{yyt}$, where $u(y,t) = 0$ for $0 \le t$ and $u(0,t) = 1$ for $t > 0$,
(a) By successive approximation (two approximations).
(b) By expansion in powers of k (up to the first power of k).
HINT. Use the Laplace transforms. Compare with Exercise 6.4.
ANS. $u = \text{erfc}\left(\dfrac{y}{2\sqrt{t}}\right) + \dfrac{ky}{4t\sqrt{\pi t}}\left(\dfrac{y^2}{t} - 1\right) e^{-y^2/(4t)}.$

10.2. Solve $\nabla^2 u + \varepsilon f(u) = 0$, $u(1, \theta) = g(\theta)$ for $f(u) = u$, u^2, $u + u^2$, respectively, and $g(\theta) = \sin \theta$, $\sin^2 \theta$, respectively. Find the first two terms in each case.

ANS. For $g(\theta) = \sin \theta$: $u \approx r \sin \theta + \dfrac{\varepsilon}{8} r(1 - r^2) \sin \theta$,

$$u \approx r \sin \theta + \frac{\varepsilon}{8} \left\{ \frac{1}{4}(1 - r^4) - \frac{1}{3} r^2 (1 - r^2) \cos 2\theta \right\},$$

$$u \approx r \sin \theta + \frac{\varepsilon}{8} \left\{ \frac{1}{4}(1 - r^4) + r(1 - r^2) \sin \theta - \frac{r^2}{3}(1 - r^2) \cos 2\theta \right\}.$$

For $g(\theta) = \sin^2 \theta$:

$$u \approx \frac{1}{2}(1 - r^2 \cos 2\theta) + \frac{\varepsilon}{8} \left\{ 1 - r^2 - \frac{r^2}{3}(1 - r^2) \cos 2\theta \right\},$$

$$u \approx \frac{1}{2}(1 - r^2 \cos 2\theta) + \frac{\varepsilon}{8} \left\{ \frac{1}{2}(1 - r^2) - \frac{r^2}{3}(1 - r^2) \cos 2\theta \right.$$
$$\left. + \frac{1}{36}(1 - r^6) + \frac{r^4}{20}(1 - r^2) \cos 4\theta \right\},$$

$$u \approx \frac{1}{2}(1 - r^2 \cos 2\theta) + \frac{\varepsilon}{8} \left\{ \frac{3}{2}(1 - r^2) - \frac{2r^2}{3}(1 - r^2) \cos 2\theta \right.$$
$$\left. + \frac{1}{36}(1 - r^6) + \frac{r^4}{20}(1 - r^2) \cos 4\theta \right\}.$$

10.3. Solve $u_{rr} + \dfrac{1}{r} u_r + \dfrac{1}{r^2} u_{\theta\theta} + \varepsilon u_r u_\theta = 0$, $u(1, \theta) = \cos \theta$.

SOLUTION. Substituting the series (10.2) into the partial differential equation and comparing coefficients of different powers of ε on both sides, we get

$$u_{0rr} + \frac{1}{r} u_{0r} + \frac{1}{r^2} u_{0\theta\theta} = 0,$$

$$u_{1rr} + \frac{1}{r} u_{1r} + \frac{1}{r^2} u_{1\theta\theta} + u_{0r} u_{0\theta} = 0,$$

$$u_{2rr} + \frac{1}{r} u_{2r} + \frac{1}{r^2} u_{2\theta\theta} + u_{1r} u_{0\theta} + u_{0r} u_{1\theta} = 0.$$

The new boundary conditions are $u_0(1, \theta) = \cos \theta$, and $u_n(1, \theta) = 0$ for $n \geq 1$. It is easy to see that

$$u_0(r, \theta) = r \cos \theta,$$

$$u_1(r, \theta) = \frac{1}{10}(r^3 - r^2) \sin 2\theta,$$

$$u_2(r, \theta) = \frac{1}{480}(r^5 - r) \cos \theta + \left(\frac{1}{35} r^4 - \frac{1}{64} r^5 - \frac{29}{2240} r^3 \right) \cos 3\theta.$$

Alternatively, we solve this problem by the method of successive approximations. Note that the solution for u_0 is $r \cos \theta$. The next approximation u_1 satisfies the given boundary condition and the partial differential equation

$$u_{1rr} + \frac{1}{r} u_{1r} + \frac{1}{r^2} u_{1\theta\theta} + \varepsilon u_{0r} u_{0\theta} = 0.$$

The solution for u_1 is

$$u_1 = r\cos\theta + \varepsilon\frac{1}{10}(r^3 - r^2)\sin 2\theta.$$

The next approximation u_2 once again satisfies the same boundary conditions, but the new partial differential equation is

$$u_{2rr} + \frac{1}{r}u_{2r} + \frac{1}{r^2}u_{2\theta\theta} + \varepsilon u_{1r}u_{1\theta} = 0,$$

whose solution is

$$u_2 = r\cos\theta + \varepsilon\frac{1}{10}(r^3 - r^2)\sin 2\theta + \varepsilon^2\Big[\frac{1}{480}(r^5 - r)\cos\theta$$
$$+ \Big(\frac{1}{35}r^4 - \frac{1}{64}r^5 - \frac{29}{2240}r^3\cos 3\theta\Big)\Big]$$
$$- \frac{\varepsilon^3}{100}\Big(\frac{r^7}{11} - \frac{r^6}{4} + \frac{2r^5}{9} - \frac{25r^4}{396}\Big)\sin 4\theta.$$

10.4. Solve $u_{rr} + \frac{1}{r}u_r + \frac{1}{r^2}u_{\theta\theta} + \varepsilon u_{r\theta} = 0, u(1,\theta) = \sin\theta$. Find the first three terms.

SOLUTION. Substituting the series (10.2) into the partial differential equation and comparing coefficients of different powers of ε on both sides, we get

$$u_{0rr} + \frac{1}{r}u_{0r} + \frac{1}{r^2}u_{0\theta\theta} = 0,$$
$$u_{1rr} + \frac{1}{r}u_{1r} + \frac{1}{r^2}u_{1\theta\theta} + u_{0r\theta} = 0,$$
$$u_{2rr} + \frac{1}{r}u_{2r} + \frac{1}{r^2}u_{2\theta\theta} + u_{1r\theta} = 0.$$

The new boundary conditions are $u_0(r,\theta) = \sin\theta$, and $u_n(1,\theta) = 0, n \geq 1$. Clearly, $u_0(r,\theta) = r\sin\theta$, and the partial differential equations for u_1 is

$$u_{1rr} + \frac{1}{r}u_{1r} + \frac{1}{r^2}u_{1\theta\theta} = -\cos\theta.$$

Its solution is

$$u_1 = \frac{1}{3}r(1 - r)\cos\theta.$$

The partial differential equation for u_2 becomes

$$u_{2rr} + \frac{1}{r}u_{2r} + \frac{1}{r^2}u_{2\theta\theta} = \frac{1}{3}(1 - 2r)\sin\theta,$$

whose solution is given by

$$u_2 = -\frac{1}{36} r(1 - 4r + 3r^2) \sin \theta.$$

10.5. In Example 10.6, choose $f(\theta) = \sin^2 \theta$, and solve the problem.
ANS.

$$u_0 = \frac{1}{2} \left(1 - r^2 \cos 2\theta \right),$$

$$u_1 = \frac{1}{2} \left(r^3 \sin 3\theta - r \sin \theta \right),$$

$$u_2 = \frac{1}{8} - \frac{3}{4} r^2 \cos 2\theta - 3 \cos 3\theta + \frac{5}{8} r^4 \cos 4\theta.$$

10.6. Solve $\nabla^2 u = 0, u(1 + \varepsilon \cos \theta, \theta) = f(\theta)$, for (a) $f(\theta) = \cos \theta$, and (b) $f(\theta) = \sin \theta$.
ANS. (a) $u_0 = r \cos \theta, u_1 = -\frac{1}{2}(1 + r^2 \cos 2\theta), u_2 = \frac{1}{2}(r^3 \cos 3\theta + r \cos \theta)$;
(b) $u_0 = r \sin \theta, u_1 = -\frac{1}{2}r^2 \sin 2\theta, u_2 = \frac{1}{2}(r^3 \sin 3\theta + r \sin \theta)$.

10.7. Find the exact solution of $y u_x + (x + \varepsilon u) u_y = 0, u(x, 1) = x$, by the method of characteristics (Chapter 2), and then find an approximate solution (up to three terms of a series solution or three approximations by the method of successive approximation) of this partial differential equation by perturbation methods.
ANS. $(1 + 2\varepsilon) u^2 - 2\varepsilon u x + y^2 - x^2 - 1 = 0$,

$$u = \frac{\varepsilon x + \sqrt{[(1 + x^2 - y^2) + 2\varepsilon(1 + x^2 - y^2) + \varepsilon^2 x^2]}}{1 + 2\varepsilon}.$$

The series solution is

$$u = \sqrt{1 + x^2 - y^2} + \left(x - \sqrt{1 + x^2 - y^2} \right) \varepsilon$$
$$+ \left(\frac{4x^2 + 3\left(1 - y^2\right)}{2\sqrt{1 + x^2 - y^2}} - 2x \right) \varepsilon^2 + \cdots.$$

10.8. Find the exact solution of

$$\frac{u_x}{1 + y^2} + (x + \varepsilon u^2) u_y = 0, \quad u(x, 0) = x,$$

by the method of characteristics (Chapter 2), and then find an approximate solution (up to three terms of a series solution or three approximations by the method of successive approximation) of the partial differential equation by perturbation techniques.

ANS. $u^2 + 2\varepsilon u^2(u - x) = x^2 - 2\tan^{-1} y$,

$$u = \sqrt{x^2 - 2\tan^{-1} y} + \varepsilon\sqrt{x^2 - 2\tan^{-1} y}\left(x - \sqrt{x^2 - 2\tan^{-1} y}\right)$$
$$+ \varepsilon^2\sqrt{x^2 - 2\tan^{-1} y}\left(4x^2 - 4x\sqrt{x^2 - 2\tan^{-1} y} - 5\tan^{-1} y\right) + \cdots.$$

10.9. Solve $\nabla^2 u = 0$, $0 < x < \pi$, $y > \varepsilon x$, $0 < \varepsilon \ll 1$, such that $u(0, y) = u(\pi, y) = 0$, and $u(x, \varepsilon x) = \sin x$.

SOLUTION. Using (10.2), we have

$$u(x, \varepsilon x) = \sum_0^\infty \varepsilon^n x^n \frac{\partial^n u}{\partial^n y}(x, 0).$$

Hence,

$$\sin x = \sum_0^\infty \varepsilon^n u_n(x, 0) + \varepsilon x \sum_0^\infty \varepsilon^n \frac{\partial u_n}{\partial y}(x, 0) + \frac{1}{2}\varepsilon^2 x^2 \sum_0^\infty \varepsilon^n \frac{\partial^2 u_n}{\partial y^2}(x, 0) + \cdots.$$

Comparing powers of ε on both sides, we get

$$u_0(x, 0) = \sin x,$$
$$u_1(x, 0) + x\frac{\partial u_0}{\partial y}(x, 0) = 0,$$
$$u_2(x, 0) + x\frac{\partial u_1}{\partial y}(x, 0) + \frac{x^2}{2}\frac{\partial^2 u_0}{\partial y^2}(x, 0) = 0,$$

and conditions on $u_n(x, 0)$ can be derived similarly. All u_n satisfy the Laplace equation. The solution for u_0 is clearly given by $u_0 = e^{-y}\sin x$, and u_1 satisfies the following conditions:

$$u_1(0, y) = u_1(\pi, y) = 0, \quad u_1(x, 0) = x\sin x.$$

Assuming a solution of the form

$$u_1 = \sum_{n=1}^\infty A_n e^{-ny}\sin nx,$$

we find that

$$A_1 = \frac{\pi}{2}, \quad A_n = -\frac{4n[1 + (-1)^n]}{\pi(n^2 - 1)^2}, \quad n \neq 1.$$

Thus, u_1 is completely determined. We continue this process and determine u_n for higher values of n.

10.10. Obtain a steady-state perturbation solution for the following problem: $u_t + ku_{yyt} = ku_{yyy} + u_{yy} + u_y$, subject to the conditions: $u(0, t) = e^{i\omega t}$, $\lim\limits_{y \to \infty} u(y, t) = 0$, and $\lim\limits_{k \to 0} u(y, t, k) = u(y, t, 0)$. Steady-state in this case implies that the initial conditions are to be ignored.

(a) Find a solution of the form $u = e^{-hy + i\omega t}$, where $\Re\{h\} > 0$ and assume that $h \approx h_0 + h_1 k + O(k^2)$.

(b) Same as in part (a) except that $h \approx h_0 + h_1 k + h_2 k^2 + O(k^3)$.

(c) Find a solution of the form $u = u_0 + ku_1 + k^2 u_2 + O(k^3)$.

ANS. Parts (a) and (b): Take $u = e^{-(a+ib)y + i\omega t}$. Then

$$a + ib = \frac{1}{2}(1 + \alpha + i\beta) + \left(\frac{A + iB}{2(\alpha^2 + \beta^2)}\right)k + \left(\frac{C + iD}{\alpha^2 + \beta^2}\right)k^2 + O(k^3),$$

$$\alpha = \sqrt{\frac{1}{2}(\sqrt{1 + 16\omega^2} + 1)}, \quad \beta = \sqrt{\frac{1}{2}(\sqrt{1 + 16\omega^2} - 1)},$$

$$A = \alpha + \alpha^2 + \beta^2 + 4\beta\omega - 2\alpha\omega^2,$$

$$B = 2\omega(\alpha^2 + \beta^2) - \beta + 4\alpha\omega + 2\omega^2\beta,$$

$$C = p\alpha + q\beta,$$

$$D = q\alpha - p\beta,$$

$$p = \frac{[(3 + 3\alpha - 2\omega\beta)A - (3\beta + 2\omega\alpha + 8\omega)B]\sqrt{1 + 16\omega^2} - A^2 + B^2}{4(1 + \omega^2)},$$

$$q = \frac{[(3 + 3\alpha - 2\omega\beta)B + (3\beta + 2\omega\alpha + 8\omega)A]\sqrt{1 + 16\omega^2} - 2AB}{4(1 + \omega^2)}.$$

Note that $C = 0 = D$ for part (a).

Part (c): $u_0 = e^{-(1+\alpha+i\beta)y/2 + i\omega t}$, $u_1 = P_1 y\, e^{-(1+\alpha+i\beta)y/2 + i\omega t}$,

$$u_2 = \left(P_2 y + P_3 y^2\right) e^{-(1+\alpha+i\beta)y/2 + i\omega t}, \text{ where}$$

$$P_1 = -\frac{(\alpha - i\beta)(1 + 4i\omega - 2\omega^2) + \sqrt{1 + 16\omega^2}\,(1 + 2i\omega)}{2\sqrt{1 + 16\omega^2}},$$

$$P_2 = \frac{P_1^2(\alpha - i\beta)}{\sqrt{1 + 16\omega^2}}, \quad P_3 = \frac{1}{2}P_1^2.$$

HINT. Assume a solution of the form $u(y, t) = F(y)\, e^{i\omega t}$. Substituting this solution into the given equation, we get $i\omega F + i\omega F'' = kF''' + F'' + F'$. This equation is an ordinary differential equation with constant coefficients and therefore has solutions of the form e^{-hy}. The characteristic equation is given by $-kh^3 + (1 - i\omega k)h^2 - h - i\omega = 0$. Its valid solutions are those for which $\Re\{h\} > 0$. Thus, assume solutions of the form $h = h_0 + kh_1 + k^2 h_2$, where $h_2 = 0$ for part (a).

10.11. Find the steady-state solution of

$$\frac{\partial u}{\partial t} - s_0 \frac{\partial u}{\partial z} = \frac{\partial U}{\partial t} + \nu \frac{\partial^2 u}{\partial z^2} - M\,(u - U),$$

subject to the conditions $u(0, t) = 0$, and $\lim\limits_{z \to \infty} u(z, t) = U(t) = U_0\left(1 + \varepsilon e^{i\omega t}\right)$, with ε as the perturbation parameter, where U_0 denotes the constant mean free-stream velocity, s_0 the constant suction velocity, ν the kinematic viscosity, and M a constant representing the influence of the magnetic field. This problem represents the fluctuating flow of a viscous fluid past a porous infinite plate in the presence of a magnetic field.

ANS. $u(y, t) = u(\zeta, t) = U_0\left[1 - e^{-m_1 \zeta} + \varepsilon\left(1 - e^{-m_2 \zeta}\right) e^{i\omega t}\right]$, where

$$\zeta = \frac{s_0 y}{\nu}\,,\; m_1 = -\frac{1}{2}\left(1 + \sqrt{1 + 4M_1}\right),$$
$$m_2 = -\tfrac{1}{2}\left(1 + \sqrt{1 + 4M_1 + 4i\lambda}\right),$$
$$M_1 = \frac{M\nu}{s_0^2}\,,\; \lambda = \frac{\omega\nu}{s_0^2}.$$

11

Finite Difference Methods

Among different numerical techniques for solving boundary and initial value problems, the finite difference methods are widely used. These methods are derived from the truncated Taylor's series, also known as Taylor's formula, where a given partial differential equation and the boundary and initial conditions are replaced by a set of algebraic equations that are then solved by various well-known numerical techniques. These methods have a significant advantage over other methods because of its simplicity of analysis and computer codes in solving problems with complex geometries. We will discuss difference schemes for first- and second-order partial derivatives, and then apply them to numerically solve boundary and initial value problems for second-order partial differential equations.

11.1. Finite Difference Schemes

Consider a single-valued, finite function $u(x)$ that belongs to the class $C^\infty(R^1)$. Then by Taylor's theorem,

$$u(x+h) = u(x) + h\,u'(x) + \frac{h^2}{2!}\,u''(x) + \frac{h^3}{3!}\,u'''(x) + \cdots, \qquad (11.1)$$

$$u(x-h) = u(x) - h\,u'(x) + \frac{h^2}{2!}\,u''(x) - \frac{h^3}{3!}\,u'''(x) + \cdots. \qquad (11.2)$$

If we subtract (11.2) from (11.1), we get

$$u(x+h) - u(x-h) = 2h\,u'(x) + O(h^3).$$

Then

$$u'(x) = \frac{u(x+h) - u(x-h)}{2h},$$ (11.3)

with a truncation error of $O(h^3)$. The approximation formula (11.3) is known as the *first-order central difference formula*. Geometrically, it represents the slope of the chord $\overline{AB}$ (Fig. 11.1).

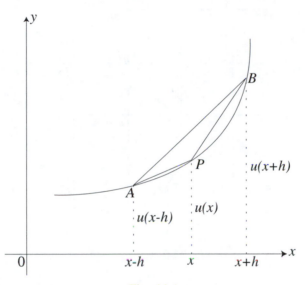

Fig. 11.1.

Similarly,

$$u'(x) = \frac{u(x) - u(x-h)}{h},$$ (11.4)

and

$$u'(x) = \frac{u(x+h) - u(x)}{h},$$ (11.5)

each with a truncation error of $O(h^2)$, are known as the *backward* and *forward difference formulas*, respectively, representing the slope of the chord $\overline{PA}$ and $\overline{BP}$, respectively (Fig. 11.1).

If we add (11.1) and (11.2), we get

$$u(x+h) + u(x-h) = 2\,u(x) + h^2\,u''(x) + O(h^4),$$

which yields the second-order central difference formula

$$u''(x) = \frac{u(x+h) - 2u(x) + u(x-h)}{h^2},$$ (11.6)

with a truncation error of $O(h^4)$. Based on the Taylor series expansion, the truncation error is approximately $h\,f'''(x)$ for both forward and backward schemes, and $\frac{1}{12}h^2 f^{(4)}(x)$ for the central scheme.

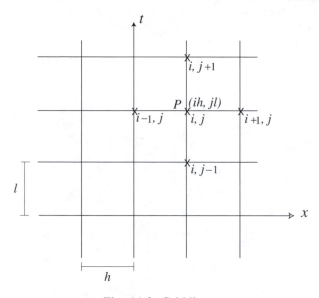

Fig. 11.2. Grid lines.

In the case of a function $u(x,t)$ of two independent variables x and t, we partition the x-axis into intervals of equal length h, and the t-axis into intervals of equal length l. The (x,t)-plane is divided into equal rectangles of area hl by the grid lines parallel to Ot, defined by $x_i = ih$, $i = 0, \pm 1, \pm 2, \ldots$, and by the grid lines parallel to Ox, defined by $y_j = jl$, $j = 0, \pm 1, \pm 2, \ldots$ (Fig. 11.2). We will use the following notation: Let $u_P = u(ih, jl) = u_{i,j}$ denote the value of the function $u(x,t)$ at a mesh point (node) $P\,(ih, jl)$. Then, in view of formula (11.6), we have the following three central difference schemes:

$$
\left.\frac{\partial^2 u}{\partial x^2}\right|_P = \left.\frac{\partial^2 u}{\partial x^2}\right|_{i,j} = \frac{u\left((i+1)h, jl\right) - 2u(ih, jl) + u\left((i-1)j, jl\right)}{h^2}
$$
$$
= \frac{U_{i+1,j} - 2U_{i,j} + U_{i-1,j}}{h^2} \equiv \frac{\delta_x^2\, U_{i,j}}{h^2}, \quad i = 1, 2, \ldots, n-1,
$$

(11.7)

with a truncation error $-\dfrac{h^2}{12}u_{xxxx}(\bar{x}, t)$, where $x_{i-1} < \bar{x} < x_i$;

$$
\left.\frac{\partial^2 u}{\partial t^2}\right|_P = \left.\frac{\partial^2 u}{\partial t^2}\right|_{i,j} = \frac{U_{i,j+1} - 2U_{i,j} + U_{i,j-1}}{l^2} \equiv \frac{\delta_t^2\, U_{i,j}}{l^2}, \quad j = 1, 2, \ldots, m-1,
$$

(11.8)

with a truncation error $-\dfrac{l^2}{12}u_{tttt}(x,t')$, where $t_{j-1} < t' < t_j$; and

$$\frac{\partial^2 u}{\partial x \partial t}\bigg|_P = \frac{U_{i+1,j+1} - U_{i+1,j-1} - U_{i-1,j+1} + U_{i-1,j-1}}{4hl}, \qquad (11.9)$$

with a truncation error $-\dfrac{h^2}{6}u_{xxxt}(\bar{x},\bar{t}) - \dfrac{l^2}{6}u_{xttt}(x',t')$. Note that the difference operators δ_x^2 and δ_t^2 are the finite difference analogs of the partial differential operators $\dfrac{\partial^2}{\partial x^2}$ and $\dfrac{\partial^2}{\partial t^2}$, respectively. Moreover, from (11.3) we have the first-order central difference formulas

$$\frac{\partial u}{\partial x}\bigg|_P = \frac{u\left((i+1)h, jl\right) - u\left((i-1)h, jl\right)}{h} = \frac{U_{i+1,j} - U_{i-1,j}}{h}, \qquad (11.10)$$

with a truncation error $-\dfrac{h^2}{6}u_{xxx}(\bar{x},\bar{t})$, and

$$\frac{\partial u}{\partial t}\bigg|_P = \frac{U_{i,j+1} - U_{i,j-1}}{l}, \qquad (11.11)$$

with a truncation error $-\dfrac{l^2}{6}u_{ttt}(x,t')$. The first-order backward and forward difference formulas for $u(x,t)$ are similarly derived from (11.4) and (11.5). The second-order forward and backward difference schemes are defined, respectively, by

$$\frac{\partial^2 u}{\partial x^2}\bigg|_{i,j} = \frac{U_{i+2,j} - 2U_{i+1,j} + U_{i,j}}{h^2}, \quad i = 0,1,2,\dots,n-2, \qquad (11.12)$$

and

$$\frac{\partial^2 u}{\partial x^2}\bigg|_{i,j} = \frac{U_{i,j} - 2U_{i-1,j} + U_{i-2,j}}{h^2}, \quad i = 0,1,2,\dots,n-2. \qquad (11.13)$$

11.2. First-Order Equations

Consider the first-order quasilinear partial differential equation of the form

$$a\,u_x + b\,u_y = c, \quad \text{or} \quad a\,p + b\,q = c, \qquad (11.14)$$

where a, b, and c are functions of x, y, and u only (see §2.3). This equation yields the auxiliary system of ordinary differential equations

$$\frac{dx}{a} = \frac{dy}{b} = \frac{du}{c}. \qquad (11.15)$$

It is shown in Chapter 2 that at each point of the solution domain of Eq (11.14) there exists a direction, known as the characteristic, along which the solution of this partial differential equation coincides with the solutions of the ordinary differential equations (11.15). We will denote by C the projection of this characteristic in the (x, y)-plane. Let us assume that the solution of u is known at every point of the characteristic C in the (x, y)-plane, and that C is distinct from the initial curve Γ, on which the initial value of u is prescribed.

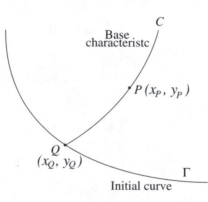

Fig. 11.3.

Let u be prescribed on the initial curve Γ, which does not coincide with any characteristic C for the partial differential equation (11.14). Let $P(x_P, y_P)$ be a point on the characteristic C, which passes through a point $Q(x_Q, y_Q)$ on Γ such that P and Q are close to each other, i.e., $|x_P - x_Q|$ is small (Fig. 11.3).

We denote the m-th approximation of u by $u^{(m)}$, and of y by $y^{(m)}$ for $m = 1, 2, \dots$. Let us assume that x_P is known. Then, in view of (11.3) and (11.4), the *first approximation* of $a\,dy = b\,dx$ is given by

$$a_Q\left(y_P^{(1)} - y_Q\right) = b_Q\left(x_P - x_Q\right), \qquad (11.16)$$

which yields $y_P^{(1)}$, whereas from $a\,du = c\,dx$ we have

$$a_Q\left(u_P^{(1)} - u_Q\right) = c_Q\left(x_P - x_Q\right), \qquad (11.17)$$

which yields $u_P^{(1)}$. For the *second approximation* we have, by using the average values,

$$\frac{a_Q + a_P^{(1)}}{2}\left(y_P^{(2)} - y_Q\right) = \frac{b_Q + b_P^{(1)}}{2}\left(x_P - x_Q\right), \qquad (11.18)$$

which yields $y_P^{(2)}$, and

$$\frac{a_Q + a_P^{(1)}}{2}\left(u_P^{(2)} - u_Q\right) = \frac{c_Q + c_P^{(1)}}{2}\left(x_P - x_Q\right), \qquad (11.19)$$

which yields $u_P^{(2)}$. Subsequent (higher) approximations are similarly obtained.

EXAMPLE 11.1. Consider the quasilinear partial differential equation

$$x^{3/2} u_x - u\, u_y = u^2,$$

where $u = 1$ on the initial line $y = 0$, $0 < x < \infty$. The auxiliary system of equations is

$$\frac{dx}{x^{3/2}} = \frac{dy}{-u} = \frac{du}{u^2}.$$

Solving $\dfrac{dx}{x^{3/2}} = \dfrac{du}{u^2}$, we find that $\dfrac{2}{\sqrt{x}} = \dfrac{1}{u} - A$, where A is constant on a particular characteristic C. Thus, at a point $(x_Q, 0)$ on the initial curve Γ, we know that $u = 1$, and that gives $A = 1 - \dfrac{2}{\sqrt{x_Q}}$, which yields the solution along the base characteristic C_Q as

$$\frac{1}{u} = 1 + \frac{2}{\sqrt{x}} - \frac{2}{\sqrt{x_Q}}. \tag{11.20}$$

Similarly, solving $\dfrac{dy}{-u} = \dfrac{du}{u^2}$, we have $y = -\ln(Bu)$, and the solution on the characteristic C_Q, with $B = 1$, is

$$u = e^{-y}. \tag{11.21}$$

Thus, the solution along a characteristic C_Q is given by (11.20) or (11.21). Eliminating u between Eqs (11.20) and (11.21), we find the equation of the characteristic C_Q as

$$y = \ln\left(1 + \frac{2}{\sqrt{x}} - \frac{2}{\sqrt{x_Q}}\right). \tag{11.22}$$

Now, we find the first and the second approximations at the point $P(1.1, y)$. Let $x_Q = 1$, $x = 1.1$. Then $dx = x_P - x_Q = 0.1$, $y_Q = 0$, and $u_Q = 1$.

FIRST APPROXIMATION. In view of (11.18) and (11.19), from $x^{3/2}\, dy = -u\, dx$ we obtain

$$x_Q^{3/2}\left(y_P^{(1)} - 0\right) = -u_Q\, dx,$$

which gives

$$y_P^{(1)} = -\frac{u_Q}{x_Q^{3/2}}\, dx = -0.1.$$

Also from $x_Q^{3/2}\, du = u_Q^2\, dx$, we get

$$x_Q^{3/2}\left(u_P^{(1)} - 1\right) = u_Q^2\, dx,$$

which yields

$$u_P^{(1)} = 1 + \frac{u_Q^2}{x_Q^{3/2}}\, dx = 1 + \frac{1}{(1)^{3/2}}\,(0.1) = 1.01.$$

SECOND APPROXIMATION. The equation $u\, dx = -x^{3/2}\, dy$ gives

$$\frac{u_Q + u_P^{(1)}}{2}\, dx = -\frac{x_Q^{3/2} + x_P^{3/2}}{2}\left(y_P^{(2)} - y_Q\right),$$

$$\frac{1 + 1.01}{2}\,(0.1) = -\frac{(1)^{3/2} + (1.1)^{3/2}}{2}\,y_P^{(2)},$$

which yields $y_P^{(2)} = -0.097507$. Also, from $x^{3/2}\, du = u^2\, dx$, we have

$$\frac{x_Q^{3/2} + x_P^{3/2}}{2}\left(u_P^{(2)} - u_Q\right) = \frac{u_Q^2 + u_P^2}{2}\, dx,$$

$$\frac{(1)^{3/2} + (1.1)^{3/2}}{2}\left(u_P^{(2)} - 1\right) = \frac{(1)^2 + (1.1)^2}{2}\, dx,$$

which yields $u_P^{(2)} = 1.10261$.

The exact values from (11.22) and (11.20) for $x_Q = 1$, $x = 1.1$ are

$$y_P = \ln\left(1 + \frac{2}{\sqrt{1.1}} - \frac{2}{1}\right) = -0.0976953,$$

$$u_P = e^{-y_P} = 0.91485.$$

Higher-order approximations are continued until the computed values differ from each other within the preassigned tolerance. ∎

11.3. Second-Order Equations

Let a region Ω in the (x, t)-plane be partitioned into a grid (x_i, t_j), $0 \le i \le n$, and $0 \le j \le m$, as in Fig. 11.2. After replacing all derivatives in a given partial differential equation

$$L\,u = f, \quad \text{for } x,\, t \in \Omega, \tag{11.23}$$

by their respective difference quotients, we obtain a finite difference equation of the form

$$D\,U_{i,j} = f_{i,j} \quad \text{for } (x_i, t_j) \in \Omega, \tag{11.24}$$

where D denotes the difference operator. Note that Eq (11.24) is the discretized form of the given equation (11.23) such that the solution $U_{i,j}$ approximates $u(x,t)$ at the grid nodes.

DEFINITION 11.1. The *local truncation error* $\varepsilon_{i,j}$ is the amount by which the solution $U_{i,j}$ fails to satisfy Eq (11.23), i.e.,

$$\varepsilon_{i,j} = D\,U_{i,j} - f_{i,j}. \tag{11.25}$$

DEFINITION 11.2. The difference equation (11.24) is said to be *consistent* with the given partial differential equation (11.23) if

$$\lim_{h,l \to 0} \varepsilon_{i,j} = 0. \tag{11.26}$$

DEFINITION 11.3. The *discretization error* $V_{i,j}$ is defined by $V_{i,j} = U_{i,j} - u_{i,j}$, where $U_{i,j}$ is the exact solution of Eq (11.24), and $u_{i,j}$ is the solution of Eq (11.23) evaluated at (x_i, t_j).

DEFINITION 11.4. The difference scheme, defined by Eq (11.24), is said to be *convergent* if

$$\lim_{h,l \to 0} |V_{i,j}| = \lim_{h,l \to 0} |U_{i,j} - u_{i,j}| = 0 \quad \text{for } (x_i, t_j) \in \Omega. \tag{11.27}$$

In some cases the difference method may not be convergent, although it may be consistent. There are examples discussing these issues in Abbott and Basco (1990). Also, an example suggested by Du Fort and Frankel (see Carrier and Pearson, 1988, p. 263) is

$$\frac{U_{i,j+1} - U_{i,j-1}}{2l} = a^2 \frac{U_{i+1,j} - U_{i,j+1} - U_{i,j-1} - U_{i-1,j}}{h^2},$$

which is always stable, where a is a constant, but is not consistent unless $(l/h) \to 0$ as $l \to 0$ and $h \to 0$.

The concept of stability of a finite difference scheme is based on the propagation of the error $E_{i,0} = V_{i,0} - U_{i,0}$ with increasing j, where $U_{i,0}$ denote the initial values for the difference equation (11.24), and $V_{i,0}$ are the initial values obtained from the solution of a perturbed difference system. For a partial differential equation with a bounded solution, the difference scheme (11.24) is said to be *stable* if the errors $E_{i,j} = V_{i,j} - U_{i,j}$ are uniformly bounded in i as $j \to \infty$, i.e.,

$$|E_{i,j}| < M \quad \text{for } j > J,$$

where M is a positive constant and J a positive integer. A theorem, known as the *Lax equivalence theorem*, states that stability of a solution is both a necessary and sufficient condition for the convergence of a finite difference problem which is consistent with a well-posed initial and boundary value problem.

To illustrate the finite difference method for second-order equations, we first consider a very simple example.

EXAMPLE 11.2. Let a one-dimensional steady-state heat conduction problem be defined by

$$u'' = -x^2, \quad 0 < x < 1, \quad u(0) = 1, \quad u(1) = 2. \tag{11.28}$$

A partition of the interval $[0, 1]$ into equally spaced points is given by

$$0 = x_0 < x_1 < \cdots < x_n = 1,$$

with the step size $h = x_{i+1} - x_i = \dfrac{1}{n}$. Using the forward difference scheme (11.12) on the problem (11.28), we have

$$\frac{U_{i+2} - 2U_{i+1} + U_i}{h^2} = -x_i^2, \quad i = 0, 1, 2, \dots, n-1.$$

In particular, say, for $n = 4$, we get the system of equations

$$16 \left(U_2 - 2U_1 + U_0\right) = 0,$$
$$16 \left(U_3 - 2U_2 + U_1\right) = -\frac{1}{16},$$
$$16 \left(U_4 - 2U_3 + U_2\right) = -\frac{1}{4}.$$

In view of the boundary conditions, we have $U_0 = 1$, and $U_4 = 2$. Then the above system yields

$$2U_1 - U_2 = 1,$$
$$U_1 - 2U_2 + U_3 = -\frac{1}{256},$$
$$U_2 - 2U_3 = -\frac{129}{4},$$

which, by using the Gauss elimination method, gives

$$U_1 = \frac{643}{512} = 1.25586, \quad U_2 = \frac{387}{256} = 1.51172, \quad U_3 = \frac{903}{512} = 1.7367.$$

Alternatively, since only u'' occurs in problem (11.28), we can use the central difference scheme (11.7). Thus,

$$\frac{U_{i+1} - 2U_i + U_{i-1}}{h^2} = -x_i^2, \quad i = 1, 2, \dots, n-1.$$

Then, for $n = 4$, we get the system of equations

$$16\left(U_2 - 2U_1 + U_0\right) = -\frac{1}{16},$$
$$16\left(U_3 - 2U_2 + U_1\right) = -\frac{1}{4},$$
$$16\left(U_4 - 2U_3 + U_2\right) = -\frac{9}{16},$$

or

$$2U_1 - U_2 = \frac{257}{256},$$
$$U_1 - 2U_2 + U_3 = -\frac{1}{64},$$
$$U_2 - 2U_3 = -\frac{521}{256},$$

which gives

$$U_1 = \frac{325}{256} = 1.29653, \quad U_2 = \frac{393}{256} = 1.53516, \quad U_3 = \frac{457}{256} = 1.7816.$$

The exact solution is

$$u(x) = 1 + \frac{13}{12}x - \frac{x^4}{12}.$$

A comparison with the exact solution shows that the central difference scheme gives a better approximation for problem (11.22). The results are shown in the following table.

x	Forward Difference	Central Difference	Exact
0.0	1.0	1.0	1.0
0.25	1.25586	1.26953	1.27051
0.5	1.51172	1.53516	1.53646
0.75	1.76367	1.78516	1.78613
1.0	2.0	2.0	2.0 ∎

Note that if the interval is $[a, b]$, then we use the transformation

$$\xi = \frac{x - a}{b - a},$$

to reduce this interval to $[0, 1]$.

11.3.1. Diffusion Equation. Consider the one-dimensional diffusion equation

$$u_t = k\,u_{xx}, \tag{11.29}$$

where k denotes the thermal diffusivity. For the grid $(x_i, t_j) = (ih, jl)$, we will discuss the following three finite difference schemes:

(a) FORWARD DIFFERENCE (EXPLICIT SCHEME):

$$\frac{U_{i,j+1} - U_{i,j}}{l} = k \frac{U_{i+1,j} - 2U_{i,j} + U_{i-1,j}}{h^2},$$

or

$$U_{i,j+1} = \left(1 + r\,\delta_x^2\right) U_{i,j}, \tag{11.30}$$

where $r = k\,l/h^2$.

(b) BACKWARD DIFFERENCE (IMPLICIT SCHEME):

$$\frac{U_{i,j+1} - U_{i,j}}{l} = k \frac{U_{i+1,j+1} - 2U_{i,j+1} + U_{i-1,j+1}}{h^2},$$

or

$$\left(1 - r\,\delta_x^2\right) U_{i,j+1} = U_{i,j}. \tag{11.31}$$

(c) CRANK-NICOLSON (IMPLICIT SCHEME):

$$\frac{U_{i,j+1} - U_{i,j}}{l} = k \frac{\delta_x^2 U_{i,j} + \delta_x^2 U_{i,j+1}}{h^2},$$

or

$$\left(1 - \frac{r}{2}\delta_x^2\right) U_{i,j+1} = \left(1 + \frac{r}{2}\delta_x^2\right) U_{i,j}. \tag{11.32}$$

Note that the Crank-Nicolson scheme is derived by averaging the finite differences at the points (i, j) and $(i, j+1)$ (see Exercise 11.2). Also, the above forward difference scheme (11.30) is conditionally stable iff $r < 1/2$, but the other two schemes (11.31) and (11.32) are always stable.

EXAMPLE 11.3. Consider the boundary value problem

$$\begin{aligned} u_t &= u_{xx}, \quad 0 < x < 1, \quad t > 0, \\ u(0, t) &= 0 = u(1, t), \quad \text{for } t > 0, \\ u(x, 0) &= f(x), \quad \text{for } 0 < x < 1. \end{aligned} \tag{11.33}$$

This problem is studied in Example 5.2 by the separation of variables method. Since the space derivative is of second order, we will use the central difference scheme (11.7), where $U_{i,j} = u(x_i, t_j)$ denotes the temperature at a point x_i at time t_j. For the time derivative we use the forward difference scheme

$$u_t = \frac{U_{i,j+1} - U_{i,j}}{l},$$

where $l = t_{j+1} - t_j$. Then Eq (11.33) is approximated by

$$r\left[U_{i+1,j} - 2U_{i,j} + U_{i-1,j}\right] = U_{i,j+1} - U_{i,j},$$
$$i = 1, 2, \ldots, n-1, \quad \text{and} \quad j = 0, 1, 2, \ldots, \tag{11.34}$$

where $r = l/h^2$, and the boundary and initial conditions become

$$U_{0,j} = 0 = U_{n,j}, \quad \text{for } j = 1, 2, \ldots,$$
$$U_{i,0} = f(x_i) = f_i, \quad \text{for } i = 0, 1, 2, \ldots, n.$$

After rearranging the terms in Eq (11.34), we get

$$U_{i,j+1} = r\, U_{i-1,j} + (1 - 2r)\, U_{i,j} + r\, U_{i+1,j}. \tag{11.35}$$

This difference equation allows us to compute $U_{i,j+1}$ from the values of U that are computed for earlier times. Note that the value $U_{i,0} = f_i$ is a prescribed value of u at time $t = 0$ and $x = x_i$.

As an example, we take $n = 4$, i.e., $h = 1/4$. The value of r in Eq (11.35) must be chosen properly so that the solution remains stable. It has been determined that for a stable solution the value of r must be such that the coefficients of u on the right side of Eq (11.35) remain nonnegative (Smith, 1985, Ch. 3). Hence, we must have $0 < r \le \dfrac{1}{2}$. Then, in view of this restriction, we must have $l \le rh^2 = \dfrac{1}{32}$. In the marginal case when $r = 1/2$, we get $l = 1/32$. With these values of r and l, the system (11.35) reduces to

$$U_{i,j+1} = \frac{1}{2}\left[U_{i-1,j} + U_{i+1,j}\right], \quad j = 0, 1, 2, \ldots.$$

Since $U_{0,j} = U_{4,j} = 0$ for all j, we find that for $j = 0$:

$$U_{i,1} = \frac{1}{2}\left[U_{i-1,0} + U_{i+1,0}\right],$$

or, successively,

$$U_{1,1} = \frac{1}{2}f_2, \quad U_{2,1} = \frac{1}{2}(f_1 + f_2), \quad U_{3,1} = \frac{1}{2}f_2;$$

and for $j = 1$:

$$U_{1,2} = \frac{1}{2}U_{2,1}, \quad U_{2,2} = \frac{1}{2}(U_{1,1} + U_{3,1}), \quad U_{3,2} = \frac{1}{2}U_{2,1},$$

and so on. The values of $U_{i,j}$ for $f(x) = \cos \pi x$ are listed below in a tabular form for some successive values of t.

$2\,t$	$x=0$	$x=0.25$	$x=0.5$	$x=0.75$	$x=1$
$2\,0$	1	$1/\sqrt{2} \approx 0.7071$	0	$-1/\sqrt{2} \approx -0.7071$	-1
$1/32$	0	0.5	0	-0.5	0
$1/16$	0	0	0	0	0
$3/32$	0	0	0	0	0
$1/8$	0	0	0	0	0

Notice that the values average out in the outer columns. This will happen if $r = 1/2$ is chosen. The solution for problem (11.33) for different values of t with $r = 0.1$ is presented in the following table:

$2\,t$	$x=0$	$x=0.25$	$x=0.5$	$x=0.75$	$x=1$
$2\,0$	1	$1/\sqrt{2} \approx 0.7071$	0	$-1/\sqrt{2} \approx -0.7071$	-1
$1/32$	0	0.665685	0	-0.665685	0
$1/16$	0	0.532548	0	-0.532548	0
$3/32$	0	0.426039	0	-0.426039	0
$1/8$	0	0.340831	0	-0.340831	0 ∎

11.3.2. Wave Equation. To approximate the solution of the wave equation

$$u_{tt} = c^2\, u_{xx}, \tag{11.36}$$

we use the difference schemes (11.30)–(11.32) discussed earlier for the diffusion equation, but more frequently used schemes are the forward difference and the Crank-Nicolson. Let $(x_i, t_j) = (ih, jl)$, $(i, j = 0, 1, \dots)$. Then

(a) EXPLICIT SCHEME (FORWARD DIFFERENCE):

$$\frac{U_{i,j+1} - 2U_{i,j} + U_{i,j-1}}{l^2} = c^2\, \frac{U_{i+1,j} - 2U_{i,j} + U_{i-1,j}}{h^2},$$

or

$$\delta_t^2\, U_{i,j} = c^2 \rho^2\, \delta_x^2\, U_{i,j}, \tag{11.37}$$

where $\rho = l/h$.

(b) IMPLICIT SCHEME (CRANK-NICOLSON):

$$\frac{U_{i,j+1} - 2U_{i,j} + U_{i,j-1}}{l^2} = \frac{c^2}{2} \Big[\frac{U_{i+1,j+1} - 2U_{i,j+1} + U_{i-1,j+1}}{h^2} + \frac{U_{i+1,j-1} - 2U_{i,j-1} + U_{i-1,j-1}}{h^2} \Big],$$

or

$$\delta_t^2 U_{i,j} = \frac{c^2 \rho^2}{2} \left[\delta_x^2 U_{i,j+1} + \delta_x^2 U_{i,j-1} \right]. \qquad (11.38)$$

Note that the central difference schemes (11.7) and (11.8) are sometimes also used (see Example 11.5).

EXAMPLE 11.4. To use a finite difference scheme for solving the wave equation (11.36) with the Neumann initial conditions, i.e., for

$$u(x,0) = f(x),$$
$$u_t(x,0) = g(x),$$

we take $U_{i,0} = f(x_i) = f_i$. Now, we find the startup value of $U_{i,1}$ as follows: Under the assumption that $f \in C^2$, we have, by Taylor's theorem,

$$u(x_i, t_0) = u(x_i, 0) + l\, u_t(x_i, 0) + \frac{l^2}{2}\, u_{tt}(x_i, 0) + O(l^3)$$

$$= u(x_i, 0) + l\, g(x_i) + \frac{l^2}{2}\, c^2 f''(x_i) + O(l^3)$$

$$= u(x_i, 0) + l\, g_i + \frac{c^2 \rho^2}{2}\, [f_{i+1} - 2f_i + f_{i-1}] + O(l^2 h^2 + l^2),$$

where $g_i = g(x_i)$, $f_i = f(x_i)$, and (11.6) is used to approximate $f''(x_i)$. The above expansion gives

$$g_i = \frac{U_{i,1} - U_{i,0}}{l} - \frac{1}{2} c^2 \rho^2 [f_{i+1} - 2f_i + f_{i-1}],$$

where $U_{i,0} = f_i$. Then the startup value $U_{i,1}$ is given by

$$U_{i,1} = f_i + l \left[g_i + \frac{1}{2} c^2 \rho^2 \left(f_{i+1} - 2f_i + f_{i-1} \right) \right]. \blacksquare$$

EXAMPLE 11.5. Consider the wave equation $u_{tt} = u_{xx}$, $0 < x < 1$, $t > 0$, subject to the boundary conditions

$$u(0, t) = 0 = u(1, t) \quad \text{for } t > 0,$$

and the initial conditions

$$u(x, 0) = f(x), \quad u_t(x, 0) = g(x) \quad \text{for } 0 < x < 1.$$

This problem is solved in Example 5.1 by the separation of variables method. We use the central difference schemes (11.7) and (11.8) for both u_{xx} and u_{tt}. Thus,

$$u_{xx} = \frac{U_{i+1,j} - 2U_{i,j} + U_{i-1,j}}{h^2},$$

$$u_{tt} = \frac{U_{i,j+1} - 2U_{i,j} + U_{i,j-1}}{l^2}.$$

Then the wave equation, and the boundary and the initial conditions reduce to

$$U_{i,j+1} = \rho^2 U_{i-1,j} + 2(1 - \rho^2)U_{i,j} + \rho^2 U_{i+1,j} - U_{i,j-1},$$
$$i = 1, 2, \ldots, n-1, \quad j = 0, 1, 2, \ldots,$$
$$U_{0,j} = 0 = U_{n,j} \quad \text{for } j = 1, 2, \ldots,$$
$$U_{i,0} = f_i, \quad U_{i,1} - U_{i,0} = l\, g_i \quad \text{for } i = 0, 1, \ldots, n,$$

(11.39)

where $\rho = l/h$, and the forward difference scheme is used for the initial condition $u_t(x, 0) = g(x)$.

Let $f(x) = \sin x$, and $g(x) = 1 - x$. Then, with $h = l = 1/4$, the initial conditions become

$$U_{i,0} = f_i = \sin x_i \quad \text{for } i = 0, 1, 2, 3, 4,$$

$$U_{i,1} - U_{i,0} = l\, g_i = \frac{i}{4}(1 - x_i) \quad \text{for } i = 0, 1, 2, 3, 4.$$

Then the first equation in (11.39) becomes

$$U_{1,j+1} = U_{0,j} + U_{2,j} - U_{0,j-1},$$
$$U_{2,j+1} = U_{1,j} + U_{3,j} - U_{2,j-1},$$
$$U_{3,j+1} = U_{2,j} + U_{4,j} - U_{3,j-1}.$$

Since the boundary conditions are $U_{0,j} = 0 = U_{4,j}$, we get the solution for successive values of $j = 1, 2, \ldots$, which is presented in the following table.

$2\,t$	$x = 0$	$x = 0.25$	$x = 0.5$	$x = 0.75$	$x = 1$
2 0	0	0	0	0	0
0.25	0.2474	0.4272	0.3421	0.5728	0.6579
0.5	0.4794	0.5895	0.6707	0.7526	0.9021
0.75	0.6816	0.7230	−0.0921	−0.0522	0.8447
1.0	0	0	0	0	0 ■

EXAMPLE 11.6. Solve $u_{tt} = 4\, u_{xx}$, $0 < x < 1$, $t > 0$, subject to the initial conditions $u(x, 0) = \sin \pi x$, $u_t(x, 0) = 0$ for $0 < x < 1$, and the boundary conditions $u(0, t) = 0 = u(1, t)$ for $t > 0$, by using (a) the explicit scheme (11.37), and (b) the implicit scheme (11.38). Note that the exact solution is

$$u(x, t) = \sin \pi x \, \cos 2\pi t,$$

and the d'Alembert's solution is

$$u(x, t) = \frac{1}{2}\left[\sin(\pi x + 2\pi t) + \sin(\pi x - 2\pi t)\right].$$

See Mathematica Project 11.1 (§11.4). ∎

11.3.3. Poisson's Equation. Consider Poisson's equation in a rectangle $\Omega = \{0 < x < a,\, 0 < y < b\}$ with boundary Γ:

$$u_{xx} + u_{yy} = f(x,y), \tag{11.40}$$

subject to the Dirichlet boundary condition

$$u = g(x,y) \quad \text{on } \Gamma. \tag{11.41}$$

For $f = 0$, Eq (11.40) becomes the Laplace equation. We will analyze the simple case when $a = b$, with the uniformly spaced grid lines of size $h = l = a/4$. Then the nodes are given by $(x_m, y_n) = (mh, nl)$, where $m, n = 0, 1, 2, 3, 4$ (see Fig. 11.4). By using the central difference scheme (11.7), Eq (11.40) reduces to the finite difference equation

$$\delta_x^2 U_{m,n} + \delta_y^2 U_{m,n} = h^2 f_{m,n}, \tag{11.42}$$

where $f_{m,n} = f(x_m, y_n)$. The boundary condition (11.41) then becomes $U_{m,n} = g_{m,n} = g(x_m, y_n)$ for $m, n = 0, 1, 2, 3, 4$.

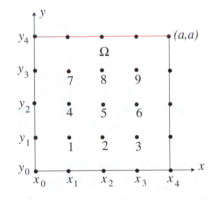

Fig. 11.4. Grid lines on the square Ω.

By reordering Eq (11.42), we get

$$4U_{m,n} - U_{m+1,n} - U_{m-1,n} - U_{m,n+1} - U_{m,n-1} = -h^2 f_{m,n}. \tag{11.43}$$

The unknown values of u are at the nodes $1, 2, \ldots, 9$ (Fig. 11.4). We use the notation:

$$U_{1,1} = U_1, \quad U_{2,1} = U_2, \quad U_{3,1} = U_3,$$
$$U_{1,2} = U_4, \quad U_{2,2} = U_5, \quad U_{3,2} = U_6,$$
$$U_{1,3} = U_7, \quad U_{2,3} = U_8, \quad U_{3,3} = U_9.$$

Then Eq (11.43) is written in the form of a system of algebraic equation

$$[A]\{U\} = \{F\}, \tag{11.44}$$

where the matrix $[A]$ is a 9×9 symmetric matrix

$$[A] = \begin{bmatrix} 4 & -1 & 0 & -1 & 0 & 0 & 0 & 0 & 0 \\ & 4 & -1 & 0 & -1 & 0 & 0 & 0 & 0 \\ & & 4 & -1 & 0 & -1 & 0 & 0 & 0 \\ & & & 4 & -1 & 0 & -1 & 0 & 0 \\ & & & & 4 & -1 & 0 & -1 & 0 \\ & & & & & 4 & 0 & 0 & -1 \\ & & & & & & 4 & -1 & 0 \\ & \text{sym} & & & & & & 4 & -1 \\ & & & & & & & & 4 \end{bmatrix}, \tag{11.45}$$

and

$$\{U\} = \begin{Bmatrix} U_1 \\ U_2 \\ U_3 \\ U_4 \\ U_5 \\ U_6 \\ U_7 \\ U_8 \\ U_9 \end{Bmatrix}, \quad \{F\} = \begin{Bmatrix} -h^2 f_{1,1} + g_{0,1} + g_{1,0} \\ -h^2 f_{2,1} + g_{2,0} \\ -h^2 f_{3,1} + g_{3,0} \\ -h^2 f_{1,2} + g_{0,2} \\ -h^2 f_{2,2} \\ -h^2 f_{3,1} + g_{4,2} \\ -h^2 f_{1,3} + g_{0,3} + g_{1,4} \\ -h^2 f_{2,3} + g_{2,4} \\ -h^2 f_{2,3} + g_{4,3} + g_{3,4} \end{Bmatrix}. \tag{11.46}$$

The difference method explained above has a truncation error of $O(h^2)$. If the boundary value problem (11.40)–(11.41) has a unique solution, h is small and $[A]$ is nonsingular, then system (11.44) has a unique solution. The system (11.44) can be easily solved.

In the case of a curved boundary Γ, consider a node in Ω, which has at least one adjacent node outside Ω (see Fig. 11.5).

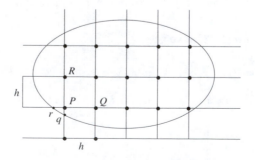

Fig. 11.5. Curved boundary Γ.

Let $P = (x_m, y_n)$ be a point inside Ω and near the boundary Γ. Then the coordinates of the adjacent points q and r on the boundary Γ are:

$$q = (x_{m+1}, y_n) = (x_m + \alpha h, y_n),$$
$$r = (x_m, y_{n+1}) = (x_m h, y_n + \beta h),$$

where $0 < \alpha, \beta < 1$. Moreover, the values of $u(q)$ and $u(r)$ are known, since u is prescribed on the boundary Γ. By Taylor's theorem

$$u(q) = u(P) + \alpha\, h\, u_x(P) + \frac{\alpha^2 h^2}{2!}\, u_{xx}(P) + O(h^3),$$

$$u(Q) = u(P) - h\, u_x(P) + \frac{h^2}{2!}\, u_{xx}(P) + O(h^3).$$

After eliminating $u_x(P)$ from these two expansions, we obtain

$$u_{xx}(P) = \frac{2[u(q) - (1+\alpha)u(P) + \alpha\, u(Q)]}{\alpha(1+\alpha)} + O(h).$$

Similarly,

$$u_{yy}(P) = \frac{2[u(r) - (1+\beta)u(P) + \beta\, u(R)]}{\beta(1+\beta)} + O(h).$$

Hence, the finite difference approximation for the Poisson's equation (11.40) defined on a region with a curved boundary Γ is

$$\frac{U(Q)}{1+\alpha} + \frac{U(R)}{1+\beta} - \left(\frac{1}{\alpha} + \frac{1}{\beta}\right) U(P) + \frac{U(q)}{\alpha(1+\alpha)} + \frac{U(r)}{\beta(1+\beta)} = \frac{1}{2} h^2 f(P). \quad (11.47)$$

EXAMPLE 11.7. To find finite difference solutions of the Laplace equation $u_{xx} + u_{yy} = 0$ on the quarter circular region $\Omega = \{x^2 + y^2 < 1, y > 0\}$, subject to the boundary conditions $u(x, y) = 10$ on $x^2 + y^2 = 1, y > 0\}$, and $u(x, y) = 0$ on $0 < x < 1, y = 0$ and $u_x = 0$ on $x = 0, 0 < y < 1$, we choose the grid with $h = 1/2$ (see Fig. 11.6).

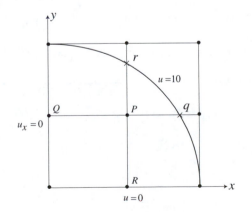

Fig. 11.6.

From the boundary conditions we have $U_{0,0} = 0 = U_{1,0}$ and $U_{2,0} = 10 = u(q) = u(r)$. The only unknown values of u are at the nodes $P = (x_1, y_1)$, and $Q = (x_0, y_1)$. From (11.43) the difference equation centered at Q is

$$4U_{0,1} - U_{1,1} - U_{-1,1} - U_{0,2} - U_{0,0} = 0. \tag{11.48}$$

Note that the coordinates of $q = (\sqrt{3}\,h, h)$, and of $r = (h, \sqrt{3}\,h)$. Hence, $\alpha = \beta = \sqrt{3} - 1$. Also, the boundary conditions yield $U_{1,1} = 0$. Thus, from (11.48) we have

$$2U_{0,1} - U_{1,1} = 5. \tag{11.49}$$

Then the difference equation (11.47) gives

$$\frac{U(Q)}{\sqrt{3}} + \frac{U(R)}{\sqrt{3}} - \frac{2U(P)}{\sqrt{3} - 1} + \frac{U(q)}{\sqrt{3}(\sqrt{3} - 1)} = 0,$$

or

$$\frac{U_{0,1}}{\sqrt{3}} + \frac{U_{1,0}}{\sqrt{3}} - \frac{2U_{1,1}}{\sqrt{3} - 1} + \frac{U(q)}{\sqrt{3}(\sqrt{3} - 1)} = 0,$$

which yields

$$(1 - \sqrt{3})\,U_{0,1} + 2\sqrt{3}\,U_{1,1} = 20. \tag{11.50}$$

By solving (11.49) and (11.50), we get

$$U_{0,1} = \frac{20 + 10\sqrt{3}}{3\sqrt{3} + 1} = 6.02317, \quad U_{1,1} = \frac{35 + 5\sqrt{3}}{3\sqrt{3} + 1} = 7.046349. \blacksquare$$

11.4. Mathematica Projects

PROJECT 11.1. We use Mathematica to solve part (a) of Example 11.6.

```
In[1]:  Clear[u,n,m,i,j,r,l,h];
        n:= 5
        m:= 5
        c:= 2
        h:= 1/(n-1)
        l:= 1/(m-1)
        r:= c l/h;
        u=Table[1,n,m];
```

In[9]: f[i_]:= Sin[Pi 1/(n-1) (i-1)]
  g[i_]:= 0

In[11]: Do[u[i,1] = f[i];
  u[i,2] = (1-r^2) f[i] + k g[i] + r^2/2 (f[i+1]+f[i-1]), {i,1,n}];

In[13]: Do[
  u[[1,j]] = 0;
  u[[n,j]] = 0;,
  {j,1,n}];

In[14]:
  Do[
  u[[i,j]] = r^2 u [[i-1,j-1]] + 2(1-r^2) u [[i,j-1]] +
  r^2u [[i+1,j-1]] - u [[i,j-2]], {i,2,n-1}, {j,3,m}];

Out[15]:

0	0	0	0	0
0.7070107	-0.12132	-0.665476	8.11418	-44.0196
1.	-0.171573	-0.941125	7.15642	-5.64069
0.7070107	-0.12132	-0.665476	0.349676	27.1931
0	0	0	0	0

(* The exact values follow *)

In[16]: Table[Sin[Pi h (i-1)] Cos[2 Pi k (j-1)],
  {i,1,n}, {j,1,m}//N//TableForm

Out[20]:

0	0	0	0	0
0.7070107	0	-0.7070107	0	0.7070107
1.	0	-1.	0	1.
0.7070107	0	-0.7070107	0	0.7070107
0	0	0	0	0

(* Note the value of r and the instability *)

PROJECT 11.2. The Notebook fd.nb is useful not only to solve problems but also generate Mathematica codes for different schemes.

11.5. Exercises

11.1. Use the central difference scheme, with $n = 4$, to solve

$$u'' - u = -2, \quad 0 < x < 1, \quad u'(0) = 0, \quad u(1) = 1.$$

HINT: $16(U_{i+1} - 2U_i + U_{i-1}) - U_i = -2$ for $i = 1, 2, 3, 4$, $h = 1/4$, and the boundary conditions give $U_{i+1} = U_i$, and $U_4 = 1$.

11.2. Derive the Crank-Nicolson scheme (11.32).

ANS. The central difference at the point (i, j) is

$$u_{xx} = \frac{U_{i+1,j} - 2U_{i,j} + U_{i-1,j}}{h^2} = \frac{\delta_x^2 U_{i,j}}{h^2},$$

and at the point $(i, j + 1)$ it is given by

$$u_{xx} = \frac{U_{i+1,j+1} - 2U_{i,j+1} + U_{i-1,j+1}}{h^2} = \frac{\delta_x^2 U_{i,j+1}}{h^2}.$$

The result follows by taking the average of these two differences.

11.3. (a) Show that the forward difference scheme (11.30) has a truncation error of $O(l + h^2)$; (b) Show that the local truncation error is of $O(l^2 + h^4)$ if $r = 1/6$.

SOLUTION. (a) Note that for $(x_i, t_j) = (ih, jl)$, we have

$$(u_t - k\, u_{xx})_{i,j} = \frac{U_{i,j+1} - U_{i,j}}{l} - k\, \frac{\delta_x^2 U_{i,j}}{h^2}$$
$$- \frac{l}{2}\, u_{tt}(x_i, \bar{t}_j) + \frac{kh^2}{12}\, u_{xxxx}(\bar{x}_i, t_j),$$

where $x_{i-1} < \bar{x} < x_i$ and $t_j < \bar{t}_j < t_{j+1}$. Then the truncation error is given by

$$\frac{l}{2}\, u_{tt}(x_i, \bar{t}) - \frac{kh^2}{12}\, u_{xxxx}(x\bar{x}_i, t_j) = O(l + h^2),$$

provided u_{tt} and u_{xxxx} are bounded.

(b) If $r = 1/6$, then $(u_t - k\, u_{xx})_{i,j} = 0$ leads to

$$\frac{U_{i,j+1} - U_{i,j}}{l} - k\, \frac{\delta_x^2 U_{i,j}}{h^2} = \left[\frac{l}{2}\, u_{tt}(x_i, \bar{t}) - \frac{kh^2}{12}\, u_{xxxx}(x\bar{x}_i, t_j) \right]_{i,j}$$
$$= O(l^2) + O(h^4).$$

11.4. Show that the forward scheme (11.30) is convergent for the problem $u_t = k\, u_{xx}$, $0 < x < 1$, $t > 0$, subject to the Dirichlet boundary conditions $u(0, t) =$

$f(t)$, $u(1,t) = g(t)$ for $t > 0$, and the initial condition $u(x,0) = F(x)$ for $0 < x < 1$.

SOLUTION. With $(x_i, t_i) = (ih, jl)$ for $i = 0, 1, \ldots, n$ and $j = 0, 1, \ldots, s$, and with $nh = 1$, $js = t_0$, where t_0 is a prescribed value, $0 \le t \le t_0$, we have

$$U_{i,j+1} = U_{i,j} + r\,\delta_x^2 U_{i,j},$$
$$U_{i,0} = F(x_i) = F_i, \quad U_{0,j} = f(t_j) = f_j, \quad U_{n,j} = g(t_j) = g_j.$$

Let $V_{i,j} = U_{i,j} - u_{i,j}$ be the discretization error, as in Definition 11.3. Then

$$V_{i,j+1} = r\,V_{i-1,j} + (1 - 2r)V_{i,j} + r\,V_{i+1,j} + \frac{l^2}{2}\,u_{tt}(x_i, \bar{t}_j)$$
$$+ \frac{klh^2}{12}\,u_{xxxx}(\bar{x}_i, t_j),$$

where $x_{i-1} < \bar{x}_i < x_i$ and $t_j < \bar{t}_j < t_{j+1}$. Since $r \le 1/2$, we have

$$|V_{i,j+1}| \le r\,|V_{i-1,j}| + (1 - 2r)|V_{i,j}| + Al^2 + Blh^2$$
$$\le \max_{0<i<n} |V_{i,j}| + Al^2 + Blh^2,$$

where

$$A = \max \left| \frac{1}{2} u_{tt}(x,t) \right|, \quad B = \max \left| \frac{k}{12} u_{xxxx}(x,t) \right|,$$

since both u_{tt} and u_{xxxx} are assumed continuous. The above inequality gives

$$\|V_{j+1}\| \le \|V_j\| + Al^2 + Blh^2,$$

where $\|V_j\| = \max_{0<i<n} |V_{i,j}|$, or since $\|V_0\| = 0$,

$$\|V_j\| \le j\left(Al^2 + Blh^2\right) \le t_0\left(Al + Bh^2\right),$$

i.e., $\|V_j\| \to 0$ uniformly in the domain of definition of the problem as $h, l \to 0$ for $0 \le t \le t_0$.

11.5. Derive the matrix form of the system of difference equations for the problem in Example 11.5 if the backward scheme (11.31) is used.

ANS. The matrix equation is $[A]\{U\} = \{B\}$, where

$$[A] = \begin{bmatrix} 1+2r & -2r & 0 & \cdots & 0 & 0 & 0 \\ -r & 1+2r & -r & \cdots & 0 & 0 & 0 \\ & -r & 1+2r & \cdots & 0 & 0 & 0 \\ & & & \cdots & & & \\ & \text{sym} & & \cdots & -r & 1+2r & -r \\ & & & & 0 & -2r & 1+2r \end{bmatrix},$$

and

$$\{U\} = \begin{Bmatrix} U_{0,j+1} \\ U_{1,j+1} \\ U_{2,j+1} \\ \cdots \\ U_{n-1,j+1} \\ U_{n,j+1} \end{Bmatrix}, \quad \{B\} = \begin{Bmatrix} U_{0,j} - 2hr f_{j+1} \\ U_{1,j} \\ U_{2,j} \\ \cdots \\ U_{n-1,j} \\ U_{n,j} + 2hr g_{j+1} \end{Bmatrix}.$$

11.6. Solve the heat equation $u_t = u_{xx}$, $0 < x < 1$, $t > 0$, with $n = 4$, $r = 1/2$, $l = 1/32$, subject to the boundary and initial conditions $u(0, t) = 0 = u(1, t)$ for $t > 0$, and $u(x, 0) = x$.

ANS. Exact solution: $u = \displaystyle\sum_{n=1}^{\infty} \frac{2(-1)^n}{n\pi} \sin n\pi x \, e^{-n^2\pi^2 t}$.

$2t$	$x = 0$	$x = 0.25$	$x = 0.5$	$x = 0.75$	$x = 1$
$2\,0$	0	0.25	0.5	0.75	1
$1/32$	0	0.25	0.5	0.75	0
$1/16$	0	0.25	0.5	0.25	0
$3/32$	0	0.25	0.25	0.25	0
$1/8$	0	0.125	0.25	0.125	0 ∎

11.7. Solve $u_t = u_{xx} + 1$, $0 < x < 1$, $t > 0$, subject to the boundary and initial conditions $u(0, t) = 0 = u(1, t)$ for $t > 0$ and the initial condition $u(x, 0) = 0$ for $0 < x < 1$, with $r = 1/2$, $l = 1/32$, and $n = 4$.

ANS.

$2t$	$x = 0$	$x = 0.25$	$x = 0.5$	$x = 0.75$	$x = 1$
$2\,0$	0	0	0	0	0
$1/32$	0	0.03125	0.03125	0.03125	0
$1/16$	0	0.046875	0.0625	0.046875	0
$3/32$	0	0.0625	0.078125	0.0625	0
$1/8$	0	0.0703125	0.09375	0.0703125	0 ∎

11.8. Solve the wave equation $u_{tt} = u_{xx}$, $0 < x < 1$, $t > 0$, with $r = 1/2$, $l = 1/32$, and $n = 4$, subject to the boundary and the initial conditions:
(a) $u(0, t) = 0 = u(1, t)$ for $t > 0$, and $u(x, 0) = 0$, $u_t(x, 0) = 1$ for $0 < x < 1$.
(b) $u(0, t) = 0 = u(1, t)$ for $t > 0$, and $u(x, 0) = \sin \pi x$, $u_t(x, 0) = 0$ for $0 < x < 1$.

Ans. (a)

2 t	x = 0	x = 0.25	x = 0.5	x = 0.75	x = 1
2 0	0.0	0.0	0.00	0.00	0.00
0.25	0.0	0.25	0.25	0.25	0.00
0.5	0.0	0.25	0.50	0.25	0.00
0.75	0.0	0.25	0.25	0.25	0.00
1.0	0.0	0.00	0.00	0.00	0.00 ∎

(b)

2 t	x = 0	x = 0.25	x = 0.5	x = 0.75	x = 1
2 0	0.0	0.7071	1.0000	0.7071	0.0
0.25	—	0.7071	1.0000	0.7071	0.0
0.5	0.0	0.2929	0.4142	0.2929	0.0
0.75	0.0	−0.2929	−0.4142	−0.2929	0.0
1.0	0.0	−0.7071	−1.0000	−0.7071	0.0 ∎

11.9. Find the system of equations $[A]\{U\} = \{F\}$ for the Neumann boundary value problem

$$u_{xx} + u_{yy} = f(x, y), \quad \text{in } \Omega, \quad \frac{\partial u}{\partial n} = g(x, y) \quad \text{on } \Gamma,$$

where Ω is the rectangle $\{0 < x < a, 0 < y < b\}$. Choose the nodes (mh, nh) for $m = 0, 1, 2, \ldots, M$ and $n = 0, 1, 2, \ldots, N$, such that $Nh = b$. Ans. Eq (11.43) for $m = 0, 1, 2, \ldots, M$ and $n = 0, 1, 2, \ldots, N$ leads to

$$U_{M+1,N} - U_{M-1,N} = 2h\, g_{m,n}, \quad n = 1, 2, \ldots, N - 1,$$
$$U_{M,N+1} - U_{M,N-1} = 2h\, g_{m,N}, \quad m = 1, 2, \ldots, M - 1,$$

and the boundary condition gives

$$U_{-1,n} - U_{1,n} = 2h\, g_{0,n}, \quad n = 1, 2, \ldots, N - 1,$$
$$U_{m,-1} - U_{m,1} = 2h\, g_{m,0}, \quad m = 1, 2, \ldots, M - 1.$$

At a corner node where the outward normal **n** is undefined, we shall take the normal derivative as the average value of the two normal derivatives at the two adjacent boundary nodes. Thus, the boundary conditions reduce to

$$U_{-1,0} + U_{0,-1} = U_{1,0} + U_{0,1} + 4h\, g_{0,0},$$
$$U_{M,-1} + U_{M+1,0} = U_{M,1} + U_{M-1,0} + 4h\, g_{M,0},$$
$$U_{M+1,N} + U_{M,N=1} = U_{M-1,N} + U_{M,N-1} + 4h\, g_{M,N},$$
$$U_{0,N+1} + U_{-1,N} = U_{0,N-1} + U_{1,N} + 4h\, g_{0,N}.$$

11.10. In Exercise 11.9, take $M = N = 3$ and $g = 0$. Determine the matrix $[A]$ and the vectors $\{U\}$ and $\{F\}$.

ANS.

$$[A] = \begin{bmatrix} 4 & -2 & 0 & -2 & 0 & 0 & 0 & 0 & 0 \\ & 4 & -1 & 0 & -2 & 0 & 0 & 0 & 0 \\ & & 4 & 0 & 0 & -2 & 0 & 0 & 0 \\ & & & 4 & -2 & 0 & -1 & 0 & 0 \\ & & & & 4 & -1 & 0 & -1 & 0 \\ & & & & & 4 & 0 & 0 & -1 \\ & & & & & & 4 & -2 & 0 \\ & \text{sym} & & & & & & 4 & -1 \\ & & & & & & & & 4 \end{bmatrix},$$

and

$$\{U\} = \begin{Bmatrix} U_1 \\ U_2 \\ U_3 \\ U_4 \\ U_5 \\ U_6 \\ U_7 \\ U_8 \\ U_9 \end{Bmatrix}, \qquad \{F\} = -h^2 \begin{Bmatrix} f_{1,1} \\ f_{2,1} \\ f_{3,1} \\ f_{1,2} \\ f_{2,2} \\ f_{3,1} \\ f_{1,3} \\ f_{2,3} \\ f_{2,3} \end{Bmatrix}.$$

11.11. Let Ω be the square region $\{0 < x, y < 1\}$ with boundary Γ. Find the finite difference equation for the boundary value problem

$$u_{xx} + u_{yy} + c\,u = f(x, y) \quad \text{in } \Omega,$$

subject to the Dirichlet boundary condition $u = g(x, y)$ on Γ.

ANS. $(c - 4)\, U_{m,n} + U_{m,n-1} + U_{m-1,n} + U_{m,n-1} + U_{m+1,n} = f_{m,n}$.

11.12. Find the finite difference equation for the boundary value problem

$$(a\,u_x)_x + (b\,u_y)_y = 0, \quad \text{in } \Omega = \{0 < x, y < 1\},$$

where a and b are positive functions of x, y, and u.

ANS. On the square grid $(x_m, y_n) = (mh, nl)$, the difference equation is given by

$$a_{m+1/2,n}\, U_{m+1,n} - \left(a_{m+1/2,n} + a_{m-1/2,n}\right) U_{m,n} + a_{m-1/2,n}\, U_{m-1,n}$$
$$+ b_{m,n+1/2}\, U_{m,n+1} - \left(b_{m,n+1/2} + b_{m,n-1/2}\right) U_{m,n} + b_{m,n-1/2}\, U_{m,n-1}$$
$$= h^2\, f_{m,n}.$$

A

Green's Identities

Green's theorems and identities play an important role in partial differential equations. We present some definitions and results.

Let Ω be a finite domain in R^2 bounded by a smooth closed curve Γ, and let w and F be scalar functions, and $\mathbf{G}$ a vector function in the class $C^0(\Omega)$. Then

$$\text{Gradient theorem:} \quad \iint_\Omega \nabla F \, dx \, dy = \oint_\Gamma \mathbf{n} \, F \, ds,$$

$$\text{Divergence theorem:} \quad \iint_\Omega \nabla \cdot \mathbf{G} \, dx \, dy = \oint_\Gamma \mathbf{n} \cdot \mathbf{G} \, ds,$$

$$\text{Stokes theorem:} \quad \oint_\Gamma \mathbf{G} \cdot d\mathbf{r} = \iint_\Omega (\nabla \times \mathbf{G}) \cdot \mathbf{n} \, d\Omega,$$

where $\mathbf{n}$ is the outward normal to the curve Γ, ds denotes the line element, and $d\mathbf{r} = \mathbf{i} \, dx + \mathbf{j} \, dy$. The divergence theorem is also known as the Gauss theorem. The gradient and divergence theorems lead to the following two useful identities in R^2:

$$\iint_\Omega (\nabla F) \, w \, dx \, dy = -\iint_\Omega (\nabla w) \, F \, dx \, dy + \oint_\Gamma \mathbf{n} \, w \, F \, ds, \qquad \text{(A.1)}$$

$$-\iint_\Omega (\nabla^2 F) \, w \, dx \, dy = \iint_\Omega \nabla w \cdot \nabla F \, dx \, dy - \oint_\Gamma \frac{\partial F}{\partial n} \, w \, ds, \qquad \text{(A.2)}$$

where $\dfrac{\partial}{\partial n} = \mathbf{n} \cdot \nabla = n_x \dfrac{\partial}{\partial x} + n_y \dfrac{\partial}{\partial y}$ is the normal derivative operator. Using the gradient theorem, the component forms of (A.1), with appropriate variables, are as follows:

$$\iint_\Omega w \frac{\partial F}{\partial x}\, dx\, dy = -\iint_\Omega \frac{\partial w}{\partial x} F\, dx\, dy + \oint_\Gamma n_x\, w\, F\, ds,$$
$$\iint_\Omega w \frac{\partial F}{\partial y}\, dx\, dy = -\iint_\Omega \frac{\partial w}{\partial y} F\, dx\, dy + \oint_\Gamma n_y\, w\, F\, ds.$$
$$\text{(A.3)}$$

The gradient and divergence theorems are valid in R^3 if the surface integral on the left side is replaced by a volume integral and the line integral on the right is replaced by a surface integral. Let the functions $M(x, y, z)$, $N(x, y, z)$, and $P(x, y, z)$, where $(x, y, z) \in \Omega$, be the components of the vector $\mathbf{G}$ in R^3. Then, by the divergence theorem

$$\iiint_\Omega \left(\frac{\partial M}{\partial x} + \frac{\partial N}{\partial y} + \frac{\partial P}{\partial z} \right) d\Omega$$
$$= \iint_{\partial\Omega} [M \cos(\mathbf{n}, x) + N \cos(\mathbf{n}, y) + P \cos(\mathbf{n}, z)]\, dS,$$
$$\text{(A.4)}$$

where dS denotes the surface element, $\partial\Omega$ denotes the boundary of Ω, and $\cos(\mathbf{n}, x)$, $\cos(\mathbf{n}, y)$, and $\cos(\mathbf{n}, z)$ are the direction cosines of $\mathbf{n}$. If we take $M = u \dfrac{\partial v}{\partial x}$, $N = u \dfrac{\partial v}{\partial y}$, and $P = u \dfrac{\partial v}{\partial z}$, then (A.4) yields

$$\iiint_\Omega \left(\frac{\partial u}{\partial x} \frac{\partial v}{\partial x} + \frac{\partial u}{\partial y} \frac{\partial v}{\partial y} + \frac{\partial u}{\partial z} \frac{\partial v}{\partial z} \right) d\Omega = \iint_{\partial\Omega} u \frac{\partial v}{\partial n}\, dS - \iiint_\Omega u \nabla^2 v\, d\Omega,$$
$$\text{(A.5)}$$

which is known as *Green's first identity*. Moreover, if we interchange u and v in (A.4), we get

$$\iiint_\Omega \left(\frac{\partial u}{\partial x} \frac{\partial v}{\partial x} + \frac{\partial u}{\partial y} \frac{\partial v}{\partial y} + \frac{\partial u}{\partial z} \frac{\partial v}{\partial z} \right) d\Omega = \iint_{\partial\Omega} v \frac{\partial u}{\partial n}\, dS - \iiint_\Omega v \nabla^2 u\, d\Omega.$$
$$\text{(A.6)}$$

If we subtract (A.5) from (A.6), we obtain *Green's second identity*:

$$\iiint_\Omega (u \nabla^2 v - v \nabla^2 u)\, d\Omega = \iint_{\partial\Omega} \left(u \frac{\partial v}{\partial n} - v \frac{\partial u}{\partial n} \right) dS, \qquad \text{(A.7)}$$

which is also known as *Green's reciprocity theorem*. This result also holds in R^2. Note that Green's identities are valid even if the domain Ω is bounded by finitely many closed surfaces; however, in that case the surface integrals must be evaluated over all surfaces that make the boundary of Ω, and in R^2 the line integrals must be evaluated over all paths that make the boundary of Ω. If f and g are real and harmonic in $\Omega \subset R^2$ and $\Gamma = \partial\Omega$, then from (A.7)

$$\int_\Gamma \left(f \frac{\partial g}{\partial n} - g \frac{\partial f}{\partial n} \right) ds = 0. \qquad \text{(A.8)}$$

Let D be a simply connected region in the complex plane with boundary Γ. Let z_0 be any point inside D, and let Ω be the region obtained by indenting from D a disk of radius ε and center at z_0, where $\varepsilon > 0$ is small (Fig. A.1 (a)). Then ∂D consists of the contour Γ together with the contour Γ_ε.

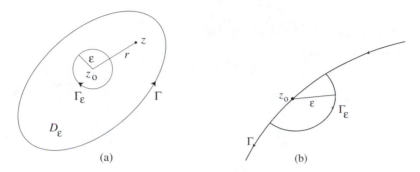

Fig. A.1.

If we set $f = u$ and $g = \log r$ in (A.8), where $z \in D$ and $r = |z - z_0|$, then since $\dfrac{\partial}{\partial n} = -\dfrac{\partial}{\partial r}$ on Γ_ε, we get

$$\int_\Gamma \left[u \frac{\partial (\log r)}{\partial n} - \log r \frac{\partial u}{\partial n} \right] ds - \int_{\Gamma_\varepsilon} \left(\frac{u}{r} - \log r \frac{\partial u}{\partial r} \right) ds = 0. \qquad (A.9)$$

Since

$$\lim_{\varepsilon \to 0} \int_{\Gamma_\varepsilon} \frac{u}{r}\, ds = \lim_{\varepsilon \to 0} \int_0^{2\pi} u(z_0 + \varepsilon\, e^{i\theta}) \frac{1}{\varepsilon}\, \varepsilon\, d\theta = 2\pi\, u(z_0),$$

$$\lim_{\varepsilon \to 0} \int_{\Gamma_\varepsilon} \log r \frac{\partial u}{\partial r}\, ds = \lim_{\varepsilon \to 0} \int_0^{2\pi} \log \varepsilon \frac{\partial u}{\partial \varepsilon}\, \varepsilon\, d\theta = 0,$$

we let $\varepsilon \to 0$ in (A.9) and obtain

$$2\pi\, u(z_0) = \int_\Gamma \left[u \frac{\partial (\log r)}{\partial n} - \log r \frac{\partial u}{\partial n} \right] ds, \qquad (A.10)$$

which is known as *Green's third identity*. Note that Eq (A.10) gives the value of a harmonic function u at an interior point in terms of the boundary values of u and $\dfrac{\partial u}{\partial n}$. If the contour Γ is continuously differentiable (has no corners) and if the point z_0 is on Γ, then instead of indenting the entire disk of radius ε, we indent from D a half disk with center at the point z_0 (Fig. A.1(b)), and Green's third identity becomes

$$\pi\, u(z_0) = \text{p.v.} \int_\Gamma \left[u \frac{\partial (\log r)}{\partial n} - \log r \frac{\partial u}{\partial n} \right] ds \qquad (A.11)$$

where p.v. denotes the principal value of the integral.

B

Orthogonal Polynomials

A set of of polynomials $\{f_i\}$ with degree i and such that $\langle f_i, f_j \rangle = 0$ for $i \neq j$ is called a set of orthogonal polynomials with respect to the inner product $\langle f_i, f_j \rangle$. Let $w(x)$ be an admissible weight function on a finite or infinite interval $[a, b]$. If we orthonormalize the powers $1, x, x^2, \ldots$, we obtain a unique set of polynomials $p_n(x)$ of degree n with positive leading coefficient, such that

$$\int_a^b w(x)p_n(x)p_m(x)\, dx = \delta_{mn} = \begin{cases} 0 & \text{if } m \neq n, \\ 1 & \text{if } m = n, \end{cases} \tag{B.1}$$

where δ_{mn} is known as the Kronecker delta. Table B.1 gives the classical polynomials corresponding to their specific weights and intervals.

Table B.1

Name	Symbol	Interval	$w(x)$
Chebyshev, 1st kind	$T_n(x)$	$[-1, 1]$	$(1 - x^2)^{-1/2}$
Chebyshev, 2nd kind	$U_n(x)$	$[-1, 1]$	$(1 - x^2)^{1/2}$
Gegenbauer (ultraspherical)	$C_n^\mu(x)$	$[-1, 1]$	$(1 - x^2)^{\mu - 1/2}, \ \mu > -1/2$
Hermite	$H_n(x)$	$(-\infty, \infty)$	e^{-x^2}
Jacobi	$P_n^{(\alpha,\beta)}(x)$	$[-1, 1]$	$(1 - x)^\alpha (1 + x)^\beta, \ \alpha, \beta > 1$
Laguerre	$L_n(x)$	$[0, \infty)$	e^{-x}
Generalized Laguerre	$L_n^{(\alpha)}$	$[0, \infty)$	$x^\alpha e^{-x}, \ \alpha > 1$
Legendre	$P_n(x)$	$[-1, 1]$	1

We provide certain important data for some of the most frequently used polynomials. Information on others is available in Abramowitz and Stegun (1965). Zeros of these polynomials can be easily computed by Mathematica.

Orthogonal polynomials satisfy the differential equation

$$g_1(x)\,y'' + g_2(x)\,y' + a_n\,y = 0, \tag{B.2}$$

where $g_1(x)$ and $g_2(x)$ are independent of n, and a_n are constants that depend on n only.

1. Chebyshev Polynomials of the First Kind, $T_n(x)$, over the interval $[-1, 1]$, such that $T_n(1) = 1$. This polynomial satisfies Eq (B.2) with $g_1(x) = 1 - x^2$, $g_2(x) = -x$, and $a_n = n^2$. The m-th zero $x_{n,m}$ of $T_n(x)$ is given by

$$x_{n,m} = \cos\frac{(2m-1)\pi}{2n}.$$

Other relevant data are as follows:

Norm:
$$\int_{-1}^{1} \left(1 - x^2\right)^{-1/2} \left[T_n(x)\right]^2 dx = \begin{cases} \dfrac{\pi}{2}, & n \neq 0, \\ \pi, & n = 0. \end{cases}$$

Series form:
$$T_n(x) = \frac{n}{2} \sum_{k=0}^{[n/2]} (-1)^k \frac{(n-k-1)!}{k!(n-2k)!} (2x)^{n-2k} = \cos(n \arccos x).$$

Indefinite and definite integrals:
$$\int T_0\,dx = T_1, \quad \int T_1\,dx = \frac{T_2}{4},$$

$$\int T_n\,dx = \frac{1}{2}\left[\frac{T_{n+1}(x)}{n+1} - \frac{T_{n-1}(x)}{n-1}\right], \quad \int_{-1}^{1} T_n\,dx = \begin{cases} \dfrac{2}{1-n^2}, & n \text{ even}, \\ 0, & n \text{ odd}. \end{cases}$$

Inequality: $\left|T_n(x)\right| \leq 1, \quad -1 \leq x \leq 1.$

Rodrigues' formula:
$$T_n(x) = \frac{(-1)^n \left(1 - x^2\right)^{1/2} \sqrt{\pi}}{2^{n+1}\,\Gamma(n+1/2)} \frac{d^n}{dx^n}\left\{(1 - x^2)^{n-1/2}\right\}.$$

2. Chebyshev Polynomials of the Second Kind, $U_n(x)$, over the interval $[-1, 1]$, such that $U_n(1) = n + 1$. This polynomial satisfies Eq (B.2) with $g_1(x) = 1 - x^2$, $g_2(x) = -3x$, and $a_n = n(n+2)$. The m-th zero $x_{n,m}$ of $U_n(x)$ is given by

$$x_{n,m} = \cos\frac{m\pi}{n+1}.$$

Other relevant data include:

Norm: $\displaystyle\int_{-1}^{1} \left(1 - x^2\right)^{1/2} \left[U_n(x)\right]^2 dx = \frac{\pi}{2}$.

Series form: $\displaystyle U_n(x) = \sum_{k=0}^{[n/2]} (-1)^k \frac{(n-k)!}{k!(n-2k)!} (2x)^{n-2k} = \frac{T'_{n+1}(x)}{n+1}$,

$$U_n(\cos\theta) = \frac{\sin(n+1)\theta}{\sin\theta}.$$

Definite integral: $\displaystyle\int_{-1}^{1} U_n\, dx = \begin{cases} \dfrac{2}{n+1}, & n = 2m, \\ 0, & n = 2m+1. \end{cases}$

Inequality: $\left|U_n(x)\right| \leq n+1, \quad -1 \leq x \leq 1$.

Rodrigues' formula: $\displaystyle U_n(x) = \frac{(-1)^n (n+1)\sqrt{\pi}}{(1-x^2)^{1/2}\, 2^{n+1}\, \Gamma(n+3/2)} \frac{d^n}{dx^n}\left\{(1-x^2)^{n+1/2}\right\}$.

3. Gegenbauer (or Ultraspherical) Polynomials, $C_n^\mu(x)$, over the interval $[-1, 1]$ such that $C_n^\mu(1) = \binom{n+2\mu-1}{n}$. This polynomial satisfies Eq (B.2) with $g_1(x) = 1 - x^2$, $g_2(x) = -(2\mu+1)x$, and $a_n = n(n+2\alpha)$. Other relevant data are the following:

Norm: $\displaystyle\int_{-1}^{1} \left(1-x^2\right)^{\mu-1/2} \left[C_n^\mu(x)\right]^2 dx = \frac{\pi 2^{1-2\mu}\, \Gamma(n+2\mu)}{n!\,(n+\mu)\left[\Gamma(\mu)\right]^2}$.

Series form: $\displaystyle C_n^\mu(x) = \frac{1}{\Gamma(\mu)} \sum_{k=0}^{[n/2]} (-1)^k \frac{\Gamma(\mu+n-k)!}{k!(n-2k)!} (2x)^{n-2k}$.

Inequality: $\displaystyle\max_{-1 \leq x \leq 1} \left|C_n^\mu(x)\right| = \begin{cases} \binom{n+2\mu-1}{n}, & \text{if } \mu > 0, \\ \left|C_n^\mu(x')\right|, & \text{if } -1/2 < \mu < 0, \end{cases}$

where $x' = 0$ if $n = 2k$; $x' = $ maximum point nearest zero if $n = 2k+1$.

Rodrigues' formula: $\displaystyle C_n^\mu(x) = \frac{(-1)^n 2^n n!\, \Gamma(\mu+n+1/2)}{\Gamma(\mu+1/2)\, \Gamma(n+2\mu)\, (1-x^2)^{\mu-1/2}} \times$

$$\times \frac{d^n}{dx^n}\left\{(1-x^2)^{n+\mu-1/2}\right\}.$$

4. Hermite Polynomials, $H_n^{(\alpha)}(x)$, over the interval $(-\infty, \infty)$ satisfies Eq (B.2) with $g_1(x) = 1$, $g_2(x) = -x$, and $a_n = n$, and has

Norm: $\displaystyle\int_{-\infty}^{\infty} e^{-x^2} \left[H_n^{(\alpha)}(x)\right]^2 dx = \sqrt{\pi}\, 2^n\, n!$.

Inequality: $\displaystyle\left|H_{2n}(x)\right| \leq e^{x^2/2}\, 2^{2n}\, n! \left[2 - \frac{1}{2^{2n}}\binom{2n}{n}\right]$, and $\quad |H_{2n+1}(x)| \leq |x|e^{x^2/2}\dfrac{(2n+2)!}{(n+1)!}$.

Rodrigues' formula: $\quad H_n(x) = (-1)^n\, e^{x^2}\dfrac{d^n}{dx^n}e^{-x^2}$.

5. Jacobi Polynomials,

$P_n^{\alpha,\beta}(x)$, over the interval $[-1,1]$ satisfies Eq (B.2) with $g_1(x) = 1 - x^2$, $g_2(x) = \beta - \alpha - (\alpha + \beta + 2)x$, and $a_n = n(n + \alpha + \beta + 1)$, and $P_n^{\alpha,\beta}(1) = \dbinom{n + \alpha}{n}$. Other relevant data are:

Norm: $\displaystyle\int_{-1}^{1}(1-x)^\alpha(1+x)^\beta\left[P_n^{\alpha,\beta}(x)\right]^2 dx$

$$= \frac{2^{\alpha+\beta+1}\,\Gamma(n+\alpha+1)\Gamma(n+\beta+1)}{(2n+\alpha+\beta)n!\,\Gamma(n+\alpha+\beta+1)}.$$

Series form: $P_n^{\alpha,\beta}(x) = \dfrac{1}{2^n}\displaystyle\sum_{k=0}^{[n/2]}\dbinom{n+\alpha}{k}\dbinom{n+\beta}{n-k}(x-1)^{n-k}(x+1)^k$.

Inequality: $\displaystyle\max_{-1\leq x\leq 1}\left|P_n^{\alpha,\beta}(x)\right| = \begin{cases} \dbinom{n+q}{n} \sim n^q & \text{if } q = \max(\alpha,\beta) \geq -1/2, \\ \left|P_n^{\alpha,\beta}\right|(x') \sim n^{-1/2} & \text{if } q < -1/2, \end{cases}$

where x' is one of the two maximum points nearest $(\beta - \alpha)/(\alpha + \beta + 1)$.

Rodrigues' formula: $P_n^{\alpha,\beta}(x) = \dfrac{(-1)^n}{2^n n!(1-x)^\alpha(1+x)^\beta}$

$$\times \frac{d^n}{dx^n}\left\{(1-x)^{n+\alpha}(1+x)^{n+\beta}\right\}.$$

6. Laguerre Polynomials,

$L_n(x)$, over the interval $[0,\infty)$ satisfies Eq (B.2) with $g_1(x) = x$, $g_2(x) = 1 - x$, and $a_n = n$, and is such that $L_n(0) = n!$ and

$$\int_0^\infty e^{-x}L_n(x)L_m(x)\,dx = \begin{cases} 0 & \text{if } n \neq m, \\ (n!)^2 & \text{if } n = m. \end{cases}$$

Its m-th zero $x_{n,m}$ is given by

$$x_{n,m} = \frac{j_m^2}{4k_n}\left(1 + \frac{j_m^2 - 2}{48k_n^2}\right) + O\left(n^{-5}\right),$$

where $k_n = n + 1/2$ and j_m is the m-th positive zero of the Bessel function $J_n(x)$.

Other relevant data are

Norm: $\displaystyle\int_0^\infty e^{-x} \left[L_n(x)\right]^2 dx = 1.$

Series form: $\displaystyle L_n(x) = \sum_{k=0}^{n} (-1)^k \binom{n}{n-k} \frac{1}{k!} x^k.$

Inequality: $|L_n(x)| = \begin{cases} e^{x/2}, & \text{if } x \geq 0, \\ \left[2 - \dfrac{1}{n!}\right] e^{x/2}, & \text{if } x \geq 0. \end{cases}$

Rodrigues' formula: $\displaystyle L_n(x) = \frac{1}{n! e^{-x}} \frac{d^n}{dx^n} \left\{x^n e^{-x}\right\}.$

7. Generalized Laguerre Polynomials, $L_n^{(\alpha)}(x)$, over the interval $[0, \infty)$ satisfies Eq (B.2) with $g_1(x) = x$, $g_2(x) = \alpha + 1 - x$, and $a_n = n$, and its m-th zero $x_{n,m}$ is given by

$$x_{n,m} = \frac{j_{\alpha,\beta}^2}{4k_n}\left(1 + \frac{2\left(\alpha^2 - 1\right) + j_{\alpha,m}^2}{48 k_n^2}\right) + O\left(n^{-5}\right),$$

where $k_n = n + (\alpha + 1)/2$, $\alpha > -1$, and $j_{\alpha,m}$ is the m-th positive zero of the Bessel function $J_n(x)$. Other relevant data are:

Norm: $\displaystyle\int_0^\infty x^\alpha e^{-x} \left[L_n^{(\alpha)}(x)\right]^2 dx = \frac{\Gamma(n + \alpha + 1)}{n!}.$

Series form: $\displaystyle L_n^{(\alpha)}(x) = \sum_{k=0}^{n} (-1)^k \binom{n + \alpha}{n - k} \frac{1}{k!} x^k.$

Inequality: $|L_n^{(\alpha)}(x)| = \begin{cases} \dfrac{\Gamma(n + \alpha + 1)}{n! \Gamma(\alpha + 1)} e^{x/2}, & \text{if } x \geq 0,\ \alpha \geq 0, \\[2mm] \left[2 - \dfrac{\Gamma(n + \alpha + 1)}{n! \Gamma(n + 1)}\right] e^{x/2}, & \text{if } x \geq 0,\ -1 < \alpha < 0. \end{cases}$

Rodrigues' formula: $\displaystyle L_n^{(\alpha)}(x) = \frac{1}{n! x^\alpha e^{-x}} \frac{d^n}{dx^n} \left\{x^{n+\alpha} e^{-x}\right\}.$

8. Legendre Polynomials, $P_n(x)$, over the interval $[-1, 1]$ satisfies Eq (B.2) with $g_1(x) = 1 - x^2$, $g_2(x) = -2x$, and $a_n = n(n + 1)$, where $P_n(1) = 1$. If $x_{n,m}$ denotes the m-th zero of $P_n(x)$, where $x_{n,1} > x_{n,2} > \cdots > x_{n,n}$, then

$$x_{n,m} = \left(1 - \frac{1}{8n^2} + \frac{1}{8n^3}\right)\cos\frac{(4m - 1)\pi}{4n + 2} + O\left(n^{-4}\right).$$

Other relevant data are:

Norm: $\displaystyle\int_{-1}^{1}\left[P_n(x)\right]^2 dx = \frac{2}{2n+1}$.

Series form: $\displaystyle P_n(x) = \frac{1}{2^n}\sum_{k=0}^{[n/2]}(-1)^k\binom{n}{k}\binom{2n-2k}{n}x^{n-2k}$.

Indefinite integral: $\displaystyle\int P_n(x)\,dx = \frac{1}{2n+1}\left[P_{n+1}(x) - P_{n-1}(x)\right]$.

Inequality: $\left|P_n(x)\right| \le 1, \quad -1 \le x \le 1$.

Rodrigues' formula: $\displaystyle P_n(x) = \frac{(-1)^n}{2^n\,n!}\frac{d^n}{dx^n}\left\{(1-x^2)^n\right\}$.

C

Tables of Integral Transform Pairs

Some basic formulas for the pairs of the Laplace, complex (exponential) Fourier, Fourier sine, Fourier cosine, finite sine, and finite cosine transforms are provided below in tabular forms. Definitions of these transforms are given in §6.1 and §6.8. The command <<Calculus`LaplaceTransform` loads the Mathematica package for the Laplace transforms.

C.1. Laplace Transform Pairs

	$f(t)$	$F(s) = \bar{f}(s)$
1.	1	$\dfrac{1}{s}, \quad s > 0$
2.	e^{at}	$\dfrac{1}{s-a}, \quad s > a$
3.	$\sin at$	$\dfrac{a}{s^2 + a^2}, \quad s > 0$
4.	$\cos at$	$\dfrac{s}{s^2 + a^2}, \quad s > 0$
5.	$\sinh at$	$\dfrac{a}{s^2 - a^2}, \quad s > 0$
6.	$\cosh at$	$\dfrac{s}{s^2 - a^2}, \quad s > 0$
7.	$e^{at} \sin bt$	$\dfrac{b}{(s-a)^2 + b^2}, \quad s > a$

	$f(t)$	$F(s) = \bar{f}(s)$
8.	$e^{at} \cos bt$	$\dfrac{s}{(s-a)^2 + b^2}, \quad s > a$
9.	$t^n \quad (n = 1, 2, \dots)$	$\dfrac{n!}{s^{n+1}}, \quad s > 0$
10.	$t^n e^{at} \quad (n = 1, 2, \dots)$	$\dfrac{n!}{(s-a)^{n+1}}, \quad s > a$
11.	$H(t-a)$	$\dfrac{e^{-as}}{s}, \quad s > 0$
12.	$H(t-a) f(t-a)$	$e^{-as} F(s) = e^{-as} \bar{f}(s)$
13.	$e^{at} f(t)$	$F(s-a) = \bar{f}(s-a)$
14.	$f(t) \star g(t)$	$F(s)\, G(s) = \bar{f}(s)\, \bar{g}(s)$
15.	$f^{(n)}(t)$	$s^n F(s) - s^{n-1} f(0) - \cdots - f^{n-1}(0)$
16.	$f(at)$	$\dfrac{1}{a} F\left(\dfrac{s}{a}\right), \quad a > 0$
17.	$\int_0^t f(t)\, dt$	$\dfrac{1}{s} F(s) = \dfrac{1}{s} \bar{f}(s)$
18.	$\delta(t-a)$	e^{-as}
19.	$t\, f(t)$	$-\dfrac{d}{ds} F(s)$
20.	$\operatorname{erf}\left(\dfrac{a}{2\sqrt{t}}\right)$	$\dfrac{1 - e^{-a\sqrt{s}}}{s}$
21.	$\operatorname{erfc}\left(\dfrac{a}{2\sqrt{t}}\right)$	$\dfrac{e^{-a\sqrt{s}}}{s}$
22.[†]	$f(t)$ with period $= T$	$\dfrac{\int_0^T e^{-st} f(t)\, dt}{1 - e^{-Ts}}$
23.	$J_0(at)$	$\dfrac{1}{\sqrt{s^2 + a^2}}$
24.	$I_0(at)$	$\dfrac{1}{\sqrt{s^2 - a^2}}$
25.	$J_n(at)$	$\dfrac{1}{\sqrt{s^2 + a^2}} \left(\dfrac{a}{s + \sqrt{s^2 + a^2}}\right)^n, \quad n > -1$

[†] $f(t)$ is continuous in $[0, T]$ and periodic with period T, $T > 0$.

	$f(t)$	$F(s) = \bar{f}(s)$
26.	$I_n(at)$	$\dfrac{1}{\sqrt{s^2 - a^2}} \left(\dfrac{a}{s + \sqrt{s^2 - a^2}} \right)^n, \quad n > -1$
27.	$t\, J_1(at)$	$\dfrac{a}{(s^2 + a^2)^{3/2}}, \quad a > 0$
28.	$t\, I_1(at)$	$\dfrac{a}{(s^2 - a^2)^{3/2}}, \quad a > 0$
29.†	$t^n\, J_n(at)$	$\dfrac{(2n)!\, a^n}{2^n\, n!\, \left(\sqrt{s^2 + a^2}\right)^{2n+1}}, \quad n > -1/2$
30.†	$t^n\, I_n(at)$	$\dfrac{2^n}{\sqrt{\pi}}\, \dfrac{\Gamma(n + 1/2)\, a^n}{\left(\sqrt{s^2 - a^2}\right)^{2n+1}}, \quad n > -1/2$
31.	$t\, J_0(at)$	$\dfrac{s}{(s^2 + a^2)^{3/2}}, \quad a > 0$
32.	$t\, I_0(at)$	$\dfrac{s}{(s^2 - a^2)^{3/2}}, \quad a > 0$
33.	$\dfrac{(t - a)^{\mu-1}}{\Gamma(\mu)}\, H(t - a)$	$\dfrac{e^{-as}}{s^\mu}, \quad \mu > 0$
34.	$\left(\dfrac{t}{a}\right)^{(\mu-1)/2} J_{\mu-1}\left(2\sqrt{at}\right)$	$\dfrac{e^{-a/s}}{s^\mu}, \quad \mu > 0$
35.	$\left(\dfrac{t}{a}\right)^{(\mu-1)/2} I_{\mu-1}\left(2\sqrt{at}\right)$	$\dfrac{e^{a/s}}{s^\mu}, \quad \mu > 0$
36.	$\dfrac{\cos 2\sqrt{at}}{\sqrt{\pi t}}$	$\dfrac{e^{-a/s}}{\sqrt{s}}, \quad a > 0$
37.	$\dfrac{\sin 2\sqrt{at}}{\sqrt{\pi a}}$	$\dfrac{e^{-a/s}}{s^{3/2}}, \quad a > 0$
38.	$\dfrac{e^{-a^2/(4t)}}{\sqrt{\pi t}}$	$\dfrac{e^{-a\sqrt{s}}}{\sqrt{s}}, \quad a > 0$
39.	$\dfrac{a e^{-a^2/(4t)}}{2\sqrt{\pi t^3}}$	$e^{-a\sqrt{s}}, \quad a > 0$
40.	$\dfrac{e^{-bt} - e^{-at}}{t}$	$\ln\left(\dfrac{s + a}{s + b}\right), \quad a, b > 0$
41.	$\delta(t)$	1
42.	$\delta(t - a)$	$e^{-as}, \quad a > 0$

† Note that $\Re\{s\} > |\Im\{a\}|$ (in 29), and $\Re\{s\} > |\Re\{a\}|$ (in 30).

C.2. Fourier Cosine Transform Pairs

	$f(x)$	$\mathcal{F}_c\{f(x)\} = \tilde{f}_c(\alpha)$
1.	$f(ax)$	$\dfrac{1}{a}\tilde{f}_c\left(\dfrac{\alpha}{a}\right)$
2.	e^{-ax}	$\sqrt{\dfrac{2}{\pi}}\,\dfrac{a}{\alpha^2 + a^2}, \quad a > 0$
3.	$x^{-1/2}$	$\dfrac{1}{\sqrt{\alpha}}$
4.	e^{-ax^2}	$\dfrac{1}{\sqrt{2a}}\,e^{-\alpha^2/(4a)}, \quad a > 0$
5.	$\dfrac{a}{x^2 + a^2}$	$\sqrt{\dfrac{\pi}{2}}\,e^{-a\alpha}, \quad a > 0$
6.	$x^2 f(x)$	$-\tilde{f}_c''(\alpha)$
7.	$\dfrac{\sin ax}{x}$	$\sqrt{\dfrac{\pi}{2}}\,H(a - \alpha)$
8.	$f''(x)$	$-\alpha^2 \tilde{f}_c(\alpha) - \sqrt{\dfrac{2}{\pi}}\,f(0)$
9.	$\delta(x)$	$\sqrt{\dfrac{2}{\pi}}$
10.	$H(a - x)$	$\sqrt{\dfrac{2}{\pi}}\,\dfrac{\sin a\alpha}{\alpha}$
11.	$\dfrac{a + \sqrt{x^2 + a^2}}{a^2 + x^2}$	$\dfrac{e^{-a\alpha}}{\sqrt{\alpha}}$
12.	$\begin{cases} (a^2 - x^2)^{1/2}, & 0 < x < a \\ 0, & x > a \end{cases}$	$\sqrt{\dfrac{\pi}{2}}\,\dfrac{aJ_1(a\alpha)}{\alpha}$
13.	$\begin{cases} (x^2 - a^2)^{-1/2}, & a < x < \infty \\ 0, & 0 < x < a \end{cases}$	$-\sqrt{\dfrac{\pi}{2}}\,Y_0(a\alpha)$
14.	$\sin\left(a^2 x^2\right)$	$\dfrac{\pi}{4a}\left(\cos\dfrac{\alpha^2}{4a^2} - \sin\alpha^2 4a^2\right)$
15.	$\cos\left(a^2 x^2\right)$	$\dfrac{\pi}{4a}\left(\cos\dfrac{\alpha^2}{4a^2} + \sin\alpha^2 4a^2\right)$

C.3. Fourier Sine Transform Pairs

	$f(x)$	$\mathcal{F}_s\{f(x)\} = \tilde{f}_s(\alpha)$
1.	$f(ax)$	$\dfrac{1}{a}\tilde{f}_s\left(\dfrac{\alpha}{a}\right)$
2.	e^{-ax}	$\sqrt{\dfrac{2}{\pi}}\dfrac{\alpha}{a^2 + \alpha^2}$
3.	$x^{-1/2}$	$\dfrac{1}{\sqrt{\alpha}}$
4.	x^{-1}	$\sqrt{\dfrac{\pi}{2}}$
5.	$\dfrac{x}{x^2 + a^2}$	$\sqrt{\dfrac{\pi}{2}}\, e^{-a\alpha}$
6.	$\arctan(a/x)$	$\sqrt{\dfrac{\pi}{2}}\dfrac{1 - e^{-a\alpha}}{\alpha}$
7.	$x^2 f(x)$	$-\tilde{f}_s''(\alpha)$
8.	$\operatorname{erfc}\dfrac{x}{2\sqrt{a}}, \quad a > 0$	$\sqrt{\dfrac{2}{\pi}}\dfrac{1 - e^{-a\alpha^2}}{\alpha}$
9.	$f''(x)$	$-\alpha^2\tilde{f}_s(\alpha) + \sqrt{\dfrac{2}{\pi}}\,\alpha f(0)$
10.	$H(a - x)$	$\sqrt{\dfrac{2}{\pi}}\dfrac{1 - \cos a\alpha}{\alpha}$
11.	$x e^{-a^2 x^2}$	$\dfrac{\pi\alpha}{4\sqrt{2}\,a^3}\, e^{-\alpha^2/(4a^2)}$
12. †	$\dfrac{1}{\sqrt{x}}\, e^{-a/x}$	$\dfrac{1}{\sqrt{\alpha}}\, e^{-\sqrt{2a\alpha}}\left(\cos\sqrt{2a\alpha} + \sin\sqrt{2a\alpha}\right),$
13.	$\dfrac{\sin bx}{x^2 + a^2}$	$\begin{cases}\sqrt{\dfrac{\pi}{2}}\dfrac{1}{a}\, e^{-ab}\sinh(a\alpha), & 0 < \alpha < b \\[2ex] \sqrt{\dfrac{\pi}{2}}\dfrac{1}{a}\, e^{-a\alpha}\sinh(ab), & b < \alpha < \infty\end{cases}$

† $|\arg a| < \pi/2$.

C.4. Complex Fourier Transform Pairs

	$f(x)$	$\mathcal{F}\{f(x)\} = \tilde{f}(\alpha)$
1.	$f^{(n)}(x)$	$(-i\alpha)^n \, \tilde{f}(\alpha)$
2.	$f(ax), \quad a > 0$	$\dfrac{1}{a} \tilde{f}\left(\dfrac{\alpha}{a}\right)$
3.	$f(x - a)$	$e^{i\,a\,\alpha} \, \tilde{f}(\alpha)$
4.	$\delta(x - a)$	$\dfrac{1}{\sqrt{2\pi}} e^{i\,a\alpha}$
5.	$e^{-a|x|}$	$\sqrt{\dfrac{2}{\pi}} \dfrac{a}{a^2 + \alpha^2}, \quad a > 0$
6.	$e^{-a^2 x^2}$	$\dfrac{1}{a\sqrt{2}} e^{-\alpha^2/(4a^2)}$
7.	$\begin{cases} 1, & |x| < a \\ 0, & |x| > a \end{cases}$	$\sqrt{\dfrac{2}{\pi}} \dfrac{\sin a\alpha}{\alpha}$
8.	$\dfrac{1}{\sqrt{|x|}}$	$\dfrac{1}{\sqrt{|\alpha|}}$
9.	$f(x) \star g(x)$	$\sqrt{2\pi}\, \tilde{f}(\alpha)\, \tilde{g}(\alpha)$
10.	$H(x + a) - H(x - a)$	$\sqrt{\dfrac{2}{\pi}} \dfrac{\sin a\alpha}{\alpha}$
11.	$x\,e^{-a|x|}, \quad a > 0$	$\sqrt{\dfrac{2}{\pi}} \dfrac{2i a\alpha}{(\alpha^2 + a^2)^2}$
12.	$\dfrac{a}{x^2 + a^2}$	$\sqrt{\dfrac{\pi}{2}} e^{-a|\alpha|}$
13.	$\dfrac{ax}{(x^2 + a^2)^2}$	$\dfrac{i}{2} \sqrt{\dfrac{\pi}{2}}\, \alpha\, e^{-a|\alpha|}$
14.	$\cos ax$	$\sqrt{\dfrac{\pi}{2}} \left[\delta(\alpha + a) + \delta(\alpha - a)\right]$
15.	$\sin ax$	$i\sqrt{\dfrac{\pi}{2}} \left[\delta(\alpha + a) - \delta(\alpha - a)\right]$
16.	$\begin{cases} \cos ax, & |x| < \pi/(2a) \\ 0, & |x| > \pi/(2a) \end{cases}$	$\sqrt{\dfrac{2}{\pi}} \dfrac{a}{a^2 - \alpha^2} \cos\dfrac{\pi\alpha}{2a}$

17. $\begin{cases} \dfrac{1}{\sqrt{a^2 - x^2}}, & |x| < a \\ 0, & |x| > a \end{cases}$ $\qquad \sqrt{\dfrac{\pi}{2}}\, J_0(a\alpha)$

18. $\begin{cases} 1 - |x|, & |x| < 1 \\ 0, & |x| > 1 \end{cases}$ $\qquad 2\sqrt{\dfrac{2}{\pi}}\left(\dfrac{\sin(\alpha/2)}{\alpha}\right)^2$

C.5. Finite Sine Transform Pairs

The tables are for the interval $[0, \pi]$. If the interval is $[a, b]$, then it can be transformed into $[0, \pi]$ by

$$y = \frac{\pi\,(x - a)}{b - a}. \tag{C.1}$$

	$f(x)$	$\tilde{f}_s(n)$
1.	$\sin mx, \quad m = 1, 2, \ldots$	$\begin{cases} \pi/2, & n = m \\ 0, & n \neq m \end{cases}$
2.	$\displaystyle\sum_{n=1}^{\infty} a_n \sin nx$	$\pi\, a_n/2$
3.	$\pi - x$	π/n
4.	x	$\dfrac{\pi}{n}(-1)^{n+1}$
5.	1	$\dfrac{1}{n}\left[1 - (-1)^n\right]$
6.	$\begin{cases} -x, & x \leq a \\ \pi - x, & x > a \end{cases}$	$\dfrac{\pi}{n}\cos na, \quad 0 < a < \pi$
7.	$\begin{cases} x(\pi - a), & x \leq a \\ a(\pi - x), & x > a \end{cases}$	$\dfrac{\pi}{n^2}\sin na, \quad 0 < a < \pi$
8.	e^{ax}	$\dfrac{n}{a^2 + n^2}\left[1 - (-1)^n e^{a\pi}\right]$
9.	$\dfrac{\sinh a(\pi - x)}{\sinh a\pi}$	$\dfrac{n}{a^2 + n^2}$
10.	$f''(x)$	$-n^2\tilde{f}_s(n) + n\left[f(0) - (-1)^n f(\pi)\right]$

C.6. Finite Cosine Transform Pairs

The tables are for the interval $[0, \pi]$. If the interval is $[a, b]$, then it can be transformed into $[0, \pi]$ by formula (C.1).

	$f(x)$	$\tilde{f}_c(n)$
1.	$\cos mx, \quad m = 1, 2, \ldots$	$\begin{cases} \pi/2, & n = m \\ 0, & n \neq m \end{cases}$
2.	$\dfrac{a_0}{2} + \displaystyle\sum_{n=1}^{\infty} a_n \cos nx$	$\pi\, a_n/2$
3.	$f(\pi - x)$	$(-1)^n\, \tilde{f}_c(n)$
4.	1	$\begin{cases} \pi, & n = 0 \\ 0, & n = 1, 2, \ldots \end{cases}$
5.	x	$\begin{cases} \pi^2/2, & n = 0 \\ \dfrac{1}{n^2}\left[(-1)^n - 1\right], & n = 1, 2, \ldots \end{cases}$
6.	x^2	$\begin{cases} \pi^3/3, & n = 0 \\ \dfrac{2\pi}{n^2}(-1)^n, & n = 1, 2, \ldots \end{cases}$
7.	$\begin{cases} 1, & 0 < x < a \\ -1, & a < x < \pi \end{cases}$	$\begin{cases} 2a - \pi, & n = 0 \\ \dfrac{2}{n} \sin na, & n = 1, 2, \ldots \end{cases}$
8.	$\dfrac{e^{ax}}{a}$	$\dfrac{(-1)^n\, e^{a\pi} - 1}{a^2 + n^2}$
9.	$\dfrac{\cosh\left(c(\pi - x)\right)}{\sinh(\pi c)}$	$\dfrac{c}{c^2 + n^2}$
10.	$f''(x)$	$-n^2 \tilde{f}_c(n) + (-1)^n\, f'(\pi) - f'(0)$

D

Bessel Functions

The Bessel functions are introduced and discussed in §4.6. We summarize here information about some of these functions for ready reference.

D.1. Bessel's Equation

The Bessel's equation is $x^2 y'' + xy' + (x^2 - n^2) = 0$, and its solution is

$$y = \begin{cases} A_1 J_n(x) + A_2 Y_n(x) & \text{for all } n, \\ A_1 J_n(x) + A_2 J_{-n}(x) & \text{for } n \neq 0, 1, 2, \ldots, \end{cases} \tag{D.1}$$

where $J_n(x)$ and $Y_n(x)$ are the Bessel functions of the first and second kind, respectively, each of order n. They are defined by

$$J_n(x) = \sum_{k=0}^{\infty} \frac{(-1)^k (x/2)^{2k+n}}{k! \, \Gamma(k+1+n)}$$

$$= \frac{x^n}{2^n \, \Gamma(1+n)} \left[1 - \frac{x^2}{2(2+2n)} + \frac{x^4}{(2)(4)(2+2n)(4+2n)} - \cdots \right],$$

$$J_{-n}(x) = \sum_{k=0}^{\infty} \frac{(-1)^k (x/2)^{2k-n}}{k! \, \Gamma(k+1-n)}$$

$$= \frac{x^n}{2^{-n} \, \Gamma(1-n)} \left[1 - \frac{x^2}{2(2-2n)} + \frac{x^4}{(2)(4)(2-2n)(4-2n)} - \cdots \right]$$

$$= (-1)^n J_n(x) \quad \text{for } n = 0, 1, 2, \ldots,$$

$$Y_n(x) = \begin{cases} \lim\limits_{k \to n} \dfrac{J_k(x) \cos k\pi - J_{-k}(x)}{\sin k\pi} & \text{for } n = 0, 1, 2, \ldots, \\[2ex] \dfrac{J_n(x) \cos n\pi - J_{-n}(x)}{\sin n\pi} & \text{for } n \neq 0, 1, 2, \ldots, \end{cases}$$

$$Y_{-n}(x) = (-1)^n Y_n(x) \quad \text{for } n = 0, 1, 2, \ldots,$$

$$Y_0(x) = \frac{2}{\pi} \left[J_0(x) \left(\ln \frac{x}{2} + \gamma \right) + \frac{2}{1} J_2(x) - \frac{2}{2} J_4(x) + \frac{2}{3} J_6(x) - \cdots \right],$$

$$Y_1(x) = \frac{2}{\pi} \left[J_1(x) \left(\ln \frac{x}{2} + \gamma \right) - \frac{1}{x} - \frac{1}{2} J_1(x) + \frac{9}{4} J_3(x) - \cdots \right]$$

$$= -\frac{d}{dx} \left[Y_0(x) \right], \quad \gamma = 0.577215665 \quad \text{(Euler's constant)}.$$

RECURRENCE RELATIONS:

$$J_{n+1}(x) = \frac{2n}{x} J_n(x) - J_{n-1}(x) \qquad\qquad Y_{n+1}(x) = \frac{2n}{x} Y_n(x) - Y_{n-1}(x)$$

$$= \frac{n}{x} J_n(x) - \frac{d}{dx} \left[J_n(x) \right] \qquad\qquad = \frac{n}{x} Y_n(x) - \frac{d}{dx} \left[Y_n(x) \right]$$

$$= -x^n \frac{d}{dx} \left[x^{-n} J_n(x) \right], \qquad\qquad = -x^n \frac{d}{dx} \left[x^{-n} Y_n(x) \right],$$

$$J_{n-1}(x) = \frac{2n}{x} J_n(x) - J_{n+1}(x) \qquad\qquad Y_{n-1}(x) = \frac{2n}{x} Y_n(x) + Y_{n+1}(x)$$

$$= \frac{n}{x} J_n(x) + \frac{d}{dx} \left[J_n(x) \right] \qquad\qquad = \frac{n}{x} Y_n(x) - \frac{d}{dx} \left[Y_n(x) \right]$$

$$= x^{-n} \frac{d}{dx} \left[x^{-n} J_n(x) \right], \qquad\qquad = x^{-n} \frac{d}{dx} \left[x^{-n} Y_n(x) \right],$$

$$J_n'(x) = \frac{1}{2} \left[J_{n-1}(x) - J_{n+1}(x) \right], \qquad Y_n'(x) = \frac{1}{2} \left[Y_{n-1}(x) - Y_{n+1}(x) \right],$$

$$\int x^{n+1} J_n(x) \, dx = x^{n+1} J_{n+1}(x), \qquad \int x^{n+1} Y_n(x) \, dx = x^{n+1} Y_{n+1}(x),$$

where the prime (') denotes differentiation with respect to x.

ODD INTEGRAL ORDERS:

$$J_{1/2}(x) = \sqrt{\frac{2}{\pi x}} \sin x, \qquad\qquad Y_{1/2}(x) = -\sqrt{\frac{2}{\pi x}} \cos x,$$

$$J_{3/2}(x) = \sqrt{\frac{2}{\pi x}} \left(\frac{\sin x}{x} - \cos x \right), \qquad Y_{3/2}(x) = -\sqrt{\frac{2}{\pi x}} \left(\frac{\cos x}{x} + \sin x \right),$$

$$J_{-1/2}(x) = \sqrt{\frac{2}{\pi x}} \cos x, \qquad\qquad Y_{-1/2}(x) = \sqrt{\frac{2}{\pi x}} \sin x,$$

$$J_{-3/2}(x) = -\sqrt{\frac{2}{\pi x}} \left(\frac{\cos x}{x} + \sin x \right), \qquad Y_{-3/2}(x) = -\sqrt{\frac{2}{\pi x}} \left(\frac{\sin x}{x} + \cos x \right).$$

HANKEL FUNCTIONS OF ORDER n:

$$H_n^{(1)}(x) = J_n(x) + i\,Y_n(x), \qquad H_n^{(2)}(x) = J_n(x) - i\,Y_n(x),$$

$$J_n(x) = \frac{H_n^{(1)}(x) + H_n^{(2)}(x)}{2}, \qquad Y_n(x) = \frac{H_n^{(1)}(x) - H_n^{(2)}(x)}{2i}.$$

ASYMPTOTIC EXPANSIONS for large x:

$$J_n(x) \sim \sqrt{\frac{2}{\pi x}} \cos\left(x - \frac{n\pi}{2} - \frac{\pi}{4}\right), \quad Y_n(x) \sim \sqrt{\frac{2}{\pi x}} \sin\left(x - \frac{n\pi}{2} - \frac{\pi}{4}\right);$$

and for large n:

$$J_n(x) \sim \frac{1}{\sqrt{2\pi n}} \left(\frac{e\,x}{2n}\right)^n, \quad Y_n(x) \sim \frac{2}{\sqrt{\pi n}} \left(\frac{e\,x}{2n}\right)^{-n}.$$

D.2. Modified Bessel's Equation

The modified Bessel's equation is $x^2 y'' + xy' - (x^2 + n^2) = 0$, and its solution is

$$y = \begin{cases} A_1 I_n(x) + A_2 K_n(x) & \text{for all } n, \\ A_1 I_n(x) + A_2 I_{-n}(x) & \text{for } n \neq 0, 1, 2, \ldots, \end{cases} \tag{D.2}$$

where $I_n(x)$ and $K_n(x)$ are the modified Bessel functions of the first and second kind, respectively, each of order n. They are defined by

$$I_n(x) = i^{-n} J_n(ix) = \sum_{k=0}^{\infty} \frac{(x/2)^{2k+n}}{k!\,\Gamma(k+1+n)}$$

$$= \frac{x^n}{2^n\,\Gamma(1+n)} \left[1 + \frac{x^2}{2(2+2n)} + \frac{x^4}{(2)(4)(2+2n)(4+2n)} + \cdots\right],$$

$$I_{-n}(x) = i^n J_{-n}(x) = \sum_{k=0}^{\infty} \frac{(x/2)^{2k-n}}{k!\,\Gamma(k+1-n)}$$

$$= \frac{x^{-n}}{2^{-n}\,\Gamma(1-n)} \left[1 + \frac{x^2}{2(2-2n)} + \frac{x^4}{(2)(4)(2-2n)(4-2n)} + \cdots\right],$$

$$I_{-n}(x) = I_n(x) \quad \text{for } n = 0, 1, 2, \ldots,$$

$$I_0(x) = 1 + \frac{(x/2)^2}{(1!)^2} + \frac{(x/2)^4}{(2!)^2} + \frac{(x/2)^6}{(3!)^2} + \cdots,$$

$$I_1(x) = \frac{x}{2} \left[1 + \frac{(x/2)^2}{2(1!)^2} + \frac{(x/2)^4}{3(2!)^2} + \frac{(x/2)^6}{4(3!)^2} + \cdots\right] = \frac{d}{dx}[I_0(x)].$$

$$K_n(x) = \begin{cases} \lim\limits_{k \to n} \dfrac{\pi}{2} \dfrac{I_{-k}(x) - I_k(x)}{\sin k\pi} & \text{for } n = 0, 1, 2, \ldots, \\[2mm] \dfrac{\pi}{2} \dfrac{I_{-n}(x) - I_n(x)}{\sin n\pi} & \text{for } n \neq 0, 1, 2, \ldots, \end{cases}$$

$$K_{-n}(x) = K_n(x) \quad \text{for } n = 0, 1, 2, \ldots,$$

$$K_0(x) = -\left[I_0(x) \left(\ln \frac{x}{2} + \gamma \right) - \frac{2}{1} I_2(x) - \frac{2}{2} I_4(x) + \frac{2}{3} I_6(x) - \cdots \right],$$

$$K_1(x) = \left[I_1(x) \left(\ln \frac{x}{2} + \gamma \right) - \frac{1}{x} - \frac{1}{2} I_1(x) + \frac{9}{4} I_3(x) - \cdots \right]$$

$$= -\frac{d}{dx} \left[K_0(x) \right], \quad \gamma = 0.577215665 \quad \text{(Euler's constant)}.$$

RECURRENCE RELATIONS:

$$I_{n+1}(x) = -\frac{2n}{x} I_n(x) + I_{n-1}(x) \qquad K_{n+1}(x) = \frac{2n}{x} K_n(x) + K_{n-1}(x),$$

$$= -\frac{n}{x} I_n(x) + \frac{d}{dx} \left[I_n(x) \right] \qquad\qquad = \frac{n}{x} K_n(x) - \frac{d}{dx} \left[K_n(x) \right]$$

$$= x^n \frac{d}{dx} \left[x^{-n} J_n(x) \right], \qquad\qquad = -x^n \frac{d}{dx} \left[x^{-n} Y_n(x) \right],$$

$$I_{n-1}(x) = \frac{2n}{x} I_n(x) + I_{n+1}(x) \qquad K_{n-1}(x) = -\frac{2n}{x} K_n(x) + K_{n+1}(x)$$

$$= \frac{n}{x} I_n(x) + \frac{d}{dx} \left[I_n(x) \right] \qquad\qquad = -\frac{n}{x} K_n(x) - \frac{d}{dx} \left[K_n(x) \right]$$

$$= x^{-n} \frac{d}{dx} \left[x^n I_n(x) \right] \qquad\qquad\quad = -x^{-n} \frac{d}{dx} \left[x^n K_n(x) \right]$$

$$I_n'(x) = \frac{1}{2} \left[I_{n+1}(x) + I_{n-1}(x) \right], \qquad K_n'(x) = -\frac{1}{2} \left[K_{n+1}(x) + K_{n-1}(x) \right],$$

$$\int x^{n+1} I_n(x)\,dx = x^{n+1} I_{n+1}(x), \qquad \int x^{n+1} K_n(x)\,dx = -x^{n+1} K_{n+1}(x),$$

where the prime ($'$) denotes differentiation with respect to x.

ODD INTEGRAL ORDERS:

$$I_{1/2}(x) = \sqrt{\frac{2}{\pi x}} \sinh x, \qquad\qquad K_{1/2}(x) = e^{-x} \sqrt{\frac{\pi}{2x}},$$

$$I_{3/2}(x) = -\sqrt{\frac{2}{\pi x}} \left(\frac{\sinh x}{x} - \cosh x \right), \qquad K_{3/2}(x) = e^{-x} \sqrt{\frac{\pi}{2x}} \left(\frac{1}{x} + 1 \right),$$

$$I_{-1/2}(x) = \sqrt{\frac{2}{\pi x}} \cosh x, \qquad\qquad K_{-1/2}(x) = e^{-x} \sqrt{\frac{\pi}{2x}},$$

$$I_{-3/2}(x) = -\sqrt{\frac{2}{\pi x}} \left(\frac{\cosh x}{x} - \sinh x \right), \qquad K_{-3/2}(x) = e^{-x} \sqrt{\frac{\pi}{2x}} \left(\frac{1}{x} + 1 \right).$$

ASYMPTOTIC EXPANSIONS for large x:

$$I_n(x) \sim \sqrt{\frac{1}{2\pi x}}\, e^x, \qquad K_n(x) \sim \sqrt{\frac{\pi}{2x}}\, e^{-x}.$$

D.3. Infinite Series and Definite Integrals

INFINITE SERIES:

$$J_n(x+y) = \sum_{k=-\infty}^{\infty} J_k(x)\, J_{n-k}(x), \quad n = 0, 1, 2, \ldots,$$

$$1 = J_0(x) + 2J_2(x) + \cdots + 2J_{2n}(x) + \cdots,$$

$$x = 2\left[J_1(x) + 3J_3(x) + \cdots + (2n+1)J_{2n+1}(x) + \cdots\right],$$

$$x^2 = 8\left[J_2(x) + 4J_4(x) + \cdots + n^2 J_{2n}(x) + \cdots\right],$$

$$\sin x = 2\left[J_1(x) - J_3(x) + J_5(x) - \cdots\right],$$

$$\cos x = J_0(x) - 2\left[J_2(x) - J_4(x) + \cdots\right],$$

$$\sinh x = 2\left[I_1(x) + I_3(x) + I_5(x) + \cdots\right],$$

$$\cosh x = I_0(x) + 2\left[I_2(x) + I_4(x) + \cdots\right].$$

DEFINITE INTEGRALS:

$$\int_0^\infty J_n(\alpha x)\, dx = \frac{1}{\alpha} \quad \text{for } n > -1, \qquad \int_0^\infty \frac{J_n(\alpha x)}{x}\, dx = \frac{1}{n} \quad \text{for } n = 1, 2, \ldots,$$

$$\int_0^\infty e^{-\alpha x} J_0(\beta x)\, dx = \frac{1}{\sqrt{\alpha^2 + \beta^2}}, \qquad \int_0^\infty e^{-\alpha x} J_1(\beta x)\, dx = \frac{1}{\beta}\left(1 - \frac{\alpha}{\sqrt{\alpha^2 + \beta^2}}\right),$$

$$\int_0^\infty \frac{J_m(x) J_n(x)}{x}\, dx = \begin{cases} \dfrac{2}{(m^2 - n^2)\,\pi} \sin\dfrac{(m-n)\pi}{2} & \text{if } m \neq n \\[2ex] \dfrac{1}{2m} & \text{if } m = n \end{cases}, \quad \text{for } m+n > 0.$$

For other Bessel functions and additional information, see Abramowitz and Stegun (1965).

E

Mathematica Notebooks and Projects

A list of Mathematica Packages and Notebooks available on the CRC website *www.crcpress.com* is given below.

bessel.nb
declare.m
declare.nb
drum.nb
eigenpair.nb
equilateral.nb
EquationType.m
EquationType.nb
Example2.1.nb
Example2.5.nb
Example2.7.nb
Example2.8.nb
Example2.15.nb
Example2.18.nb
Example2.19.nb
Example2.22.nb
Example2.23.nb
Example2.25.nb
Example5.1.nb

Example5.2a.nb
Example5.2b.nb
Example5.3.nb
Example6.nb
Example6.5.nb
Example7.18.nb
Exercise2.4.nb
fd.nb
galerkin.nb
greens.nb
Intro2Mma.nb
InverseOperator.nb
Orthonormality.nb
PackagesAndNotebooks.nb
perturbationmethods.nb
plotfourier.nb
proj7.3.nb
proj7.4.nb
ReadMe.nb

We reiterate the suggestion made on page 7: After downloading the above files, the user should first create a copy of these originals and then only modify a copy. After opening a Notebook, the user will be presented with a sequence of cells, which may be open or closed. To open a group of cells, double-click a closed group's cell bracket at the right.

The file mu.pdf, mentioned on page 135, is available on the above CRC website.

A glossary of Mathematica functions, which was appended to the first edition, is removed from the current edition because on-line help is available in the Mathematica environment under the Help Browser. However, this glossary is now available as the file Glossary.pdf on the CRC website.

MATHEMATICA PROJECTS. A list of Mathematica projects presented in the book is given below.

Project	Page	Project	Page
1.1	24	5.2	143
1.2	24	5.3	144
2.1	59	5.4	145
2.2	59	6.1	178
2.3	59	6.2	179
2.4	59	6.3	179
2.5	59	6.4	180
2.6	59	6.5	180
2.7	60	6.6	182
2.8	60	6.7	201
2.8	60	6.8	202
2.10	60	6.9	202
2.11	60	6.10	203
3.1	81	7.1	246
4.1	106	7.2	247
4.2	106	7.3	247
4.3	107	7.4	248
4.4	108	7.5	249
4.5	109	9.1	329
4.6	109	10.1	351
4.7	109	11.1	376
5.1	142	11.2	377

Bibliography

M. B. Abbott and D. R. Basco, *Computational Fluid Dynamics*, Longman Scientific & Technical, Essex, UK, and Wiley, New York, 1990.

M. Abramowitz and I. A. Stegun (Eds.), *Handbook of Mathematical Functions*, Dover, New York, 1965.

H. Bateman, *Partial Differential Equations*, Dover, New York, 1944.

————, *Partial Differential Equations of Mathematical Physics*, Cambridge University Press, Cambridge, 1959.

M. Becker, *The Principles and Applications of Variational Methods*, MIT Press, Cambridge, MA, 1964.

W. E. Boyce and R. C. DiPrima, *Elementary Differential Equations*, 5th ed., Wiley, New York, 1992.

G. Carrier and C. Pearson, *Partial Differential Equations*, 2nd ed., Academic Press, New York, 1988.

————, M. Krook, and C. E. Pearson, *Functions of a Complex Variable: Theory and Technique*, McGraw-Hill, New York, 1966.

H. S. Carslaw and J. C. Jaeger, *Conduction of Heat in Solids*, 2nd ed., Oxford University Press, New York, 1959.

R. V. Churchill, *Operational Methods*, 3rd ed., McGraw-Hill, New York, 1972.

————and J. W. Brown, *Fourier Series and Boundary Value Problems*, McGraw-Hill, New York, 1978.

J. J. Connor and C. A. Brebbia, *Finite Element Techniques for Structural Engineers*, Butterworths, London, 1973.

R. Courant and D. Hilbert, *Methods of Mathematical Physics, Vol. 1, 2*, Interscience, New York, 1963, 1965.

A. J. Davies, *The Finite Element Method*, Clarendon Press, Oxford, 1980.

B. Davies, *Integral Transforms and Their Applications*, Springer-Verlag, New York, 1978.

H. F. Davis, *Fourier Series and Orthogonal Functions*, Allyn and Bacon, Boston, 1963.

G. F. D. Duff, *Partial Differential Equations*, Univ. Toronto Press, 1956.

D. G. Duffy, *Transform Methods for Solving Partial Differential Equations*, CRC Press, Boca Raton, 1994.

B. Epstein, *Partial Differential Equations*, McGraw-Hill, New York, 1962.

A. Erdélyi, W. Magnus, F. Oberhettinger and F. G. Tricomi, *Tables of Integral Transforms*, vol. 1, McGraw-Hill, New York, 1954.

S. J. Farlow, *Partial Differential Equations for Scientists and Engineers*, Wiley, New York, 1982.

B. A. Finlayson, *The Method of Weighted Residuals and Variational Principles*, Academic Press, New York, 1972.

M. J. Forray, *Variational Calculus in Science and Engineering*, Academic Press, New York, 1972.

F. G. Friedlander, *An Introduction to the Theory of Distributions*, Cambridge Univ. Press, Cambridge, 1982.

P. R. Garabedian, *Partial Differential Equations*, Wiley, New York, 1964.

I. M. Gelfand and G. E. Shilov, *Generalized Functions and Operations*, vol. I (Translation from Russian), Academic Press, New York, 1964.

C. Gordon, L. Webb and S. Wolpert, *One cannot hear the shape of a drum*, Bull. Am. Math. Soc. (July 1992), 134–138.

M. D. Greenberg, *Application of Green's Functions in Science and Engineering*, Prentice-Hall, Englewood Cliffs, NJ, 1971.

R. Haberman, *Elementary Applied Partial Differential Equations*, 2nd ed., Prentice-Hall, Englewood Cliffs, NJ, 1987.

F. B. Hildebrand, *Methods of Applied Mathematics*, 2nd ed., Prentice-Hall, Englewood Cliffs, NJ, 1965.

E. Hille, *Ordinary Differential Equations in the Complex Domain*, Wiley, New York, 1976.

K. Hoffman, *Analysis in Euclidean Space*, Prentice-Hall, Englewood Cliffs, NJ, 1975.

K. M. Humi and W. B. Miller, *Boundary Value Problems and Partial Differential Equations*, PWS-KENT Publishing Company, Boston, MA, 1992.

F. John, *Partial Differential Equations*, Springer-Verlag, New York, 1982.

M. Kac, *Can one hear the shape of a drum?*, Am. Math. Monthly **74** (1966), 1–23.

P.N. Kaloni, *Fluctuating flow of an elastico-viscous fluid past a porous flat plate*, Phys. Fluids **10** (1967), 1344–1346.

L. V. Kantorovitch and V. L. Krylov, *Approximate Methods of Higher Analysis*, Interscience, New York, 1958.

R. Kanwal, *Generalized Functions: Theory and Technique*, Academic Press, New York, 1983.

J. Kevorkian, *Partial Differential Equations*, Wadsworth & Brooks/Cole, Belmont, 1990.

E. Kreiszig, *Introductory Functional Analysis with Applications*, Wiley, New York, 1978.

P. K. Kulshrestha and P. Puri, *An exact solution of hydromagnetic rotating flow*, Developments in Mech. **5** (1969), Proc. 11th Midwestern Conf., Iowa State Univ. Press, 265–271.

_____ and P. Puri, *Wave structure in oscillatory Couette flow of a dusty gas*, Acta Mech. **46** (1983), 127–135.

P. K. Kythe, *An Introduction to Boundary Element Methods*, CRC Press, Boca Raton, 1995.

_____, *Fundamental Solutions for Differential Operators and Applications*, Birkhäuser, Boston, 1996.

_____, *Computational Conformal Mapping*, Birkhäuser, Boston, 1998.

_____ and P. Puri, *Unsteady MHD free-convection flows with time-dependent heating in a rotating medium*, Astrophys. Space Sc. **135** (1987), 219–228.

_____ and P. Puri, *Unsteady MHD free-convection flows on a porous plate with time-dependent heating in a rotating medium*, Astrophys. Space Sc. **143** (1988), 51–62.

_____ and P. Puri, *Unsteady MHD free-convection oscillatory flow on a porous plate with time-dependent heating in a rotating medium*, Astrophys. Space Sc. **149** (1988), 107–114.

———— and P. Puri, *The effects of induced magnetic field on unsteady hydromagnetic flows in a rotating medium*, Astrophys. Space Sc. **174** (1990), 121–133.

———— and P. Puri, *Computational Methods for Linear Integral Equations*, Birkhäuser, Boston, 2002.

———— ,P. Puri, and M. R. Schäferkotter, *Partial Differential Equations and Mathematica*, CRC Press, Boca Raton, 1997.

N. Levinson and R. M. Redheffer, *Complex Variables*, Holden-Day, San Francisco, 1970.

D. R. Lick, *The Advanced Calculus of One Variable*, Appleton-Century Crofts, New York, 1971.

C. R. MacCluer, *Boundary Value Problems and Orthogonal Expansions*, IEEE Press, New York, 1994.

A. G. Mackie, *Boundary Value Problems*, Scottish Academic Press, Edinburgh, 1989.

R. McOwen, *Partial Differential Equations*, Prentice-Hall, Englewood Cliffs, NJ, 1996.

S. A. S. Messiha, *Laminar boundary layers in oscillatory flow along an infinite flat plate with variable suction*, Proc. Camb. Phil. Soc. **62** (1966), 329–337.

S. G. Mikhlin, *Variational Methods in Mathematical Physics*, Pergamon Press, New York, 1964.

P. M. Morse and H. Feshbach, *Methods of Theoretical Physics*, Parts 1 and 2, McGraw-Hill, New York, 1953.

F. Oberhettinger, *Tables of Fourier Transforms and Fourier Transforms of Distributions*, Springer-Verlag, Berlin, 1990.

M. N. Özişik, *Heat Conduction*, Wiley, New York, 1980.

L. E. Payne, *Improperly Posed Problems in Partial Differential Equations*, SIAM, Philadelphia, PA, 1975.

I. G. Petrovskii, *Partial Differential Equations*, W. B. Saunders Co., Philadelphia, 1967.

M. H. Protter, *Can one hear the shape of a drum? Revisited*, SIAM Rev. **29** (1987), 185–197.

M. Protter and H. Weinberger, *Maximum Principles in Differential Equations*, Springer-Verlag, Berlin, 1984.

P. Puri, *Fluctuating flow of viscous fluid on a porous plate in a rotating medium*, Acta Mechanica **21** (1975), 153–158.

———— and P. K. Kulshrestha, *Rotating flows of non-Newtonian fluids*, Applicable Analysis **4** (1974), 131–140.

———— and P. K. Kulshrestha, *Unsteady hydromagnetic boundary layer in a rotating medium*, J. Appl. Mech., Trans. ASME **98** (1976), 205–208.

———— and P. K. Kythe, *Wave structure in unsteady flows past a flat plate in a rotating medium*, Proc. SECTAM XIV, Developments in Theor. Appl. Mech. **14** (1988), 207–213.

———— and P. K. Kythe, *Some inverse Laplace transforms of exponential form*, ZAMP **39** (1988), 150–156; 954.

J. N. Reddy, *An Introduction to the Finite Element Method*, McGraw-Hill, New York, 1984.

G. F. Roach, *Green's Functions: Introductory Theory and Applications*, 2nd ed., Cambridge University Press, Cambridge, 1982.

C. C. Ross, *Differential Equations*, Springer-Verlag, New York, 1995.

S. L. Ross, *Differential Equations*, Blaisdell, Waltham, 1964.

Z. Rubenstein, *A Course in Ordinary and Partial Differential Equations*, Academic Press, New York, 1969.

T. Schücker, *Distributions, Fourier Transforms and Some of Their Applications to Physics*, World Scientific, Singapore, 1991.

V. I. Smirnov, *A Course in Higher Mathematics*, Vol. IV: Integral Equations and Partial

Differential Equations, Pergamon Press, London, 1964.

G. D. Smith, *Numerical Solutions of Partial Differential Equations: Finite Difference Methods*, 3rd ed., Clarendon Press, Oxford, 1985.

I. N. Sneddon, *Partial Differential Equations*, McGraw-Hill, New York, 1957.

———, *Fourier Transforms and Their Applications*, Springer-Verlag, Berlin, 1978.

A. Sommerfeld, *Partial Differential Equations in Physics*, vol. VI, Academic Press, New York, 1964.

V. M. Soundalgekar and P. Puri, *On fluctuating flow of an elastico-viscous fluid past an infinite plate with variable suction*, J. Fluid Mech. **35** (1969), 563–573.

I. Stakgold, *Boundary Value Problems of Mathematical Physics*, Vol. II, Macmillan, New York, 1968.

———, *Green's Functions and Boundary Value Problems*, Wiley, New York, 1979.

W. A. Strauss, *Partial Differential Equations*, Wiley, New York, 1992.

R. L. Street, *The Analysis and Solution of Partial Differential Equations*, Brooks/ Cole, Monterey, CA, 1973.

J. T. Stuart, *A solution of Navier-Stokes and energy equations illustrating the response of skin friction and temperature of an infinite plate thermometer to fluctuations in the stream velocity*, Proc. Roy. Soc. A **A 231** (1955), 116–130.

G. I. Taylor, *The determination of stresses by means of soap films. The Mechanical Properties of Fluids*, Blackie & Son, Ltd., London, Glasgow, 1937, pp. 136.

J. F. Treves, *Linear Partial Differential Equations with Constant Coefficients*, Gordon and Breach, New York, 1966.

D. W. Trim, *Applied Partial Differential Equations*, PWS-Kent, Boston, MA, 1990.

A. N. Tychonov and A. A. Samarski, *Partial Differential Equations of Mathematical Physics*, Vols. I and II, Holden-Day, Inc., San Francisco, 1964, 1967.

V. S. Vladimirov, *Equations of Mathematical Physics*, English Translation, Mir Publishers, Moscow, 1984.

D. Vvedensky, *Partial Differential Equations with Mathematica*, Addison-Wesley, Workingham, UK, 1993.

J. S. Walker, *Fourier Analysis*, Oxford University Press, Oxford, 1988.

G. N. Watson, *A Treatise on the Theory of Bessel Functions*, 2nd ed., Cambridge University Press, Cambridge, 1944.

H. F. Weinberger, *A First Course in Partial Differential Equations*, Xerox, Lexington, MA, 1965.

R. E. Williamson, R. H. Crowell and H. F. Trotter, *Calculus of Vector Functions*, 2nd ed., Prentice-Hall, Englewood Cliffs, NJ, 1968.

S. Wolfram, *The Mathematica Book*, 4th ed., Wolfram Media, Champaign, IL, and Cambridge University Press, Cambridge, UK, 1999.

E. C. Young, *Partial Differential Equations*, Allyn and Bacon, Boston, 1972.

E. C. Zachmanoglou and D. W. Thoe, *Introduction to Partial Differential Equations with Applications*, Williams & Wilkins, Baltimore, 1976.

E. Zauderer, *Partial Differential Equations of Applied Mathematics*, Wiley, New York, 1983.

Index